EARTH IN CRISIS

EARTH IN CRISIS

an introduction to the earth sciences

Thomas L. Burrus

Herbert J. Spiegel

Both of the Department of Chemistry and Earth Science,
Miami-Dade Community College, North Campus,
Miami, Florida

with 459 *illustrations*
original drawings by George Ondricek, Jr.

The C. V. Mosby Company

Saint Louis 1976

Printed in the United States of America

Distributed in Great Britain by Henry Kimpton, London

Library of Congress Cataloging in Publication Data

Burrus, Thomas L 1937-
Earth in crisis.

Includes index.
1. Earth sciences. 2. Astronomy. I. Spiegel, Herbert, joint author. II. Title.
QE26.2.B87 550 75-22029
ISBN 0-8016-0918-6

GW/CB/B 9 8 7 6 5 4 3 2 1

To the students whose use of the information contained herein will benefit mankind.

To Tommy and Scott, Karen and Hilary, who will inherit the earth.

PREFACE

To see the earth as it truly is,
small and blue and beautiful in that eternal silence where it floats, is to see ourselves as riders on the earth together, brothers on that bright loveliness in the eternal cold—brothers who know now they are truly brothers.

A Reflection
Archibald MacLeish

These words summarize a fact of which we on earth are only now becoming aware: the earth at any moment is nearly a closed system, that is, a system constantly changing in response to factors from within but influenced only to a minute degree by factors from without. Our own destiny is linked to these changes, and we must learn to adapt to them or perish. Our recent history unfortunately indicates that we are not adapting but rather attempting to force nature to adapt to us. This attitude is already creating severe problems. The next several decades hopefully will see attempts to repair the damage we have done to our earth in such a short time. The success of these efforts will depend largely on the awareness of the general public of the issues involved.

With this in mind, *Earth in Crisis* deliberately stresses the problems that remain to be solved rather than those already behind us, the difficulties we have created rather than those we have conquered, the harm we have done to the environment rather than the cases in which we have protected fragile ecosystems. This is not meant as negativism but rather to provide a basis for future positive action.

The title, format, and content of the book were chosen to suit its audience: freshmen and sophomore general education students. We feel that the purpose of an earth science text should no longer be merely to provide an introduction for future scientists but rather to afford a basis for intelligent choices made by average citizens. Our readers are the future voters, taxpayers, employees, employers, and politicians as well as scientists. To this end, we have striven to build a text that is readable and that covers, in addition to physical phenomena, the social, economic, and environmental considerations and implications of our attempt to deal with our world.

To turn a dream into a book required the combined efforts of many people. Thanks should go to our families and friends who have seen all too little of us while our energies went into manuscript preparation. Thanks are due also to our students who often acted as sounding boards for the ideas and manner of presentation that comprise this book. Finally, we wish to express our gratitude to all the people who helped in their own individual ways: George Ondricek, Jr., who prepared the artwork; Barbara Hinsz, our typist, grammarian, critic, and morale booster; Joe Zawodny and Jim McWhorter, who helped with the oceanography section; Anita Friedman, who proofread and criticized portions of the manuscript; Louis King, who helped select the introductory quotations; Jane Morganroth and Louise Arnoth for their assistance with the paperwork and miscellaneous correspondence; Chris Burrus and Toby Spiegel, who read and typed various portions of the manuscript; and Frank Groselle, Dennis Hipple, Willis Holland, Jim Kilps, George Winston, and Lyle Joyce for assistance, criticism, suggestions, and moral support.

Thomas L. Burrus
Herbert J. Spiegel

CONTENTS

part two **GEOLOGY**

part three **OCEANOGRAPHY**

EARTH IN CRISIS

PROLOGUE

So, nature calls through all her systems wide,
Give me thy love, O Man, so long denied.

The Symphony
Sidney Lanier

A GRAND TOUR

Traveling at the speed of light and with the freedom of a god from some remote part of the universe, billions of years elapse before we approach a cluster of 19 galaxies known as the Local Group. After we pass the largest member of the group, the Great galaxy in Andromeda (Fig. 1), we travel for more than 2 million years before we come to the Local Group's second largest member, which is simply named the Galaxy. It is a gigantic spiral pinwheel 100,000 light-years in diameter and 10,000 light-years thick, and it consists of more than 100 billion stars. The Galaxy is one of the largest aggregate forms of matter in the universe, just as the atom is the smallest. It consists not only of stars but also of vast quantities of gas and dust. Rotating on its axis once in 200 million years, the Galaxy plunges through the universe at a speed of 50 miles/sec relative to other members of the Local Group.

We enter the Galaxy and pass through great clouds of interstellar gas and dust, nonuniform in density but everywhere less dense than the best vacuum that could be produced in a laboratory.

An additional 30,000 years of travel takes us past a triple star system called Alpha Centauri. Then, after another 4 years elapse, we approach a feeble yellow star shining in the perpetual darkness of space. On closer examination we see that this star, called the sun, holds a system of nine major bodies in its firm gravitational grasp. These nine objects, known as planets, have no light of their own and reflect only that which their giant master sends them.

Passing the planet furthest from the sun, we are able to see that it is a small sphere with a diameter of only 4000 miles. Its temperature of −400° F is much too cold to support human life. This planet, known as Pluto, has no satellites and completes one orbit around that distant yellow star every 248 years.

Moving on through this solar system, after 75 minutes of travel we arrive at another planet. Neptune, as it is called, is 28,000 miles in diameter. Its atmosphere is composed chiefly of hydrogen and methane gas, and this, together with a temperature of −350° F, makes Neptune an inhospitable place for man. Carrying its two satellites with it, Neptune completes one revolution around the sun every 165 years.

After another 92 minutes we come to the next planet, Uranus. This green planet and its five satellites orbit the sun every 84 years. Uranus, which is 31,000 miles in diameter, has a temperature of −300° F and an atmosphere similar to that of Neptune. Uranus also is a most unfriendly environment.

Leaving Uranus, 75 minutes pass before we approach an incredibly beautiful ringed sphere called Saturn, which is 75,000 miles in diameter (Fig. 2). Saturn's ring system, with four distinct parts, surrounds the planet's equatorial region like the brim of a straw hat. Saturn rules 10 satellites; one of them, Titan, has an atmosphere of poisonous methane gas. With a temperature of −230° F and a banded atmosphere composed of methane and hydrogen, this planet is not environmentally suited to human habitation. Saturn orbits the distant sun once every 29½ years.

Leaving Saturn, we arrive 42 minutes later at Jupiter, an even larger planet. This giant master of 13 satellites has a diameter of 88,000 miles and orbits the sun once every 12 years. Like Saturn, its atmosphere is composed of methane, hydrogen, and ammonia. Because of this and its temperature of −220° F, Jupiter is a planet incapable of supporting human life.

A few minutes away from this great planet we encounter thousands of small minor planets orbiting the sun. With diameters ranging in size from 488 miles to less than 1 mile, none of these small objects is large enough to retain an

Fig. 1. The Great galaxy in Andromeda is a separate star system 2 million light-years from our galaxy. With the naked eye, binoculars, or even a small telescope, only the bright central area can be seen. M32 at left and NGC 205 at right are this galaxy's two companion galaxies. (Courtesy University of Chicago Yerkes Observatory, Williams Bay, Wisc.)

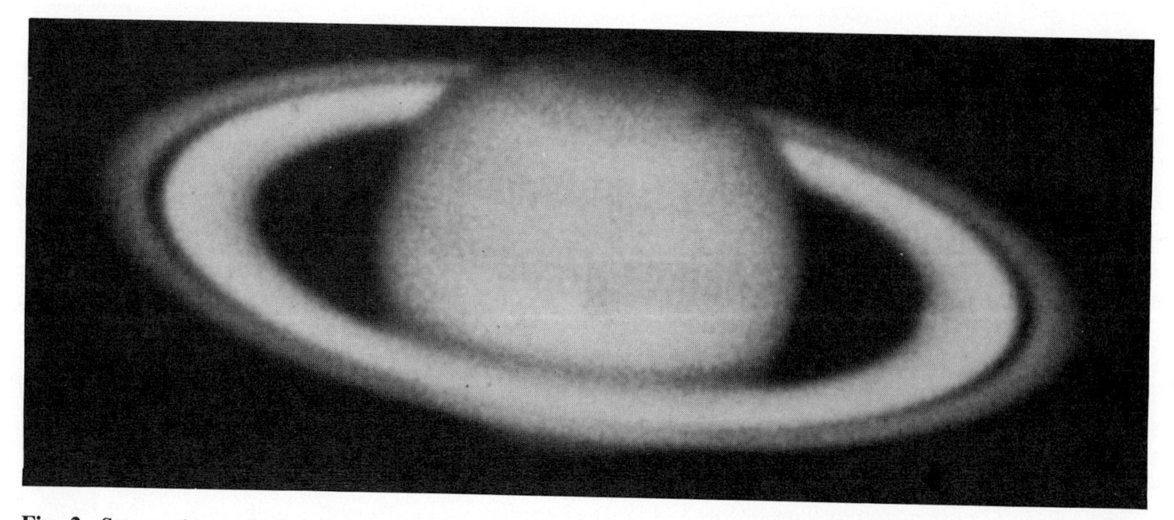

Fig. 2. Saturn. Atmospheric bands and three regions of the rings (the outer ring, Cassini's division, and the bright ring) show clearly in this photograph taken with the 60-inch reflecting telescope at Mt. Wilson Observatory. (Courtesy University of Chicago Yerkes Observatory, Williams Bay, Wisc.)

atmosphere, a prerequisite for the natural survival of man.

The next stop on our exciting journey brings us in 30 minutes to the small red planet Mars, which, accompanied by its two small satellites, completes one orbit of the sun every 687 days. We can actually see the prominent polar ice caps of frozen carbon dioxide at both the north and south poles. Mars is 4200 miles in diameter and has a thin atmosphere composed chiefly of carbon dioxide with traces of oxygen and water vapor. While the planet has a pleasant temperature of about 80° F on the sunlit hemisphere, its nighttime reading plunges to a bitter-cold −100° F. Although Mars comes close to having the capacity to sustain human life, its environment is still far from adequate.

Racing toward the sun from Mars, in 4 minutes we are approaching a breathtakingly beautiful blue planet that has a rather large satellite revolving about it. Both this planet and its barren satellite complete one trip around the sun in 365¼ days. This planet has a diameter of 7927 miles and an atmosphere composed mostly of nitrogen, oxygen, and argon, together with some carbon dioxide, water vapor, and other gases. Its mild average temperature of 60° F makes it a likely place for the survival of man. From a distance we can see white clouds and brownish surface markings on its otherwise bluish face. Ice caps of frozen water appear to mark the polar regions of this most majestic of all planets, Earth.

Making a note to return soon to this very hospitable planet, we continue our tour of the solar system. Two and one-half minutes away from Earth, we come to a planet with an atmosphere so thick that we can see no surface features at all. Venus, as this planet is called, has a diameter of 7700 miles and revolves in a 225-day trip around the sun. Possessing no satellites, Venus is much closer than Earth to the sun, and its atmosphere of carbon dioxide literally traps large quantities of heat. Because of these conditions, the planet has an average temperature of 900° F, not at all suitable for human life.

Leaving Venus, another 2½ minutes pass before we arrive at Mercury, the last and smallest planet in the nine-member planetary family. With the sun just another 3 minutes away, Mercury completes its revolution about this star in 88 days. Mercury has a diameter of about 3000 miles, has no satellites, and broils at a temperature only slightly less than that of Venus. It has a very thin atmosphere of helium and because of this and its proximity to the sun, Mercury is uninhabitable by us.

In order to get a closer look at a star, we must now proceed onward in the direction of the sun. This gigantic ball of glowing turbulent gases is a truly awe-inspiring sight. The visible part, or photosphere, is 864,000 miles in diameter and has a temperature of 11,000° F. The sun's interior temperature is estimated to be as high as 30,000,000° F, and it is truly the most massive body in this solar system. With an interior density estimated at 11 times that of lead, the sun rotates on its axis from west to east, completing one turn in 24 days, 16 hours. The sun is composed primarily of hydrogen and helium, plus smaller quantities of other elements. This giant master of the solar system is consuming itself at a fantastic rate, chiefly through the process of hydrogen fusion, or the conversion of hydrogen atoms into helium atoms. This process releases enormous quantities of energy that flood the entire solar system.

Leaving the vicinity of the sun before we come too close and are consumed, we return to Earth, the small blue sphere lost in the vastness of space. Of all the members of this solar system, only Earth is capable of supporting human life. Of the billions of galaxies we have seen throughout our long journey through the universe, each consisting of billions of stars, could there in fact be other planets like Earth that might develop or sustain life as we know it? From a purely statistical standpoint there should be millions of such planets, but unfortunately we know of only one. Earth, then, is truly an oasis in space.

THE CHALLENGE

Millions of years ago, long before our ancestors appeared on earth, there were natural crises within, on, and above the planet. Still changing, the earth will produce crises as long as its internal processes continue and until the sun's fires are quenched billions of years from now. As passengers on the earth, we have enough natural crises with which to concern ourselves without worsening our lot by creating more problems.

The earth has a finite size and capacity. The entire geo-ecosystem, however, is in a delicate state of balance that can be easily upset, as we are just now beginning to realize. Just as the death of a canary taken into a coal mine alerted miners to the presence of toxic gas and an impending disaster, the earth's "canaries," the small biologic systems, are also beginning to die; this should be a warning to us. Lake Erie (Fig. 3) and the Hudson River are perfect examples of environments that we have killed biologically. The crisis in the petroleum industry is another symptom. If such disasters are allowed to continue unabated as they are at present, the crises produced by overpopulation will undoubtedly spread to larger systems on earth, if indeed they have not already. When the critical land, water, and atmospheric interrelationships are destroyed, human life will then cease on this planet.

The grave problems facing us in the latter part of the

Fig. 3. Sign on the beach at Lake Erie warns that fish and other water species are not the only ones endangered by human habits. (Courtesy United States Environmental Protection Agency, Washington, D.C.)

twentieth century are complex and most of all critical. They must be solved, however, if we are to survive, which is not an easy task for creatures who apparently can neither live in peace with their neighbors nor balance their own budgets. We are so deeply involved with the demands of everyday living that we cannot see ourselves dying in the process. We think that we are at the zenith of evolution—all knowing, all seeing, all powerful. It would appear that for us to solve our problems we must first see ourselves in the proper perspective.

TOWARD THE SOLUTION

The study of the total environment of the earth and its setting in space is called earth science. This is not one specific area of study but rather a group of interrelated fields. The earth sciences encompass astronomy, the study of the universe; geology, the study of the earth; oceanography, the study of the oceans; and meteorology, the study of the atmosphere. Each of these specific areas of study is further subdivided into areas that have become individual fields in themselves. Also involved in the study of the earth sciences are the newer environmental sciences, and, of course, the classic sciences: chemistry, physics, and biology.

A study of the earth sciences not only enable us to be aware of our place in the scheme of nature but also creates awareness of our limitations and responsibilities. Equipped with a basic understanding of the problems seen in their proper environmental setting, we can then begin the task of making earth a spaceship that will sustain us on our voyage through time.

part one **ASTRONOMY**

One might think the atmosphere was made transparent
with this design, to give man, in the heavenly bodies, the
perpetual presence of the sublime. Seen in the streets of the cities
how great they are! If the star should appear one night
in a thousand years, how would men believe and adore;
and preserve for many generations the remembrance
of the city of God which had been shown. . . .
He who knows the most, he who knows what sweets and virtues
are in the ground, the waters, the planets, the heavens,
and how to come at these enchantments, is the rich and noble man.

Nature
Ralph Waldo Emerson

Earth as it appears from 22,500 miles in space. This photograph was taken by the ATS 3 weather satellite on July 4, 1973; continental outlines, latitude, and longitude are superimposed. (Courtesy Environmental Science Services Administration, Washington, D.C.)

1 *widening horizons*

A brief history of astronomical thought

. . . but I rather say
That you remain
A world above man's head,
To let him see
How boundless might his
Soul's horizons be . . .

A Summer Night
Matthew Arnold

Astronomy, the study of the universe, is the oldest of all the sciences, for people have always looked to the heavens with wonder and curiosity. In fact, speculation about astronomy coverges with philosophy and religion in questions about the origins of the world and the place of people in it. Written records from as long ago as 3000 B.C. reveal the fascination of early man with celestial phenomena (Fig. 1-1).

ANCIENT ASTRONOMY

Ancient astronomy encompasses the period from prehistory to the year 1543. This year saw the publication of *De Revolution Orbium Coelestium,** the great work of Nicolas Copernicus that revolutionized astronomical thought.

The early Chinese

The Chinese culture is not only one of the oldest of all contemporary civilizations, it also has the longest continuous recorded history, which dates back to about 3000 B.C. It is here that the record of ancient astronomy begins.

The early Chinese made systematic observations and kept good records. They noted the movements and positions of the five planets visible with the naked eye—Mercury, Venus, Mars, Jupiter, and Saturn—even though the true nature of these celestial bodies was unknown to them. The Chinese, like virtually every culture throughout history, developed and used a calendar, a system of fixing the division of time in such a way as to suit their needs. With their calendar the Chinese predicted the start of the seasons. They cataloged 284 constellations, star groups that appear to belong together to form a figure; each consists of about five stars. The appearance of sunspots was recorded, but it was believed that they were in the earth's atmosphere. Comets, meteors and meteorite falls, and eclipses of the sun and moon were observed and recorded. The Chinese developed a catalog of stars containing more than 800 entries, and from time to time they observed the appearance of a "guest star," now known as a nova or supernova. The most famous supernova explosion ever observed was recorded by the Chinese in 1054 A.D.; its remains are still visible today as the Crab nebula in the constellation of Taurus the Bull (Fig. 1-2).

The Babylonians

Babylonia, an ancient near-eastern kingdom prominent in the second and first millennia B.C., was located in an area corresponding approximately to modern Iraq. This early culture developed on the floodplains of the Tigris and Euphrates Rivers, the area historically known as the "Fertile Crescent." It is generally believed that the roots of western civilization lie close to ancient Babylonia, and some theologians believe this area was the site of the biblical Garden of Eden. Some people also believe that Babylonia was the location of the famous flood of Noah.

*The title is translated as *The Revolution of Celestial Bodies*.

Fig. 1-1. Babylonian clay brick found near Larsa shows this early civilization's concept of constellation of Leo the Lion. (Courtesy Adler Planetarium, Chicago, Ill.)

Fig. 1-2. Crab nebula in Taurus is all that remains after the explosion of a star that was observed by Chinese astronomers in 1054 A.D. (Courtesy University of California Lick Observatory, Santa Cruz, Calif.)

The Babylonians made great advances in astronomy. Among the oldest Babylonian astronomical records are observations of the planet Venus. Mention is also made of the zodiacal belt and its constellations, through which the sun was observed to move during the year. The Babylonians developed a 12-month, 360-day lunar calendar and every 6 years added an extra 30-day month to make up the discrepancy that they knew existed between their observations of the motion of the moon and of the sun. The Babylonians are credited with the invention of the sundial, even though the oldest remains of such an instrument were found in Egypt. The Babylonians knew the positions of the sun, moon, and the five visible planets, and by 250 B.C. they had discovered the 18-year eclipse cycle, the *saros,* which enabled them to predict with great accuracy the dates of future solar and lunar eclipses. Astrology, of such great interest in the past and again today, apparently had its beginning in Babylonia. Many of the constellations we refer to today, such as Orion the Hunter and Leo the Lion, and many familiar stars in twentieth-century skies such as Aldeberan in the constellation of Taurus and Sirius in the constellation of Canis Major, the Big Dog, were named by these ancient people.

The Egyptians

The Greek historian Herodotus (ca 484-ca 425 B.C.) described Egypt as "the gift of the Nile." This statement accurately characterizes Egypt's total dependence on the great river that formed and nourished its land and without which the Egyptian culture would not have come into being.

The Egyptians' dependence on the Nile led to many of their early astronomical discoveries. The early Egyptian astronomers, priests in the pharaoh's court, were the first to discover the 365-day year, a finding that was due primarily to the flooding of the Nile. One day each year Sirius, the brightest star, rises at the same time as the sun. The early Egyptians knew that this event was followed closely by the Nile flood. So regular was the flooding of the river that the average interval between successive floods was computed to be 365 days. (The average of not more than 50 years' observation would have given the true length of the year—365¼ days—but as it was the Egyptians were very close.) Their calendar was divided into 12 months, each of 30 days' duration. At the end of the year, 5 days were added to be consistent with their observations.

Other important indirect astronomical contributions of the early Egyptians were in the areas of mathematics and surveying. They developed concepts of arithmetic and geometry and applied them to measurements of the land and the construction of temples and pyramids.

The early Greeks

Ancient astronomy reached its zenith under the early Greeks (ca 600 B.C.– ca 400 A.D.), who applied to astronomy the advances they had made in mathematics. So dominant was Greek thought that vestiges of the Greek concept of the universe survived virtually unchallenged into medieval times. Since scientific experimentation was almost nonexistent, the Greek scholars worked with logic, trying to arrive at valid conclusions based on observations. Because they had no reliable method by which to test their conclusions, many errors became accepted as true explanations of the universe. Nonetheless, many brilliant and useful astronomical ideas did emerge from the work of the early Greeks.

Among these was the concept of the universe called the celestial sphere. In this scheme the earth was a fixed, flat body occupying the center of the universe. Surrounding the earth was a hollow, spherical shell—the celestial sphere. Some Greeks envisioned a sphere composed of crystalline material with tiny jewel-like stars embedded in it, while others believed that the sphere contained holes through which light from a more distant region of fire shone through. Even though the celestial sphere concept is false, the illusion is so convincing that an imaginary sphere of stars is still useful in understanding astronomy (Chapter 2).

Through observations of solar eclipses, the early Greeks realized that the stars were present during the day as well as at night. They were familiar with the motion of the sun on the celestial sphere, the annual path of which they called the *ecliptic,* for they knew that in order for eclipses to occur, the moon and sun had to be on this path. The sky visible from Greece over the course of a year was divided into constellations named after gods in the Greek mythology. These early scholars further developed the Babylonian idea of the zodiac, dividing the narrow zone on either side of the ecliptic into 12 parts, or signs, each occupied by a constellation. The Greeks differentiated between fixed stars and wandering stars; the latter they named planets. Some individual contributions are so important that they merit closer study.

Thales and Anaxagoras

Thales (ca 600 B.C.) introduced geometry and surveying from Egypt and used old Babylonian records to predict eclipses of the sun and moon. Anaxagoras (ca 500 B.C.), on the other hand, concluded that the sun was not a god but was rather a lump of incandescent stone larger than the moon, even though the two appeared to be the same size. Because of his theories, which contradicted the accepted beliefs of his time, Anaxagoras was prosecuted and exiled.

Pythagoras and Philolaus

Pythagoras (570-490 B.C.) believed that the earth was spherical instead of flat. Philolaus (ca 350 B.C.), a student of Pythagoras', believed that the earth revolved around a central fire that was not the sun. In Philolaus' scheme the earth was balanced by a counterearth, and the sun, moon, and all the planets revolved around the central fire. This was the first known concept of the universe that was not earth centered, although the idea of a heliocentric (sun-centered) universe did not gain widespread acceptance until well into the sixteenth century A.D.

Aristotle

Aristotle (384-322 B.C.) was primarily responsible for developing the logical though incorrect concept that a fixed, spherical earth occupied the center of the universe. Although several Greek scholars took exception to the idea, so convincing were Aristotle's arguments that the geocentric (earth-centered) concept of the universe became the accepted model for almost 2000 years. Aristotle also offered explanations of the phases of the moon and the occurrence of eclipses.

Aristarchus and Eratosthenes

Aristarchus (310-230 B.C.) developed methods of measuring the relative distances from the earth to the moon and sun as well as their relative sizes. His methods were based on the construction of triangles. Aristarchus concluded from his calculations that the moon was 10 earth diameters away from the earth and that the sun was 200 earth diameters distant. By studying eclipses and the sizes of shadows cast by the earth and moon, Aristarchus concluded that the diameter of the moon was one-third that of the earth, while the diameter of the sun was seven times the diameter of the earth. Although the results of his calculations were incorrect, his methods and basic approach to the problem are consonant with those in use today. Aristarchus was also among the first to believe that the earth rotated on its own axis and revolved around the sun, which he thought was the center of the universe. Aristarchus also correctly explained the changing of the seasons—the 23½° tilt of the earth on its axis (Chapter 9). Eratosthenes (276-194 B.C.) first determined the earth's circumference by measuring an arc of longitude between Alexandria and Syene, Egypt (Fig. 9-1).

Hipparchus

The greatest of the early Greek astronomers was Hipparchus (190-125 B.C.). By observing the sun's motion, he was able to determine the exact length of the seasons, and he made a table giving the sun's position for every day of the year. He also catalogued and gave positions for more than 800 stars, which he divided into six categories of brightness. Comparing the positions of these stars with positions that had been given by other observers 150 years earlier, he found that the stars' locations had changed. He thereby discovered the phenomenon known today as *precession,* the slow wobbling of the earth's rotational axis caused by the gravitational pull of both the moon and the sun (p. 21). Because of his expanded knowledge of the motions of the sun and moon, Hipparchus was able to predict eclipses more accurately than his predecessors. He is also credited with the further development of trigonometry, which he used in his calculations, and with devising the system of latitute and longitude by which points on the earth's surface are still located (p. 17).

Ptolemy

Ptolemy (100-170 A.D.) is considered the last of the great Greek astronomers, even though he was an Egyptian by birth. Ptolemy's major contribution was his great work called *The Almagest,* a compilation of both his own findings and earlier observations, especially those of Hipparchus.

The Almagest consists of 13 books, the first of which contains the principles of astronomy and spherical trigonometry. The second book lists the zones into which Ptolomy divided the earth and explains the rising and setting of the sun, while the third book gives the length of the year and sets forth a theory of solar motion. The fourth book deals with the length of the month and explains lunar motion, and book five lists the distances of the earth from the sun and moon as well as details of the instruments Ptolemy used in making his observations. The sixth book explains the positions of the sun and moon and the occurrence of eclipses. The seventh and eighth books, based on the work of Hipparchus, contain a catalog of more than a thousand fixed stars and include a discussion of precession. The remaining books contain an explanation of planetary motion that has come to be called the Ptolemaic system.

The Ptolemaic system is a geocentric theory of planetary motion in which each planet moves uniformly in a small circle called the *epicycle*. At the same time the center of this small circle moves along the circumference of a larger circle called the *deferent,* which in turn is centered on the earth (Fig. 1-3). The sun and moon, at that time also considered to be planets, were thought to orbit the earth in circular paths called *eccentrics*.

Such combinations of circular paths were intended to duplicate the observed courses of the five known planets around the celestial sphere and to explain the variable

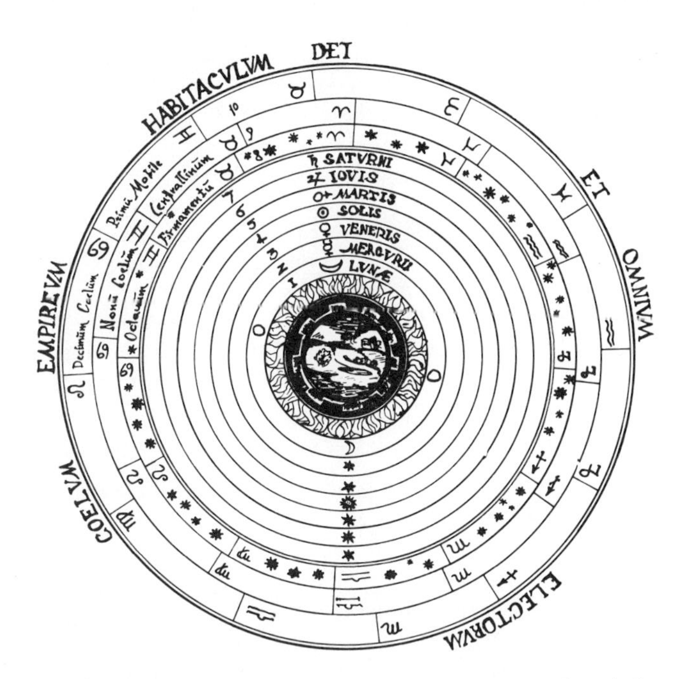

Fig. 1-3. Ptolemaic concept of the universe showing the relative placement of celestial bodies. This concept prevailed until the Renaissance. The original drawing was published in Copernicus' *Cosmographia* in 1543.

speeds of the sun and moon. The main purpose of the Ptolemaic system was to predict the positions of these bodies for any desired date. So convincing was Ptolemy's work that his system remained the accepted explanation for the workings of the universe until the publication of Copernicus' work more than 1300 years later.

Hindu and Arabian astronomy

The 1300 years from the death of Ptolemy to the time of Copernicus was a very uneventful period in the history of astronomy. The ancient civilizations had either disappeared or were on the decline, while European cultures were entering the Dark Ages. The only important contributions to astronomy during this period were made by the Hindu and Arabian astronomers. The Hindus, ancestors of present-day inhabitants of modern India and Pakistan, developed our system of numbers, using them in calculating such astronomical data as the distance from earth to the moon, which they measured as 64½ earth radii (the actual figure is about 60).

The Arabs primarily provided continuity between ancient and modern astronomy. They brought the Hindu numbering system to Europe and developed the new mathematical tool called algebra. Some of the stars, such as Betelgeuse, Mizar, and Dabiha, bear Arabic names today.

Early astronomy of the Western Hemisphere

We often tend to forget that while the cultures of the Babylonians, Egyptians, Greeks, Hindus, and Arabs were flourishing on one side of the world, other civilizations were developing in the Western Hemisphere. Isolated physically and intellectually from the rest of the world by the great oceans, the civilizations of North and South America seem to have gone through ages and phases of cultural development as though they were on a separate planet bound on an independent course.

Most pre-Columbian North American cultures did not have comprehensive written records. Traces of recent achievements were wiped out by conquerors from across the eastern seas. Never before had entire civilizations been so suddenly and thoroughly destroyed. As a result, the details of early American cultures lay buried until nearly 100 years ago, when archeologists and historians began to explore this part of the past. As data trickled in, it became apparent that the Olmecs, Chavins, Moches, Mayans, Nazcans, Aztecs, and Incas had an amazing interest in and knowledge of astronomy. Today, unfortunately, the true meaning of much of their knowledge is still unclear.

The Mayans (ca 1000 B.C.–ca 300 A.D.), for example, have been called the "Greeks of the New World" for theirs were the highest intellectual achievements of pre-Columbian America. They had a calendar more precise than that used in Europe and discovered and applied the integer 0 before the Arabs. They had a vast astronomical knowledge and were highly interested in measuring time.

The people of the Nazcan culture (ca 500 A.D.), on the other hand, laid out huge geometric figures in the desert of sourthern Peru. The lines of these figures coincide in direction with the position of the sun at the start of the seasons, and these constructions have been called the "largest astronomy book in the world." The Aztecs (ca 1200 A.D.) made a calendar stone, a circular rock 13 feet (4 meters) in diameter and weighing 20 tons. This huge stone was not only an intricate calendar but also a synthesis of the Aztec view of the universe with the sun at the center. A replica of this calendar stone is the national emblem of Mexico. The Incas (ca 1400 A.D.) constructed sundials chiseled from solid rock. It is thought that they were used for the observation of the sun's position on the first day of summer and winter.

MODERN ASTRONOMY

The modern period of astronomy extends from 1543 to the present, and it is a story of men, not cultures.

Nicolas Copernicus

Nicolas Copernicus (1473-1543), a Polish mathematician and astronomer, is acknowledged as the founder of modern astronomy. After a visit to Italy where the Pythagorean speculations were known, Copernicus began observations that became the basis for sun-centered theory of the universe. Copernicus showed that this theory explained the facts more simply than the deferents and epicyles of Ptolemy.

Copernicus explained the motion of the celestial bodies by assuming that the earth and other planets revolve around the sun while rotating at the same time on their own axes (Fig. 1-4). The earth's rotation explained the rising and setting of the sun and stars, while the sun's changing position among the stars was explained by the movement of the earth in its orbit. As Copernicus stated:

> First and above all lies the sphere of fixed stars, containing itself and all things, for that very reason immovable; in truth the frame of the universe, to which the motion and position of all other stars are referred. Though some men think it to move in some way, we assign another reason why it appears to do so in our theory of the movement of the earth. Of the moving bodies first comes Saturn, who completes his circuit in 30 years. After him, Jupiter, moving in a twelve year revolution. Then Mars, who revolves biennially. Fourth in order an annual cycle takes place, in which we have said is contained the earth. With the lunar orbit as an epicycle. In the fifth place Venus is carried round in nine months. Then Mercury holds the sixth place, circulating in the space of 80 days. In the middle of all dwells the sun. . . . We find, therefore, under this orderly arrangement, a wonderful symmetry in the universe, and a definite relation of harmony in the motion and magnitude of the orbs, of a kind it is not possible to obtain in any other way.

Copernicus assumed that the planets had circular orbits, however, and because their observed motions were not uniform, he had to retain some of Ptolemy's epicycles. It remained for Johannes Kepler in the seventeenth century to correct this assumption by introducing the concept of elliptical orbits. Galileo, also in the seventeeth century, furnished the first empirical evidence for the validity of Copernicus' theory. By discrediting the belief that the celestial sphere

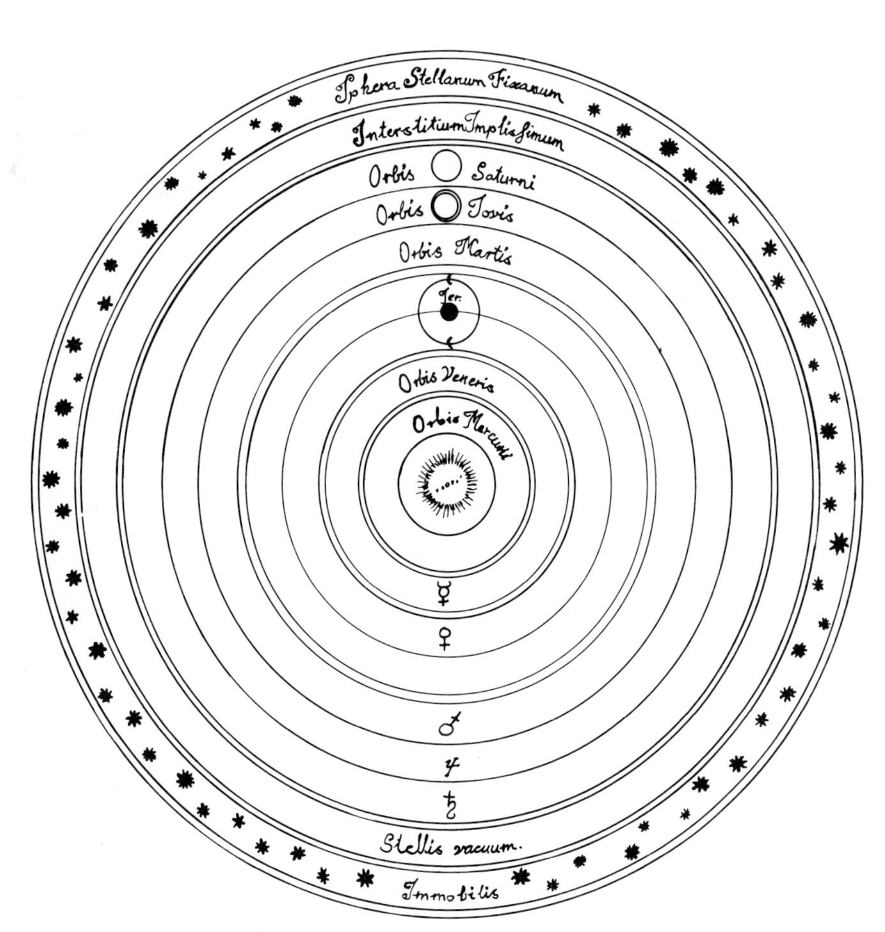

Fig. 1-4. Copernican concept of the universe. This drawing must be compared with Fig. 1-3 for a full appreciation of the progress astronomers made during this one revolutionary century. The original drawing was first published in 1647 in Galileo's *Selenographia*.

must be small enough to rotate daily around the earth, the Copernican theory enlarged man's concept of the universe and thus became the basis of modern astronomy.

Tycho Brahe

Tycho Brahe (1546-1601), the most notable of Copernicus' immediate successors, was a Danish astronomer. He is remembered as the greatest astronomer before the advent of the telescope because of his many precise observations of the positions of stars and planets, especially Mars. At this time the Copernican theory was being put to a critical test and was still not widely accepted, even by Tycho.

Tycho was the observer *extraordinaire,* so it is perhaps not surprising that his critical objection to the Copernican system was based on a simple observation, the absence of stellar parallax (Fig. 7-3). If the earth moved around the sun as Copernicus had said, some displacement of the stars should then be visible due to the motion of the observer, a concept called *parallax*. The lack of observed parallax led Aristotle and Ptolemy to the conclusion that the earth was the center of the universe and prevented Tycho from becoming an advocate of the heliocentric concept. Had there been a single star whose parallax could be detected with the unaided eye, Tycho would have discovered it and the triumph of the Copernican system would have been much swifter.

Not being able to reconcile the Copernican system and his observations, Tycho developed his own system, which was really a compromise between those of Copernicus and Ptolemy. In Tycho's system the five known planets orbited the sun, while the sun and its five "satellites" orbited a central, fixed earth. The starry sphere meanwhile rotated around the earth.

If this compromise system had been Tycho's only contribution to astronomy, he would not be so significant a figure. As it was, he provided two new pieces of evidence so important that the traditional picture of the heavens had to be altered. This evidence is of great interest for two reasons. Apart from the impact on man's traditional beliefs, Tycho's discoveries also showed how an observer tends to notice only those things that he is prepared to see. Events such as those Tycho observed in the sky during the 1570s had occurred before and since without creating the same excitement.

The first observation was related to the immutability of the heavens. In the year 1572 a bright light appeared in the constellation of Cassiopeia and grew rapidly in intensity until it was as bright as the brightest stars in the sky. We now know that what Tycho observed was a *supernova,* a faint star that suddenly blew up and gradually burned out. As the finest observer of his time, Tycho was able to establish that, despite all the rumors to the contrary, the light was stationary and in all respects indistinguishable from a normal bright star. A "new" star had thus been found on the supposedly unchanging celestial sphere, and it became known as Tycho's Star.

Tycho's second critical observation was made in 1577, the year in which a great comet appeared in the sky and remained visible for several months. He followed its movement carefully, comparing its changing position with the positions of the stars, moon, and other celestial bodies. Tycho concluded that, contrary to popular belief, the comet was not in the earth's atmosphere but was moving on a course far beyond the moon, a theory that he confirmed trigonometrically. This alone was a significant contribution, but even more important, the comet of 1577 cast doubt on the existence of solid planetary spheres that were commonly believed to hold the planets in place. The presence of comets in the world beyond the moon at once reopened the question of the mechanism of the planetary system; this time the question remained open for more than a century until Newton's theory of gravitation.

Tycho was also one of the best instrument makers ever to work in the field of astronomy. His quadrants and other direction finders were made with great precision. Because he was an honest and systematic observer, Tycho was capable of the most advanced work that it was possible to achieve so long as observations were limited to what was visible to the naked eye. The next step in the destruction of the medieval concept of the universe came some 30 years later as a result of Galileo's work with the telescope.

Galileo Galilei

Galileo (1564-1642), an Italian, was the founder of the science of physics and observational astronomy. A list of his excursions into science includes almost every topic in the physical sciences, but two subjects preoccupied him: Copernican astronomy and the mathematical theory of motion. Galileo perceived these as separate areas of study, and this was unfortunate, for if he had tied the two together he could have carried his work even farther.

Although Galileo made only one serious, limited excursion into astronomy, its direction was well chosen and he exploited his discoveries to the maximum. Let his own words of 1610 tell the story:

> About ten months ago a report reached my ears that a Dutchman had constructed a telescope, by the aid of which visible objects, although at a great distance from the eye of the observer, were seen distinctly as if near; and some proofs of its most wonderful performances were reported, which some gave credence to, but others contradicted. A few days after, I received confirma-

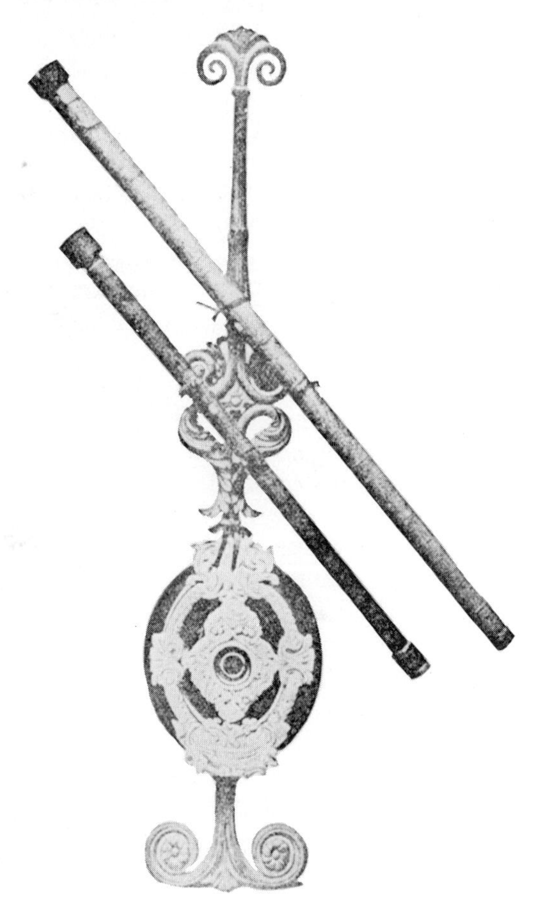

Fig. 1-5. Galileo's delicately ornamented telescopes show few visible features that would relate them to the huge equipment later scientists developed based on Galileo's principle. (Courtesy University of Chicago Yerkes Observatory, Williams Bay, Wisc.)

tion of the report in a letter written from Paris by a noble Frenchman, Jacques Badovere, which finally determined me to give myself up first to inquire into the principle of the telescope, and then to consider the means by which I might compass the invention of a similar instrument, which after a little while I succeeded in doing, through a deep study of the theory of refraction; and I prepared a tube, at first of lead, in the ends of which I fitted two glass lenses, both plane on one side, but on the other side one spherically convex, and the other concave. Then bringing my eye to the concave lens I saw objects satisfactorily large and near, for they appeared one-third of the distance off and nine times larger than when they are seen with the natural eye alone.

Contrary to popular belief, Galileo is not credited with the invention of the telescope but with its application to astronomy (Fig. 1-5). Galileo's contributions to astronomy are impressive:

1. He discovered the true nature of the moon's features.
2. He discovered the four largest satellites of Jupiter.
3. He discovered that the Milky Way was composed of stars.
4. He discovered the phases of Venus, a strong argument for the Copernican system.
5. He described Saturn as a three-body system but could not explain its appearance.
6. He found that sunspots were actually on the sun and not in the earth's atmosphere.

In summary, Galileo did two things for astronomy. First, he made the telescope an indispensable instrument for future observations, since for the first time man became aware of heavenly objects that he could not see with his naked eye. Second, he firmly established the Copernican heliocentric system.

Johannes Kepler

Johannes Kepler (1571-1630) was a German mathematician and astronomer whose self-appointed mission in life was to reveal the new inner coherence of the Copernican system. His main ambition was to produce a "celestial physics" of a new kind, in which the forces responsible for observed phenomena would be brought to light. By 1596 Kepler's ideas were little more than possibilities. To proceed, he needed to check his ideas against the best astronomical records, and this meant that he needed access to Tycho Brahe's work. He contacted Tycho and was hired as an assistant. Thus were joined the incomparable records of Tycho and the magnificent mind of Kepler.

The results of this union were Kepler's three laws of planetary motion, which were based almost entirely on observations of the planet Mars.

FIRST LAW: The planetary orbits take the geometric form of an ellipse, with the sun at one focus (Fig. 1-6).

SECOND LAW: The speed of any planet varies in inverse proportion to its distance from the sun in such a way that the line joining the planet to the sun sweeps out equal areas in equal times (Fig. 1-6).

THIRD LAW: The square of the planetary year (P) is equal to the cube of its average distance from the sun (A):

$$P^2 = A^3$$

Kepler realized that the critical problem of the planetary system was one of dynamic balance, but he did not have the means to discover the true forces that kept the planets in their orbits. As he noted:

If the earth and moon were not kept in their respective orbits by a spiritual or some other equivalent force, the earth would ascend towards the moon one fifty-fourth part of the distance, and the moon would descend the remaining fifty-three parts of the inter-

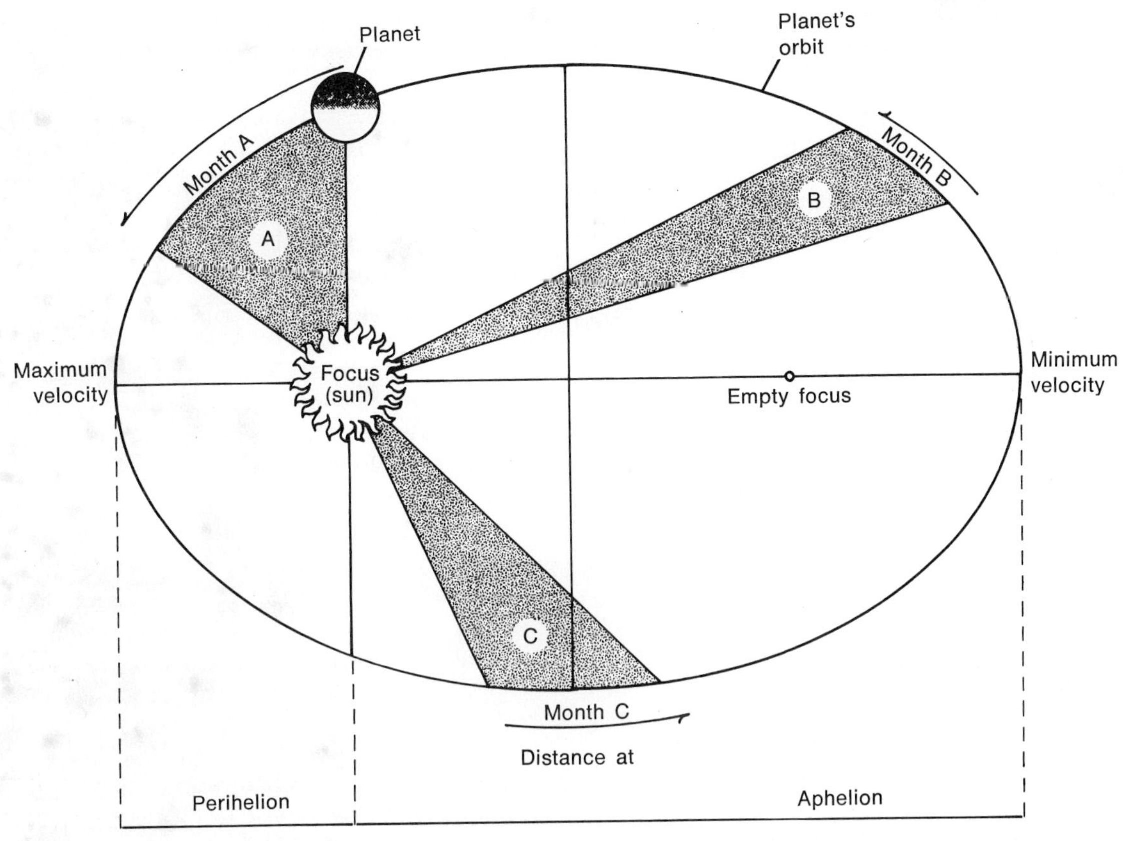

Fig. 1-6. Diagrammatic representation of Kepler's first and second laws. Kepler's first law states that a planet's orbit must be in the shape of an ellipse. Kepler's second law states that the speed of this planet will vary in such a way that during month *A* it will cover exactly the distance required to make the area of gray section *A* equal to the areas of gray sections *B* and *C*.

val, and thus they would unite—assuming that they are both of the same density.

The final combination of astronomy and dynamics was still 60 years away.

Isaac Newton

Isaac Newton (1642-1727) was an Englishman who has been described as "the greatest genius that ever lived" and as "easily the greatest man of physical science in historic times." The passage of more than 200 years has not diminished his reputation. Newton had the imagination and the mathematical capacity first to understand the whole system of ideas that had gone before and then to put the pieces together to form a single picture. His contributions to science may be grouped in three major areas of study: motion and force, light, and calculus.

Motion and force. Newton formulated the basic laws of mechanics, the study of which is now known as Newtonian physics. He showed that these laws applied to celestial objects as well as to objects on earth. Newton's laws of motion, stated in their simplest form, are as follows:

> FIRST LAW: In the absence of an external force, a body at rest will remain at rest or a body in motion will remain in motion at constant speed in a straight line.

Kepler's first law, on the other hand, suggested that motion was the natural state in the universe but that the paths followed by celestial objects were not straight lines and their speeds were not constant.

> SECOND LAW: When a force acts on a body, its motion is changed in the direction of the applied force and the rate of change is proportional to the magnitude of the force.
>
> THIRD LAW: All forces occur in pairs. If a force is exerted on a body, that body reacts with an equal and opposite force.

Assuming that the planets obeyed the same laws of motion as objects on earth, Newton wondered what kept the moon, which moves in an approximate circle around the earth, from flying off into space. He came to the conclusion that it must be the pull of the earth that controlled the moon. He then assumed that every body in the universe is attracted to every other body by a force directly proportional to the masses of the bodies multiplied together and inversely proportional to the square of their distance from each other. Doubling the distance would reduce the attraction to a fourth, tripling the distance would reduce the attraction to a ninth, and so on. This became the *law of universal gravitation*.

By combining this simple law with his laws of motion, Newton proved mathematically that the attraction of the sun would cause the planets to trace out elliptical orbits with exactly the characteristics given by Kepler's three laws, which accurately expressed the facts of observation but offered no reason for them. Now there was an explanation —universal gravity.

Light. Through his work with the prism in 1665, Newton demonstrated that visible light is actually made up of light of various colors, which he listed as red, orange, yellow, green, blue, indigo, and violet. He had thus discovered the continuous spectrum of light. He also established that light of a definite color will always be affected in the same way by the prism, which explained the color aberration of lenses, a common fault of refracting telescopes. To correct the problem he constructed the first reflecting telescope, which contained a mirror rather than a lens. Today all the largest telescopes in the world are of this type. Much of our knowledge about the universe has been gained through the study of spectra (Chapter 5).

Calculus. Lacking an existing mathematical tool to do his work, Newton proceeded to invent one, which he called "methods of fluxions." Today it is known as infinitesimal calculus, which consists of two branches, *differential* calculus and *integral* calculus. Calculus involves the rates of variations of quantities, and since celestial objects are in constant change, it has been an indispensible tool to the astronomer.

The launching of the first unmanned and manned earth satellites of the late 1950s, the lunar flights and manned landings of the 1960s and 1970s, and the interplanetary unmanned flights all required application in Newton's laws. Despite Newton's extraordinary achievements, he was aware that much remained to be accomplished. Shortly before his death in 1727 he said:

> I do not know what I appear to the world; but to myself I seem to have been only like a boy, playing on the seashore, and diverting myself in now and then finding a smoother pebble or a prettier shell than ordinary, while the great ocean of truth lay all undiscovered before me.

• • •

It is impossible in such a brief chapter to give the history of all developments in astronomy. These highlights of important beliefs, discoveries, and events show the evolution and maturation of our concept of the universe, a process that continues today and will no doubt continue forever. Other historically important contributions are noted throughout the chapters that follow.

2 *the sky as seen from earth*

The celestial sphere

When a soft and purple mist
Like a vaporous amethyst,
Or an air-dissolved star
Mingling light and fragrance, far
From the curved horizon's bound
To the point of Heaven's profound,
Fills the overflowing sky . . .

Lines Written Among the Euganean Hills
Percy Bysshe Shelley

The celestial sphere, our view of the sky from earth, is one of the most convincing optical illusions that we will ever see. The earth appears to be stationary while the sky appears to turn. Celestial objects appear to rise in the east, move across the sky, and set in the west. The sky appears to be curved, and all celestial bodies appear to be at the same distance from the observer. The sun appears to slowly shift its position from north to south and back again during the course of the year. The sun and moon appear to be the same size in the sky but each appears to increase in size when it rises or sets. All are optical illusions.

As mentioned in Chapter 1, the ancient Greeks were primarily responsible for developing the concept of the celestial sphere, and to them the illusion was real. The earth, or terrestrial sphere, was a fixed body occupying the center of a great spherical hollow shell, the celestial sphere, in which were embedded all celestial objects. The Greeks did not attempt to discover how far above the surface of the earth the celestial sphere was located. However, they reasoned that the celestial sphere was of great size and at a great distance from the earth. Otherwise, as people moved about from place to place on the earth's surface, they would see an apparent displacement in the direction of the stars. The Greeks also reasoned that the celestial sphere rotated on an axis that passed through the earth, for the rotation of the earth was unknown to them. The rotation of the celestial sphere then caused the stars to rise in the east, move across the sky, and set in the west.

Today, with nearly 2000 years of experience behind us, we know that the celestial sphere does not really exist. The sky only looks like a sphere because of the curvature of the spherical earth. Celestial objects only appear to be embedded in the sphere because they all are at such great distances from us that they look as though they are at the same distance. The sky only appears to move because of the earth's rotation (Chapter 9). Even though the celestial sphere is an optical illusion, the concept is still very useful to navigators and astronomers and an unending source of pleasure to the amateur observer.

POINTS, CIRCLES, AND POSITIONS ON EARTH'S SURFACE

As people learned more about the earth on which they lived and began to travel across longer distances, their need for a system that could enable them to locate their position on the earth's surface and to tell time became increasingly apparent. The problem was solved by devising an imaginary coordinate system of grids that consisted of great and small circles. The present system of grids was developed during the seventeenth century at the Royal Greenwich Observatory just outside London.

An imaginary plane passes through the earth's center, intersecting the surface along a path known as a great circle. Stated another way, a *great circle* is a circle drawn on the surface of a sphere in such a way that the geometric center of the circle and the center of the sphere coincide. A great circle always divides a sphere such as the earth into two equal halves. All other circles constructed on a sphere are called *small circles*. There can be no straight lines on a sphere, only circles or parts thereof called *arcs*.

The coordinate system used today consists of a series of imaginary great and small circles arranged on the earth's surface so that they intersect each other at 90° angles (Fig. 2-1). One series runs west to east, the other runs north to south. The west-east system begins with a great circle called the *equator,* which encircles the earth at its exact midpoint. The equator divides the planet into two equal Northern and Southern Hemispheres. Arranged north and south of the equator and running parallel to it are a series of small circles known as *parallels,* which have smaller and smaller circumferences as they approach the poles. The angular separation between the equator and any parallel is called *latitude,* which is reckoned in *degrees, minutes,* and *seconds* northward or southward from the equator. The equator is numbered 0° latitude and the north and south geographic poles are numbered 90° N and 90° S, respectively. A degree of latitude is equal to about 70 miles (112.7 km), a minute of latitude is about 1.2 miles (1.9 km), and a second of latitude is about 0.2 miles (0.32 km), or approximately 100 feet (30.5 meters).

Intersecting the system of parallels of latitude at right angles is the system of *meridians of longitude*. A series of 180 great circles run north to south through the poles, completely encircling the earth. Each circle is divided by the poles into two halves known as meridians. The total number of meridians, then, is 360, corresponding to the number of degrees in a circle or around a sphere.

The meridian that passes through the Greenwich Observatory is named the *prime meridian*. The angular separation of any meridian from the prime meridian is called *longitude,* which is reckoned in degrees, minutes, and seconds east and west of the prime meridian to a maximum of 180°.

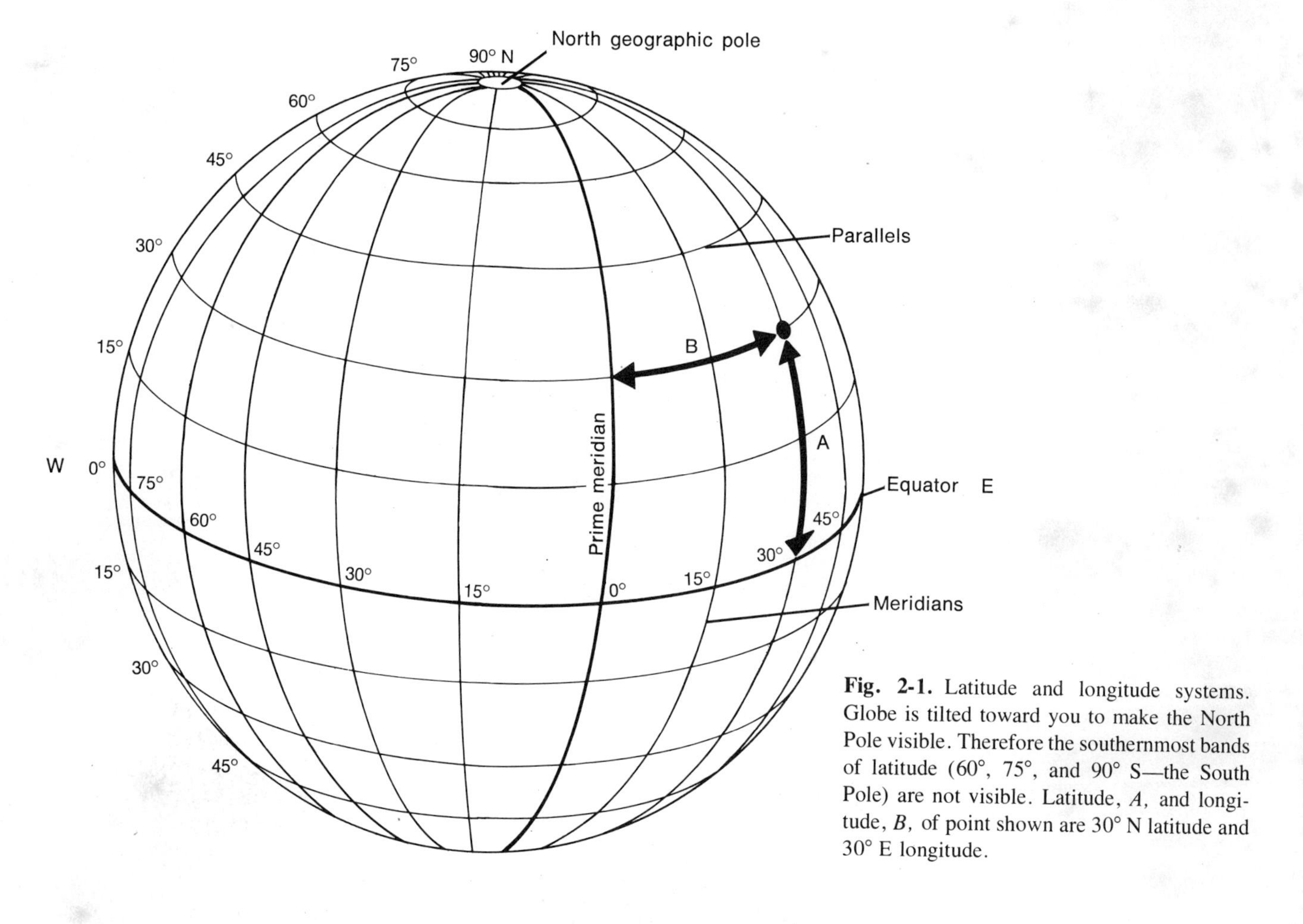

Fig. 2-1. Latitude and longitude systems. Globe is tilted toward you to make the North Pole visible. Therefore the southernmost bands of latitude (60°, 75°, and 90° S—the South Pole) are not visible. Latitude, *A,* and longitude, *B,* of point shown are 30° N latitude and 30° E longitude.

The prime meridian, like the equator, is numbered 0° longitude. All meridians converge at the poles and reach their maximum separation, about 70 miles (112.7 km), as they cross the equator. Therefore the separation in miles between meridians of longitude is a function of the latitude and is not a fixed and constant distance, as is the distance between parallels of latitude.

A simple analogy may help you to understand the system more clearly. The earth's coordinate system is almost exactly analogous to a street-avenue grid network used in many towns and cities. Streets generally run from west to east, while avenues are usually oriented north to south. A main street usually divides the city into northern and southern sectors, while a main avenue separates the city into western and eastern parts. All streets will run approximately parallel to the main street and all avenues will run nearly parallel to the main avenue. If you wished to know how far away 62nd Street was from the main street, you would immediately say 62 blocks. Even though the streets run west to east, you would almost instinctively go north or south of the main street to find this location. Likewise, if you wanted to find 18th Avenue, you would go either west or east of the main avenue, which runs north to south.

The same logic may be applied to locations on earth. To find the latitude of west-east parallels, one must go north or south of the equator. To obtain the longitude of north-south running meridians, one must go west or east of the prime meridian. The latitude and longitude of Miami, Florida, for example, is approximately 26° N and 80° W. This city is therefore found 26 parallels north of the equator and 80 meridians west of the prime meridian.

POINTS AND CIRCLES ON THE CELESTIAL SPHERE

We have learned that a series of imaginary points and circles have been drawn on the surface of the earth for the

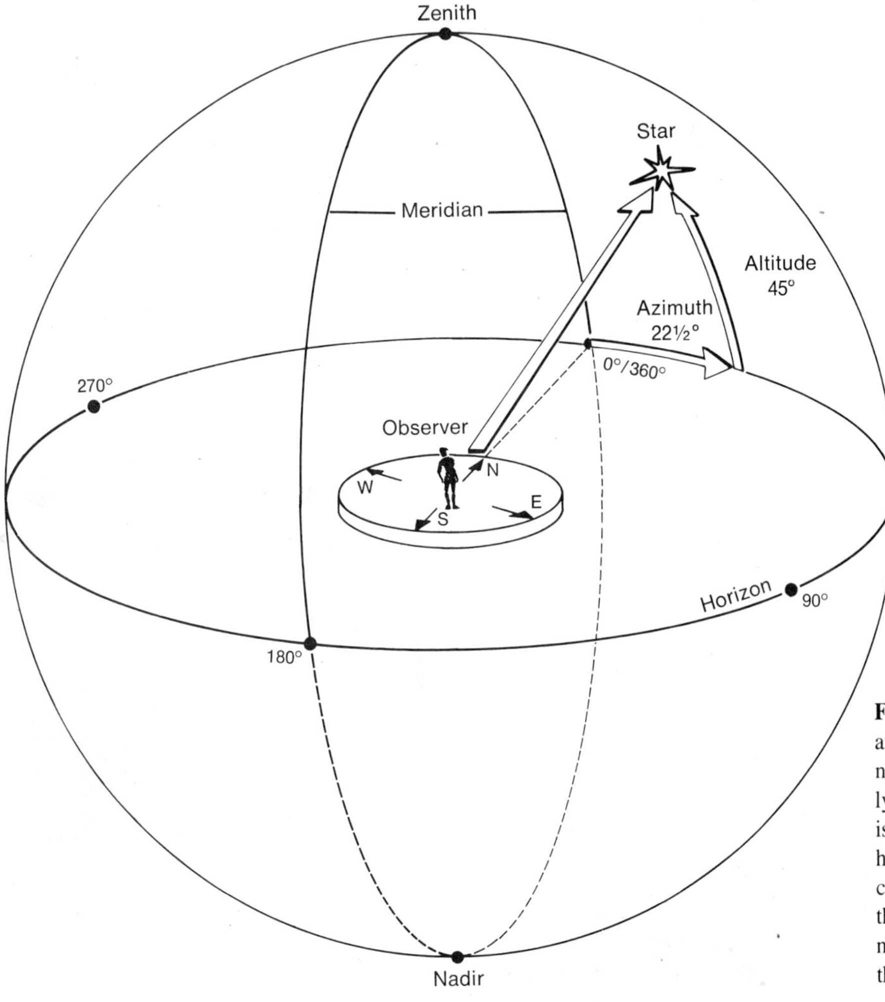

Fig. 2-2. Points of observer-oriented altitude-azimuth system. Circle that connects the zenith, nadir, and the two points on the horizon lying due north and due south of the observer is called the meridian. Once the meridian and horizon are established, any celestial object can be located by measuring the arc between the object and one or the other of these imaginary circles. Azimuth (horizontal distance) in this case is 22½° and altitude is 45°.

purposes of navigation and time keeping. The celestial sphere also has imaginary points and circles that enable the observer to locate the position of celestial objects. Two groups of points and circles are in common use today; one is observer oriented and thus variable, while the other is sky oriented and therefore fixed.

Observer-oriented points and circles

For the observer standing on its surface, the earth always fills half the sky. This immediately reduces the observer's view of the celestial sphere by 50%. To help observers orient themselves, a number of imaginary points and a circle have been devised. Their location depends on the observer's position (Fig. 2-2).

Zenith and nadir

The point on the celestial sphere directly above the observer's head is called the zenith. Everyone has a different zenith; as the observer moves about on the surface, so does this important point. The point on the celestial sphere 180° from the zenith is the nadir. The nadir is always directly under the observer's feet, obscured by the earth.

Horizon

Ninety degrees from the zenith or nadir is a point on the celestial sphere called the *true* or *astronomical* horizon. To a geographer, the horizon is anywhere the sky meets the land, a building, or any object. The astronomer calls this the *apparent* horizon. For an observer on the earth's surface the true and apparent horizons never coincide because of the curvature of the earth. The true horizon, then, is always below the apparent horizon; as a result, it is never seen but is rather determined mathematically. The true and apparent horizons come closer to coinciding at sea than at any other location.

Meridian

The meridian is an imaginary great circle on the celestial sphere that runs from north to south through the observer's zenith, northern and southern horizon points, and nadir. The meridian completely encircles the celestial sphere, even though observers in their mind's eye can only see half of the meridian at any given time. We call the part of the meridian that is above the horizon at all times the *meridian arc;* this always divides the observer's visible sky into two halves—eastern and western. The effect seen by observers is that their meridian remains stationary while the celestial sphere sweeps by every 24 hours.

The meridian is a concept that is at least in part well known to almost everyone. We know that A.M. means morning and P.M. means evening. Actually the abbreviation A.M. is an abbreviation for *ante meridian* (before the meridian) and P.M. is the abbreviation for *post meridian* (after the meridian). The instant the sun appears to cross the meridian during daylight hours is called *noon*. Twelve hours later the sun again crosses the meridian, but this time we cannot see its passage because that part of the meridian below the horizon is involved. The instant of this passage is called *midnight*.

Altitude-azimuth coordinate system

The altitude of a celestial object is its "height" above the true horizon (Fig. 2-2). Altitude is measured in angular degrees, minutes, and seconds along an imaginary arc from the true horizon (through the object to be measured) to the zenith. The true horizon in this case is said to be 0° and the zenith 90°. If a star, for example, were located on this arc midway between the true horizon and the zenith, its altitude would be 45°.

Azimuth, on the other hand, is the number of degrees, minutes, and seconds of angular measurement along the horizon from the north point to the altitude arc of the object (Fig. 2-2). The north horizon point is called 0° and 360°. East is 90°, south is 180°, and west is 270°. For example, if the star with an altitude of 45° were north-northwest of the observer, its azimuth would be 22½°.

Combining the aspects of observer-oriented points, circles, and measurements produces the altitude-azimuth coordinate system that may be used to locate celestial objects. The main shortcoming of the observer-oriented system is that the altitude-azimuth coordinates continually change as the celestial sphere (earth) turns. For example, a star rising in the southeast would have an altitude of 0° and an azimuth of 135°. Six hours later its altitude would be 30° and its azimuth would be 205°.

Sky-oriented points and circles

The earth-fixed points and circles, consisting of the north and south geographic poles, the equator, parallels, and meridians, have been extended to the celestial sphere to form sky-oriented points and circles (Fig. 2-3). While the observer-related points and circles change with the movement of the observer or are different for different observers, sky-fixed points and circles remain constant except over long periods of time.

Celestial poles

The points on the celestial sphere immediately above the earth's geographic North and South Poles are the north and south celestial poles. Just as the earth rotates about the

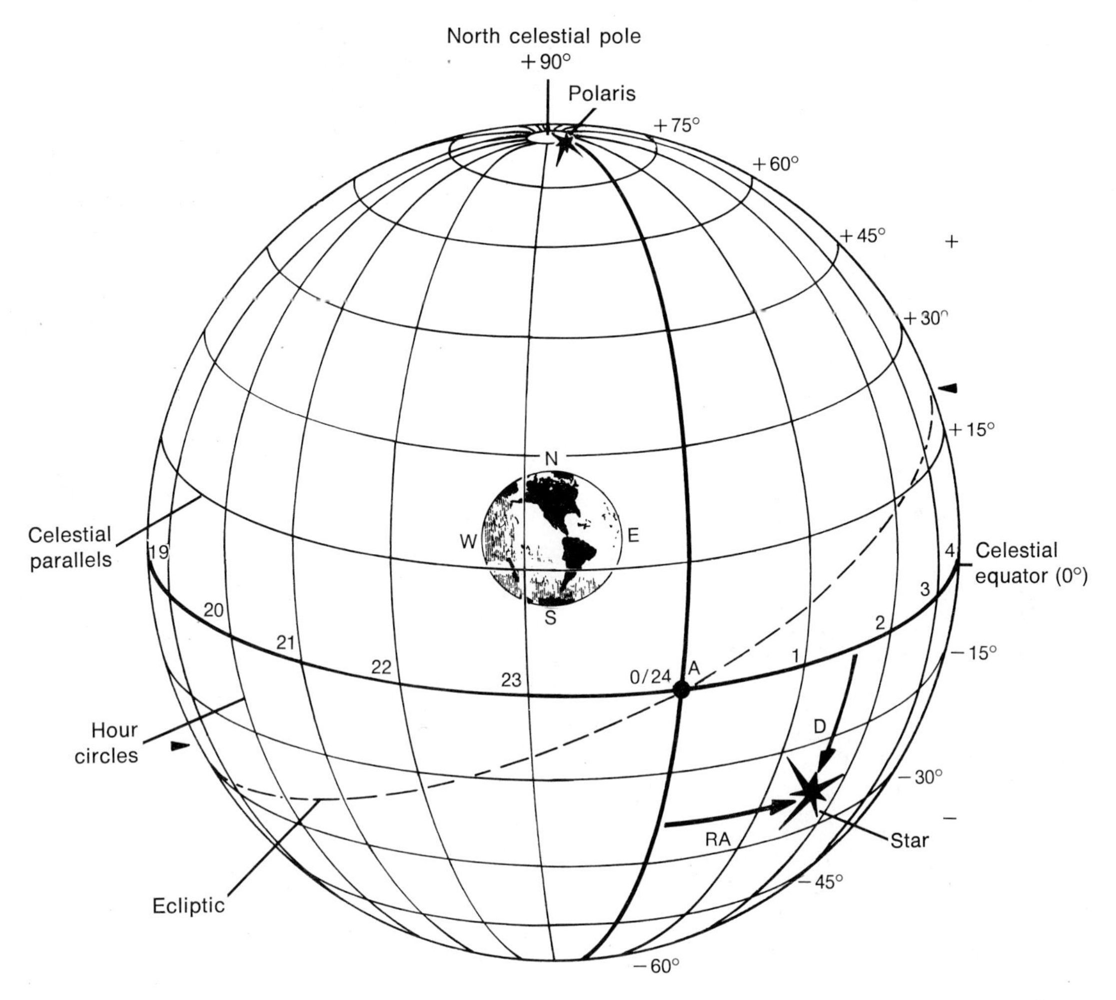

Fig. 2-3. Sky-oriented points and circles. This sytem works in same manner as observer-oriented system but has the advantage of permanently fixed points. This makes it possible for two observers to communicate about features without being in exactly the same position. Tilt of sphere makes −75° and −90° (south celestial pole) invisible. *A* indicates vernal equinox; *D* indicates declination; *RA* indicates right ascension.

geographic poles, the celestial sphere appears to turn about the celestial poles. Within about 1° of the north celestial pole today is the star named *Polaris, the North Star,* making the location of the north celestial pole a simple matter. Unfortunately for observers in the Southern Hemisphere, no such star visible to the unaided eye marks the south celestial pole.

Precession

The gravitational effect of the moon and to a lesser extent the effect of the sun cause the rotational axis of the earth to wobble slowly like a spinning top. This motion, called precession, causes the celestial poles to transcribe a circle westward among the stars, returning to their original position 26,000 years later (Fig. 2-4). The net effects are that the present north star, Polaris, will not always be the north star, and the sun's apparent position in space on the first day of spring, known as the *vernal equinox,* will move westward among the stars. The next bright north star in about 12,000 years will be Vega, which is located in the constellation Lyra, the Harp. The vernal equinox does not remain fixed but slips westward completely around the ecliptic, spending 866 years in each sign of the zodiac. In the time of the ancient Greeks the vernal equinox was in the constellation Aries. It is now in Pisces preparing to enter Aquarius—hence the popular saying "Age of Aquarius." The first day of spring, now occurring on or about March 21, will continue to move forward in the calendar until

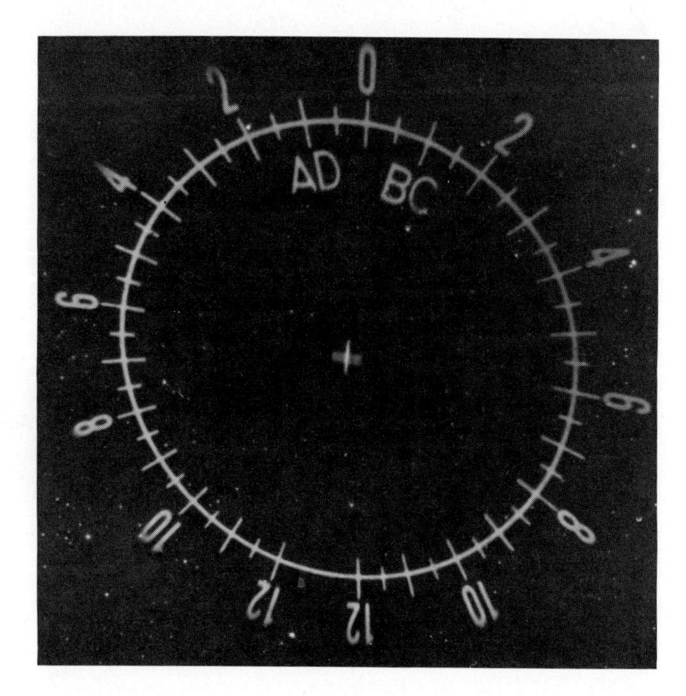

Fig. 2-4. Precession of north celestial pole. The present North Star (Polaris) is found nearly at the 2 A.D. (2000) mark. The next bright pole star will be Vega, located just outside the circle near 12 A.D. (12,000) mark. (Constellation near 4 B.C. mark is Ursa Major, the Big Bear, which contains the Big Dipper.) Time exposure was taken at the Miami Museum of Science Space Transit Planetarium, and slight distortion is due to curvature of dome. (Courtesy Miami Space Transit Planetarium, Miami, Fla.)

spring occurs in our present winter months, necessitating a change in thinking if not calendar revision itself. The precession of the vernal equinox will also require major changes in thinking on the part of astrology buffs, for eventually all sun signs, or the position of the sun with respect to the stars on one's birthday, will also move westward. Someone born between June 21 and July 20, although considered a Cancer astrologically, has his sun in Gemini astronomically. Astrology for the most part has disregarded precession.

Celestial equator

Immediately above and running parallel to the earth's equator is the celestial equator, an imaginary great circle completely encircling the celestial sphere in an east-west direction halfway between and at right angles to the celestial poles. The celestial equator divides the celestial sphere into equal northern and southern halves, just as the earth's equator divides our planet into the Northern and Southern Hemispheres.

Celestial parallels and hour circles

Each of the parallels and some of the meridians belonging to the earth-fixed coordinate system have also been extended to the celestial sphere. Running from east to west parallel to the celestial equator are celestial parallels—180 imaginary small circles—90 north of the celestial equator and 90 south. The circumference of the celestial parallels grows progressively smaller toward the celestial poles.

Running from north to south on the celestial sphere are 24 great circles called hour circles, which are analogous to meridians on earth. The hour circles converge at the celestial poles and completely encircle the celestial sphere. The logic of dividing the celestial sphere into hour circles is based on the fact that the earth turns through a total of 360° at an angular rate of 15° each hour, so that the period of the earth's rotation, the day, is 24 hours long. The celestial sphere naturally reflects the earth's rotation, so in 1 hour a celestial object will move westward 15°.

Ecliptic and zodiac

The sun appears to move eastward on the celestial sphere about 1° each day, completing one circuit of the sky each year. This apparent annual path of the sun on the celestial sphere is called the ecliptic. Actually it is the earth's eastward revolution around the sun that produces this optical illusion. The ecliptic may also be thought of as the plane of the earth's orbit around the sun. If the orbit of the earth were a flat sheet extended inward and out to encompass the entire solar system, it would be found that the orbits of all the planets and their satellites lie virtually in the same plane. The fact that the earth, moon, and sun must be aligned in this plane for eclipses to occur is the origin of the term "ecliptic." To better comprehend the ecliptic plane, imagine the sun at the center of a dime with all planets and their satellites orbiting within the thickness of the coin.

From earth the sun appears to move against a background of more distant stars. These background stars can only be seen during a total eclipse of the sun, which occurs when the moon passes directly between the earth and the sun. An eclipse of the sun can only occur during daylight hours, so that the light from the sun can be blocked by the moon. When this occurs the sky becomes dark and stars that would otherwise not be seen at that hour become visible. During the course of 1 year the sun appears to move through 12 constellations, groups of stars that appear to belong together (Fig. 2-5). The early Greeks referred to the narrow zone containing these 12 constellations as the zodiac, or zone of animals, because the star groups reminded them of different creatures in their mythology. The zodiac consists today as it did then of Aries the Ram, Taurus the

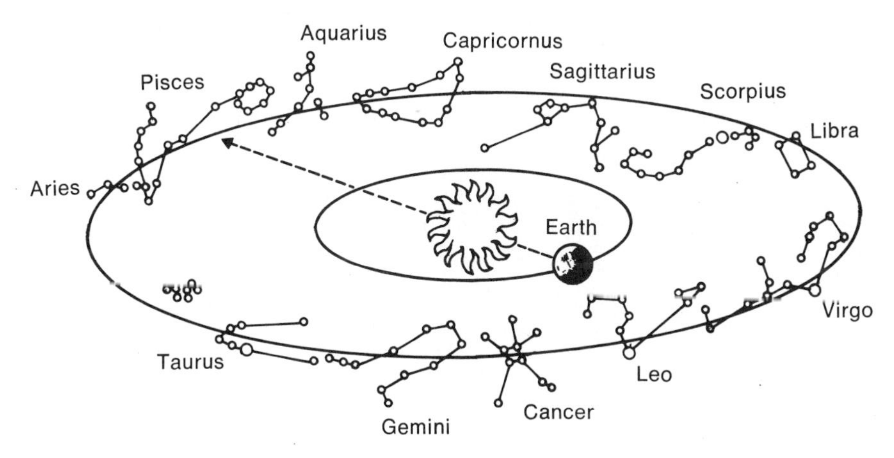

Fig. 2-5. Constellations of the zodiac as seen from outside the celestial sphere. Astrologers say the sun is "in" a constellation when the sun lies between the constellation and the earth. In this case, the sun is "in" Pisces.

Bull, Gemini the Twins, Cancer the Crab, Leo the Lion, Virgo the Virgin, Libra the Scales, Scorpio the Scorpion, Sagittarius the Archer, Capricorn the Goat, Aquarius the Water Bearer, and Pisces the Fish (Fig. 2-6). The star group in which the sun appears at the moment of one's birth determines one's sun sign, a basic aspect of astrology.

The zodiac is 16° wide and 360° long. The constellations of the zodiac are arranged along the ecliptic, with each occupying 30° or 2 hour circles. Even though the zodiac consists of twelve equal parts, the stars visible with the naked eye that make up each of the twelve signs occupy different percentages of their assigned areas and are of different brightnesses. Taurus and Scorpio, for example, are bright constellations that fill much of their designated space, whereas Cancer and Libra are dim and do not fill their part of the zodiac.

Equinoxes and solstices

The rotational axis of the earth lies tilted to the plane of its orbit by an angle of 23½°. As a result, the ecliptic is also inclined to the celestial equator at the same angle. The ecliptic and the celestial equator cross each other at only two points on the celestial sphere—the equinoxes (equal nights), when day and night are of equal length over the entire earth (Fig. 2-8). The sun can thus never be more than 23½° away from the celestial equator. This maximum separation likewise occurs at only two points—the solstices (the sun stands still) (Fig. 2-8). For 6 months of the year, from about March 21 until September 21, the ecliptic is in the Northern Hemisphere of the celestial sphere. From about September 21 until March 21, on the other hand, the ecliptic is in the Southern Hemisphere.

On or about March 21 each year the sun appears to be located on the celestial equator, preparing to enter the Northern Hemisphere. This point is called the *vernal equinox* (*A* in Fig. 2-8). When the sun is at the vernal equinox, the spring season begins for those living north of the equator.

Three months later, on or about June 21 each year, the sun reaches the point on the celestial sphere where it is as far north of the celestial equator as it can ever get. This point is known as the *summer solstice,* the start of summer for the Northern Hemisphere of the earth (*B* in Fig. 2-8). The summer solstice is now in the constellation Gemini.

On or about September 21, 3 months after it reaches the summer solstice, the sun is again on the celestial equator, but this time it is preparing to enter the Southern Hemisphere. This point at which the sun crosses the celestial equator going south is called the *autumnal equinox,* the first day of fall for the Northern Hemisphere (*C* in Fig. 2-8). The constellation Virgo is now the location of the autumnal equinox.

Three months after the autumnal equinox the sun reaches a point at which it is as far south of the celestial equator as it can get. The sun reaches this point, called the *winter solstice,* on or about December 21 each year, the start of winter for the Northern Hemisphere (*D* in Fig. 2-8). The winter solstice is now in the constellation Sagittarius.

The seasons are reversed in the Southern Hemisphere. People living south of the equator are accustomed to calling March 21 the first day of fall, June 21 the first day of winter, September 21 the first day of spring, and December 21 the first day of summer.

Fig. 2-6. The constellation of Gemini the Twins. If you connect the stars indicated by the arrows it is possible to envision the heads, bodies, and feet of Castor and Pollux standing arm in arm. (Photograph by John Clement, Jr.)

Right ascension–declination coordinate system

The right ascension (RA) of a celestial object is its distance east of the vernal equinox, and it is measured in hours, minutes, and seconds (Fig. 2-3).* Right ascension is roughly analogous to longitude on earth. The main difference between the two is that longitude is measured both east and west of the prime meridian, whereas right ascension is measured only eastward from the vernal equinox. The hour circle that passes through the vernal equinox is called the *equinoctal colure* and is designated both 0^h and 24^h. The hour circles passing through the summer solstice, autumnal equinox, and winter solstice are designated 6^h, 12^h, and 18^h, respectively. Thus a celestial object could have a right ascension of 14^h, 36^m, 21^s and could be found 14^h, 36^m, 21^s east of the vernal equinox somewhere between the north and south celestial poles.

Just as longitude alone will not specifically locate a city on the earth's surface, neither will right ascension pinpoint the location of a celestial object. Another coordinate is needed. In the case of the earth-fixed system the necessary coordinate is latitude. For the sky-fixed system the required coordinate is called declination. The declination of a celestial object is its distance north or south of the celestial equator (Fig. 2-3). If the object is north of the celestial equator, it is said to have a plus (+) declination, if south, a minus (−) declination. Declination is analogous to latitude on the earth's surface. The celestial equator has a declination of 0°, while the north and south celestial poles have a declination of +90° and −90°, respectively. A celestial object could have a declination of −37°, 18^m, 53^s and would be found 37°, 18^m, 53^s south of the celestial equator. Combining the declination of −37°, 18^m, 53^s with the right ascension of 14^h, 36^m, 21^s mentioned earlier would specifically locate this object.

THE OBSERVER'S VIEW OF THE CELESTIAL SPHERE

The orientation and apparent motion of the celestial sphere depends on the observer's location on the surface of the earth. The observer's view of the celestial sphere from the poles, equator, and midlatitudes is explored in the following discussion.

An observer at the poles

If an observer is at either geographic pole of the earth, the corresponding celestial pole would be at his zenith. The celestial equator would coincide with his entire true horizon, and he would be able to see only those celestial objects in his hemisphere, that is, in the Northern Hemisphere if he were at the North Pole or in the Southern Hemisphere if he were at the South Pole. All stars would be circumpolar; that is, they would never rise or set but would transcribe circles around the celestial pole.

For the observer at the geographic poles, day would last 6 months and night would last 6 months. Let us take an observer at the North Pole to illustrate the positions and movement of the sun as seen from the poles. It should be remembered that the dates will be reversed but the effect would be the same from the South Pole.

The observer at the North Pole would see the sun on the horizon on or about March 21. For the next 6 months the sun would slowly gain altitude, reaching a maximum of 23½° on or about June 21. Meanwhile, each day during this 6-month period the sun would not rise or set but would completely circle the sky above the horizon. After June 21, the sun would begin to lose altitude, and on or about September 21, it would again be on the horizon. For the next 6 months the sun would be below the horizon, and darkness would prevail.

An observer at the equator

If an observer is at the earth's equator, the celestial equator passes through his zenith from east to west. The north and south celestial poles would be at his north and south horizon points, respectively. Unlike an observer at the poles, at which stars never rise or set, the observer at the equator would see all stars rise and set. During the course of 1 year he could theoretically see all objects in both hemispheres. Geographically speaking, the equator is thus the perfect location for an observatory.

An observer at the midlatitudes

The observer between the poles and the equator, where the bulk of the world's population lives, would see some of the effects visible to both the polar and equatorial observer (Fig. 2-7). The altitude of the north celestial pole for a Northern Hemisphere observer and the south celestial pole for a Southern Hemisphere observer would be equal to his latitude; stars within this distance of the poles would be circumpolar. For example, from New York City, latitude 41° N, the north celestial pole would be found 41° above the northern horizon and all objects within 41° of the north celestial pole would never rise or set.

Since the celestial equator is an extension of the earth's equator, the location of the celestial equator for an observer at the surface of the earth, like the altitude of the celestial pole, is also a function of his latitude. For the observer in

*The hours, minutes, and seconds used in measuring right ascension are commonly expressed as in the following: 4^h, 12^m, 18^s.

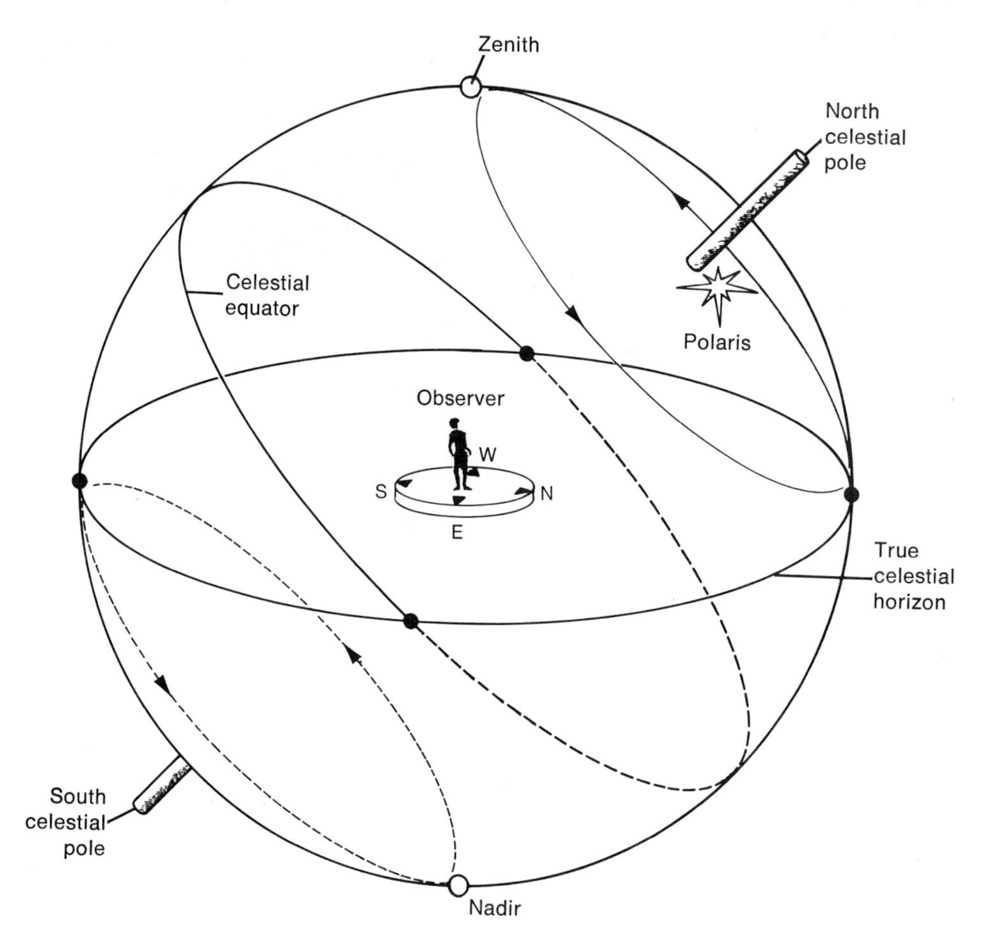

Fig. 2-7. The celestial sphere as seen from midnorthern latitudes. To an observer north of the equator, the altitude of Polaris above the northern horizon is equal to his latitude. Stars within this distance are circumpolar. The distance from the observer's zenith to the celestial equator is also equal to his latitude. The distance from the celestial equator to the observer's southern horizon is equal to his co-latitude (90° minus observer's latitude equals co-latitude).

the Northern Hemisphere, the celestial equator is always south from his zenith at a distance equal to his latitude. From New York City the celestial equator is 41° S of the observer's zenith. In the Southern Hemisphere the same principle applies, except the direction of the celestial equator is always north.

SIDEREAL TIME

We are all familiar with time based on the movement of sun—solar time. There is apparent solar time that is measured by a sundial, mean solar time measured by clocks and watches, zone time, local time, and daylight savings time —all based on the apparent movement of the real or an "imaginary" sun. Some further aspects of time are explored in Chapter 9, but for the sake of the point at hand, apparent and mean solar time are discussed here.

The *apparent* solar day is the period of time required by the earth to rotate once with respect to the sun. It is measured by two successive meridian passes of the real sun. Because the earth is orbiting the sun in an elliptic path at varying speeds and distances, all apparent solar days are of a different length. To make order out of chaos, all apparent solar days are averaged to make a *mean* solar day based on the *apparent* motion of a "fictitious" sun that is always on time. The mean solar day is 24 hours long.

Because the stars are much more distant than the sun, the rotation of the earth with respect to the stars (the sidereal day) is much more accurate and constant. The sidereal day is measured by sidereal time and is divided into sidereal hours, minutes, and seconds. Sidereal time is vernal equinox time because it is a measure of the rotation of the earth with respect to the vernal equinox rather than the sun. The

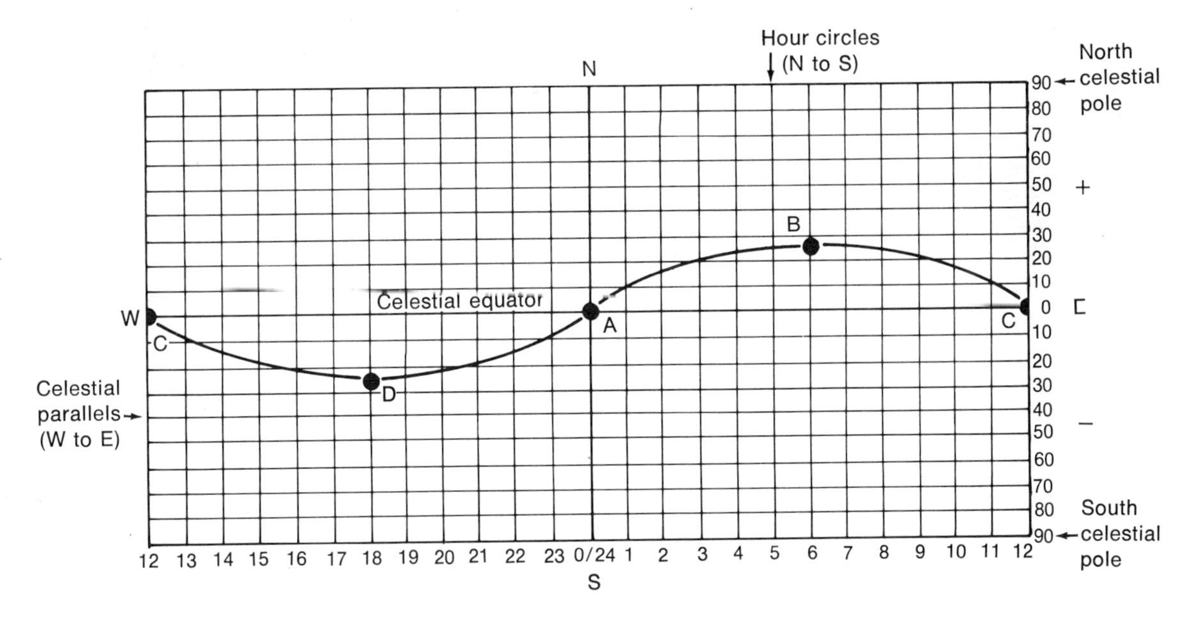

Fig. 2-8. Celestial sphere chart. *A,* Vernal (spring) equinox (March 21); *B,* summer solstice (June 21); *C,* autumnal (fall) equinox (September 21); *D,* winter solstice (December 21).

sidereal day is 23^h, 56^m, 4.091^s long and begins at sidereal noon, 0^h, 0^m, 0^s sidereal time, with the vernal equinox on the observer's meridian.

Sidereal time is used primarily in the astronomical observatory or by individual observers. The main function of sidereal time is to indicate when a particular hour circle is crossing the observer's meridian. This allows an observer to locate celestial objects with the unaided eye, binoculars, or telescope. In other words, sidereal time will tell an observer which part of the celestial sphere is visible on any day of the year or at any hour of the night. For the remainder of this chapter, sidereal time may be considered simply as the hour circle crossing the observer's meridian.

Sidereal time charts are available to the observer and are usually based on an 8:00 P.M. standard clock time, a convenient hour for observing. There is a different sidereal time for each day of the year. In other words, at 8:00 P.M. each night throughout the year, a different part of the celestial sphere is crossing the observer's meridian. A sidereal time chart that can be used for many years is found in Appendix B. If observing before 8:00 P.M., subtract (after 8:00 P.M., add) the correct amount of time, so that the sidereal time will be accurate. For example, if the sidereal time is 4^h, 0^m, 0^s at 8:00 P.M. on January 20, then the sidereal time will be 3^h, 0^m, 0^s at 7:00 P.M. and 5^h, 0^m, 0^s at 9:00 P.M.

Sidereal time, on the average, increases by approximately 4 minutes each day. This is explained by the fact that in one day the earth has orbited $^1/_{365}$ of its way around the sun—about 1°. It takes the earth 4 minutes to rotate through 1°, which brings the same star to the horizon or meridian 4 minutes earlier than the night before. It should also be remembered that the earth's 1°/day revolution around the sun causes the 1°/day apparent motion of the sun eastward on the ecliptic.

LOCATING OBJECTS ON THE CELESTIAL SPHERE

Novice observers have difficulty locating celestial objects on the celestial sphere for they do not know when or where to look, and with the exception of the sun and moon, all celestial objects tend to look the same to them. They also have difficulty differentiating the various constellations. With practice, however, celestial objects become friends that are recognizable on sight. To help the novice become familiar with the location of objects on the celestial sphere, a celestial sphere chart has been developed. Its organization and use are described in the remainder of this chapter.

Organization of the celestial sphere chart

The celestial sphere chart takes the form of a rectangle that covers 180° from top to bottom and 24 hours (or 360°) from side to side (Fig. 2-8). North is at the top, south is at the bottom, east is to the right, and west is to the left. Celestial sphere charts designed for outdoor use have the east-west directions reversed so that when the chart is held over the observer's head it will coincide with what can be seen in the sky. The celestial sphere chart shown in Fig. 2-8 has been designed for use at the observer's desk or on the ground; the information can then be applied to the sky.

The celestial equator is found in the center of the chart; it runs from east to west and is numbered 0°. Celestial parallels also run east to west parallel to the celestial equator. There are in reality 180 celestial parallels but on the chart only every tenth parallel is shown. Even though celestial parallels are imaginary circles, they will appear as straight lines on the celestial chart. The celestial poles are imaginary points represented on the chart by the lines at the top and bottom.

The 24 hour circles, shown on the chart running from north to south, are numbered consecutively from 0 to 24 hours eastward from the vernal equinox. The twelfth hour circle is given twice on the chart, once to the right and once to the left, to remind the observer that the chart represents the real sky. Each hour circle, representing the rotation of the earth through 15°, is divided into minutes and seconds like a clock.

The curving heavy line is the ecliptic, the annual path of the sun or the plane of the earth's orbit. The ecliptic space between each hour circle represents about 15 days of apparent solar motion; the space between every two hour circles represents a month. Four points along the ecliptic in Fig. 2-8 have been lettered *A*, *B*, *C*, and *D*. Point *A* is called the vernal equinox, and the hour circle passing through this point, labeled 0/24 hours, is called the equinoctal colure. Point *B* is called the summer solstice, the point at which the sun reaches its greatest northern declination, +23½°, and the sixth hour circle runs through it. Point *C* is the autumnal equinox and, like the twelfth hour circle that runs through it, has been shown twice on the chart. Point *D* is the winter solstice, where the sun has its greatest southern declination, −23½°.

The observer's location and view

The celestial sphere is the sky as seen from earth, so that the observer is always below a point on the chart called the zenith, whose declination is always equal to the observer's latitude. The meridian runs from north to south through the observer's zenith. It appears to remain stationary while the celestial sphere turns by once every 24 hours. From true horizon to true horizon in all directions is 180°, or half of the celestial sphere. From the meridian to the eastern or western horizon is 90° or 6 hour circles. Remember that the earth, turning at the rate of 15°/hr, takes one fourth of a day or 6 hours to complete one fourth of a turn or 90°.

Use of the celestial sphere chart

Now that you are familiar with the nature of the celestial sphere chart, your location on it, and your view of it, you are ready to make use of it. To properly use the chart you must first do the following:

1. Assume you will observe at 8:00 P.M. local time. (If you observe at a different time, be sure to calculate sidereal time.)
2. Find the sidereal time for the date of your observation. This may be obtained from the sidereal time chart found in Appendix B.
3. Obtain the right ascension and declination of the object to be located. The coordinates of many celestial objects may be found in Appendix A. Additional coordinates may be found in the *American Ephemeris and Nautical Almanac*.
4. Determine your latitude to the nearest whole degree. The latitudes of many cities may be found in the *World Almanac and Book of Facts* or in any good almanac. If you live in or near a city not listed you may obtain your approximate latitude from a map.
5. Sketch a celestial sphere chart similar to the one shown in Fig. 2-8 and follow steps 6 through 15 below.
6. Mark your meridian on the chart. Remember that your meridian is equal to the sidereal time.
7. Place your zenith point on the meridian. The zenith is equal to your latitude.
8. Mark your horizons on the chart. The true horizon is always 90° or 6 hours from the zenith or meridian.
9. Plot the position of the celestial object to be found using its right ascension and declination.
10. Draw an arrow on the chart from your zenith toward the plotted location of the object.
11. Note the direction of the arrow on the chart and notice approximately how far in degrees north or south of the zenith and in hours east or west of the meridian the object is located.
12. If on the chart you observe that the celestial object is not located between the horizons, then it cannot be seen at 8:00 P.M. If this is the case, determine when it will rise during the night by counting hours from the object westward until the eastern horizon is reached. If the object is calculated not to reach the eastern horizon before daylight, you will have to observe at a different time of the year.
13. Locate the sun on the chart. The sun always appears to cross the meridian at noon. At 8:00 P.M., then, the sun will have appeared to move 8 hour circles

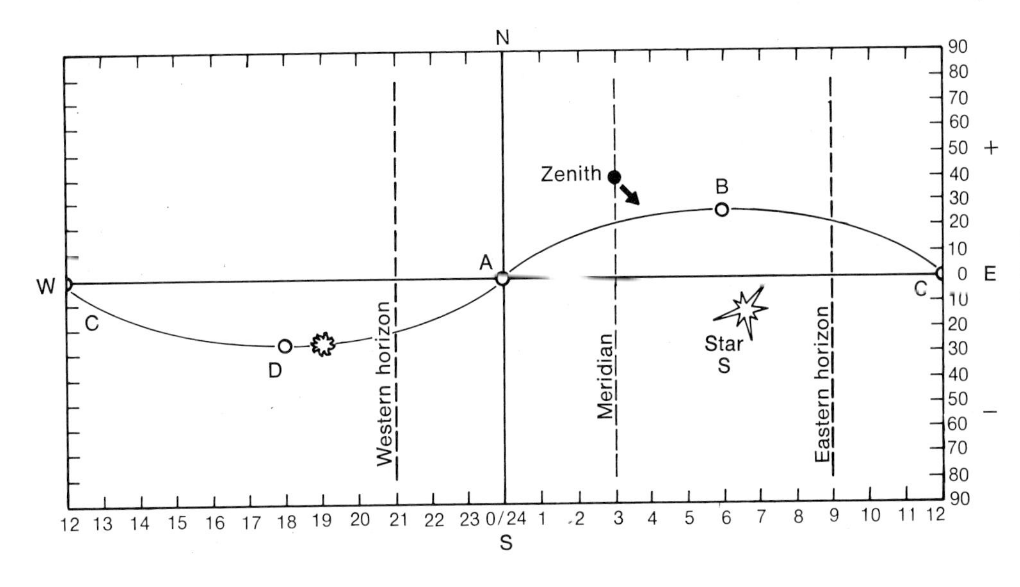

Fig. 2-9. Celestial sphere chart to be used in solving the problem below.

west of the meridian. To locate the sun on the chart, count 8 hour circles west of the meridian and place a point on the ecliptic.

14. Confirm the correct date of the year by using the known dates for points *A, B, C,* or *D*. One hour circle equals about 15 days; two hour circles equal approximately 1 month.
15. With the location of the celestial object in hand, go outside and find the object using your unaided eye, binoculars, or a telescope.

To help you understand more fully the processes involved in locating a specific celestial object, complete the following problem by using Fig. 2-9 and the steps just listed.

Given:

1. Clock time, 8 P.M.
2. Sidereal time, 3^h, 0^m
3. Zenith, +40°
4. Right ascension of star S, 6^h, 39^m
5. Declination of star S, −16°, 43^m

Find:

1. The position of star S
2. The month and date of the year

Solution:

1. Star S is 3½ hours east of the meridian and 56° south of the zenith.
2. The sun is at the 19 hour circle, making the date January 5.

LOCATION OF THE MOON AND PLANETS

You have now learned how to locate celestial objects using their "fixed" coordinates in the sky. However, there are other objects of interest that do not have such a fixed position because of their proximity to us. The celestial objects that appear to move rapidly on the celestial sphere are the moon and the planets.

The moon, because it is our nearest celestial neighbor, appears to move more rapidly in the sky than any other major celestial body. The moon moves eastward on the celestial sphere, completing one circuit every 27 days, 7 hours, 43 minutes, 11.5 seconds—a time interval known as the sidereal month. Completing 360° in 27⅓ days means that each day the moon moves 13° eastward, just less than 1 hour circle (360° ÷ 27⅓ days = 13°/day). During the sidereal month the moon moves through all 12 signs of the zodiac, spending just over 2 days in each.

The moon's orbit is inclined to the ecliptic plane by 5°. As a result the moon can never be found more than 5° away from the ecliptic. The moon can thus be as far as −28½° away from the celestial equator; over a period of time you may see the moon change its north-south location by as much as 57°. In other words, sometimes the moon will be high in the sky and at other times it will be low.

The orbits of all the major planets are virtually in the same orbital plane as the earth. Only Pluto and Mercury, with orbital inclination of 17° and 7°, respectively, are more than 3° out of the ecliptic plane. As a result, Pluto is the only major planet that can ever be found out of the zodiac.

Information concerning the location of planets may be obtained from world almanacs, the *American Ephemeris and Nautical Almanac,* publications such as *Sky and Telescope* magazine, and sometimes newspapers. To find a planet, look near the ecliptic and find a bright starlike object

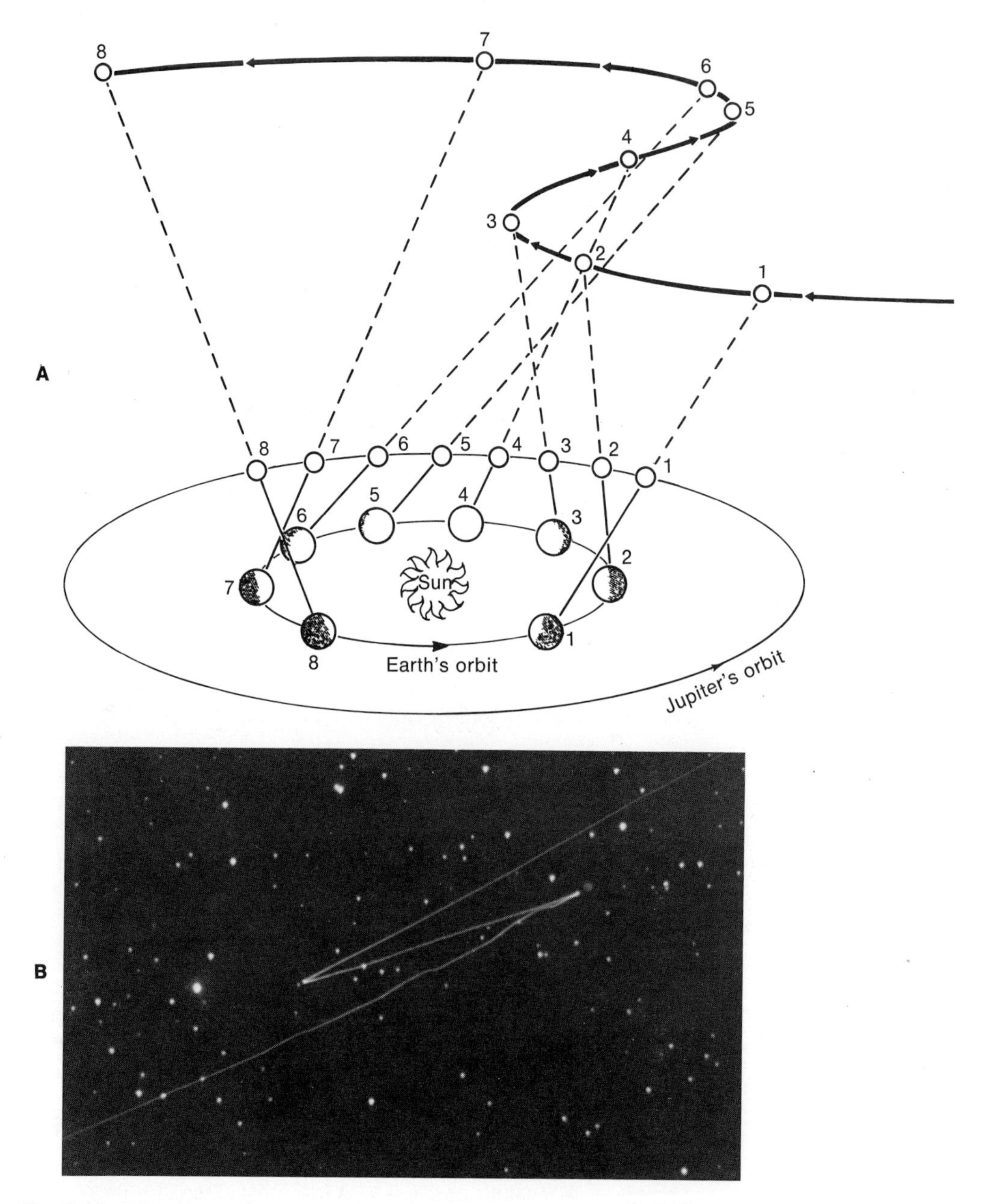

Fig. 2-10. When the earth overtakes and passes a slower, more distant planet, that planet appears to reverse its path in a phenomenon known as the retrograde loop. **A,** Diagrammatic view. **B,** Photographic time exposure. (**B** courtesy Miami Space Transit Planetarium, Miami, Fla.)

that tends not to twinkle except when rising and setting.

Because of either to their own orbital motions or the movement of the earth, planets appear to change their positions rather rapidly on the celestial sphere. Jupiter, for example, moves slowly eastward near the ecliptic, spending 1 year in each of the 12 constellations of the zodiac. Once each 399 days, however, it appears that Jupiter stops moving eastward, moves back toward the west, then finally resumes an eastward motion. This phenomenon is called a *retrograde loop* and is caused by the earth overtaking and passing a more distant, slower planet (Fig. 2-10).

3 *the ruler of the night*

The moon

. . . The moon, whose orb through
Optic glass the Tuscan artist views
At evening from the top of Fiesole,
Or in Valdarno, to descry new lands,
Rivers or mountains in her spotty globe.

Paradise Lost
John Milton

People have always regarded the moon with almost unending fascination. It was worshiped by ancient people as a god and was the subject of much of the work of early astronomers (Fig. 3-1). Lovers have whispered and sighed over it, and its beauty has been praised time and again in both poetry and song. The moon is the celestial object most often noticed by people in general and viewed telescopically by amateur astronomers. Furthermore, its exploration was the first major goal of the United States space program. Astrologers believe that the fortunes of those born between June 21 and July 20 are supposed to be ruled by the moon, and Monday was named in its honor. It supposedly influences crime rates, hurricane development, and fishing, among other things. Although much has been discovered about the moon, there is still a great deal more to be learned because this strange yet familiar apparition in the sky apparently contains many more secrets than have yet been unveiled.

ORIGIN OF THE MOON

Although the actual origin of the moon still remains a puzzle, many years of lunar studies have produced three popular theories to explain the existence of the earth's natural satellite.

Earthoon or moorth?

Late in the nineteenth century Sir George Darwin, son of the famous naturalist, proposed that the moon was at one time a part of a rapidly rotating earth. He theorized that the moon had spun off the earth early in its formative period and that the scar left behind became the Pacific Ocean basin. Darwin's lunar theory still has some advocates today, but it has lost its former popularity. One bit of evidence often given by proponents of this theory is the measurable recession of the moon away from the earth. As the speed of the earth's rotation is braked by the gravitational tidal forces of the moon at the rate of 13 seconds each century, the moon's orbital speed increases, causing it to recede approximately 400 feet (122 meters). At this rate of recession the moon and the earth might possibly have been in contact with each other about 4 to 5 billion years ago, which ironically is also the estimated age of the earth, moon, sun, and other members of our solar system. On the other hand, opponents of the theory argue that if the moon had separated from the earth it would have broken into pieces before it had a chance to escape. How then could these pieces have formed the moon as we know it today?

Space piracy

A second attempt to explain the moon's origin could be called the ''capture'' theory. According to this theory, the moon was formed elsewhere in the solar system and then became trapped by the earth's gravitational pull during a later stage of its development. Difficulty arises in explaining how the earth could have captured a body the size of the moon, for the speed and angle of capture would have been

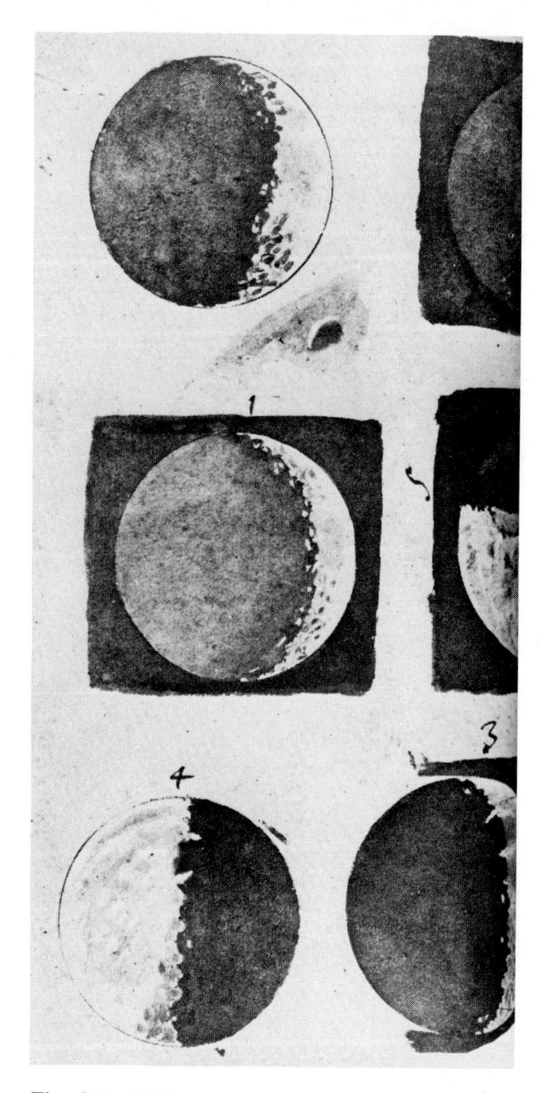

Fig. 3-1. Galileo was one of the early astronomers fascinated by our nearest astronomical neighbor. Here are some of his sketches of the earth's moon. (Courtesy University of Chicago Yerkes Observatory, Williams Bay, Wisc.)

critical, with overwhelming odds that the moon should have gone into orbit around the sun to become a planet. An alternate proposal is that the earth trapped many smaller bodies that later coalesced to form the moon.

Separate but not equal

The final and probably most widely held theory of lunar origin suggests the separate but not equal formations of the moon and the earth. Within this theory, which is interpreted in many different ways, the moon and the earth could have both been formed within the same approximate region of space at nearly the same time and in the same manner. Within the framework of this basic belief two popular schools of thought have developed; these may be called the "hot moon" and "cold moon" theories. The hot moon theory suggests that the moon and the earth could have formed separately during the cooling of a hot, gaseous cloud. The cold moon theory postulates that the origins of the moon and the earth could have been initiated by the accretion of cold, solid material that was subsequently heated. Evidence that supports the separate but not equal idea, although it is not necessarily conclusive, can be listed as follows:

1. The rocks of both the earth and the moon appear to have been formed by heat but have for the most part cooled off.
2. The densities of the earth and the moon are similar.
3. The earth and the moon are of approximately the same age.
4. The earth and the moon are similar but not identical in composition.

Every new discovery that is made through continued space exploration brings us that much closer to unlocking the mysteries of lunar origin.

THE MOON'S PLACE IN TIME AND SPACE

The moon is classified astronomically as a *satellite*—a body in orbit around a planet. Although it is the earth's only natural satellite and closest celestial neighbor, the moon is 1 of 33 satellites in our solar system (Table 1). Its size is surpassed only by Ganymede and Callisto of Jupiter, Titan of Saturn, and Triton of Neptune. In one very important respect, however, the moon does rank first. No other satellite is so large relative to its ruling planet, or so close to it. The moon's size and proximity actually involve it intimately in the environment of the earth, making it a great deal more than just another pretty sight on the celestial sphere.

Distance

The center-to-center distance from the earth to the moon averages 238,857 miles, but because the moon's orbit is not circular, the distance to the moon varies from instant to instant. At *apogee,* the closest point in its orbit to the earth, the moon is 221,463 miles away. At *perigee,* its most distant point, the moon is 253,710 miles from earth. This means that the moon is about 30 earth diameters away, a fact that explodes the notion, perpetuated by models of the earth-moon system, that the moon is very close to the earth. Although nearer than any other major celestial body, the moon is still distant by human standards of measure. At an

Table 1. Natural satellites

Planet and satellite	Average distance from planet (thousands of miles)	Orbital period (days)	Diameter (miles)*
Earth			
Moon	240	27	2,200*(5)
Mars			
Phobos	5.8	0.32	±10
Deimos	25	1.3	±5
Jupiter			
V	110	0.5	±100
I (Io)	260	1.8	2100* (6)
II (Europa)	420	3.6	1800* (7)
III (Ganymede)	670	7.2	3100* (1)
IV (Callisto)	1200	17	2900* (3)
VI	7100	266	±70
VII	7400	277	±20
X	7400	253	±10
XII	13,000	631	±10
XI	14,000	692	±10
VIII	15,000	735	±20
IX	15,000	758	±10
XII	?	?	?
Saturn			
Mimas	115	0.94	±300
Enceladus	148	1.4	350
Tethys	183	1.9	600
Dione	234	2.7	600
Rhea	327	4.5	800* (8)
Titan	758	16	3000 (2)
Hyperion	919	21	300
Iapetus	2210	79	1100
Phoebe	8040	550	130
Janus	98	0.7	190
Uranus			
Miranda	81	1.4	±100
Ariel	120	2.5	±400
Umbriel	170	4.1	±200
Titania	270	8.7	±600
Oberon	360	13	±500
Neptune			
Triton	220	5.9	2500* (4)
Nereid	3400	360	±200

*Numbers in parentheses indicate order of decreasing size of satellites.

average speed of 4000 miles/hr, astronauts require more than 2 day's travel for a one-way trip to the moon.

Several early Greek astronomers calculated the relative distance to the moon with various degrees of success. Modern calculations were initially made with the use of radar waves that were beamed from the earth to the moon's surface and back again. Within recent years, however, laser beams transmitted through earth-based telescopes have been bounced from laser retroreflectors which had been left on the moon's surface by Apollo astronauts. As a result of these laser experiments, we now know the distance to the moon within an accuracy of less than 1 mile.

Size

Many years ago it was believed that the moon was only as large as it appeared to be, which is less than the size of a dime held at arm's length. Today we know that the lunar diameter covers ½° on the celestial sphere. By knowing the actual distance to such a celestial object, it becomes a simple mathematical problem to calculate its true diameter from its apparent angular size. The diameter of the moon, thus computed, is 2160 miles, nearly one-fourth the diameter of the earth. Even with sophisticated telescopic instruments and knowledge gained from lunar exploration, this measurement of the diameter of the moon has been refined only in fractions of a mile.

Rotation

Many people still believe that the moon does not rotate on its axis because they see the same side of the moon always facing the earth. The moon, however, does spin on its own axis, rotating from west to east at an average speed of 10 miles/hr, which is considered a very fast walking speed here on earth. One side of the moon appears to face the earth at all times because over the billions of years since the formation of the moon, the earth's gravitational pull has slowed down the moon's rotation rate to the point where its rotation period equals its period of revolution—27⅓ days. This slow, synchronized rotation of the moon causes both the lunar night and day to equal approximately 14 earth days.

Because the moon is a sphere, half of it is always facing toward the earth. However, due to a combination of the motions of both the earth and the moon, a careful observer on earth is able with the aid of a telescope to view 59% of the moon's total area over a period of time. These movements of the earth and moon are called *librations,* and they are of three different types—latitudinal, longitudinal, and diurnal. The moon's rotational axis is inclined to the plane of its orbit at an angle of 6½° (as compared to the earth's tilt of 23½°), and the angle remains constant as the moon orbits the earth. As a result, each month observers are first able to see 6½° beyond the moon's north pole; 2 weeks later it is possible to see 6½° beyond the south pole. This effect is called *libration in latitude*.

Even though the moon's rotation period equals its or-

Fig. 3-2. East-to-west libration of the moon. Note the changing position of Mare Crisium, the small, dark, round patch along the right-hand edge of the photographs. (Courtesy University of California Lick Observatory, Santa Cruz, Calif.)

bital period, the two get out of step as seen from earth. This is because the moon's rotational speed stays almost constant, whereas its revolution speed varies over wider limits. We on earth are thus able to initially view approximately 8° around the eastern edge of the moon. Later on, an equal amount around the western edge is visible. This effect is known as *libration in longitude* (Fig. 3-2).

The third motion produces a *diurnal* or *daily libration* as the rotation of the earth carries an observer back and forth over a distance of nearly 8000 miles each day. At moonrise we see the moon 4000 miles east of the earth's center and have an opportunity to look beyond its western edge, which would not be possible 6 hours later when it crosses the meridian. At moonset the moon is viewed 4000 miles west of the earth's center. As a consequence, we are able to look beyond the moon's eastern edge. Diurnal librations amount to only about 1° either way.

Therefore 41% of the moon's total area is always facing us, 41% is never facing us, and 18% is sometimes facing us—changes all due to librations.

Revolution

The moon revolves in a nearly circular orbit around the earth, completing one circuit every 27 days, 7 hours, 43 minutes, and 11.5 seconds, at an average speed of 2287 miles/hr, a speed equal to the muzzle velocity of a high-powered rifle. However, the true shape of the moon's orbit is an *ellipse,* with the earth at one focus. As is true of its distance from the earth, the moon's orbital speed is a constantly changing quantity.

The orbital plane of the moon is inclined to the ecliptic plane at an angle of 5° and actually intersects it at two points called *nodes*. The point at which the moon crosses the ecliptic moving northward is called the *ascending node*. The moon's crossing point moving southward is called the *descending node*. Because of the gravitational pull of the earth and to a lesser extent of the sun, the nodal points are gradually slipping westward around the ecliptic. This motion, discovered by the Babylonians, is called the *regression* of the nodes and requires 18 years, 11 days, and 8 hours to complete one circuit of the ecliptic.

Moonrise delay

Because of the earth's eastward rotation the moon, like all other celestial objects viewed from earth, appears to rise above the eastern horizon, cross the meridian 6 hours later, and finally set below the western horizon. If the moon were not orbiting the earth, these events would occur at the same time each day. As it is, however, the moon orbits eastward 13°/day while the sun, because of the revolution of the earth, appears to move 1° in the same direction. With respect to the sun, then, the moon changes its position toward the east at a rate of 12°/day. In order to bring the moon back to the eastern horizon, the earth must rotate the extra 12° each day. To do this the earth, rotating at a rate of 15°/hr, requires an average of 50 minutes. Therefore moonrise, meridian passage, and moonset all occur approximately 50 minutes later each day.

Months

There are actually two types of months, both based on the time required by the moon to revolve around the earth. The length of time required for the moon to orbit the earth with respect to the stars is called the *sidereal* month, which is equal to 27 days, 7 hours, 43 minutes, and 11.5 seconds. As seen from earth, the moon each month appears to move through each of the 12 constellations of the zodiac, remaining in each about 2¼ days. If the moon were located in the constellation of Pisces the Fish on a given night this month, then 1 sidereal month later the moon would be back in approximately the same location in the same constellation. The sidereal month is a measure of the moon's true orbital period (Fig. 3-3).

The length of time required for the moon to orbit the earth with respect to the sun instead of the stars is known as the *synodic* month, or month of phases. The synodic month lasts 29 days, 12 hours, 44 minutes, and 2.8 seconds and is the elapsed time between two consecutive appearances of the same phase of the moon. It will be observed, for instance, that full moon does not always occur in the same constellation. If the full moon is in the constellation of Pisces the Fish this month, then next month's full moon will appear in the constellation of Aries the Ram, which is one constellation eastward (Fig. 3-3).

Why are the sidereal and synodic months of different lengths? If the earth did not revolve around the sun while the moon orbited the earth, the sidereal and synodic months would both be 27⅓ days. The earth, however, does orbit the sun, and as the moon completes one trip around the earth, the earth has moved one twelfth of the way around the sun, a distance equivalent to 30° on the celestial sphere. Full moon, our example here, can only occur when the earth lies between the moon and the sun. When the moon returns to its original position after one sidereal period, the earth and sun are out of line by 30°. In order to produce a realignment with the earth and sun to produce full phase, the moon must orbit an additional 30° to the east (Fig. 3-3). At its orbital rate of 13°/day the moon requires an extra 2 days and 5 hours to cover the 30° necessary to carry it into the adjacent eastward constellation.

Phasing

The most obvious aspect of the moon's appearance is its phasing. Like the earth, the moon is a spherical, opaque body with no light of its own except what it receives from the sun and reflects. As a result, half of the moon is always in sunlight and half is in darkness. However, the same half is not always illuminated, since the moon is forever changing its position with respect to the sun and earth. The constant change in the moon's appearance is the result of its turning different portions of its illuminated half toward us; this phenomenon is called *phasing*.

At any given time the moon is said to be in a certain phase. Actually, the moon has 29½ different phases, since each day during the month it is in a slightly different position with respect to the sun and earth. The positions are geometrically exact, allowing accurate predictions to be made as to when each phase will rise, cross the meridian, and set (Fig. 3-4). Rather than naming all 29½ phases, the progression of phases is given in terms of age, beginning with new phase when the moon is said to be zero days old. Only eight phases are actually named.

New phase can only occur when the moon is between the earth and the sun. At this position the half of the moon that is facing the earth is not receiving direct light from the sun, while the half of the earth facing the moon is. Therefore, while an observer on earth is experiencing daylight, an observer on the half of the moon facing the earth would be in darkness. The almost direct alignment of the moon between the earth and the sun each month causes the new moon to rise at sunrise, cross the meridian with the sun at local noon, and set at sunset. The new moon, then, is always above the horizon during daylight hours. At new phase the half of the moon facing the earth is illuminated

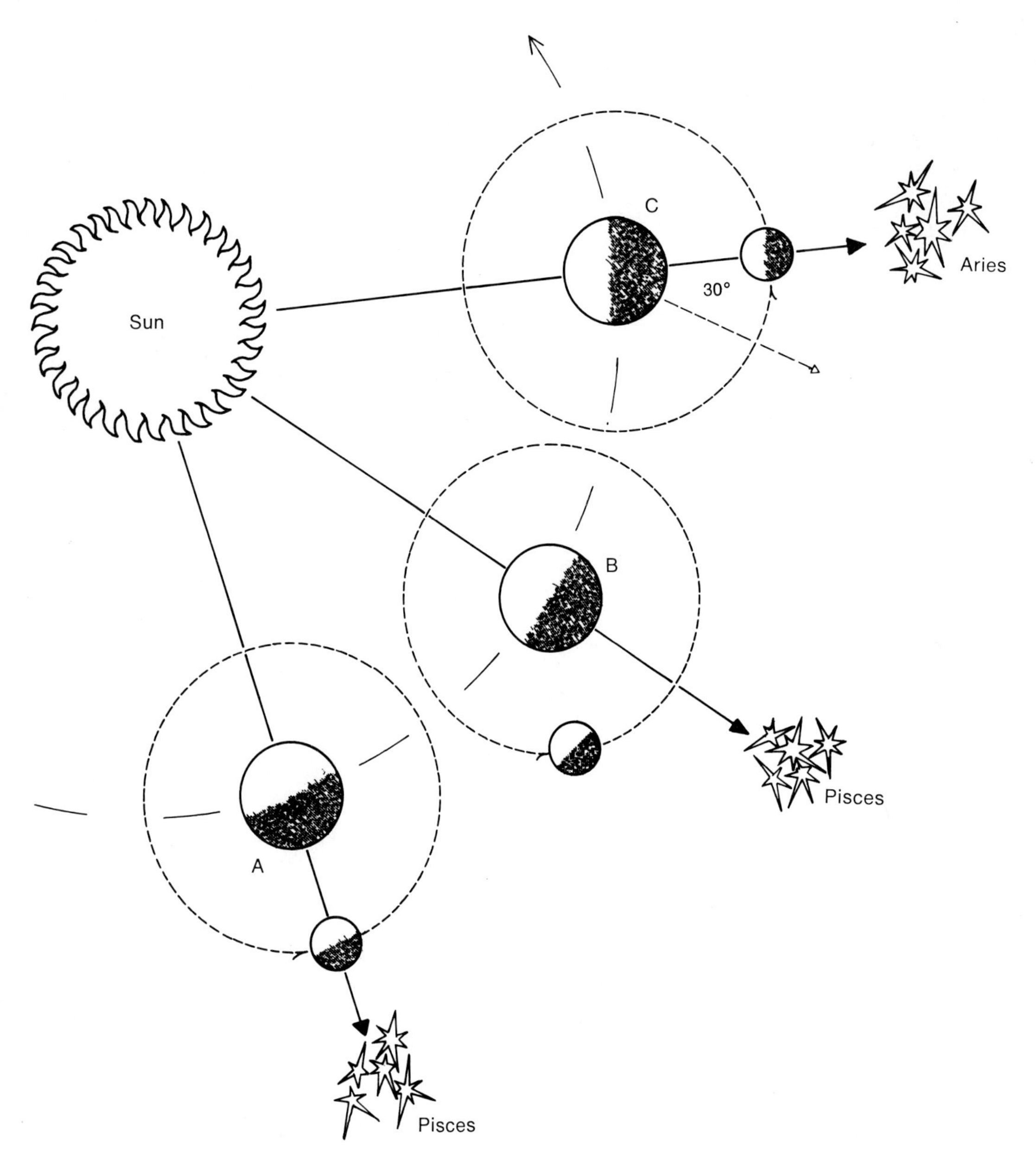

Fig. 3-3. Position of the moon at the end of the sidereal and synodic months. *A*, Position of the earth and moon at first observation (full moon in Pisces); *B*, position of earth and moon 27⅓ days later (moon at waxing gibbous phase in Pisces); *C*, position of earth and moon 2 days after *B* (moon again full but this time in Aries).

only by light reflected from the earth, which is called earthlight. Although 5½ times brighter than moonlight, earthlight is not intense enough to be seen through the overwhelming brightness of sunlight. As a result, new phase can never be seen.

Following new phase the moon, orbiting eastward around the earth, appears to move further away from the sun. The moon is first viewed as a thin, right-handed crescent in the west after sunset. To see the moon when it is less than 2 days old is a challenge for any observer because of its nearness to the sun. The French astronomer Andre Danjon (1890-1967) apparently holds the record for sighting the moon in 1931 when it was only 16 hours, 13 minutes from new phase. At 4 days the moon is said to be at *wax-*

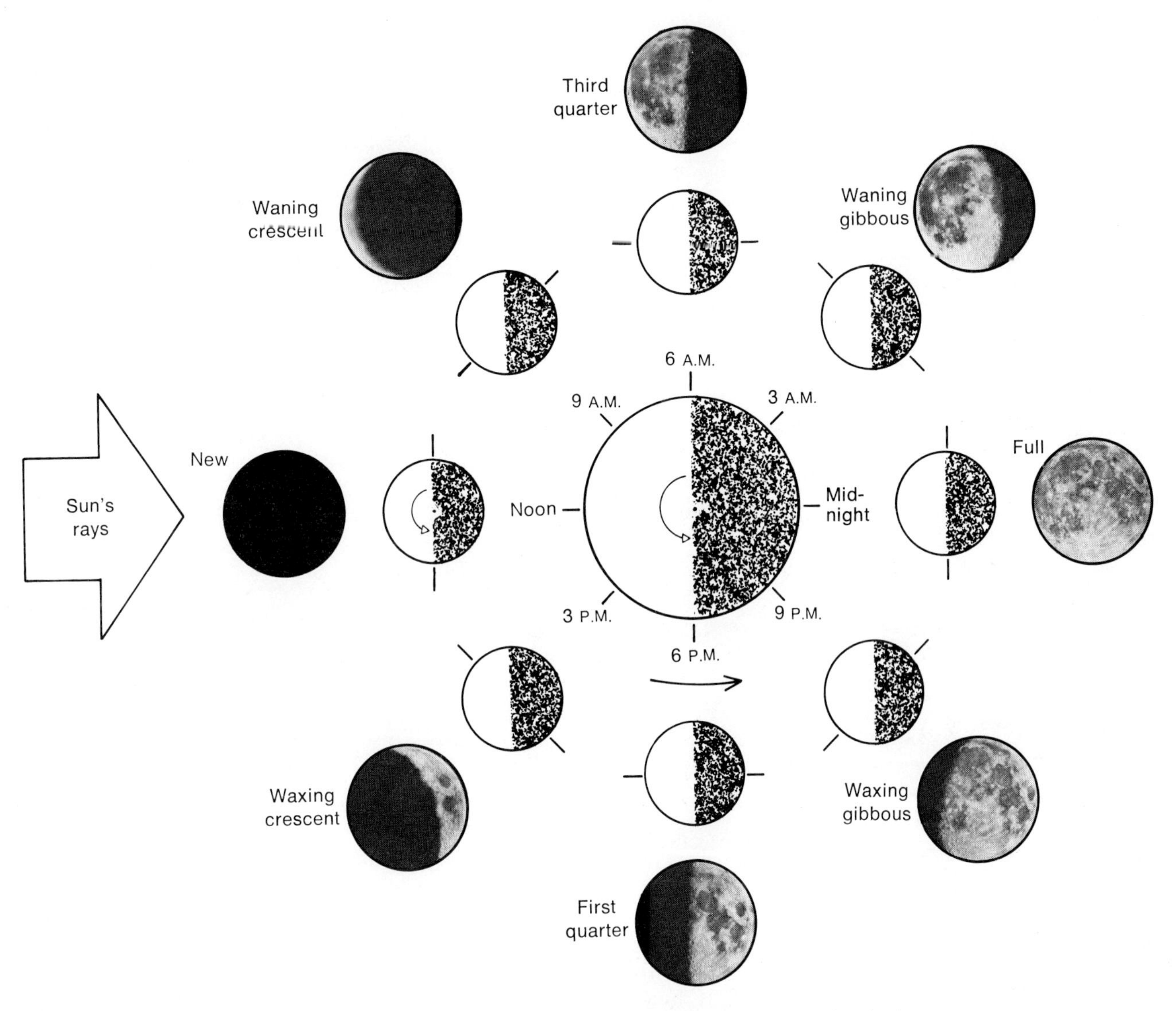

Fig. 3-4. Time-phase relationships of the eight major phases of the moon. The appearance of each phase as seen from earth is shown in insets. (Insets courtesy University of California Lick Observatory, Santa Cruz, Calif.)

ing crescent phase, in which the right-hand fourth of the moon's earth-facing hemisphere is illuminated. The moon in this phase rises at 9:00 A.M. local time (LT), crosses the meridian at 3:00 P.M. LT and sets at 9:00 P.M. LT.

At 7 days the moon reaches a point east of the sun at which it makes a 90° angle with the earth and the sun. This phase is called *first quarter* because the right-hand half of the moon's hemisphere facing the earth is illuminated. The first-quarter moon must rise at local noon, cross the meridian at sunset, and set at local midnight.

The moon at 10 days of age is at *waxing gibbous* phase, in which approximately the right-hand three fourths of the side facing the earth is illuminated. The waxing gibbous moon must rise at 3:00 P.M. LT, cross the meridian at 9:00 P.M. LT, and set at 3:00 A.M. LT.

Fourteen days after new phase the moon, the earth, and the sun are again aligned. This time, however, the moon and sun are in opposite positions with the earth in between. This is called *full* phase because the entire half of the moon facing the earth is fully illuminated. Being opposite the sun causes the full moon to rise at sunset, cross the meridian at local midnight, and set at sunrise.

Just as the new moon is above the horizon all day, the full moon is above the horizon all night. The full moon is second only to the sun in apparent brightness, thus interfering drastically with nighttime observation of other celestial bodies.

Waning gibbous phase occurs 18 days after new moon. At this phase observers on earth see the left-hand three quarters of the portion of the moon facing the earth in sunlight. A waning gibbous moon rises at 9:00 P.M. LT, crosses the meridian at 3:00 A.M. LT, and sets at 9:00 A.M. LT.

At 21 days the moon reaches *third-quarter* phase, a position in its orbit in which it forms a 90° angle with the sun and earth. This time, however, the moon is west rather than east of the sun, as it was at first-quarter phase. In third-quarter phase the left-hand half of the earth-facing hemisphere of the moon is illuminated. The third-quarter moon rises at local midnight, crosses the meridian at sunrise, and sets at local noon.

Again approaching the sun, the moon at 24 days of age is at *waning crescent* phase. In this phase the left-hand quarter of the portion of the moon facing the earth is illuminated. The moon at waning crescent phase must rise at 3:00 A.M. LT, cross the meridian at 9:00 A.M. LT, and set at 3:00 P.M. LT.

As the waning moon continues to orbit eastward, it appears to move closer to the sun. It is now seen as a thin, left-hand crescent visible in the east for a short time before sunrise. Finally, the moon disappears into the morning twilight to return again to new phase, completing its cycle of 29½ days.

Dark side or back side?

The back side of the moon, the side not visible from the earth, is often erroneously called the dark side. It is, however, no darker than the front side because all parts of the lunar surface are illuminated at some time during the month. The phases of the back side are exactly opposite those of the front side. The earth as seen from the moon also has opposite phases. For example, when the moon is at first-quarter phase as seen from the earth, the back side of the moon and the earth as seen from the moon are at third-quarter phase.

PHYSICAL CHARACTERISTICS OF THE MOON

We learned earlier that the moon has a diameter approximately one-fourth that of earth's. The moon's mass, or the quantity of matter it contains, is measured by its resistance to changes in motion. It is only one-eightieth the earth's mass, or 8.1×10^{19} tons. The resultant surface gravity of the moon is only one-sixth that of earth's, so that an astronaut who weighs 180 pounds (81.12 kg) on earth weighs only 30 pounds (13.62 kg) on the lunar surface.

Escape velocity is the velocity that a moving body must attain in order to free itself from the influence of the gravity of a celestial body and move out into space. For example, escape velocity from the earth had to be reached to enable the astronauts to leave the earth to reach the moon; escape velocity from the moon had to be reached so that the astronauts could return to earth. The velocity of escape for the moon is 1.5 miles (2.4 km)/sec, or 5400 miles (8694 km)/hr as compared to the escape velocity of earth, which is 7 miles (11.27 km)/sec, or 25,200 miles (40.572 km)/hr. From an infinite distance a freely falling body such as a spacecraft, without deceleration, would strike the surface of the moon traveling at the moon's escape velocity. This ultimate velocity of free fall for any celestial body is called *terminal velocity*.

The moon's low escape velocity makes it much easier for a spacecraft to escape from the lunar surface than from the earth's and explains why the moon lost whatever atmosphere it may have had at one time. The molecular energy of the gases of an atmosphere would undoubtedly have exceeded the moon's escape velocity and have been lost into space. If the moon does possess an atmosphere today, it is too slight to be measurable. It has been said that the astronauts left more atmosphere on the moon by venting their oxygen tanks than was present before, and even this minute amount would be lost very rapidly. Likewise, any volcanic eruptions or meteorite impacts on the moon would also throw dust and debris into space, whereas on earth such material would first be dispersed into the atmosphere and finally settle back to the surface again.

From the moon's diameter its volume can be computed, and by dividing its volume into its mass, the lunar density can be obtained. The average density of the moon is thus computed to be 3.34 gm/cm^3, compared to the earth's average density of 5.5 gm/cm^3. The moon's overall mean density is only slightly more than the 2.7 gm/cm^3 density for surface rocks on earth.

The moon, like the other satellites and planets in the solar system, has no light of its own except what it receives from the sun. As sunlight strikes either the surface or atmosphere of a celestial body, some of the light is absorbed and turned into heat and the rest is reflected. The ratio of total light reflected from a satellite or planet compared to the total light that strikes it is called *albedo*. More reflected light produces a higher albedo. The albedo of the moon is 7%; therefore it absorbs 93% of the sunlight that strikes it. The albedo of the earth, on the other hand, is 35%, and the planet in our solar system with the highest albedo (76%) is Venus.

Due to the large quantity of sunlight that it absorbs, the moon's surface gets very hot during the long lunar day; without an atmosphere to prevent the rapid loss of this accumulated heat, the moon's surface becomes very cold during the lunar night. The surface temperature of the moon varies from a daytime high of 214° F to a nighttime low of −280° F.

OBSERVING THE MOON

Even though the moon can be seen during daylight hours, it is at its magnificent best between sunset and sunrise. Unlike viewing the sun, the observer on earth may look at the moon indefinitely with either the naked eye or an optical instrument without any harmful effects. This is true because moonlight is cold, reflected sunlight that does not contain the damaging wavelengths found in light directly from the sun.

A variety of lunar surface features are easily visible with the use of a small telescope such as a 2.4-inch refractor. To do justice to the moon, however, a 3- or 4-inch refracting telescope or a 6-inch reflecting telescope should be used. There are many types of lunar surface features: maria, craters, mountains, valleys, rilles, and rays.

Maria

The large dark markings that form the famous ''man in the moon'' are called maria—the plural of the Latin *mare,* meaning sea. They are large lowland plains with relatively smooth, flat floors (Fig. 3-5). To the best of our knowledge

Fig. 3-5. Full moon showing maria, rays, and bright impact craters. Bright crater at bottom of photograph is Tycho. (Courtesy University of California Lick Observatory, Santa Cruz, Calif.)

there is no water on the moon, but if there were, the maria would indeed be seas because they are on the average 6000 feet (1830 meters) below the level of the brighter highlands. The ocean basins on earth are an average of more than 10,000 feet (3050 meters) below the level of the continents. The similarity between the lunar maria and the earth's ocean basins does not end there, however. They are also both composed of the igneous rock basalt, which is darker in color and more dense than continental or highland rocks.

It is now generally accepted that the moon's maria are younger than the highlands. They were formed by the impact of large meteorites or minor planets and were subsequently filled by outpourings of basaltic lava from below the surface. There are 30 maria on the moon—16 on the back side and 14 on the front side. Many of the maria on the earth-facing hemisphere, however, are very large. For example, Oceanus Procellarum (Ocean of Storms) is the largest of all lunar maria, occupying the greatest part of the eastern region of the front side.

Craters

The rounded depressions in the lunar surface, smaller than maria, are called craters. There are more than 30,000 craters on the moon, making them by far its most numerous surface feature. The entire surface, but especially the southern half of the front side, is studded with craters. There are fewer craters in the maria, which may be another indication of their relative youth and may also be the result of the diminishing frequency of meteorite showers over time.

There are several types of craters on the moon. The largest are *mountain-walled plains,* ranging from 60 to 140 miles (96.6 to 225 km) in diameter. Surrounded by mountains, these craters have floors that are depressed below the general level of the enclosing walls. Of these, Calvius is the largest; its diameter is 140 miles (225.4 km). Other examples of mountain-walled plain craters are Grimaldi, Ptolemacus, and Archimedes.

Mountain-ringed plains, 10 to 60 miles (16.1 to 96.6 km) in diameter, are the most numerous type of large craters. They are almost circular and have broad walls and steep inner slopes that are terraced in a series of steps to the floor. Mountain-ringed plains have floors that are considerably depressed below the level of the surrounding walls. The floors of these craters are often characterized by the presence of central mountain peaks that are believed to have been thrust up by meteorite impact (Fig. 3-6). Copernicus, Tycho, Hershel, and Theophilus are good examples of this type of crater.

Crater *rings,* a third type of crater, are 3 to 10 miles (4.8 to 16 km) across. Their shape is circular, and their walls are only slightly elevated above the outside surface, while the inside is well depressed. Thousands of these small craters are scattered over the moon's surface. Craterlets are of the same nature as crater rings but are smaller in size.

Crater *pits* are shallow depressions that do not usually have surrounding rims. Crater *chains* and chains of walled plains are also present, as are ruined ring plains, almost surely the result of erosion by micrometeorite impact, the solar wind, and lava flooding. Also visible are *twin* craters, *multiple* craters, and *hilltop* craters, the latter of which are very similar to terrestrial volcanoes and may be of igneous origin.

Mountains

Just as the lowlands of the moon are called maria, lunar highlands are referred to as *terrae,* Latin for "land." These highlands are for the most part mountain lands with bumps, hollows, peaks, and valleys. They often appear to be rough and jagged when seen from our distant earth, but actually they are much smoother when viewed from the lunar surface.

The mountains of the northern and southern parts of the moon that are visible from earth are quite different. Mountains in the southern portion are almost always associated with craters, while in the northern portion there are several mountain ranges surrounding maria. These lunar mountain ranges are usually named for such terrestrial mountain ranges as the Alps, Caucasus, Apennines, and Carpathians. Across the entire front side of the moon there are numerous isolated mountains such as the central mountain peaks in mountain-ringed plains and curious Piton and Pico in Mare Imbrium (Fig. 3-7). Lunar mountains, like virtually all other surface features of the moon, are believed to have been formed as a result of meteorite impact.

There are many mountains on the moon whose peaks rise to an elevation of 5000 to 12,000 feet (1525 to 3660 meters) above mare level. Some are considerably higher, such as the 18,000-foot (5490-meter) Mount Huygens in the lunar Apennines. At least one peak in the Leibnitz mountains near the lunar south pole was once believed to exceed the height of Mount Everest on earth. This now appears to have been a miscalculation due to the visual distortion of the moon's rim. Remembering that the moon's diameter is only one-fourth that of earth, we realize that these lunar mountains would still reach proportionally greater elevations than their terrestrial counterparts.

Mountains such as the Leibnitz range along the lunar edge give an irregular appearance to the moon's rim, a

Fig. 3-6. Photograph of Tycho, one of our moon's larger craters, was taken by Lunar Orbiter V in 1967. Note the peaks in the center of the crater. These were probably thrown up by meteor impact. (Courtesy National Aeronautics and Space Administration, Washington, D.C.)

Fig. 3-7. Mare Frigoris (above) and Mare Imbrium are separated by the Alps, a mountain range thrust up as the maria were formed by meteor impact. The prominent valley in the center of the photograph is the famous Alpine Valley. At the bottom of the photograph are Piton and Pico, two isolated mountain peaks. Perhaps they were the central peaks of craters that were subsequently filled with lava. Note rille in Alpine Valley. (Courtesy National Aeronautics and Space Administration, Washington, D.C.)

phenomenon that is especially noticeable during an eclipse of the sun. At that time, the last rays of the sun, streaming through these lunar valleys just seconds before the moon completely covers the solar disc, produce the famous "Baily's beads" (Fig. 3-11, *A*).

Valleys, rilles, and rays

There are not many well-defined lunar valleys. The best-known examples are the Rheita valley, 15 miles (24 km) wide and 100 miles (161 km) long, and the Alpine valley, 5 miles (8.05 km) wide and 75 miles (120.8 km) long, cutting through the lunar Alps (Fig. 3-7). Unlike most terrestrial valleys that resulted from erosion, lunar valleys are thought to be grabens formed by normal faulting (Chapter 18). The famous straight wall in Mare Nubium is a cliff that is 60 miles (96.6 km) long and 800 feet (244 meters) high with a 30° slope, also the result of normal faulting.

Hundreds of winding, narrow valleys called *rilles* can be found on the moon (Fig. 3-7). While some are broad and flat, others are narrow and steep. Some rilles actually appear to emanate from lunar craters, a phenomenon that has led scientists to the conclusion that many of these features were formed from the collapse of subsurface lava tubes. Other rilles are believed to be grabens, while still others are of unknown origin. Examples are Schroeter valley and Hadley rille, the latter being the now-famous site at which Apollo 15 landed on July 31, 1971.

Extending outward from nearly 100 lunar craters are

brilliant narrow white streaks known as *rays* (Fig. 3-5). These features are always best seen at full moon; some stretch for hundreds of miles. They seem to be unaffected by changes in topography, for they pass over mountains, craters, and maria alike. The largest of all the bright ray systems can be seen stemming from the craters Tycho, Copernicus, Aristarchus, and Kepler. Tycho holds the record, however, for some of its rays exceed 1000 miles in length, extending as far as the boundary of Mare Nectaris. The rays from Tycho cause the moon at full phase to resemble a peeled orange.

For a long time scientists were puzzled as to the origin of rays, although they appeared to be material splashed out of the moon by meteorite impact. In hopes that the solution could be found to the mystery surrounding these strange lunar features, a ray from the crater Copernicus was selected as the Apollo 12 landing site. The study of the rock and soil samples that were gathered by this successful lunar flight, combined with knowledge of the albedo, location, and other characteristics of rays, enabled scientists to conclude that their original guesses were basically correct. Rays are indeed formed by meteorite impact and consist of ejecta rich in the small spheres of glass so abundant in all lunar samples. The brightness of lunar craters and rays, apparently determined by the percentage of glass spherules in their constituent material, is now thought to be indicative of age—the brighter, the younger. Tycho, then, must be a relatively young crater.

ECLIPSES

The sun is not only about 400 times the diameter of the moon, but at the same time it is about 400 times further away. This means that, as seen from earth, the moon and the sun appear to be about the same angular size on the celestial sphere, ½°. The orbits of both the moon and earth are almost in the same plane, actually crossing each other at two points called nodes. The moon and earth are also both solid bodies. These three facts make possible one of nature's most spectacular events—eclipses.

The fact that the moon and sun appear to be the same angular size on the celestial sphere is significant to students of eclipses. During a total solar eclipse, the moon completely covers the disc of the sun, giving observers a view of the outer solar atmosphere that is not usually visible. The study of the outer atmospheric layers of the sun is important if we are to gain additional knowledge of both the structure of our ruling star and the processes that it is constantly undergoing.

In order for an eclipse to occur, the moon must be at one of the two nodal points in its orbit. It was pointed out earlier that the Babylonians discovered that the nodes re-

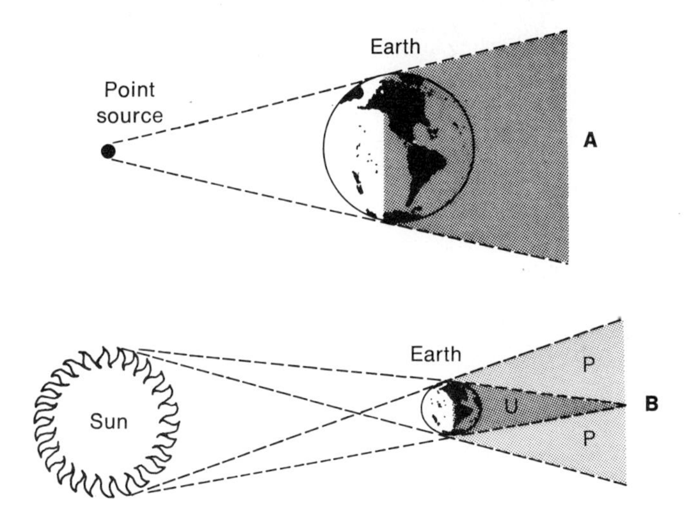

Fig. 3-8. A, Shadow the sun would cast on the earth if it were smaller or farther away. **B,** Double shadows the sun actually does cast. *U* is umbra; *P* is penumbra.

gress westward around the ecliptic and that they require 18 years, 11 days, and 8 hours to complete one circuit. At about the same time the Babylonians also found that eclipses followed a cycle of the same duration, which they named the *saros*. After an interval of 1 saros any eclipse, lunar or solar, will be repeated. The coincidence of position is only approximate, so that the following eclipse will not necessarily be of exactly the same type. A total eclipse of the sun, for example, may be followed by a partial solar eclipse. An eclipse of the moon occurred on January 30, 1972, and an eclipse of the sun followed on July 10, 1972. An eclipse of the moon, then, can be expected to occur on February 11, 1990, with an eclipse of the sun to follow on July 22. Awareness of the saros makes it possible to predict eclipses centuries in advance or calculated backward to fit ancient records and thus date historic events with which they were associated.

The moon and earth each cast two long shadows into space in a direction that is always away from the sun. There are two shadows of each body because the sun is so close to the earth and moon that it is more than just a point of light like other stars. If this were the case, only a single divergent shadow would be produced (Fig. 3-8, *A*). As it is, however, a convergent total shadow called the *umbra* and a divergent partial shadow known as the *penumbra* are cast by both the moon and the earth (Fig. 3-8, *B*). The shadows are present at all times but are visible only when an eclipse occurs as the moon moves into the shadow of the earth or vice versa.

Eclipses are of two basic types: lunar and solar. Due to the dynamics of the earth-moon system, the number of lunar eclipses occurring in any one year may vary from zero to three, while solar eclipses may occur from two to five times annually. Eclipses of all types, however, may not exceed seven in any given year.

Lunar eclipses

In order for an eclipse of the moon to occur, the moon must first move into the earth's shadow. This happens only when the moon and earth are in approximately the same orbital plane, possible only at a node and when the earth is also between the sun and moon. In other words, a lunar eclipse can only occur when the moon, at full phase, is at or near a node of its orbit (Fig. 3-9).

There are three types of lunar eclipses—total, partial, and penumbral. A *total* eclipse of the moon occurs when the moon passes directly into the umbral shadow of the earth. Because of the moon's distance from the earth, the umbral shadow of the earth is 6000 miles in diameter, and the penumbral shadow is 10,000 miles across. The moon initially moves into the penumbra, where only a slight dimming of the moon takes place. Then as the moon enters the umbra it is possible to see the curvature of the earth's shadow being cast on the moon (Fig. 3-9, *B*). During the total phase of the eclipse, when the moon is completely within the earth's total shadow, its color changes from white to copper, a phenomenon explained by the refraction of sunlight through the water vapor and dust contained within the earth's atmosphere. A total lunar eclipse, from the instant the moon touches the earth's penumbra on one side until it exits through the other, may last almost 6 hours. Total phase, however, may last no more than 1 hour and 40 minutes.

A *partial* lunar eclipse occurs when the moon moves partially but not completely into the umbra of the earth. *Penumbral* eclipses, on the other hand, occur when the moon passes either partially or completely through the penumbra but does not touch the umbra.

Regardless of type, eclipses of the moon take place only at night, so that everyone with clear weather on the night side of the earth should easily be able to view some portion of the event.

Solar eclipses

Although viewing a lunar eclipse can be a spectacular experience, witnessing a solar eclipse can be virtually awe inspiring. Ancient peoples stood in mortal fear of solar

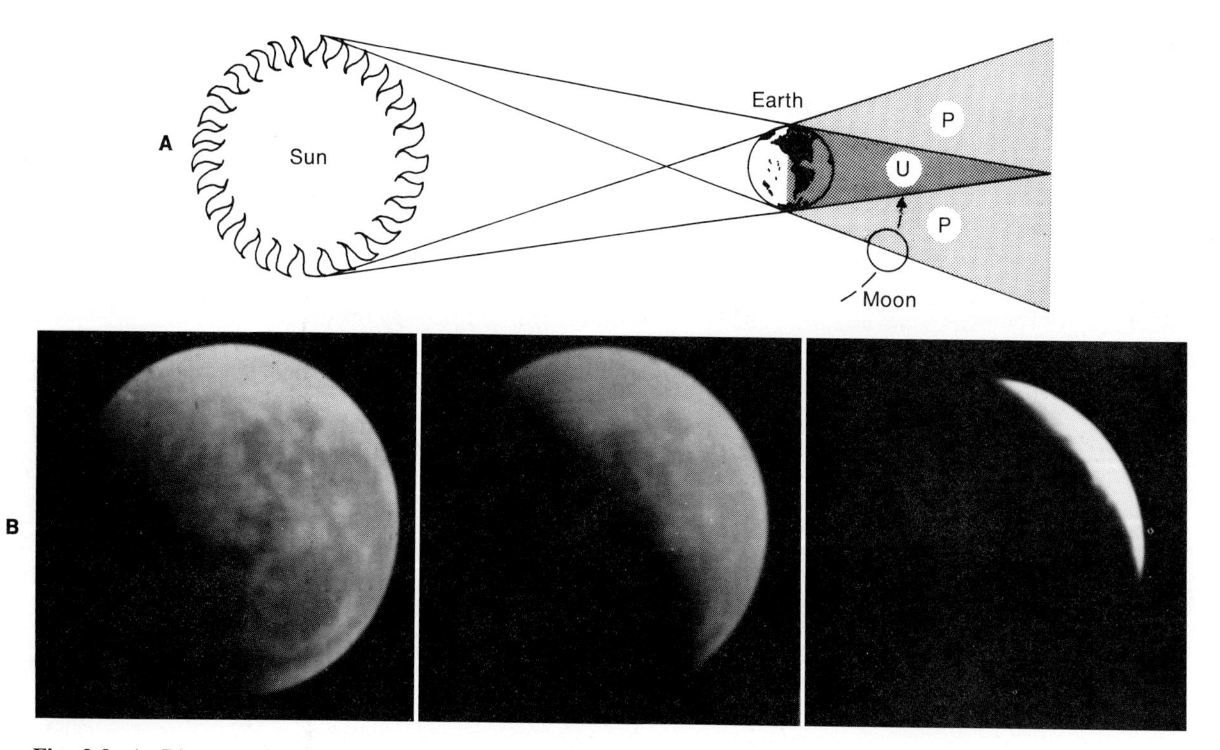

Fig. 3-9. A, Diagram of moon about to enter the earth's shadow, an event we call an eclipse. Only if it enters the umbra will the eclipse be total. **B,** Photographs of actual eclipse. (**B** courtesy University of Chicago Yerkes Observatory, Williams Bay, Wisc.)

eclipses, believing them to be the work of angry gods. Although we now know that an eclipse of the sun is the result of celestial geometry, we still cannot help but marvel at this phenomenal occurrence.

In order for a solar eclipse to occur, the moon at new phase between the earth and the sun must be at or near a node (Fig. 3-10). The earth moves into the lunar shadow at about the moon's minimum distance from the planet, causing the shadow of the moon to be just long enough to reach the earth's surface. Like the earth's shadow, the shadow of the moon also consists of a penumbra and an umbra. Where the moon's shadow touches the earth's surface, the umbra can at most be about 170 miles in diameter, while the penumbra may be as wide as 4000 miles (Fig. 3-10). The moon's eastward orbital speed of nearly 2400 miles/hr, coupled with the earth's eastward rotation speed of about 1000 miles/hr, makes the maximum eastward ground speed of the shadow of the moon 1400 miles/hr. At this speed the umbra passes an observer in no more than 7½ minutes—usually much less. Those in the path of the umbra will witness a total eclipse of the sun, while those located within the path of the penumbra will view only a partial eclipse.

A total solar eclipse starts quietly. You barely notice a dimming of the sun; the penumbra rushes over you. Then a small nick appears on the western edge of the solar disc, marking the arrival of the moon's umbra. As the minutes pass the nick enlarges, diminishing the visible portion of the sun. One instant before the moon completely covers the sun, the last rays of sunlight, streaming through lunar valleys, produce the beautiful Baily's beads phenomenon mentioned earlier (Fig. 3-11, *A*). The umbra now engulfs you from the west and the sky grows dark, even though your watch reads only about 1:00 P.M. The moon is ringed by a faint light from the outer atmosphere of the sun—the *corona*—which is not ordinarily visible (Fig. 3-11, *B*). The temperature of the air drops, night insects begin to chirp, night birds take wing, chickens go to roost, and streetlights and signs come on. The entire horizon is lighted by an eerie orange glow. Then, almost as quickly as it came, the umbra

Fig. 3-11. Phenomena of solar eclipse. **A,** Baily's beads are caused by sunlight streaming through the valleys of the moon. **B,** Sun's corona is shown during the total solar eclipse. (Courtesy University of Chicago Yerkes Observatory, Williams Bay, Wisc.)

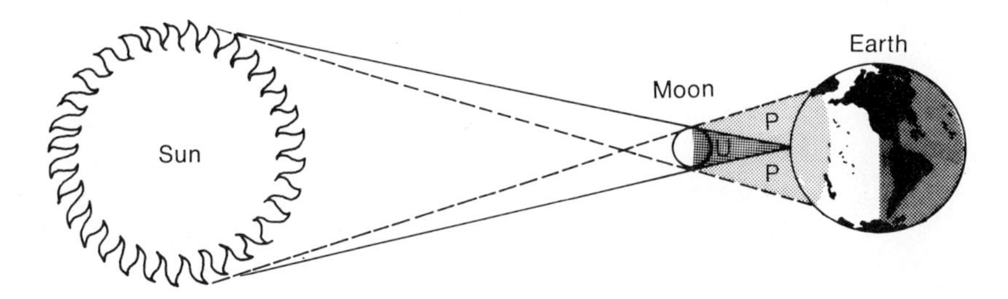

Fig. 3-10. Arrangement of sun, moon, and earth necessary to produce a solar eclipse.

speeds away towards the east and sunlight returns. The moon slowly receeds from the sun's disc until only a nick remains on the eastern edge. Except for the slight penumbral darkening, the eclipse is over.

Solar eclipses may also be of three types—total, partial, and annular. A *total* eclipse of the sun, just described, takes place whenever the umbra of the moon reaches the earth's surface. A *partial* solar eclipse, on the other hand, occurs when the penumbral shadow of the moon makes only partial contact with the surface of the earth while the umbra just fails to make contact. As noted, a partial eclipse of the sun is also seen by observers outside the umbral path of a total solar eclipse. When the moon is too far from the earth for its umbra to touch and the penumbra makes full contact, an *annular* or *ring* solar eclipse takes place. During an annular eclipse, the moon's disc is seen surrounded by the outer edge of the sun.

A word of caution must be added at this point. Never look directly at the sun, especially during a total eclipse. The result of direct observation may be severe burns of the exterior of the eye caused by infrared (heat) radiation, causing total blindness or the more subtle and often irreversible damage done to the retina by ultraviolet radiation. The iris of the eye, which is highly sensitive to light, controls the size of the pupil and this allows just the right amount of light to enter. During a total eclipse of the sun, the natural protective function of the iris ceases because the moon blocks the light from the sun. This leaves the eye vulnerable to the more dangerous, invisible ultraviolet radiation. The only safe way to ever view the sun, whether during an eclipse or not, is indirectly. Safe indirect techniques include eyepiece projection on a screen with binoculars or telescope, projection by pinhole camera, and reflection from water. Viewing the sun with sunglasses, smoked glass, and exposed film can be highly dangerous.

Since the sun is visible almost every day except during cloudy weather, you have many opportunities to observe its varying seasonal positions and daily telescopic appearance. Such observation offers one of the greatest spectacles that we can ever hope to witness.

Occultations

When the moon eclipses any body other than the sun, the event is called an *occultation*. Moving along in its orbit against the background of more distant celestial bodies, the moon frequently moves in front of a star or planet, obscuring it from the view of observers on earth. The object initially disappears behind the moon's eastern edge, then reappears at its western edge.

Occultations, which must be viewed telescopically, provide spectacular proof that the moon has no atmosphere, for the disappearance and reappearance of the occulted object is instantaneous. If the moon had an appreciable atmosphere, however, the object would dim slightly an instant before appearing to pass behind the moon. The data obtained by precisely timing occultations are of value to a more accurate understanding of the workings of the solar system.

TIDES

The word "tide" immediately brings to mind the ocean, but actually the earth would have tides even if it possessed no water. A tide is the deformation of a body by the gravitational forces exerted upon it by other bodies. Tide-producing forces on earth arise because of the gravitational pull of the moon and the sun.

Gravity, you will remember, is directly proportional to the sums of the attracting masses and inversely proportional to the square of their distance from each other. The moon is roughly 240,000 miles away from the earth, while the sun is 93 million miles distant. As a result, the sun's gravitational field is virtually nondifferential; that is, there is little change in solar gravity over the 8000 mile distance of the earth's diameter. The sun's gravitational pull, then, is almost the same on one side of the earth as it is on the other.

Lunar gravity, on the other hand, decreases rapidly through the 8000-mile distance between a point on the earth's surface facing the moon and the corresponding point on the opposite side of the earth. Tidal forces, then, occur not only as a result of gravitational pull alone but also as a result of its different effects over the earth's diameter. The sun's mass is some 27 million times greater than the moon's, but because the moon is so close to the earth, lunar tidal forces are 2½ times greater than solar tidal forces.

Tidal forces influence the solid earth, the water, and the atmosphere. The ocean tides are a good illustration of the effect of these tidal forces. Tidal bulges are produced within the oceans by the moon's tidal forces and, to a lesser extent, by the sun's. These bulges are located on the side of the earth facing the moon as well as on the opposite side. As the earth rotates, all coastal regions are carried into and out of these bulges twice each day, thereby producing high and low tides (Fig. 3-12).

In order to understand the formation of the tidal bulges, it is important to remember that the effect of the moon's gravity decreases with distance. Imagine a point A on the side of the earth facing the moon. The moon exerts a stronger pull at A than at a point B located on the far side of the earth facing away from the moon, just opposite point A.

Because of the moon's pull at point A, it is easy to

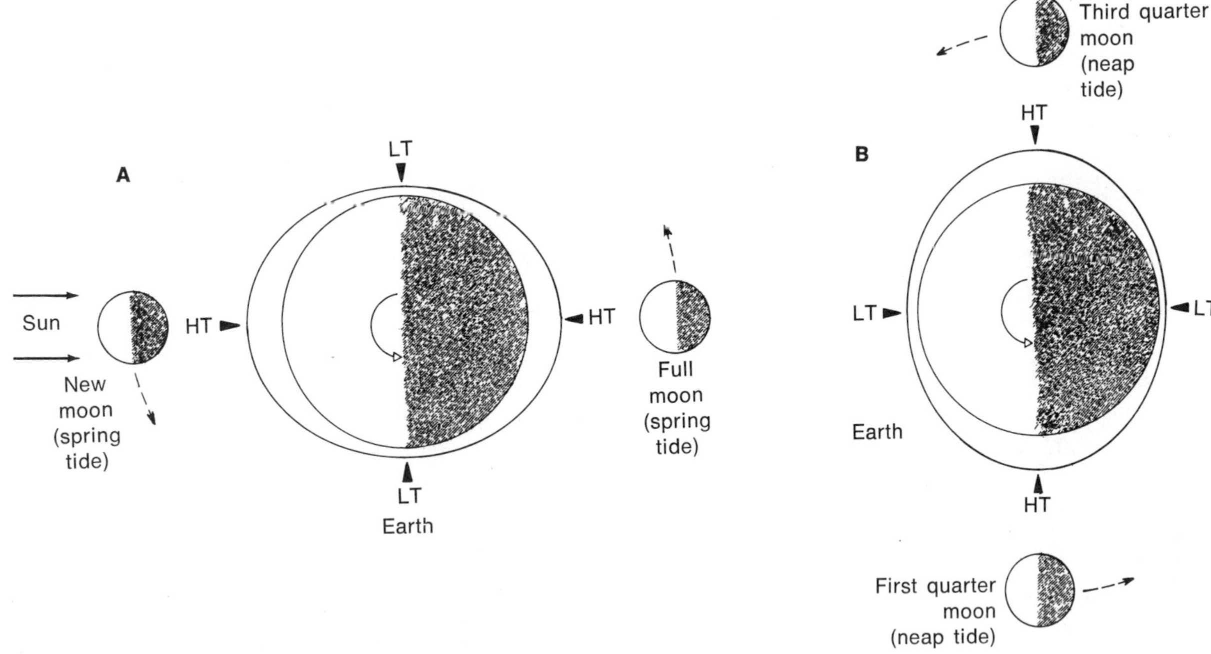

Fig. 3-12. Positions of the moon during spring tide, **A,** and neap tide, **B.** *HT* is high tide; *LT* is low tide.

understand that as the earth rotates and the oceans come to that point, the waters will rise up, or bulge, to form high tide. Actually, the moon pulls on the solid earth as well as on the water. Since the ocean waters are fluid, however, they are more responsive than the solid rock of the earth itself.

Nonetheless, the earth itself is displaced slightly toward the moon, and this helps to explain why there are high tides on the side of the earth facing away from the moon at the point we called B in our imaginary example. The pull of the moon on the earth at our point A is stronger, but the earth is a solid object, so that one part of it cannot move away from the rest. Therefore the whole body of the earth, even that part at point B on the far side, is displaced slightly toward the moon. Because the waters are not a part of the earth itself, but instead lie on its surface, when the earth moves toward the moon the waters on the earth's surface at point B tend to remain where they were, creating the bulge, or high tide, at point B.

Essentially, then, we can say that at high tide the effects of the moon's tidal forces are (1) to pull the water at our point A away from the earth's surface toward the moon and (2) to pull the earth away from the water at point B. As the earth rotates and carries coastline areas into and through points A and B, we see high tide. Then, as the earth continues to rotate we see low tides as the waters sink back to their previous levels and the waters pass out of points A and B.

When the sun, earth, and moon are aligned, which is the case both at full and new moon, the gravitational pull of the sun combines with the moon's tidal forces to produce spring tides, which are higher than average high tides (Fig. 3-12). When the sun and moon form a right angle with the earth, as they do at both the first- and third-quarter phase of the moon, their tidal forces cancel each other out, thereby producing *neap* tides, or tides that are lower than average.

The arrival times of high and low tides and the difference in height between them can vary at different locations or even in the same general area because of the variable water depths and basin and coastline shapes. One of the greatest tidal ranges in the world can be found in some parts of the Bay of Fundy, Nova Scotia, where there may be a difference of more than 50 feet (15 meters) between high and low tide. In contrast, normal tidal ranges in most parts of the world are only between 2 to 3 feet (less than 1 meter). The unusual tidal range in the Bay of Fundy is due to its V-shaped inlet, which narrows back away from the sea.

Earlier in this chapter it was pointed out that moonrise

is delayed by an average of 50 minutes each day because of the revolution of the moon. Since the tidal bulges follow the moon as it orbits the earth, there is also a daily average delay of 50 minutes in the arrival of high and low tide at any given location.

COMPOSITION, INTERIOR, AND AGE OF THE MOON

Although the six Apollo landings on the lunar surface have greatly increased our knowledge of the moon, a great deal remains to be learned. Hundreds of pounds of lunar material have been returned to the earth for analysis, and sensitive instruments have been left on the moon to relay information regarding radioactivity, magnetic fields, the solar wind, moonquakes, and temperature back to earth. Generalities about the moon are difficult to make because the different landing sites have yielded different information.

Composition

Lunar surface material may be classified into broad categories—mare and highland. The *mare* material is either basalt-like rock or the loose debris derived from it that is called lunar "soil." *Highland* material, on the other hand, is composed of aluminum-rich basalt, gabbro, or a rock called anorthosite. These rocks usually contain twice as much aluminum oxide and about half as much iron oxide as the mare rocks, although their composition varies greatly. The high aluminum content indicates that the highlands of the moon represent areas of light original lunar crust, whereas the maria are composed of material originating from the depths of the moon where the iron content is higher.

Interior

Exploration of the interior of the moon is proceeding through several different methods. Moonquakes are being recorded by seismometers left on the moon by the Apollo astronauts; the data is transmitted back to earth. Lunar magnetic fields are being studied with the use of magnetometers that were also left on the moon's surface. Measurements of heat flow from the lunar surface have been taken by astronauts using sensitive thermometers, and in work closely related to the heat flow experiments there are studies of the moon's electrical properties. Finally, earth-based telescopic measurements of the physical librations of the moon are providing indications of interior mass distribution. Librations are measured to within 1 to 4 inches (3 to 15 cm) by laser beams bounced off retroreflectors left on the lunar surface by the astronauts.

A model of the moon's interior is beginning to take shape as a result of the many experiments that have been conducted by the astronauts and with the aid of instruments they left behind, but this model is by no means the final answer. The interior of the moon is now thought to consist of at least two basic parts. The presence of a heterogeneous, solid outer layer, extending from near the surface to a depth of about 600 miles (1000 km), has been indicated by the way in which seismic waves are scattered in this region. A homogeneous, semiviscous inner zone that extends perhaps to the moon's center is suspected because of the efficient way in which seismic waves travel below a depth of 600 miles (1000 km). The absence of a general magnetic field and the low density of the moon indicate either lack of a heavy iron core such as the earth's or at best only a small one. Heat flow experiments, however, have provided evidence of high internal temperatures that are indicative of a hot core. The lunar interior appears to be seismically active because of the occurrence of several moonquakes whose foci were located 600 miles below the surface—evidence of repeated slippage in solid rock. Scientists have therefore concluded that the moon is still geologically active but only at great depths. Its exterior evolution is virtually complete.

Age

For a long time people have thought that the moon should be approximately the same age as the earth and, for that matter, as the other members of the solar system. The oldest known rocks on earth are 3.6 billion years old, while some meteorites have been radioactively dated at 4.6 billion years. What about the age of the moon? Samples of basaltic rock from the maria all yield an age of between 3.2 and 3.9 billion years. Maria "soil" samples are generally older—up to 4.6 billion years. The oldest known anorthosite from the highlands is 4.1 billion years of age. Therefore we may say that even though both the moon and the earth are very nearly the same age, most of the geologic evolution of the moon occurred before the oldest known rocks on earth were formed. It must not be concluded from this that the moon is older than the earth, but only that the moon is a "fossil planet" that completed its major geologic evolution quickly and then retained its ancient appearance for billions of years.

4 *runners of the firmament*

The solar system

Nature, by magnetic laws,
Circle unto circle draws,
But they only touch when met,
Never mingle—strangers yet.

Strangers Yet
Lord Houghton

There were five bright objects moving about among the "fixed" stars, but they were different from the stars in several ways. Not only did they wander within the constellations of the zodiac, they also shone with a clear steady light that was somehow different from the twinkling of the stars. What were they? Gods?

An object appeared in the sky. Battles ceased as warriors looked in bewilderment and fear at the strange sight above their heads. Proclamations were issued, the populations of entire cities were thrown into panic, kings abdicated their thrones, and people died of fear. What were these apparitions that roused such terror? Could they be omens of impending death and destruction?

INVENTORY OF THE SOLAR SYSTEM

Men eventually learned the true nature of these celestial objects. The wanderers of the heavens were not gods but planets, and the strange objects that appeared from time to time were comets, not bad omens. We now know that planets and comets are children of the sun and brothers, sisters, and near relatives to the earth and moon. They are all members of the solar system—a family of small celestial bodies held firmly in the gravitational grasp of their giant father, the sun. Although the solar system encompasses a region of space extending for billions of miles in all directions from the sun, members of the solar family must be considered close to one another when compared by the colossal "yardstick of the universe."

The solar system consists of a star called the sun, nine known major planets, 33 satellites, thousands of minor planets or asteroids, and an apparently inexhaustible supply of comets and meteoroids. The sun alone virtually *is* the solar system for several reasons. First, it comprises 99.86% of the system's total mass; second, its overall gravity is dominant; and finally, its brilliant light illuminates all other members of the solar system, and its enormous energy makes life possible on earth.

BIRTH OF THE SOLAR SYSTEM

If in our imaginations we could go back in time about 4.5 billion years, we would see a very strange sight, a large, red, young sun, newly formed of an interstellar cloud composed mostly of gaseous hydrogen, with heavier elements making up only a small fraction of the total. The young sun is unaccompanied by the familiar planets we know today but instead is surrounded by a huge, disc-shaped mass of dust and gas, cold material remaining after the sun's formation. This gas and dust is in turbulent motion, with the result that large changes in density occur from point to point in the disc. Occasionally some of this material collects and develops a density and mass so high that its own gravitational attraction prevents it from breaking up. Such masses of material are permanent condensations called *protoplanets.* As each of the protoplanets moves around the sun, it increases its mass and, as a result, its temperature, by sweeping up more matter. The heavier, solid material sinks toward the center, leaving the protoplanet enveloped by a halo of gas—its primordial atmosphere. As interplanetary space becomes more transparent, the protoplanet atmospheres are warmed by the sun's heat. The warming effect is most intense on those evolving planets nearest the sun, causing them to lose virtually all their atmosphere and leaving only a solid core behind. In the case of the more distant protoplanets the heating effect is so minute that they are still close to their ancestral stage even today.

Gas and dust, from which all galaxies, stars, planets, and satellites are presumably made, is not uniformly distributed in space; it is found in more dense masses in some places than in others. The greater the original concentration of matter in a particular region, the larger the mass of the protoplanet. Whether the original condensation zone was rich in gas or dust also plays a role in the ultimate formation of the planet. A small mass rich in gas and relatively poor in dust could have produced a type of planet called *Jovian* (like Jupiter). Jovian planets have low densities and are represented in our solar system by Saturn, Uranus, and Neptune, as well as Jupiter. Very small regions that were rich in dust but poor in gas could have formed a second type of planet referred to as *terrestrial* (like the earth). Mercury, Venus, Earth, Mars, and Pluto are the terrestrial planets in our sun's family. No planets of our solar system, because of their small masses, have ever generated light of their own, although Jupiter came close. It has been estimated that if Jupiter's mass were only 10% greater than it is today, it would be a faintly shining star.

The protoplanet hypothesis, discussed above, is presently the most popular theory regarding the formation of our solar system. Other theories that at one time were seriously considered include the random capture, encounter, and nebular hypotheses. The *random capture* theory was abandoned because it failed to explain the solar system as it is today, and it is mentioned only to show the manner in which human thinking has matured with regard to the origin of the solar family. The random capture theory assumes that planets wandered through space and from time to time were captured by the gravitational attraction of a star such as the sun. If this had been the case, the orbital directions and inclinations of the planets today should be random, and they are not.

The encounter and nebular hypotheses differ primarily in their account of the manner in which the planetary matter was separated from the sun. The *encounter* theory stipulates that the sun made a close approach to another star, and that as a result, ribbons of hot gas were torn away from each star by tidal forces. The gas then cooled and condensed into small solids that in turn collided, cohered, swept up more material, and finally grew into planets. There are four major objections to the encounter hypothesis.

1. The hot gaseous ribbons should have dispersed rather than condensed.
2. Tidal forces would not have been strong enough to eject material unless the approach was very close, an extremely unlikely event.
3. It is highly improbable that the sun ever came close to another star at all.
4. The encounter could have occurred at any time after the formation of the sun, so the planets should therefore be younger than their ruling star. Our best evidence to date indicates that the solar system and the sun are about the same age.

The *nebular* theory proposed that a rapidly rotating primordial sun, itself formed from a gaseous nebula (cloud), spun off rings of hot gas that later condensed to form the planets. This theory was rejected because such an origin should have left the sun with most of the angular momentum in the solar system, when in reality it has only $^1/_{200}$ of the total. As in the encounter hypothesis, it is not clear why condensation into solids rather than dispersion into space would have occurred.

Theories about the formation of the solar system are still only theories. More pieces of an existing model may be added or an entirely new explanation may be developed as our knowledge of the universe continues to increase at an ever-accelerating rate.

MAJOR PLANETS

The wandering stars that so intrigued the ancients were called "planets" by the Greeks. Although early astronomers recognized the existence of five planets, little was known about their nature until Galileo introduced the use of the telescope to astronomy. Since that time, however, at a rate of approximately one each 100 years, astronomers have been discovering new planets. Today the number of planets stands at nine, including the earth (Table 2). Are there other major undiscovered planets? Although astronomers from time to time have reported such sightings, these reports are still unconfirmed. It is also interesting that astrologers have long predicted and now anticipate the impending discovery of two new planets.

Inferior planets

Inferior planets are those whose orbits are smaller than that of the earth. There are only two inferior planets—Mercury and Venus. From Mars, however, the earth would be considered inferior, and from Pluto, the planet most distant from the sun, all other planets would be so classified.

Mercury: the elusive planet

Mercury, one of the five planets known to the ancients, was named by the Romans for the mythologic messenger of the gods (Fig. 4-1). Its diameter of 3030 miles makes

Table 2. Major celestial bodies in our solar system

Celestial body	Symbol	Year of discovery	Discoverer
Sun	☉	Unknown	Unknown
Moon	☾	Unknown	Unknown
Mercury	♂	Unknown	Unknown
Venus	☿	Unknown	Unknown
Mars	♃	Unknown	Unknown
Jupiter	♀	Unknown	Unknown
Saturn	♄	Unknown	Unknown
Uranus	⛢	1781	Herschel
Neptune	♆	1846	Adams and Le Verrier
Pluto	♇	1930	Tombaugh

Table 3. Densities of selected celestial bodies*

Celestial body	Density (gm/cm^3)
Earth	5.50
Mercury	5.20
Venus	5.10
Pluto	4.85†
Mars	4.00
Neptune	2.00
Uranus	1.50
Jupiter	1.33
Saturn	0.70
Moon	3.34
Sun	1.40

*So that you can compare these densities with a familiar substance, remember that the density of water is 1 gm/cm^3.
†This figure is approximate.

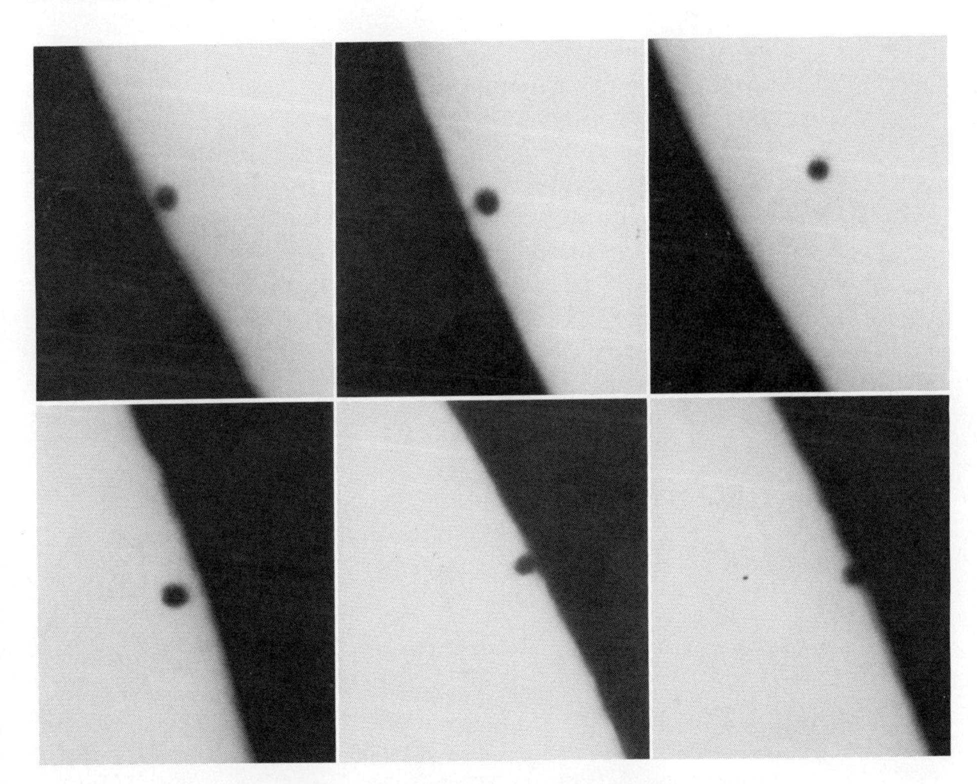

Fig. 4-1. November 10, 1973, transit of Mercury as photographed at Orcines, France. Ingress occurred at 7:49 universal time (UT) (upper left) and egress (lower right) almost 6 hours later at 13:17 UT. Mercury is seen as a small black dot against the bright photosphere of the sun. (Photograph by J. Dragesco.)

it the smallest planet; in fact, its diameter is only 870 miles larger than that of the moon. Mercury is the least massive planet, with a mass of 0.056 relative to the earth's mass of 1.000. It is the planet closest to the sun, from which its mean distance is only 36 million miles, or 0.39 astronomical units (AU).* With an average orbital speed from west to east of 30 miles/sec (108,000 miles/hr), Mercury is the fastest planet and as a result has the shortest revolution period—88 earth days are required for it to travel once around the sun. Its escape velocity is 2.6 miles/sec (9360 miles/hr), its surface gravity is 0.36 *g* (36% of earth's 1 *g*) and its albedo is 7%; all are lower than comparable data for any other major planet. Mercury also experiences the greatest and most drastic surface temperature variations.

Its density of 5.2 gm/cm^3 is similar to that of the earth. Therefore Mercury, in addition to several other planets with high densities, is classified as a terrestrial planet (Table 3).

Like all planets in the solar system, Mercury rotates on its axis. For many years its rotation was thought to be synchronized with its revolution—both equal to 88 earth days. This would have caused the planet to always keep the same side turned toward the sun, similar to the way in which one hemisphere of the moon always faces the earth. Recent earth-based radar studies of Mercury's rotation, however, indicate that the planet turns once each 59 days. The fact that 59 is almost exactly two thirds of 88 means that Mercury spins three times on its axis for every two revolutions it makes about the sun, and thus turns all portions of its surface toward the sun. Mercury's rotation is direct, that is, from west to east, which is the dominant direction of rotation as well as revolution in our solar system.

In March of 1974, instruments on Mariner 10 discovered a very thin atmosphere of helium extending as far as 300 miles from the surface of Mercury. The source of this atmosphere, only one-hundred billionth as dense as the earth's, is probably the solar wind. An alternate source could be the decay of radioactive materials within the planet.

The extreme temperature variations that occur on Mercury's surface and the planet's low albedo are also indicative of the presence of very little atmosphere. Mariner 10 detected ferociously high temperatures of 800° F (426° C) bathing the planet during the daytime, whereas temperatures at night plunged to a bitter cold −280° F (−137° C), the widest range in the solar system. If it had an appreciable atmosphere, Mercury would undoubtedly have a much higher temperature at night and an albedo greater than the 7% we have been able to measure. This would be the case because a thicker atmosphere would trap some of the daytime heat as the planet turned and would also reflect more sunlight, thus increasing the planet's albedo.

*The distance between the earth and sun is said to measure 1 AU, or an average of 93 million miles. Distances between other celestial bodies are often expressed in astronomical units.

The first flyby of Mercury by Mariner 10 also detected a magnetic field only one thousandth as strong as earth's. Mercury's magnetic field is believed to be produced by an electric current induced in the planet as Mercury cuts through solar magnetic fields.

Telescopically, Mercury is at first glance a disappointing planet, for it has neither discernible surface features nor any known satellites. Through a small telescope it is visible only as a small, reddish disc, which over a period of time passes through phases like the moon. With proper mental preparation on the part of the observer, however, Mercury can be a source of esthetic pleasure. For example, it should be remembered that few people have ever seen Mercury either with the naked eye or a telescope, including the great astronomer Copernicus.

Mariner 10 showed Mercury to be much like the moon in appearance, complete with craters, maria, and cliffs. The similarity ends there, for calculations based on Mariner 10's data show Mercury to be more than 100 times more massive than previous estimates. Its lightweight, moonlike surface encloses a heavy, earthlike iron core.

The reason for its elusiveness is that when it is viewed from earth, Mercury can never be more than 28 angular degrees from the sun. With the earth rotating at the rate of 15°/hr, the tiny planet can never be seen more than 1 hour, 52 minutes before sunrise or after sunset. Mercury, then, can never be seen in a completely dark sky.

The orbital plane of Mercury is inclined 7° to the ecliptic, giving it the second highest orbital inclination in the solar system. The planet is so nearly aligned with our line of sight of the sun, however, that we do not directly observe Mercury's revolution. Instead, the planet appears to be moving almost in a straight line, first to the east of the sun and then to the west, spending much of its period of revolution either lost in the solar glare or hidden below the earth's horizon.

Venus: planet of clouds

The planet Venus was named for the goddess of love in Roman mythology, and like Mercury, it is an inferior planet that was known to the ancients (Fig. 4-2). Earth and Venus are often called twin or sister planets because they are so nearly alike both in diameter and mass. Every 19 months Venus comes within 25 million miles of the earth, also making Venus our nearest planetary neighbor.

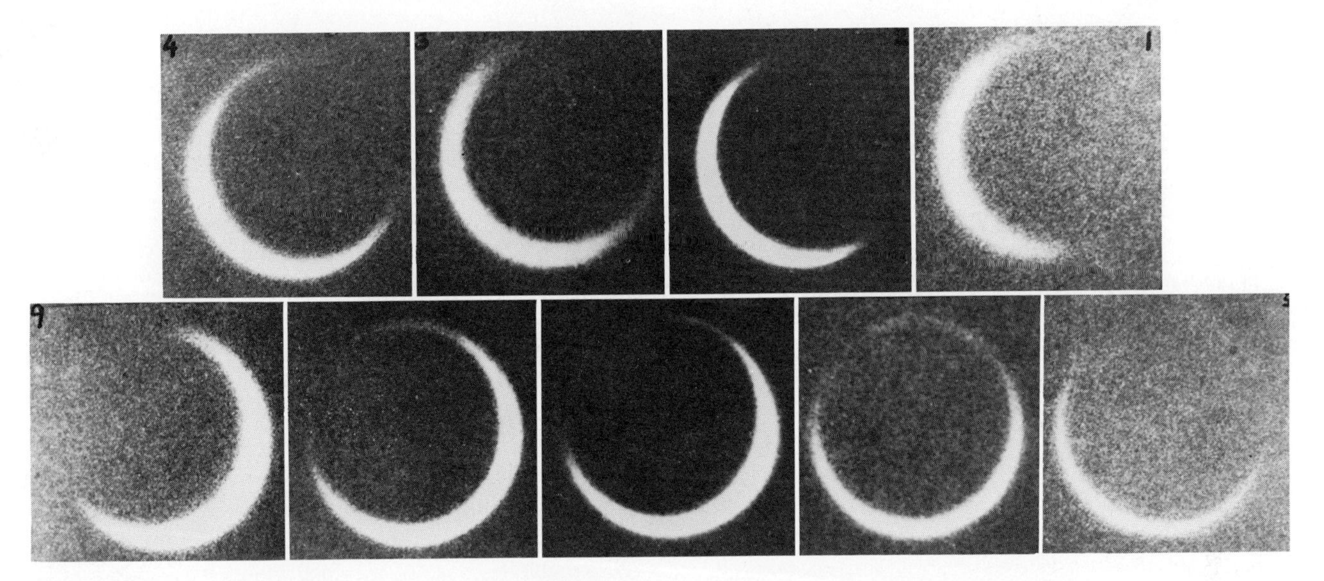

Fig. 4-2. Venus passing inferior conjunction. It was Galileo who first discovered that Venus passes through phases similar to the moon's. Photographs were taken with a small reflector and show the planet in crescent phases. Photograph No. 6 shows the planet's atmospheric ring as well. (Courtesy University of Chicago Yerkes Observatory, Williams Bay, Wisc.)

The orbit of Venus, more nearly circular than the path of any other planet, is inclined to the ecliptic at an angle of 3½°. The average distance of the planet from the sun is 67 million miles, or 0.72 AU. Venus completes its west-to-east revolution of the sun in 225 earth days at an average speed of 22 miles/sec (79,200 miles/hr).

For centuries it was impossible to determine the rotation period of Venus because its surface was obscured by heavy clouds. It was later decided that Venus, like Mercury, must have a synchronous rotation period. The matter was settled recently by sensitive, earth-based radar scans that were able to penetrate the shrouding cloud layers of the planet. The results were startling: Venus rotates slowly from east to west, a direction opposite that of all other planets except Uranus, and completes one turn in 243 earth days on an axis inclined 32° to its orbital plane. It has been proposed that this unusual rotational direction and period could have been produced by the gravitational effect of the earth, which strongly affects Venus at its closest approach.

Venus has a diameter of 7700 miles and a mass 77% that of the earth's. Its density of 5.1 gm/cm^3 classifies it as a terrestrial planet. The surface gravity of Venus is 0.86 g, and its escape velocity is 6.5 miles/sec (23,400 miles/hr). It has no known satellites.

Venus is usually the third brightest object seen in the sky. Not only is it so bright that on a clear, dark night it casts a shadow, but the planet may also be seen telescopically in the daytime. Its brightness can be explained by its proximity to the earth and its high albedo of 76% that is produced by a very dense atmosphere.

The atmosphere of Venus, very different from that of the earth, is a dense envelope of carbon dioxide as thick as 35 miles; its atmosphere may also contain traces of other gases. The atmospheric pressure at the surface of Venus is hundreds of times greater than the atmospheric pressure at the surface of the earth. This overwhelming pressure is believed to have contributed in part to the premature failure of several Russian probes that successfully landed on the surface of Venus. The heavy atmospheric cover of carbon dioxide gas creates a most efficient greenhouse effect on the planet, trapping large quantities of the heat that radiates from the planet's surface. The convection currents thus produced cause the atmosphere of Venus to circulate in great spiral patterns. Thus Venus, although nearly twice as far from the sun as Mercury, maintains the highest surface temperature of any planet, often exceeding 800° F (537° C). Venus clearly is an awesome world of constant, unbearable heat and crushing pressure.

Venus is also disappointing when viewed telescopically. It appears as a small disc; no surface features are visible because of the dense atmospheric cover. However, Venus may be seen passing through phases like the moon. When Venus is closest to the earth and is therefore at its brightest, it is oddly enough at a thin crescent phase (Fig. 4-2).

With an orbit almost twice as large as that of Mercury, Venus, as seen from earth, may appear as far as 47° away from the sun. As a result, Venus may sometimes be seen against a dark sky and may remain above the horizon for as long as 3 hours, 6 minutes. Like Mercury, Venus may also be seen swinging in an almost straight line back and forth on either side of the sun. It first appears in the east before sunrise. Then gradually, with each passing day, the planet appears to approach the sun, becoming lost in the solar glare. As if by magic, however, Venus reappears several weeks later in the west after sunset. This pattern of movement was confusing to the ancients, who believed that Venus was two different planets.

Periods and positions of inferior planets

Like the moon, planets have two types of revolution periods—sidereal and synodic. The *sidereal* period of a planet is the length of time it requires to complete one full orbit of the sun with respect to the stars—its true revolution period. The *synodic* period, on the other hand, is the elapsed time between the occurrence of two consecutive orbital positions of an inferior planet with respect to the earth and sun. If the earth were stationary while the planet revolved around the sun, both periods would be equal. In reality, however, the earth does move in its orbit, which makes the sidereal and synodic periods of unequal duration.

For example, Mercury's sidereal period is 88 days, while that of Venus is 225 days. Their synodic periods, however, are 115.88 and 583.92 days, respectively. The reason for the difference is that the earth, like a runner on the outside of a track, moves more slowly in a larger orbit than Mercury and Venus. Thus additional time is required for the earth to return to the same relative position with respect to either one of the inferior planets.

Inferior planets have four basic positions: inferior conjunction, eastern elongation, superior conjunction, and western elongation (Fig. 4-3). When either one of the inferior planets is aligned between the sun and the earth and thus invisible because of the solar glare, it is said to be at *inferior conjunction*. When either Mercury or Venus forms a right angle with the earth east of the sun, they are at *eastern elongation* and may only be seen in the west after

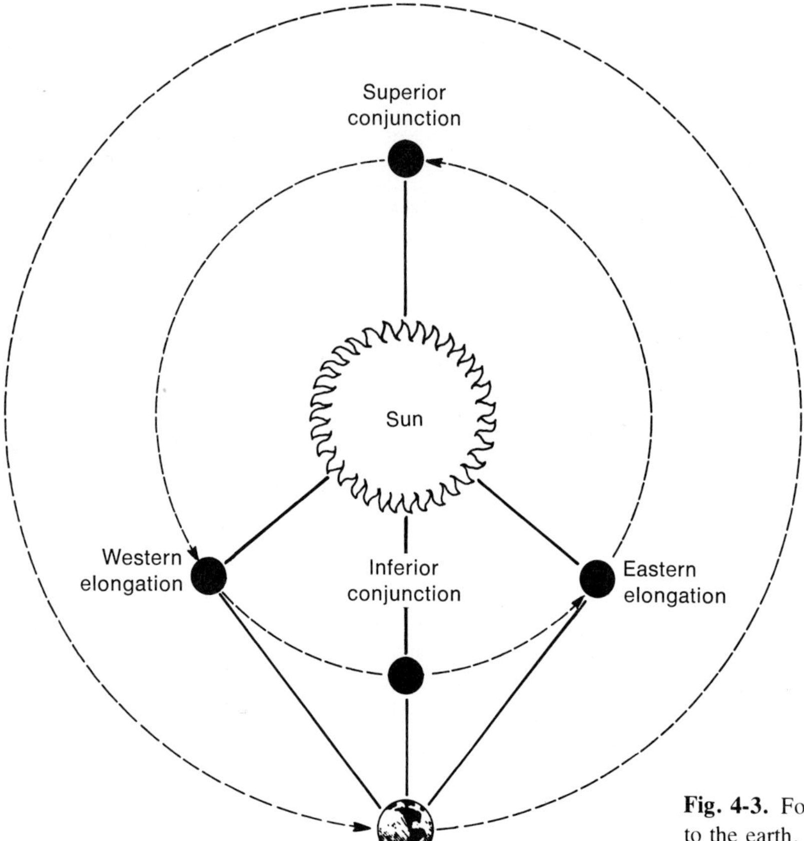

Fig. 4-3. Four basic positions of the inferior planets with respect to the earth.

sunset. When an inferior planet again aligns itself with the earth so that the sun is between the earth and the planet, the planet is at *superior conjunction* and invisible once more due to the glare of the sun. As an inferior planet makes a 90° angle with the earth west of the sun, it is at *western elongation* and can only be viewed in the eastern sky before sunrise. The maximum angular separation of the inferior planets from the sun, as seen from earth, always occurs at either eastern or western elongation.

Transits. The inferior planets are involved from time to time in an interesting phenomenon. When either Mercury or Venus aligns itself directly between the earth and the sun, a position possible only at inferior conjunction, the observer on earth sees a tiny black dot moving slowly westward across the solar disc. Such an event is known as a transit (Fig. 4-1).

Transits of Mercury always occur in May or November, for it is in those months that the earth is at or near a node of Mercury's orbit. There are, on the average, 13 transits of Mercury each century at intervals of 10, 3, 13, and 7 years. The most recent transit of Mercury took place on November 9, 1973, and the next one is scheduled for May of 1986. Transits of this small planet may only be viewed with a telescope.

Transits of Venus are much rarer than those of Mercury because our sister planet has a much larger orbit and requires more time to return to a node. Transits of Venus always take place in pairs during the months of June or December, separated by an interval of 8 years. The pairs themselves, however, are separated by either 113 or 130 years. The last transits of Venus occurred on December 9, 1874, and on December 6, 1882. The next pair is predicted for June 8, 2004, and June 6, 2012. Transits of Venus may be seen without a telescope if proper safety precautions are used for viewing the sun.

The earth

Earth is classified as neither an inferior planet nor a superior planet, since it is the platform from which we as observers view the entire universe. The earth is the planet whose composition, structure, and processes are important enough to us to be the subject of the remaining three-fourths of this book.

Superior planets

Superior planets are those whose orbits are larger than the earth's. From the sun all planets would be classified as superior, while from Pluto there is no known superior planet. From the earth there are six superior planets: Mars, Jupiter, Saturn, Uranus, Neptune, and Pluto.

Mars: the red planet

Mars, the first superior planet, was named for the Roman god of war. The planet was one of the five wandering stars known to the ancients, and because of its blood-red color, it quickly became associated with death and disaster. No other planet has so seized our imaginations and teased our curiosity. Until recently, for example, little green men were suspected of piloting flying saucers from the red planet, there was talk of Martian invasions and war between the worlds, and many believed in the possibility of intelligent life on the planet capable of constructing sophisticated irrigation systems (the Martian "canals") for watering abundant plant life.

Unmanned satellite flybys and orbital missions of the 1960s and 1970s, however, have done much to puncture these myths about Mars. The facts that have replaced the fantasies indicate that Mars is a desolate place (Fig. 4-4). The planet seems incapable of supporting any higher forms of life as we know it, and even the presence of the

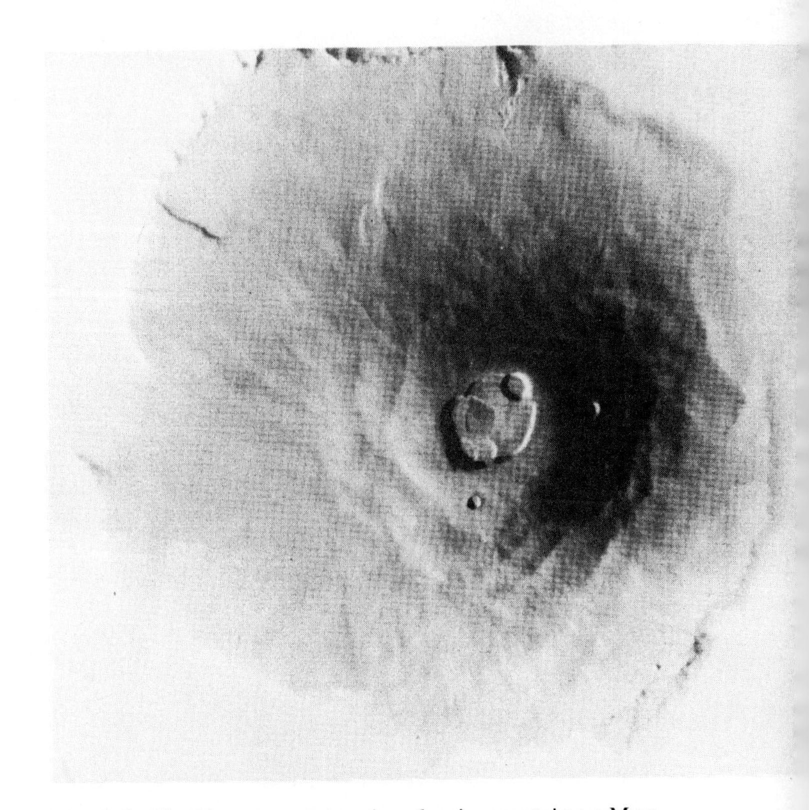

Fig. 4-4. Nix Olympica, a gigantic volcanic mountain on Mars, is more than three times as broad as the most massive volanic pile on earth. (Photograph was taken by Mariner 9 in late January, 1972.) (Courtesy Jet Propulsion Laboratory, California Institute of Technology, Pasadena, Calif.)

simple plants is in doubt. Future manned landings hopefully will answer the question of life on Mars once and for all.

Mars is located at a mean distance of 141 million miles, or 1.52 AU, from the sun. Its orbit is inclined to the ecliptic at an angle of 1.5°, and it completes its revolution around the sun from west to east in 687 earth days at an average speed of 15 miles/sec (54,000 miles/hr). Mars may approach to within 34 million miles of the earth, a proximity second only to that reached by Venus. However, unlike Venus, Mars has a transparent atmosphere, thus making the red planet the most thoroughly studied of all planets in our solar system.

With a diameter of 4200 miles, Mars is nearly half the size of the earth and almost twice the size of the moon. Its albedo of 15% is indicative of a planet with many moonlike characteristics. Mars is classified as a terrestrial planet because it has a density of 4 gm/cm^3. The planet has a mass 11% that of the earth, a surface gravity of 0.38 *g*, and an escape velocity of 3.2 miles/sec (11,520 miles/hr).

Mars has two tiny satellites that appropriately bear the names of the war god's attendants—Phobos (Fear) and Deimos (Terror). The two Martian satellites at one time became involved in a bit of astronomical intrigue. In *Gulliver's Travels,* published in 1726, Jonathan Swift spoke through his fictitious characters to report:

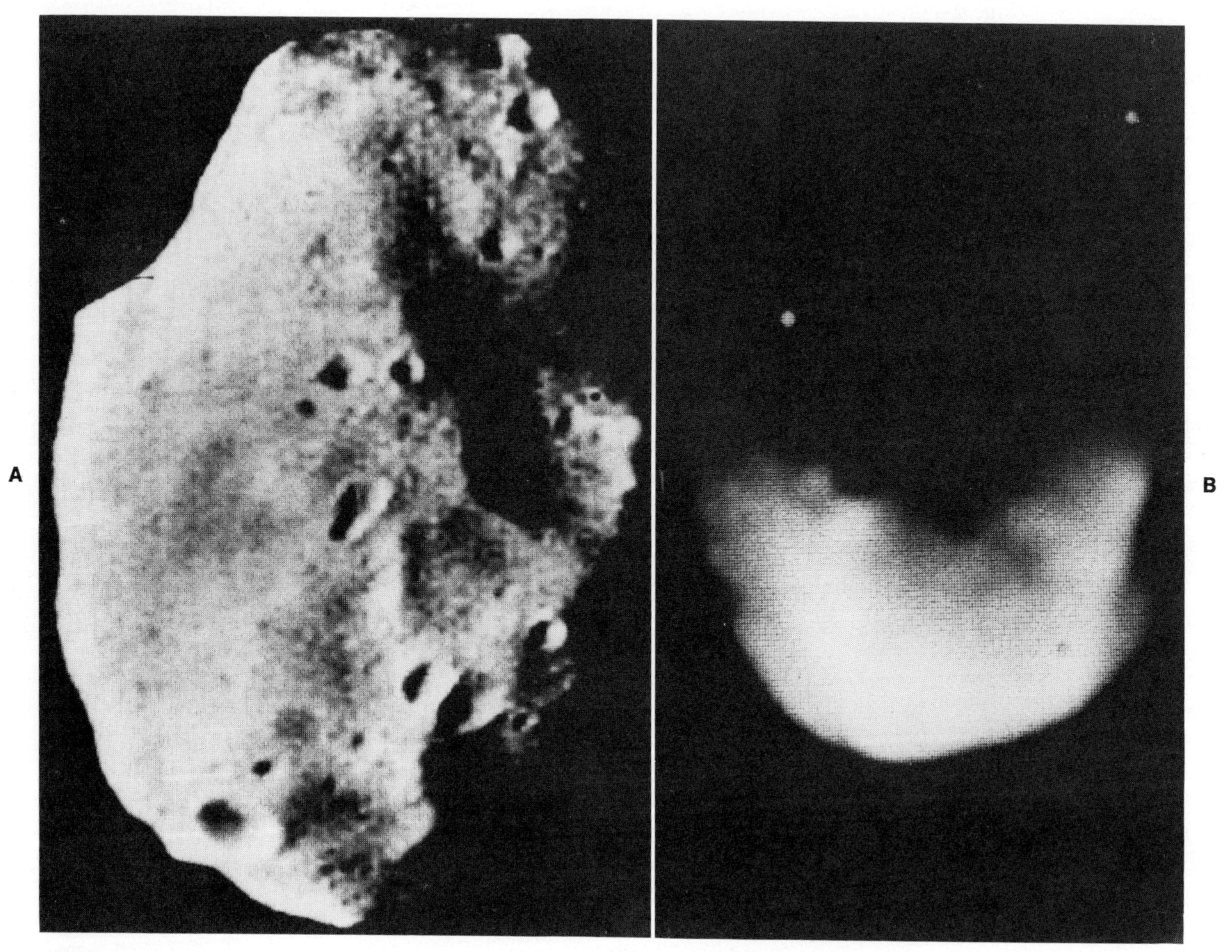

Fig. 4-5. A, Variety of craters in this computer-enhanced photograph of Phobos. Mars' innermost moon, suggests both great age and great structural strength. **B,** Deimos, Mars' other moon, shows highly irregular surface features due to its extremely low gravity. On earth or other bodies with higher gravities such projecting features are crushed, resulting in more regular, spherical shapes. (Courtesy Jet Propulsion Laboratory, California Institute of Technology, Pasadena, Calif.)

> ...two satellites, which revolve about Mars, whereof the innermost is distant from the center of the primary planet exactly 3 of the diameters, and the outermost, 5. The former revolves in a space of 10 hours and the latter in 21½, so that the square of their periodic times is equal to the cubes of their distance from Mars, which evidently shows them to be governed by the same law of gravitation that influences the other heavenly bodies.

In 1877, 150 years later, Phobos and Deimos were discovered by Asaph Hall, who was working at the United States Naval Observatory. Since no telescope in existence in 1726 was large enough to have seen the satellites, how could Jonathan Swift have known of their existence? How could he have located them with such exactness? Although his descriptions of the orbital data of the satellites were not precise, Swift was close enough to preclude guessing. Unfortunately, the source of his information died with him in 1745.

Neither Phobos nor Deimos is spherical; both are shaped somewhat like an Irish potato (Fig. 4-5). Phobos is the closer of the two to Mars, with a center-to-center distance from the red planet of about 6000 miles. It revolves in a period of 7 hours, 39 minutes, from west to east. The orbital period of Phobos is shorter than the rotation period of Mars, and therefore this satellite appears to rise in the west and set in the east no fewer than three times during each Martian day. Phobos is about 16 miles long and 13 miles wide. Deimos, about 8½ miles long and 7½ miles wide, is at an average distance of 14,500 miles from Mars and completes its revolutionary period from west to east in 30 hours, 18 minutes. These two tiny satellites are among the darkest bodies in the solar system, having albedoes of only 0.05%. They both have numerous craters that were probably formed by meteorite impact. One crater on Phobos is 3.3 miles in diameter; the impact that produced it must have been the greatest the satellite could sustain without breaking up. With so many craters, the satellites must also be very old.

Mars rotates from west to east each 24 hours, 37 minutes, 23 seconds, on an axis inclined 25° to its orbit. The Martian day, then, is only 42 minutes longer than a day on earth. Seasonal variations in the angle of incidence of the sun's rays caused by the planet's axial tilt are also much like earth's. The seasons on Mars, however, last nearly twice as long as those on earth because the orbit of the red planet is almost twice as large.

For many years observers have watched the polar caps on Mars increase in size during the Martian winter and decrease in size during its summer. The polar caps were once thought to consist of ice like those on earth. Recent investigations, however, have shown that they are composed of dry ice, or frozen carbon dioxide. Some of the dry ice changes into gaseous carbon dioxide during the Martian summer and enters the atmosphere. During the Martian winter, on the other hand, some of the atmospheric carbon dioxide changes to dry ice and is deposited on the planet's surface.

Another noticeable seasonal effect on Mars is that some of its surface features change color. Always reddish in color when viewed telescopically and by the naked eye, Mars does have some regions on its surface that change from green to brown in the Martian winter and brown to green in the Martian summer. Until recently, many believed that the changing colors on the planet were due to the seasonal growth of vegetation. The discovery of extensive dust storms on Mars, however, indicates that the effects of such atmospheric disturbances, combined with chemical changes in the planet's surface, could be the source of the observed color changes.

The atmosphere of Mars is composed almost entirely of carbon dioxide; it is suspected that there are traces of nitrogen and water vapor. The atmospheric pressure and density at the Martian surface is only about 1% that of the earth's at sea level. Thus when we make our initial visit to Mars, it will be necessary to take along an environmental system, since the red planet will offer us no life support whatsoever.

The thin Martian atmosphere results in extremely variable surface temperatures that change with the seasons, geographic locations, and time of day. At the Martian equator the daytime temperature is maintained at a pleasant 80° F (27° C) but plunges to −163° F (−73° C) at night. The average temperature at the north pole of Mars remains at approximately −163° F (−72° C) all year; at the south pole the temperature is even lower, at −212° F (−100° C). In contrast, the lowest temperature ever recorded on earth was −190° F (−88° C) at Vostok Station, Antarctica.

In many respects Mars seems to be at a point in its evolution between the stages reached by the earth and moon. For example, Mars has many impact craters similar to those on the moon but also has volcanic mountains, some of which are larger than the largest ones on earth. There are Martian chasms that dwarf the earth's Grand Canyon. Sinuous valleys are reminiscent of lunar rilles, but they possess branching tributaries that have no counterpart on the moon. It would be premature at this time to state in which direction the planet is evolving. Manned exploration of the Martian surface is necessary if we are someday to unravel the story of the evolution of the red wanderer of the night.

Jupiter: the giant planet

Located at a mean distance of 484 million miles, or 5.2 AU, from the sun, Jupiter is the largest planet of the solar system. Named for the king of the gods in Roman myth, Jupiter revolves around the sun in an orbit inclined at 2½° to the ecliptic and completes one revolution period in 11.96 earth years at an average speed of 8.1 miles/sec (29,260 miles/hr).

This giant planet has a diameter of 88,000 miles, slightly more than one-tenth the diameter of the sun. Its mass, 315 times that of the earth's, is greater than the combined masses of all the other planets. The surface gravity of Jupiter is 2.64 times that of earth, and its escape velocity is 37 miles/sec (133,200 miles/hr). With a density of 1.33 gm/cm^3, Jupiter is the first and most important member of a type of planet called Jovian. This group of planets has densities much lower than the terrestrial planets and are thus thought to be composed chiefly if not entirely of gases.

Jupiter has 13 known satellites, the greatest number of any planet in the solar system. The largest four satellites—Ganymede, Callisto, Io, and Europa—were discovered by Galileo in 1610 but named by the German astronomer Simon Marius. The satellites are also referred to by Roman numerals in order of their increasing distance from Jupiter. Io (I), with a diameter of 2000 miles, is the closest of the four to Jupiter; its mean distance is 262,000 miles, similar to the distance of the moon from the earth. Europa (II) has a diameter of 1754 miles and is found at an average distance of 416,000 miles from Jupiter. Ganymede (III), the largest of the four and the largest satellite in the solar system, is 3038 miles in diameter at a mean distance of 663,400 miles from Jupiter. Callisto (IV) is 2933 miles in diameter and at an average distance of 1,166,840 miles from Jupiter. When viewed through binoculars or a small telescope, the changing positions of these four Galilean satellites can be a source of utmost fascination for observers on earth. The other nine satellites, visible only through large telescopes, are given Roman numerals to indicate the order in which they were discovered. Satellite V is closer to Jupiter than Io and was discovered in 1892. Outward from Callisto are VI (1904), VII (1905), X (1938), XII (1951), XI (1938), VIII (1908), IX (1914), and XIII (1975). Unlike the inner eight satellites, the outer five satellites show a retrograde (east-to-west) revolution.

Jupiter, the largest planet in the solar system, also has the shortest rotation period, completing a turn from west to east in 9 hours, 55 minutes, on an axis inclined 3° to its orbit. This rapid rate of rotation has caused Jupiter to become flattened at the poles and to bulge at the equator. Its oblateness, the difference between the polar and equatorial diameters, is 1 part in 15. The oblateness of the earth, in contrast, is 1 part in 297.

Whether Jupiter has a solid surface like earth is not known, but if there is such a surface it probably consists of solid hydrogen buried below thousands of miles of frozen atmosphere. The outer gaseous atmosphere of Jupiter, sometimes referred to mistakenly as the planet's "surface," is composed of 80% hydrogen gas with smaller percentages of helium, ammonia, and methane. The temperature of this outer atmosphere is −270° F (−140° C), and its albedo is 51%. The atmosphere of Jupiter has several dark belts and bright zones lying roughly parallel to the equator, which give the planet a prominent banded appearance when it is viewed telescopically. These bands are clouds of different colored hydrogen compounds that form at different temperatures and therefore arrange themselves at different latitudes of Jupiter.

Various spots are seen from time to time in the atmospheric bands of Jupiter. The best known is the Great Red Spot, discovered by Jean Cassini (1625-1712) in 1666 (Fig. 4-6). The Great Red Spot seems to be a semipermanent component of the Jovian atmosphere. It has been under almost constant observation for well over 200 years; for some years subsequent to 1878 it was dark red in color and as large as 30,000 miles long and 7000 miles wide. Its strong color has since faded, but it is still decidedly pink.

Fig. 4-6. Jupiter's gigantic red spot is large enough to swallow not one, but several, earths. Scientists speculate that this characteristic area may be simply a large "permanent hurricane." (Courtesy National Aeronautics and Space Administration, Washington, D.C.)

At intervals the spot disappears, but it always returns; its position, although not fixed, is always in the southern hemisphere of the planet. The Great Red Spot appears to be a high cloud structure that behaves like an ascending column, flowing upward through the jovian atmosphere and out at the level of its top, somewhat like thunderstorms or hurricanes on earth.

Jupiter is the only planet in the solar system that produces more energy of its own than it receives from the sun. For example, in 1955 it was discovered that Jupiter produces radio waves; the only other known source of radio waves in our solar system is the sun. It is possible that Jupiter is so massive that it could be considered a low-grade star; its evolution stopped just short of the point at which it would be capable of producing light and the other types of short-wavelength radiation. Regardless, Jupiter is certainly a remarkable planet in many respects.

Late in 1973 an unmanned spacecraft, Pioneer 10, made the first flyby of the planet Jupiter, taking photographs and recording data on such factors as magnetic fields, radiation belts, and atmosphere. Continuing past the giant planet, Pioneer 10 will eventually become the first man-made object to leave the solar system. In November of 1974 Pioneer 11 came within 1 million miles of Jupiter and sent back superior photographs and additional data. Pioneer 11 will continue its flight to the planet Saturn, where it is expected to arrive in September of 1979.

Saturn: the ringed planet

The most distant of the planets known to the ancients was Saturn, named for the Roman god of agriculture. Jupiter may claim fame for its starlike qualities and its unrivaled system of satellites, but Saturn, also a Jovian planet, has a set of rings that is unequaled in the entire known universe. The rings make Saturn a most spectacular and beautiful celestial object.

The ring system of Saturn lies in the equatorial plane of the planet. The rings completely encircle Saturn like the brim of a hat but never touch the planet. At their broadest they are 171,000 miles in diameter, 41,500 miles wide, and only about 10 miles thick, making them the flattest known feature in the universe. The rings are inclined to Saturn's orbit at a constant angle of 27°, a phenomenon that allows observers on earth to see them in various positions —apparently from above, below, or edge on—depending on the planet's orbital position with respect to the earth (Fig. 4-7). At edge-on presentation, which occurs every 15 years or twice during Saturn's revolution of the sun, the rings are invisible in smaller telescopes and barely visible in the largest.

The ring system of Saturn consists of four rings with at least two known divisions or spaces produced by the gravitational sweeping action of Saturn's inner satellites. The outer ring, 10,000 miles wide, contains a gap called Encke's division. Next is the well-marked space called Cassini's division, which is easily seen even in small telescopes. Inside Cassini's division are the bright ring, which is 16,000 miles wide; the crepe ring, which is 10,000 miles wide; and finally the D ring, which was discovered in 1969.

The exact nature and origin of Saturn's ring system are not known. The rings appear to consist of innumerable small, solid particles, each in an individual orbit around the planet. The particles that make up the rings are thought to be either rocklike materials coated with frozen gases or chunks of frozen gases. The debate over the origin of Saturn's rings continues. The two main schools of thought are (1) that the rings are the remains of a satellite that approached Saturn so closely that it broke up or (2) that

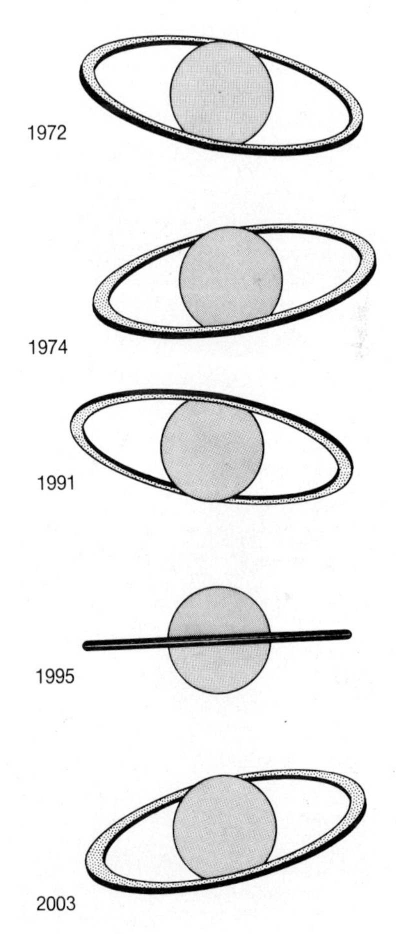

Fig. 4-7. Sketches of Saturn showing the changing aspects of the ring system along our line of sight during a 29½-year cycle.

the rings are residual material left over from the formation of Saturn. When Pioneer 11 reaches Saturn in September of 1979, plans call for the spacecraft to pass between the D ring and the planet itself.

Saturn is at a mean distance of 887 million miles from the sun (9.5 times the earth's distance); even at its closest approach the ringed planet is still 740 million miles from earth. Its orbit is inclined at 2½° to the ecliptic, and Saturn requires 29½ years at an average speed of 6 miles/sec (21,600 miles/hr) to complete one revolution around the sun.

The diameter of Saturn is 75,000 miles and its mass is 94 times that of the earth. Its density of 0.7 gm/cm^3, the lowest of any planet in the solar system, would make Saturn float in water if enough water could be found. Saturn has a surface gravity 1.17 times that of the earth's and an escape velocity of 22 miles/sec (79,200 miles/hr). It rotates once on its axis each 10 hours, 38 minutes and shows an oblateness of 1 part in 10, the highest in the solar system.

Without its rings, Saturn would look like a slightly smaller version of Jupiter. Like Jupiter, Saturn is thought to be composed chiefly of gases, perhaps with a core of solid hydrogen. The visible "surface" of Saturn is its atmosphere, consisting primarily of hydrogen and hydrogen compounds such as methane and ammonia. The planet has a "surface" temperature of −230° F (−110° C) and an albedo of 50%, characteristics again very much like Jupiter's. Dark atmospheric belts extend across the disc of Saturn roughly parallel to its equator, but they are less distinct than the belts of Jupiter. Spots are rather rare but sometimes they become quite striking. Conspicuous white spots, for example, were seen in 1933 and again in 1960. The presence of spots on Saturn may indicate that the ringed planet is undergoing processes closely akin to those on Jupiter.

Saturn has a complement of 10 satellites, the last of which was discovered in 1966 when the rings were last presented edge-on to the earth. Titan, Saturn's largest satellite, has a diameter of 2976 miles and is larger than the moon; it is the only satellite in the solar system known to have an atmosphere (methane). Titan is easily seen in small telescopes.

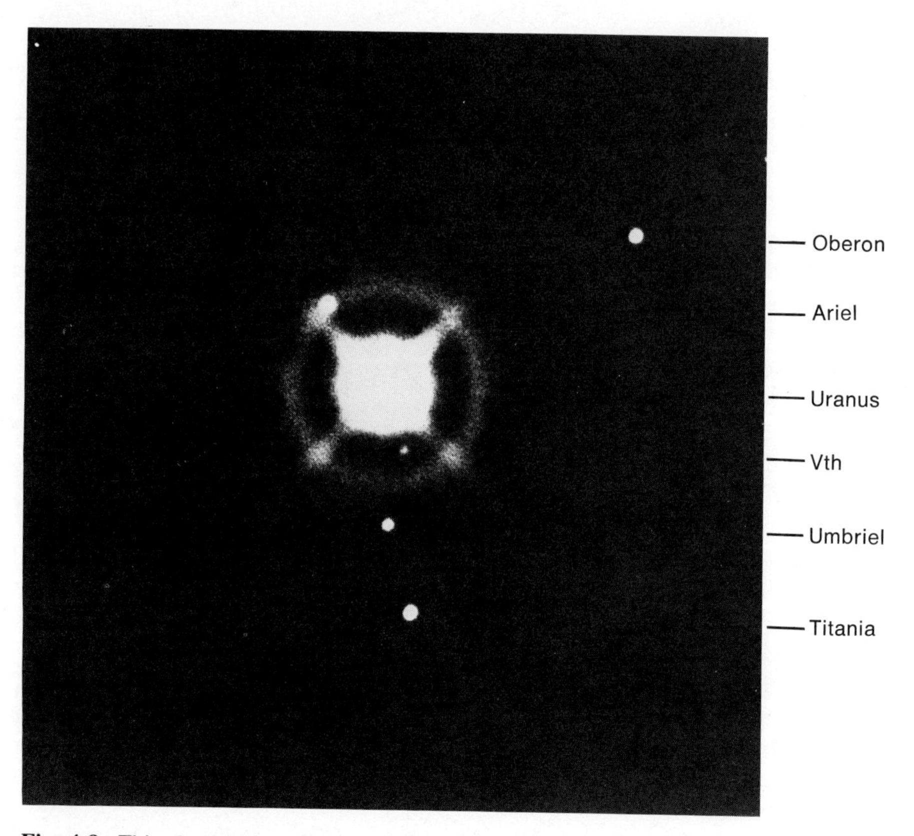

Fig. 4-8. This photograph of the satellite system of Uranus is unique in that it includes the moon Miranda, labeled simply *Vth* within the overexposed ring of the planet. (Courtesy University of Chicago Yerkes Observatory, Williams Bay, Wisc.)

Uranus: the strange planet

Uranus, named for the god of the sky in Greek mythology, was the first planet discovered after the development of the telescope (Fig. 4-8). The planet is visible to the unaided eye, but because it resembles a very faint star it was probably seen by the ancients but not discovered officially until 1781. In that year William Herschel (1738-1822) found the planet more or less by accident. The English astronomer was making a systematic survey of the stars when he noticed one that refused to remain in one place. He recognized it immediately as a planet and, on checking older records, found that Uranus had been charted at least 20 times as a star.

At a mean distance of 1.783 billion miles from the sun, Uranus is twice as far as Saturn and 19 times farther than the earth. At that distance, the planet requires 84 earth years at an average speed of 4.5 miles/sec (16,200 miles/hr) to complete one revolution of the sun. The orbit of Uranus is inclined to the ecliptic by only 0.46^{s}, the smallest inclination in the solar system.

Uranus has a diameter of 31,000 miles, a mass 14½ times that of the earth, and a density of 1.5. Uranus, then, is a Jovian planet. Its surface gravity is 1.13 times that of earth and its escape velocity is 20 miles/sec (72,000 miles/hr). Uranus spins from east to west once every 10 hours, 45 minutes on an axis tilted 82° to its orbit, the greatest of any planet. Therefore the poles rather than the equator, as is the case with the other planets, alternately face the earth as Uranus orbits the sun.

Uranus has an albedo of 66% and a "surface" temperature of −300° F (−148° C). Like the other Jovian planets, Uranus is thought to be composed chiefly of gases. The planet has a thick atmosphere composed primarily of hydrogen and methane; methane gives Uranus a definite greenish telescopic appearance. Other than some indefinite bands near the equator, no surface features have been noted.

Five satellites revolve from east to west around Uranus, all near the planet's equatorial plane. The smallest and closest is Miranda, which was discovered in 1948. Its diameter is 200 miles and its mean distance from Uranus is 81,000 miles. Next in order of distance are Ariel (372 miles in diameter) and Umbriel (248 miles in diameter), discovered in 1951 by William Lassell. Titania, Uranus' largest satellite, has a diameter of 620 miles, and Oberon is 558 miles in diameter; both were discovered by William Herschel in 1787. Oberon revolves at an average distance of 364,000 miles from Uranus.

Neptune: the twin planet

Neptune was discovered in 1846 by mathematical prediction. By 1846 the orbit of Uranus was well established but the seventh planet was definitely not following its predicted path. It was decided that Uranus' orbital variations were caused by the gravitational attraction of an unseen mass—a new planet. Neptune, the eighth planet from the sun and the last of the Jovian planets, was discovered independently by the Englishman John Adams and the Frenchman Joseph Leverrier and was named for the mythologic god of the sea.

At a mean distance of 2.793 billion miles from the sun, Neptune is 1 billion miles farther from the sun than Uranus and 30 times farther than the earth. Neptune moves in an orbit inclined at 1.77° at an average speed of 3 miles/sec (10,800 miles/hr), completing a revolution in 165 earth years. Since its discovery in 1846, Neptune has yet to complete one revolution.

Neptune may be regarded as an almost perfect twin of Uranus, but its diameter of 28,000 miles is slightly smaller and its density of 2 is slightly greater. Neptune's mass is equal to 17 earth masses, its surface gravity is 1.41 *g*, and its escape velocity is 24 miles/sec (86,400 miles/hr). Neptune's albedo is 62%, and its surface temperature is −350° F (−217.7° C). On an axis inclined to its orbit at 29°, Neptune rotates from west to east in 15 hours, 48 minutes. Its oblateness is 1 part in 40. Like Uranus, Neptune, when viewed telescopically, appears as a small greenish disc whose color has been found to be due to methane; hydrogen is also present in its atmosphere. No surface markings have yet been found on the planet (Fig. 4-9).

Neptune has two known satellites. The larger, named Triton, is 2480 miles in diameter and orbits its master planet from east to west every 5 days, 21 hours, at an average distance of 220,000 miles. Triton was discovered by the English astonomer William Lassell in 1846, shortly after the discovery of Neptune itself. The other satellite, Nereid, is less than 22 miles in diameter and was discovered by G. P. Kuiper in 1949. Nereid has the most eccentric orbit of any satellite in the solar system, and its distance from Neptune varies from 1 to 6 million miles. Its average distance, however, is 3 million miles. The tiny satellite orbits from west to east, completing its revolutionary period in 359 earth days.

Pluto: the distant planet

After the discovery of Neptune and calculations of its orbit, Percival Lowell, an American astronomer, found that the new planet could not account for all of the variations in the orbit of Uranus. Calculations were made and the search was begun for a ninth planet. Although Lowell began searching in 1905, it was not until 1930, 14 years after his death, that the new planet was found near its predicted location by Clyde W. Tombaugh, working at the observatory founded by Lowell near Flagstaff, Arizona.

Fig. 4-9. Neptune and its two satellites. Triton (larger, clear spot on left) and Nereid (arrow). Planet is purposely overexposed. Fuzzy spots surrounding the planet are galaxies. (Courtesy University of Chicago Yerkes Observatory, Williams Bay, Wisc.)

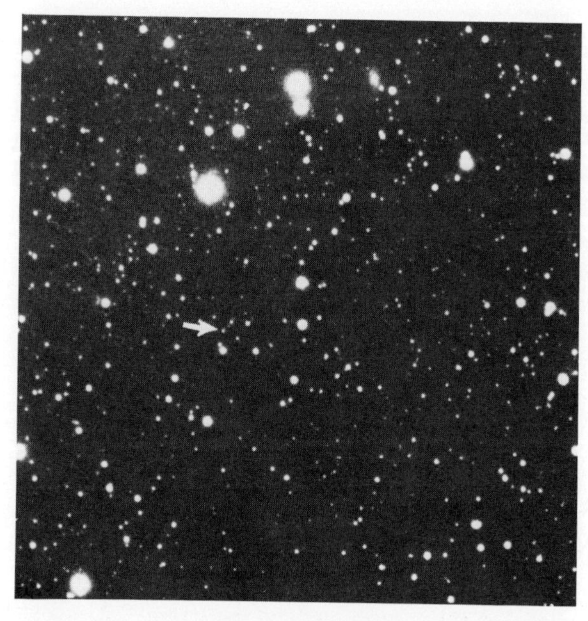

January 23, 1930

January 29, 1930

Fig. 4-10. Copies of small sections of the discovery plates showing images (arrows) of Lowell's mathematically predicted trans-Neptunian planet. This planet, later named Pluto, was found on February 18, 1930, on examination of these plates. (Courtesy Lowell Observatory, Flagstaff, Ariz.)

Pluto, named for the Roman god of the lower world, was discovered by the use of a newly developed device called a blink-comparing machine (Fig. 4-10). The device compares two photographs of the same region of space taken some weeks apart. Any object that has changed position during the time interval appears to jump back and forth. Pluto, moving much faster than the background stars, thus appeared to jump, and another new planet had been discovered.

Pluto, the planet with the largest orbit in the solar system, is generally regarded as the outermost planet. Strictly speaking, however, this is not always true. Pluto's mean distance from the sun is 3.675 billion miles, or 39 times farther than the earth, but because its orbit is more eccentric than that of any other planet, its distance from the sun varies between 2.766 and 4.566 billion miles. Therefore at perihelion, the point at which it is closest to the sun, Pluto is closer than Neptune; from 1979 to 1998 Neptune will be the most distant planet.

The orbit of Pluto is inclined to the ecliptic at an angle of 17°, the highest in the solar system. The planet requires 248 earth years at an average speed of 2.9 miles/sec (10,440 miles/hr) to complete its eastward orbit of the sun. It moves so slowly among the stars that it remains for approximately 20 years in each of the 12 constellations of the zodiac. Since its discovery in Gemini in 1930, Pluto has since moved eastward through Cancer and Leo and is now located in Virgo.

Pluto apparently has a diameter of 4000 miles, a mass 11% that of the earth, and a density of 4.85 gm/cm^3, which makes it a terrestrial planet. Its period of rotation is 6.4 days from west to east. Pluto has an albedo of 16%, a surface temperature of −400° F (−240° C), and no measurable atmosphere. This distant planet also has no known satellites.

Even through the largest telescopes, Pluto is not visible as an appreciable disc but appears as a starlike object. Hence data concerning the planet are subject to constant

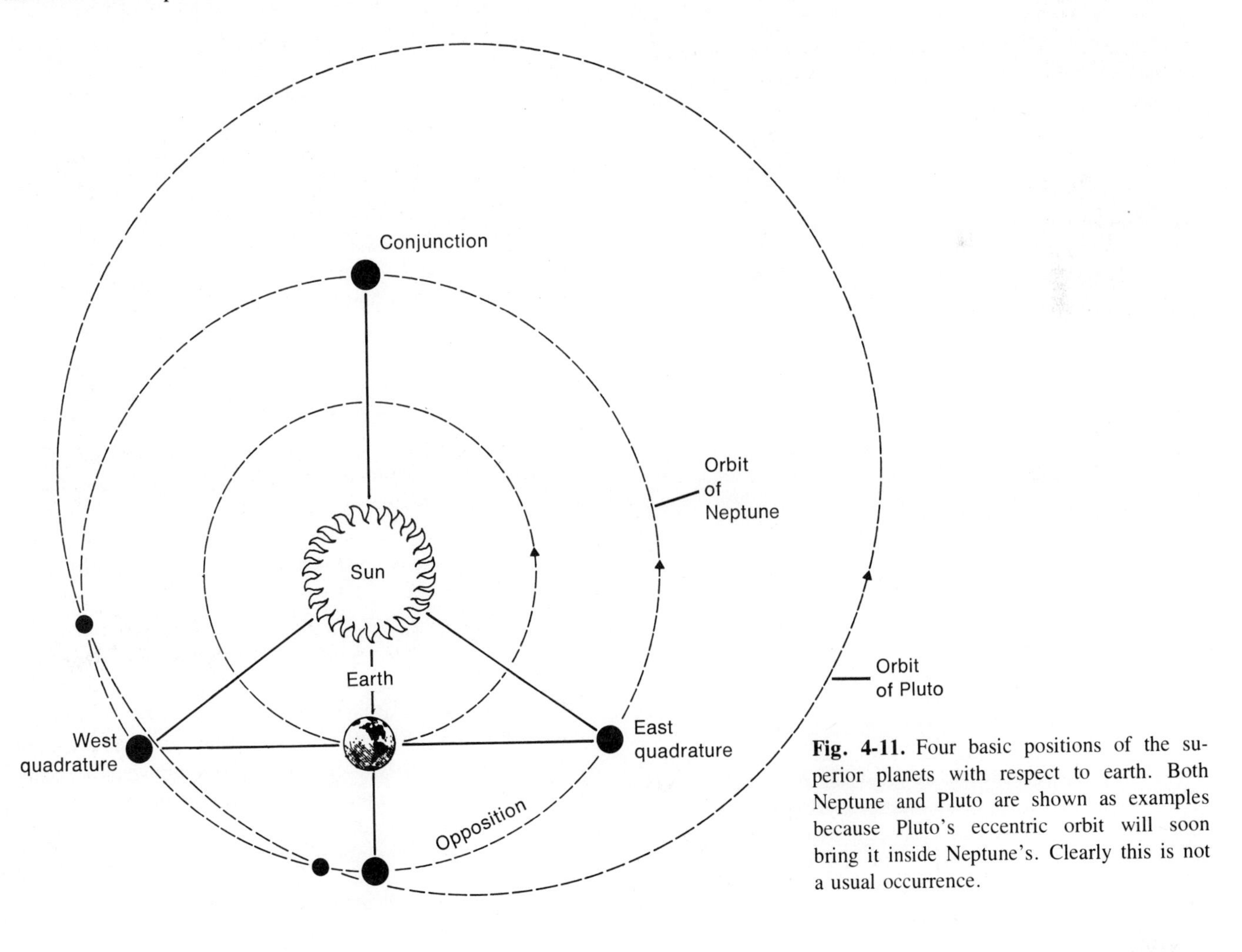

Fig. 4-11. Four basic positions of the superior planets with respect to earth. Both Neptune and Pluto are shown as examples because Pluto's eccentric orbit will soon bring it inside Neptune's. Clearly this is not a usual occurrence.

revision. Pluto is beyond the range of small telescopes and even taxes the power of a 15-inch reflector. The chances are that no person reading this book will ever see Pluto except in photographs.

Positions and periods of superior planets

Like inferior planets, superior planets may have four basic orbital positions with respect to the sun and earth: opposition, west quadrature, conjunction, and east quadrature (Fig. 4-11). When the earth is between the sun and a superior planet, the planet is said to be at *opposition*. The sun and the superior planet in this case are always 180° apart and appear in opposite parts of the sky as seen from earth. A superior planet at opposition always rises at sunset, crosses the meridian at midnight, and sets at sunrise.

When the sun is between a superior planet and the earth, the planet is at *conjunction*. A superior planet at conjunction is lost in the solar glare and cannot be seen.

When a superior planet forms a 90° angle with the earth and sun, the planet is said to be at *quadrature*. If the planet is west of the sun, it is at *west quadrature;* if it is east of the sun, it is at *east quadrature*. At west quadrature, the superior planet rises at local noon, crosses the meridian at sunset, and sets at local midnight. At east quadrature, the planet rises at local midnight, crosses the meridian at sunrise, and sets at local noon.

The synodic periods of the inferior planets are longer than their respective sidereal periods because the earth has a larger orbit. The orbits of the superior planets, on the other hand, are larger than the earth's and therefore the superior planets have synodic periods shorter than their sidereal periods. For example, Pluto has a sidereal period of 248 years but a synodic period of only about 367 days, at which time it will again be back in the same position with respect to the sun and earth.

MINOR PLANETS

Johann Titius' German translation of Charles Bonnet's *Contemplation de la Nature* was published in 1766. The book contained a rather insignificant footnote added by the translator.

> Divide the distance from the sun to Saturn into 100 parts; then Mercury is separated by 4 such parts from the sun; Venus by 4 + 3 = 7 such parts; the Earth by 4 + 6 = 10; Mars by 4 + 12 = 16. But notice that from Mars to Jupiter there comes a deviation from this exact progression. After Mars, there follows a distance of 4 + 24 = 28 parts, but so far no planet or satellite has been sighted there. . . . Let us assume that this space without doubt belongs to the still undiscovered satellites of Mars. . . . Next to this for us still-unexplored space there rise Jupiter's sphere of influence at 4 + 48 = 52 parts; and that of Saturn at 4 + 96 = 100.

In 1772 almost exactly the same words appeared in *Anleitung zur Kenntnis des gestirnten Himmels,** written by Johann Bode (1747-1826). Like Titius, Bode could not believe that there was only empty space between Mars and Jupiter. The difference between the two men's lines of thought was that Bode did not assign that empty space to some hypothetical satellite of Mars. Both men did imply, however, that there should be a yet-undiscovered body at 28 parts of Saturn's distance from the sun. The relationship became known as the Titius-Bode law and the search was on for the missing planet.

Ironically, on January 1, 1801, a planet was found at 2.676 times the earth's distance or 27.67 parts of Saturn's orbital distance. The planet, however, was only 488 miles in diameter, the first of more than 2650 small bodies that would in subsequent years be discovered in the zone 340 million miles wide between Mars and Jupiter. These small bodies were called minor planets, or asteroids.

Ceres, 488 miles in diameter, is the largest of the minor planets; then in descending order of size are Pallas (304 miles), Vesta (240 miles), and Juno (120 miles). A few other minor planets have diameters greater than 100 miles, a few hundred at from 25 to 100 miles in diameter, but the vast majority have diameters of 1 mile or less. Their albedos are usually 1% or less, and none is known to have an atmosphere. The surface gravities of minor planets are very low because of their small mass. Even on Ceres, a man weighing 175 pounds (80 kg) on earth would weigh only 6 pounds (3 kg). The escape velocity of a minor planet would also be so low that a major league pitcher could easily pitch a ball into space.

Some minor planets have orbits that are so eccentric that from time to time they come inside the orbit of the earth. Such close approaches of minor planets are welcomed by astronomers, for changes in the orbit of a minor planet can be used to more accurately determine the mass of the earth-moon system. Prophets of doom, however, usually predict a collision with catastrophic consequences for the earth, but the odds against such a collision are so huge as to be hardly worth a moment's worry.

Just as there are spaces in Saturn's rings caused by the gravitational effects of its inner satellites, so are there spaces in the minor planet zone that are swept clear by the gravity of Jupiter. The narrow zones that are free of minor planets are called *Kirkwood's gaps*. The gaps are located at positions in which the orbital period of any minor planet would be a simple fraction (for example, $^1/_2$, $^1/_3$, $^2/_5$, $^3/_5$, etc.) of Jupiter's period.

In 1772 several minor planets were discovered revolv-

*The title is translated *Introduction to the Study of the Starry Heavens*.

ing around the sun in Jupiter's orbit, 60° east and 60° west of the giant planet. There are 14 of these minor planets called the *Trojans;* nine are to the east and five are to the west. The Trojans occupy a type of "no man's land." If they were any closer to Jupiter they would become satellites of that planet. If they were any further away they would be independently orbiting the sun outside the sphere of Jupiter's influence. It is now believed, in fact, that many of Jupiter's smaller satellites and the two small satellites of Mars are trapped minor planets.

The origin of the minor planets is not known at this time, but there are two popular schools of thought. The first suggests that the minor planets are the remnants of a major planet that broke up subsequent to a close approach to Jupiter. The second theory views the minor planets as space debris that never formed into a major planet in the first place. A third theory, a variation of the first, is based on the possibility of a collision and subsequent breakup of two major planets. Of the three theories, the first is held in highest esteem.

METEOROIDS, METEORS, AND METEORITES

Minor planets are not the only natural space junk in the solar system, for uncountable numbers of small bodies called meteoroids also orbit the sun. Some meteoroids are immense, weighing tons, and many are the size of a few grains of sand. Most, however, are very tiny and are called *micrometeoroids*. Most meteoroids are apparently distributed at random throughout the solar system in orbits of all eccentricities and inclinations. On the other hand, some appear to move in swarms associated with the orbits of comets.

When a meteoroid enters the earth's atmosphere it becomes known as a meteor. Most meteors are vaporized by the heat produced by friction as they pass through the atmosphere. Therefore they usually appear for only a brief instant as a streak of light in the night sky (Fig. 4-12) and are sometimes referred to as "shooting stars." It is estimated that more than 100 million meteors enter the atmosphere every day in an almost constant stream; if it were not for our protective blanket of atmosphere they would

Fig. 4-12. If our atmosphere did not protect us, an almost constant stream of meteors like these would fall on the earth's surface. Note how this shower appears to radiate from a point off the photograph to upper right. (Photograph by Dennis Milon.)

pepper the ground in a virtually ceaseless barrage. Although meteors usually look as though they are quite close to an observer, most become visible at from 60 to 80 miles (97 to 129 km) above the earth's surface and disappear at an altitude of 40 miles (64 km). It is therefore unlikely that you will ever see a flaming stone crash to earth a short distance away, although this has happened from time to time. You may think that a meteor has fallen nearby, but more probably it will have simply vanished below the horizon. Very bright meteors are called *fireballs* and may be visible for several seconds. A meteor seen breaking up into smaller pieces is called a *bolide*. The speed of meteors sometimes actually is faster than the speed of sound.

When the earth orbits through a swarm of meteoroids (which are often associated with the paths of comets), observers on earth see a natural fireworks display called a meteor *shower*. The earth enters no fewer than 28 meteor swarms each year, but of that number only 11 are worthy of note (Table 4). Because of the great distances involved, the parallel meteor paths in a swarm seem to radiate from a central point called the *radiant* (Fig. 4-12). Meteor showers are named for the constellation in which the radiant is located. In order to have the best view of a shower, the observer should go outside after midnight and face eastward, the direction of both the earth's rotation and revolution. During an extensive meteor shower as many as 100 meteors may be seen each hour.

The earth is constantly overtaking and being overtaken by meteoroids that enter the atmosphere at any time from any direction to become sporadic meteors. Seeing them is a matter of looking in the right direction at the right time. Like meteor showers, individual meteors are more frequently seen toward the east because the earth's orbital speed often exceeds that of the meteoroids.

Occasionally a meteor survives the plunge through the earth's atmosphere and lands on the surface—a true visitor from outer space called a meteorite (Fig. 4-13). There are three basic types of meteorites: stony meteorites, or *aerolites;* iron meteorites, or *siderites;* and a combination of the two known as stony-irons, or *siderolites*. There are several interesting subclasses such as *carbonaceous chrondrites,* a type of stony meteorite that contains carbon, and *tektites,* a type of glassy aerolite. The stony variety makes up 92% of all known meteorites; irons make up 5% to 6%, and stony-irons account for the remaining 1% to 2%.

Fig. 4-13. Stone meteorite from Arkansas. This huge meteorite, which weighs 745 pounds, was the largest single stone meteorite whose fall was actually observed. (Courtesy University of Chicago Yerkes Observatory, Williams Bay, Wisc.)

Table 4. Major meteor showers

Name of shower	Date of best display each year	Approximate duration (days)	Meteors expected/hr	Associated comet	Period of comet (years)
Quadrantid	January 3	0.5	40	—	7.0
Lyrid	April 21	4	8	1861-I	415.0
Eta Aquarid	May 4	18	20	Halley	76.0
Delta Aquarid	July 30	20	20	—	3.6
Perseid	August 11	25	70	1862-II	105.0
Draconid	October 9	0.1	100	Giacobini-Zinner	6.6
Orionid	October 20	8	25	Halley	76.0
Taurid	October 31	30	10	Encke	3.3
Andromedid	November 14	—	—	Biela	6.6
Leonid	November 16	4	15	1866-I	33.0
Geminid	December 13	6	50	—	1.6

The total number of known meteorites is about 1600, and 35 of these have weighed more than 1 ton. The world's largest known meteorite, named the Hoba West, was discovered in southwest Africa; it has a volume of 9 cubic yards and weighs 66 tons. The largest meteorite on display in the United States is the 34-ton Ahnighito, at the American Museum–Hayden Planetarium in New York City.

Meteorite craters are formed by the falls of large meteorites. Two such falls in the twentieth century were observed from a distance and their crater fields, containing hundreds of small meteorite craters, were later investigated. Both occurred in Siberia—one on June 30, 1908, and the other on February 12, 1947. Larger meteorites have hit the earth and formed huge craters, but no reports of these ancient events have been found in the records of early civilizations. The largest of these craters is the New Quebec Crater, 2 miles in diameter, while Barringer Meteor Crater in Arizona is 4200 feet across and 570 feet deep. The Wolf Creek Crater in Australia has a diameter of 2800 feet. Approximately a dozen other craters are the result of meteorite impact, while it is suspected that hundreds of others, such as an arc-shaped segment of shoreline on the east side of Hudson Bay and the Carolina Bays, were formed in this way. Wabar Crater in Saudi Arabia is an astrobleme, a meteorite crater virtually obliterated by erosion (Fig. 4-14).

Fig. 4-14. Wabar Crater in Saudi Arabia is a good example of an astrobleme. It appears here as the faint, shadow-like circle in the center of the photograph. (Courtesy United States Department of the Interior, Geological Survey, Reston, Va.)

COMETS

More excitement may have been caused by comets than by any other objects that appear in the sky. People at one time actually wore charms as protection against them, which is not so surprising, considering some of the spectacular comets that have been seen. A comet seen in 344 B.C. "looked like a flaming torch," and one in 146 B.C. was "as bright as the sun." The comet of 530 A.D. reached from horizon to zenith, and a comet that appeared in 1811 had a head nearly 1 million miles in diameter and a tail 130 million miles long. The comet of 1744 had six tails forming a great fan, while the Great Comet of 1843 had a tail 200 million miles in length.

A comet has been described as "the nearest thing to nothing that anything can be and still be something." Comets in fact are flimsy masses of gas and dust circling the sun in highly elongated orbits. Comets are not thought to be native members of the solar system. It is now generally believed that as the sun orbits through the Galaxy, it encounters clouds of interstellar gas and dust that are attracted gravitationally into orbits around our star. The number of potential comets, then, appears to be unlimited.

The overall size of comets is in most cases far larger than the dimensions of planets, and a few are even comparable in volume with the sun. On the other hand, the masses of even the largest comets are negligible by planetary standards. It has been estimated that 10,000 million average comets would be needed to equal the mass of the earth.

Comets have two basic parts, a *head* and a *tail*. The head often consists of two parts: a *coma* and a *nucleus*. When the comet first appears it usually consists only of a *coma,* a swarm of loose particles and frozen gases that appears as a rounded, nebulous region. As the comet comes within about 250 million miles of the sun, it seems to undergo excitation. The coma brightens perceptibly, and in many comets a brilliant *nucleus* appears within the coma. It is believed that the nucleus forms volitilization of frozen gases within the comet. Not all comets have nuclei, but a great many do; if nuclei are present they range from less than 100 miles to 50,000 miles across. The total head region of comets can be as much as 1 million miles in diameter.

The tail is by far the most spectacular of a comet's features. It may be millions of miles in length and extend over a larger area than any other celestial body except a giant star, a star cluster, or a galaxy. The formation of the tail occurs when the outer region of the coma is whirled out by solar excitation, forced back by radiation pressures from the sun, and finally dissipated into microscopic fragments. Scattered over a tremendous area, these tiny particles are perhaps 1 mile apart, so that the tail is almost a vacuum, more tenuous than the earth's atmosphere. It is well known that the tail of a comet points approximately away from the sun—a phenomenon that is produced by the solar wind and radiation pressure.

Comets are classified in two groups on the basis of the shapes of their orbits and their periods of revolution. *Short-period* or *periodic* comets follow orbits that are elliptical in shape; approximately 100 comets of this type have been cataloged. The longest established period for this type of comet is 151 years for Rigollet's comet, and the shortest is 3.3 years for Encke's comet. Most short-period comets orbit from west to east in orbits only slightly inclined to the ecliptic. The best known short-period comet is Halley's comet, which has a period of 76 years; it will next appear in 1986.

About 45 of the short-period comets have had their orbits shortened by the gravitational pull of Jupiter. These comets are thus referred to as Jupiter's family of comets. They all have periods that range from 5 to 10 years, and at aphelion, their greatest distance from the sun, they are near the orbit of the giant planet.

Long-period comets, the second class, are more numerous and may approach the sun from all angles and directions. The long-period comets, of which 500 have been identified, travel in greatly elongated parabolic or in some cases hyperbolic orbits that are virtually straight lines. *Parabolic* comets have periods often measured in thousands of years, while *hyperbolic* comets may not return to the sun for millions of year, if at all. Comet Kohoutek, which was visible to us in 1973, is a hyperbolic comet (Fig. 4-15).

Comets lose mass each time they make a perihelion passage of the sun; therefore individually they cannot last forever. They also are easily affected by the gravitational

Fig. 4-15. The comet Kohoutek received much front-page publicity in 1973 and 1974 when it was visible from one place or another for several weeks. It disappointed amateur astronomers, however, and very few good photographs were taken. (Photograph by John Clement, Jr.)

pull of other bodies in the solar system, especially the sun and Jupiter. It is now generally believed that some long-period comets may become short-period comets with the passage of time, and that short-period comets may eventually become members of Jupiter's family of comets. Finally, after about 10,000 years short-period comets disintegrate, never to be seen again. The loss of mass and eventual disintegration of a comet is indicated by the meteor swarms associated with known comets. The Aquarid meteor shower, for example, is known to be associated with Halley's comet, and the Taurid shower is associated with comet Encke.

There are also meteor swarms associated with no observable comet. An especially interesting case is the Andromedids that move in the orbit of comet Biela, although this comet has not been seen since 1852. At its return to the sun in 1846, comet Biela was observed to split into two comets. In 1852 the two comets could be seen widening their separation. In 1858, the year of Biela's expected return, only a meteor shower was seen.

Many new comets are usually found each year—most by amateur astronomers. New comets are designated by the year in which they are found and a letter of the alphabet representing the order of their discovery in that year. The first new comet of 1976, for example, is listed as comet 1976A. In addition, comets are named for their discoverers. Comet 1969G, for example, is also called comet Tago-Sato-Kosaka after the three Japanese who discovered it.

The record number of comets observed in one year was 22 in 1947. This total included eight comets that had been found in 1946 and were still observable, eight newly discovered long-period comets, four previously recorded short-period comets that returned, one newly discovered short-period comet, and one object whose nature was not determined. In 1938, by contrast, not only were no new comets noted but only one previously recorded comet was observed.

5 *a glimmer of starlight*

Astronomical instruments and the analysis of stellar radiation

No unregarded star contracts its light
Into so small a character removed
But if we steadfast looke
We shall discerne
In it, as in some holy booke,
How man may heavenly knowledge learne.

1634
William Habington

For centuries man has dreamed of reaching out and touching the beauty of the heavens above him. Unlike the biologist, geologist, chemist, or physicist, who deal with materials and forces that can readily be obtained or tested here on earth, astronomers must study most of their subjects from afar. Although an occasional meteorite survives a trip through the earth's atmosphere, and although we have recently obtained samples of the moon, the vast bulk of astronomical information rides in on beams of energy produced by the sun and other stars. It has therefore been necessary for us to devise instruments as extensions of our eyes, ears, and hands in order to collect data and interpret this information in the light of known natural laws.

ELECTROMAGNETIC RADIATION

Stars produce a wide range of energy types that are collectively referred to as the electromagnetic spectrum (Fig. 5-1). Each type of energy on the spectrum is called electromagnetic radiation because it has both electrical and magnetic properties. All forms of electromagnetic energy travel at the same speed—186,000 miles/sec, and have certain properties in common. Electromagnetic radiation, for example, travels in the form of waves having a wavelength, frequency, and amplitude. The *wavelength* is the distance from the crest of one wave to the crest of the next. *Frequency,* on the other hand, is the number of wavelengths that pass a given point in a given amount of time. Finally, *amplitude* is defined as half the vertical distance from crest to trough of the wave (Fig. 5-2). One form of electromagnetic energy differs from another only in these properties, so one type of electromagnetic energy can be converted into another type with relative ease. Although electromagnetic energy arrives on earth from uncountable numbers of stars, only the energy reaching us from the sun is of any major importance in our daily lives. The other stars are so far away that the effect of their energy on the earth is negligible.

Gamma and X-rays

Gamma rays have the shortest wavelength and the highest frequency and are thus the most energetic and potentially dangerous type of electromagnetic radiation. Although generated within the cores of stars, most gamma rays are converted into other forms of energy, primarily X-rays, by absorption and reemission long before they reach the star's surface. Of those few gamma rays that do arrive in the vicinity of the earth, virtually all are absorbed by atoms and molecules of nitrogen and oxygen in the upper atmosphere. Therefore gamma rays of solar origin are of little consequence. Gamma rays are produced on earth in small quantities by nuclear reactors and are emitted by radioactive substances.

X-rays stream outward into space from the sun and other stars but, like gamma rays, most are absorbed by

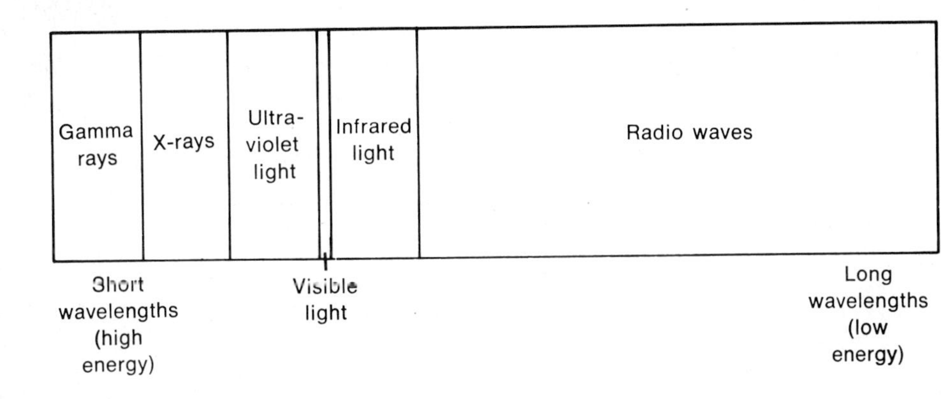

Fig. 5-1. Diagram of electromagnetic spectrum.

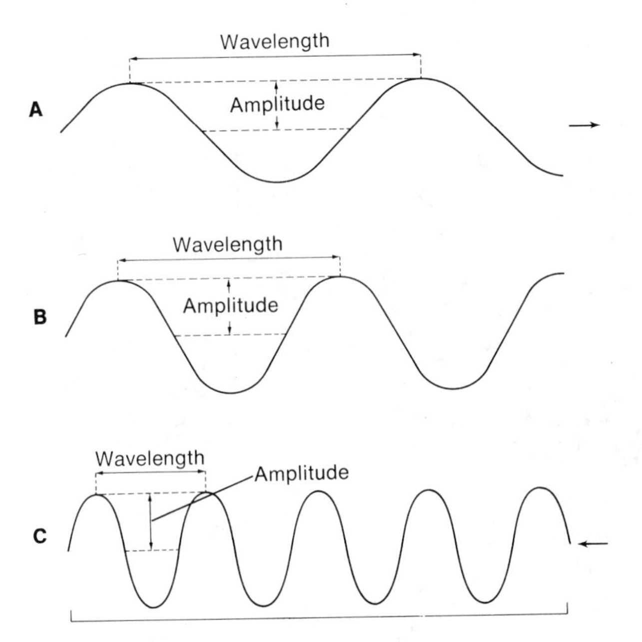

Fig. 5-2. Relationship between wavelength, frequency, and amplitude. **A,** Source of waves receding from observer (Frequency = 1.5 wavelengths/sec). **B,** Source stationary with respect to observer (Frequency = 2.25 wavelengths/sec). **C,** Source approaching observer (Frequency = 4.5 wavelengths/sec).

nitrogen and oxygen atoms and molecules to form the *ionosphere* (Chapter 13).

Ultraviolet radiation and visible light

Ultraviolet radiation, which arrives in the vicinity of the earth in potentially harmful quantities from the sun, is absorbed primarily by oxygen in the stratospheric layer of the atmosphere, forming the gas ozone. Some ultraviolet energy, however, does penetrate to the earth's surface. It is this energy that produces the common radiation burns called sunburn. Overexposure to ultraviolet radiation is thought to be a leading cause of skin cancer, and such exposure is especially dangerous to the human eye. The earth's atmosphere is more transparent to visible light radiation than to any other type. Except when it is reflected back into space by clouds, visible light from the sun penetrates to the earth's surface.

Infrared radiation and radio waves

Infrared radiation, or heat energy directly from the sun, is absorbed almost entirely by carbon dioxide and water vapor in the earth's troposphere. Thus most of the heat we feel in the atmosphere is produced by the absorption of visible light and its subsequent reemission as infrared radiation.

Radio waves, the type of electromagnetic radiation with the longest wavelengths, are harmless in every known respect. They are so weak, in fact, that special instruments are needed to detect and record them (p. 79).

LIGHT AND OPTICAL TELESCOPES

The sun, with a surface temperature of 10,000° F, radiates primarily visible light energy. The same is also true of virtually all other stars, although most have surface temperatures higher or lower than the sun. Planets and satellites in our solar system shine by reflected sunlight, while galaxies are made up of stars not vastly different from the sun. It is natural then that we, whose eyes evolved in response to sunlight, have accumulated most of our knowledge of the universe from a study of the visible light from celestial bodies.

Although marvelously efficient instruments in bright sunlight, our eyes are very poor receptors of the faint light that reaches earth from distant objects. We have thus found it necessary to build telescopes that serve as gigantic superhuman eyes, gathering millions of times more light than human eyes are capable of perceiving. From Galileo's first "optik glass" to today's enormous observatory reflectors, telescopes have been the basic tools used by astronomers in their exploration of the universe.

Law of refraction and refracting telescopes

When a beam of light passes from one transparent medium such as air into a different transparent medium such as water or glass, the beam is always bent at the boundary between the two media (Fig. 5-3, *A*). This bending is known as *refraction,* and it is the result of the effects that different transparent media have on the speed of light.

In a vacuum, light, like other forms of electromagnetic radiation, travels in straight paths at a speed of 186,000 miles/sec (670 million miles/hr). As light enters a medium of greater density, however, its speed is reduced. The ratio of the speed of light in a vacuum to that in a given medium is called the *index of refraction* of the medium. If the index of refraction for a vacuum is 1, then the index for air is 1.00029, for water, 1.3; for glass, 1.5; and for diamond, 2.4.

The law of refraction states that if light passes from one transparent substance into another that has a different index of refraction, the angle the light makes with the vertical (or normal) to the boundary between the two media is always *less* in the medium of higher index. In other words, if light goes from air into glass or water, it is bent toward the vertical, but if it goes from glass or water into air, it is bent away from the vertical.

The first telescopes, most commercially available telescopes for amateurs, and some of the finest observatory telescopes in the world operate on the law of refraction and are thus called *refractors*. The refracting telescope is basically a hollow tube that contains two lens systems—the larger objective lens at one end and the smaller eyepiece lens at the other. Parallel beams of light entering the objective lens from a distant celestial body are refracted in such a way that they converge at a point of *principal focus* to form a very tiny image of the object (Fig. 5-3, *B*). The purpose of the eyepiece-lens system is to magnify the image that is formed some distance behind the lens at the principal focus. The distance from the center of the lens to the focal point is called the *focal length* of the telescope; this is determined by the diameter of the objective lens and is thus a fixed quantity.

Small refractors are popular with amateurs, and the minimum useful size recommended for serious work is an instrument with an objective lens 2.4 inches in diameter

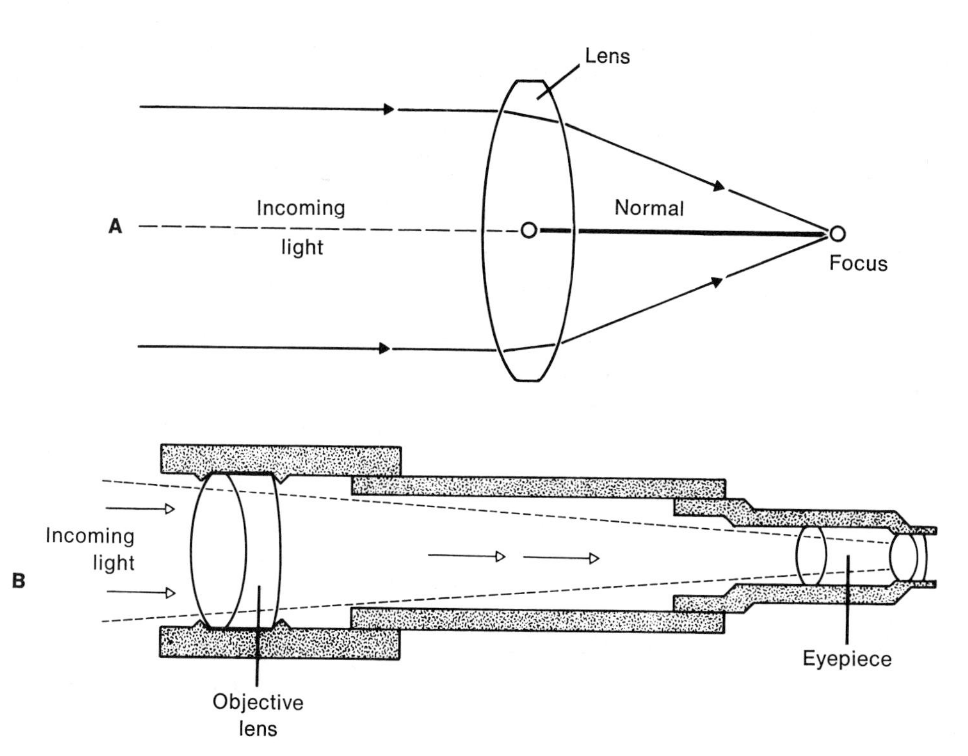

Fig. 5-3. Law of refraction. **A,** When passing from one medium to another (for example, from air to glass), light is bent because different substances have different indexes of refraction. **B,** Refracting telescope that operates on this principle.

and a focal length of 900 mm (Fig. 5-4). Such a telescope, completely equipped, retails today for approximately $365; 3- and 4-inch models retail for about $614 and $1054, respectively. Less expensive models that lack desirable equipment are available; used, well-maintained telescopes may occasionally be found by the careful shopper.

The primary disadvantages of refractors are their size and chromatic aberration. First, their mode of operation requires that the light come through the objective lens and then straight down the tube to the principal focus. This arrangement requires that refractors be very long relative to other types of telescopes. The world's largest refracting telescope, for example, has an objective lens 40 inches in diameter and a focal length of 760 inches (63⅓ feet). This huge instrument is located at Yerkes Observatory in Williams Bay, Wisconsin. The locations and diameters of other large observatory refractors are listed in Table 5.

Chromatic or color aberration is the second and more serious problem affecting refractors, for not only does a lens refract light, it also disperses light into its individual wavelengths, or colors. This occurs because of the inability of a single lens to bring light waves of various lengths to the same focal point. Short-wavelength blue light, for example, is refracted more than long-wavelength red light, so it comes to a focus closer to the lens. Because of the chromatic problem, observatory refractors are used primarily for studying members of the solar system. Reflecting telescopes are invariably used for the study of stars and deep-space objects such as star clusters, nebulae, and

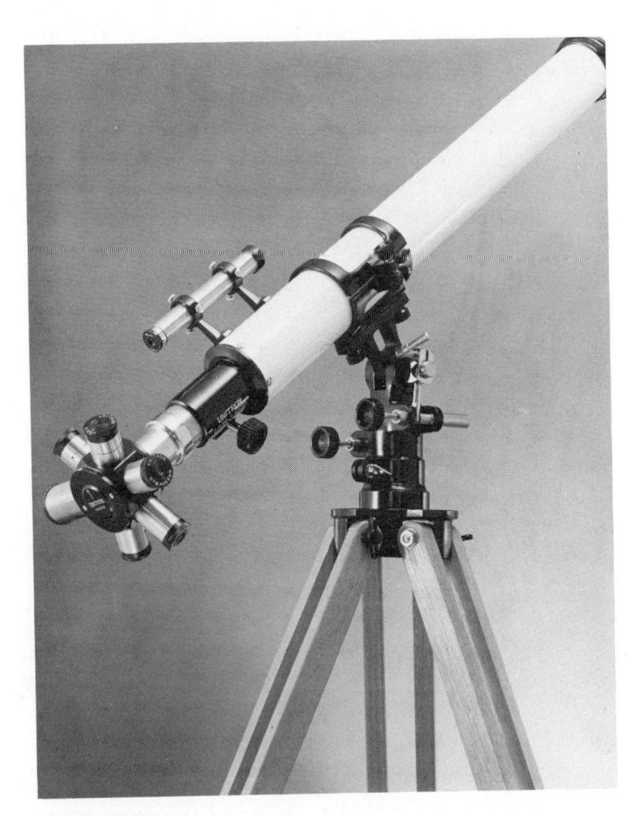

Fig. 5-4. A 2.4-inch Unitron refracting telescope with an altitude-azimuth mount. (Courtesy Unitron Instrument Co., Newton Highlands, Mass.)

Table 5. World's largest optical telescopes

Refractors	Diameter (inches)	Reflectors	Diameter (inches)
Yerkes Observatory, Williams Bay, Wisc.	40	Special Astrophysical Observatory, U.S.S.R.	236
Lick Observatory, Mt. Hamilton, Calif.	36	Hale Observatory, Mt. Palomar, Calif.	200
Astrophysics Observatory, East Germany	32	Lick Observatory, Mt. Hamilton, Calif.	120
Paris Observatory, France	32	McDonald Observatory, Fort Davis, Texas	107
Allegheny Observatory, Pittsburgh, Pa.	30	Crimean Astrophysical Observatory, U.S.S.R.	104
University of Paris, France	30	Hale Observatory, Mt. Wilson, Calif.	100
Royal Greenwich Observatory, England	28	Kitt Peak National Observatory, Tucson, Ariz.	84
Union Observatory, South Africa	26.5	McDonald Observatory, Fort Davis, Texas	82
Universitats-Sternwarte, Austria	26.5	Saint Michel Observatory, France	77
University of Virginia, Charlottesville, Va.	26	Tokyo Observatory, Japan	74
Observatory, Academy of Sciences, U.S.S.R.	26	David Dunlap Observatory, Canada	74
Astronomical Observatory, Yugoslavia	26	Helwan Observatory, Egypt	74
Leander McCormick Observatory, Charlottesville, Va.	26	Astrophysical Observatory, Japan	74
Observatory Mitaka, Japan	26	Radcliffe Observatory, South Africa	74
U.S. Naval Observatory, Washington, D.C.	26	Dominion Astrophysical Observatory, Canada	73
Mt. Stromlo Observatory, Australia	26	Perkins Observatory, Flagstaff, Ariz.	72

galaxies, where so much of our knowledge is based on color.

Law of reflection and reflecting telescopes

Another basic property of light is its ability to be reflected from a shiny surface such as a mirror. A beam of light strikes a mirror at an angle that is measured with respect to the vertical; this angle is called the *angle of incidence*. The beam is then reflected from the surface at an angle to the vertical known as the *angle of reflection* (Fig. 5-5, *A*). The law of reflections states that the angle of reflection is always equal to the angle of incidence and that the incident beam, the reflected beam, and the vertical to the surface all lie in the same plane.

Mirrors with curved surfaces may be used like lenses to converge or diverge light. Mirrors that converge light have concave curvatures, while those that diverge light are convexly curved. Converging mirrors are either *spherical* or *parabolic*. A spherical converging mirror is one whose surface is part of a sphere, and a parabolic converging mirror is one whose surface is part of a parabola.

The principal *axis* of a concave mirror is an imaginary line drawn through the center of curvature and the center of the mirror. The principal *focus* of such a mirror is that point at which incoming beams of light parallel to the principal axis intersect after they are reflected.

The largest astronomical telescopes and most of the smaller telescopes built by amateurs all have parabolic mirrors instead of lenses. Such a telescope, which operates according to the law of reflection, is called a *reflecting*

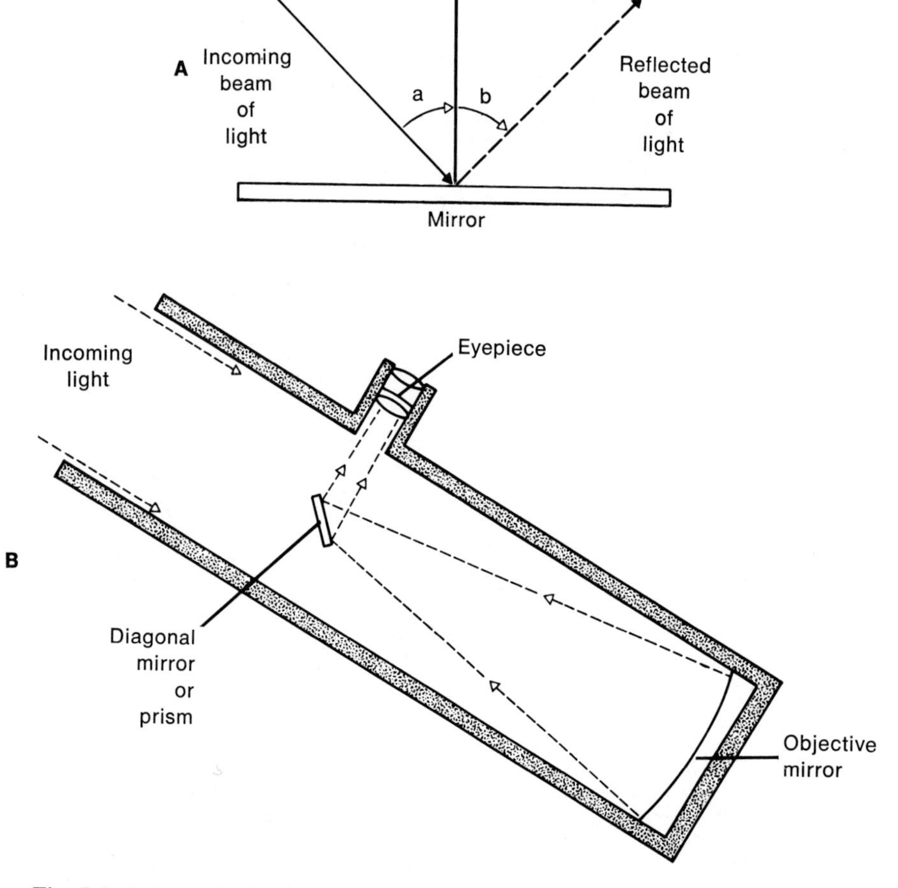

Fig. 5-5. A, Law of reflection. The incoming beam, striking the mirror at angle *a* (angle of incidence), is reflected back off at angle *b* (angle of reflection). **B,** Cutaway view of telescope based on this principle. The image is reflected between the two mirrors and magnified by the objective mirror.

telescope, or simply a *reflector* (Fig. 5-5, *B*). The first reflecting telescope was built by Isaac Newton in 1687, although the concept was developed 5 years earlier by James Gregory. It was not until the time of William Herschel in the middle of the eighteenth century, however, that the reflecting telescope became an important astronomical tool.

A reflecting telescope is basically a hollow tube, open at one end, with a parabolic mirror bolted to the other (Fig. 5-6). Light from a distant celestial object enters the tube and is reflected from the mirror to a point of prime focus, where a tiny image of the object is formed. Unlike a refractor, which produces an image behind the lens outside the tube, a reflector forms an image in front of the mirror inside the telescope itself.

A number of ingenious focal configurations have been devised to gain access to the image formed by a reflector. The arrangement used for a particular telescope is determined by the size of the instrument and the purpose for which it is intended. Popular focus configurations include the prime, Newtonian, Cassegrain, and Coudé foci (Fig. 5-7).

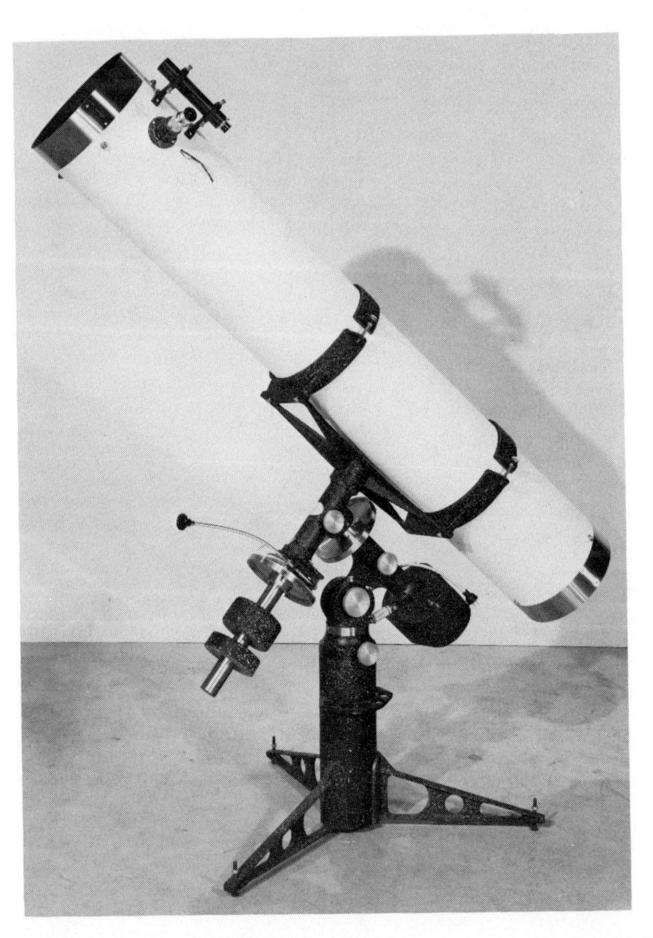

Fig. 5-6. A 10-inch deluxe reflecting telescope with equatorial mount. (Courtesy Star-Liner Co., Tucson, Ariz.)

Prime focus

The prime focus is formed within the telescope near the open end of the tube. Direct access to the prime focus is available only in the larger instruments such as the 200-inch reflecting telescope of the Hale Observatory on Mt. Palomar, California. This huge instrument has an observer's cage at the prime focus so the astronomer actually does his work inside the moving telescope.

Newtonian focus

The Newtonian focus, developed by the great English genius, is the most popular arrangement for small reflectors. A flat mirror is placed in the telescope at the prime focus. This mirror intercepts the light from the objective mirror and reflects it to a focus just outside the tube near the open end.

Cassegrain foci

There are at least two types of Cassegrain foci. The first type uses a small convex mirror at the prime focus. This mirror intercepts the light from the objective mirror and reflects it back down the tube and through a hole in the objective mirror to a focus. If the objective mirror has no hole, then another Cassegrain configuration is used. This type also involves the use of a small, flat mirror placed just above the objective mirror to intercept the light. Light is then reflected out the side of the instrument near the base to a focus.

Coudé focus

It is often necessary to bring the light from a telescope to accessory equipment rather than vice versa. Heavy, bulky equipment that is too massive to be attached to the telescope is kept in a separate room. First a convex mirror at the prime focus of the telescope intercepts the light and reflects it back down the tube. Then a small, flat mirror placed at a pivot point of the instrument again intercepts the light, reflecting it out of the tube to a focus in the room where the auxiliary equipment is located. Such an arrangement is called the Coudé focus. The Coudé room of the 200-inch reflecting telescope, for example, is immediately below the instrument.

• • •

To be minimally useful, a reflector must have an objective mirror at least 6 inches in diameter. Such an instru-

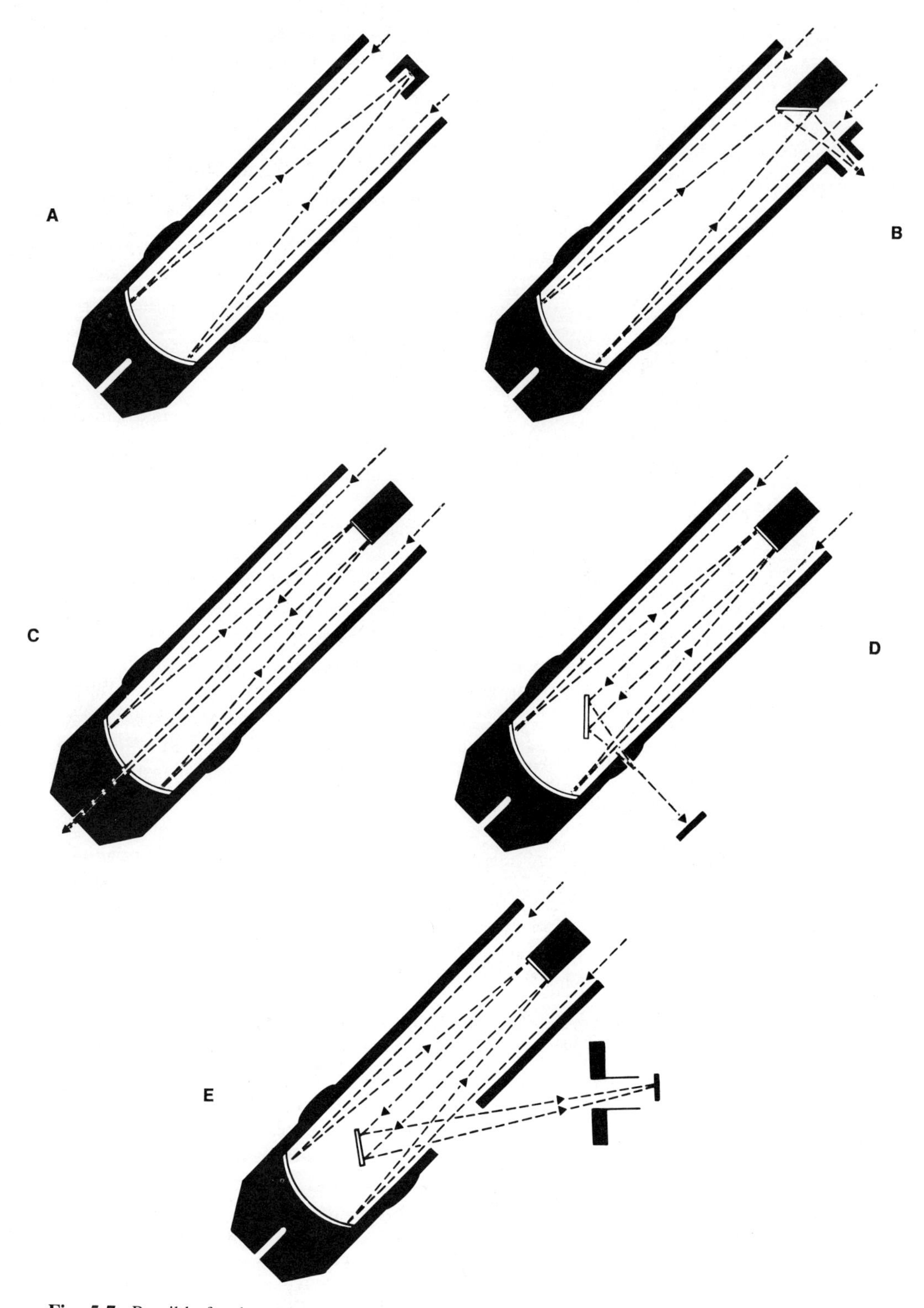

Fig. 5-7. Possible focal arrangements of reflecting telescopes. **A,** Prime focus; **B,** Newtonian focus; **C,** Cassegrain A focus; **D,** Cassegrain B focus; **E,** Coudé focus. Arrows indicate paths of light rays through each telescope.

ment can be made from a kit or purchased commercially for approximately $200 fully equipped. The world's largest reflector is a Russian instrument that is 236 inches in diameter and 82 feet in length, with a weight of 850 tons. The second largest reflector, and the instrument used in many of the important astronomical discoveries of the twentieth century, is the Hale reflector, which has a diameter of 200 inches (more than 16½ feet). This instrument weighs 500 tons and has a focal length of 660 inches (55 feet). If the earth were flat and if we could mount the 200-inch telescope mirror on top of the 1450-foot Sears Tower in Chicago and look over the surface of the earth as far as the power of the instrument would allow, we could see images of fireflies in Tokyo, Japan. The sizes and locations of other great reflectors are listed in Table 5.

Unfortunately, reflecting telescopes are not without problems, either. The main fault of a reflector is optical aberration resulting from imperfections in grinding and polishing the mirror. Although ground to within two millionths of an inch of accuracy, the 200-inch telescope still cannot be considered perfect. Because of the optical problem, reflecting telescopes are best utilized for the study of stars and deep-space objects, many of which appear only as pinpoints of light.

Inverse square law of light

The inverse square law of light states that the intensity of light decreases by an amount equal to the square of the distance between the luminous object and the observer. To illustrate this relationship between distance and light intensity, we can consider the distances from the sun to the earth, Jupiter, and Saturn, respectively. The earth, for example, is 1 AU from the sun, while Jupiter and Saturn are approximately 5 and 10 AU away respectively. As seen from Jupiter, the sun will appear only $^{1}/_{25}$ as bright as it appears from earth, while from Saturn the sun would appear $^{1}/_{100}$ as intense as from earth. The inverse square law is an important tool in determining astronomical distances (p. 104).

Doppler effect

The apparent wavelength and frequency of sound or electromagnetic energy depends on the relative motion of the emitting source with respect to the observer (Fig. 5-2). This effect was first described in 1842 by Christian Doppler (1803-1853), an Austrian physicist. The first correct explanation and proper deductions, however, were given by A. H. Pizeau (1819-1896) in 1848. Regardless, the phenomenon is called the Doppler effect.

The Doppler effect occurs when a source of energy and an observer are in rapid motion relative to one another. The frequency with which the sound, light, or radio waves from the source reach the observer increases or decreases according to the speed at which the source and the observer are moving closer together or farther apart. This effect means that an observer may not get a true impression of the source unless the Doppler effect is taken into account. For example, if a source is emitting light waves and the observer and the source are moving toward one another, the effect of this motion will change the frequency of the waves by shifting them toward the violet end of the visible light spectrum. Therefore the source will appear more violet than it actually is. Conversely, when a source of light waves and an observer are moving away from one another, this motion will have the effect of shifting the light waves toward the red end of the visible light spectrum (p. 105 and Fig. 5-2).

Schmidt telescope

The disadvantages of both reflectors and refractors can virtually be eliminated by the use of a telescope that employs a combination of a mirror and a lens. The Schmidt telescope, utilizing the optical system developed by Bernard Schmidt of the Hamburg Observatory, is basically a reflector with a parabolic objective mirror. Located at the center of curvature of the mirror is a thin lens, which corrects for optical aberration. The diameter of a Schmidt telescope, unlike basic reflecting telescopes, is determined by the diameter of this correcting lens. Any chromatic aberration produced by the lens is also corrected as the light reflects from the mirror.

The Schmidt telescope, also known as a *Schmidt camera* or *astrograph,* is the most efficient photographic instrument ever built. Light enters the telescope-camera through the correcting lens and is reflected from the surface of a parabolic mirror to a point of prime focus. Here the light falls on a curved photographic plate instead of entering an eyepiece. Rarely used as an optical instrument, the Schmidt telescope is able to photograph in one single exposure an area as wide as the entire constellation of Orion. Photographic detail is also clear, rich, and sharp, as if the pictures were a composite of a dozen or more ordinary photographs.

More and more commercially available small telescopes are being made with the highly folded Schmidt optics combined with a Cassegrain focal arrangement. The main advantage of such a configuration is that it permits the construction of a narrow field telescope with a large diameter and long focal length in a compact size. Such an instrument has a wide field, yet it maintains the end-focus versatility of a refractor. Small Schmidt-Cassegrain telescopes are available in 3½-, 5-, 7-, 8-, and 14-inch diameters with prices ranging from $865 to $3600 (Fig. 5-8).

The world's largest Schmidt telescope-camera is the 53-inch instrument located at the Tautenburg Observatory in East Germany. The largest Schmidt in the United States is the 48-inch astrograph belonging to the Mt. Wilson Observatory. Completed in 1949, its first task was to produce a photographic atlas of the sky. Financed by the National Geographic Society and thus called "the National Geographic–Palomar Sky Survey," the photographic examination required 7 years for completion. Had the survey been conducted by using the 200-inch telescope, however, at least 10,000 years would have been required to complete the task.

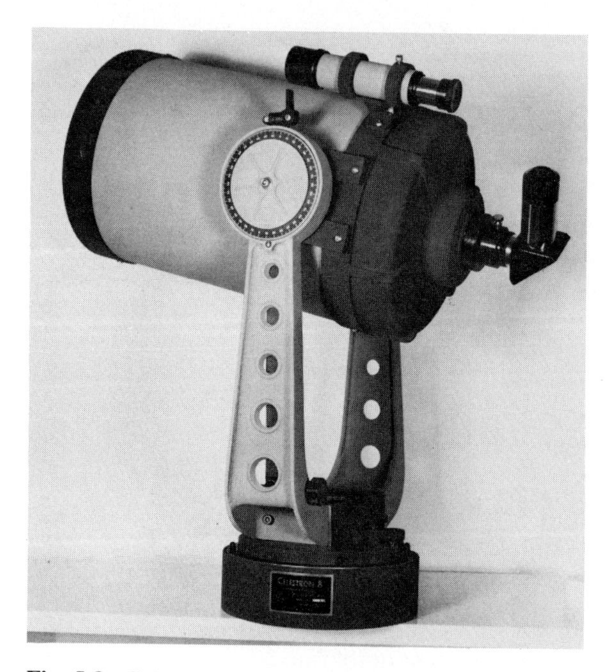

Fig. 5-8. Celestron 8, a Schmidt-Cassegrain telescope. (Courtesy Celestron Pacific, Torrance, Calif.)

A

B

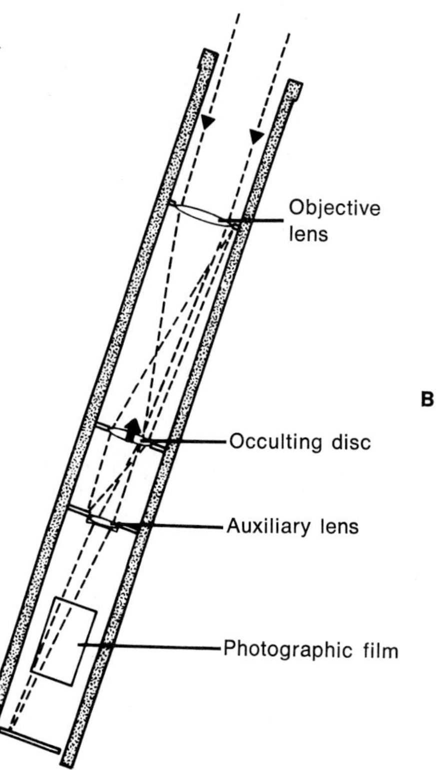

Fig. 5-9. A, Coronagraphs like this are used to measure aspects of the sun's atmosphere by simulating eclipses. **B,** Light enters the instrument, passes through the objective lens, around the occulting disc, and through auxiliary lens to photographic film. (**A** courtesy Miami Museum of Science and Space Transit Planetarium, Miami, Fla.)

Coronagraphs

The coronagraph, invented in 1930 by Bernard Lyot of the Paris Observatory, enables astronomers to artificially eclipse the sun at will in order to study the solar atmosphere (Fig. 5-9 and p. 98). The instrument is basically a reflecting telescope whose objective lens focuses the image of the main body of the sun onto an occulting disc in the focal plane. The light from the body of the sun is thus reflected out of the way by the occulting disc, and the faint light from the sun's outer atmosphere passes around the obstruction. The image is finally sent by an auxiliary lens behind the disc to the camera lens and then on to photographic film.

Magnifying powers and limits of resolution of telescopes

People often ask, "What is the power of this telescope?" "Power" in this sense could mean the total angular magnification of the tiny image formed at the principal focus by the eyepiece or the limit of resolution, often incorrectly called the resolving power. The magnifying power of a telescope is a function of the eyepiece and can be determined easily by dividing the focal length of the objective lens by the focal length of the eyepiece.

$$\text{Power} = \frac{\text{Focal length of objective}}{\text{Focal length of eyepiece}}$$

$$\text{Power} = \frac{1200 \text{ mm}}{25 \text{ mm}}$$

$$\text{Power} = 48\times$$

Consequently, a telescope has not one but many magnifying powers, which can be changed at will by viewing the focal point image with eyepieces of various focal lengths. Too great an increase in power reduces the clarity and brightness of the image because of the severe refraction through the shorter eyepiece. Therefore higher powers should be reserved for astrophotography, in which light can accumulate on photographic film over a period of time.

Resolution is the fineness of detail present in the image that is formed at the focus of a telescope. Even a lens or mirror of perfect optical quality cannot produce an image that is perfect in fineness of detail. This inability is caused by a phenomenon known as *diffraction,* effects produced by interference between the rays of light emanating from the same point. When viewing a star, for example, we do not see the geometrical image of the star itself but only its diffraction pattern. The limit of resolution of a telescope, therefore, is the smallest angle between two close-appearing stars for which two distinct images are produced. It is measured in seconds of arc and, like the focal length, is fixed for any one lens or mirror.

Telescope mounts and drives

Regardless of the type of telescope used, the mount is an important consideration. The mount is the arrangement whereby the telescope is set up for observation or photography. There are two basic types of mounts, equatorial and altitude-azimuth. The *equatorial* mount has two axes set at right angles to each other—one parallel to the earth's polar axis, the other parallel to the equator (Fig. 5-5). Thus when the telescope is properly aligned with the rotational axis of the earth through the use of the celestial pole, an object may be tracked westward by adjustment along the equatorial axis. In fact, with an electric or weight-driven clock drive that automatically turns the telescope westward at the same rate of the earth's rotation eastward (15°/hr), an object can be tracked all night and only an occasional fine adjustment will be necessary.

Altitude-azimuth mounts are usually found on inexpensive telescopes or very large instruments such as the 236-inch Russian reflector. Observing and photography are difficult because adjustments are constantly needed along two axes. Clock drives for such mounts are highly intricate and expensive.

RADIO TELESCOPES

In 1932 a young American physicist named Karl Jansky became immortal when he accidentally discovered cosmic radio noise. Working for the Bell Telephone Laboratories, he was assigned the task of studying static and other forms of interference that were causing trouble in telephone communications. Jansky designed and constructed a large directional antenna that operated at 20 megacycles/sec. The antenna was mounted on a kind of merry-go-round that enabled it to turn in any direction to pinpoint sources of interference (Fig. 5-10). Jansky kept careful records of unusual signals for a year and discovered that the noise was coming from the direction of the Milky Way.

Early in 1938, the first true radio telescope was constructed by Grote Reber and placed into operation. Located in Wheaton, Illinois, the instrument was 31 feet in diameter and took the form of a parabolic reflector, consisting of a sheet metal skin attached to a wooden framework. It was not until April of 1939, however, that Reber detected radio energy at a frequency of 162 megacycles/sec coming from the Milky Way. He then initiated a long series of observations that ultimately led to the publication in 1944 of a complete map of cosmic radio noise at that particular frequency.

Today a great many types of equipment are used for studying radio signals of extraterrestrial origin. All are called *radio telescopes.* Although they are often strikingly different in appearance, they all contain the same basic components: an antenna to capture or pick up the wave-

Fig. 5-10. The late Karl Jansky of Bell Telephone Laboratories used these rotating antennae to investigate the causes of strange noises he discovered in telephone equipment. This discovery of radiation noise emanating from the center of the Milky Way has grown into the field known as radio astronomy. (Courtesy Bell Telephone Laboratories, Inc., Murray Hill, N.J.)

lengths, a receiver to convert them into an electric current that can activate a recorder, and the recorder itself.

A radio telescope does not form an image, and therefore it cannot be used to "look through," or for photography. It measures instead the intensity of radio waves coming from a certain part of the sky. The limit of resolution of a radio telescope is a product of the diameter of the instrument and the wavelength at which it is working.

The largest radio telescope in the world is the 1000-foot stationary spherical instrument located in a natural bowl-shaped area near Arecibo, Puerto Rico (Fig. 5-11). Its movable receiver reaches to a height of 435 feet above the ground. The world's largest fully movable radio telescope is the 328-foot instrument of the Max Planck Institute for Radio Astronomy near Bonn, West Germany. Other well-known instruments include the 300-foot radio telescope at the National Radio Astronomy Observatory, Green Bank, West Virginia; the 250-foot Jodrell Bank radio telescope near Manchester, England; and the 85-foot radio telescope at Goldstone, California. The latter is famous for studies of Venus and Mars in the early 1960s and for its use in communicating with American astronauts in space.

Interferometers

The primary problem in the use of a radio telescope is the difficulty in distinguishing one radio source from another—in other words, the radio telescope's *limit of resolution*. Radio astronomers have found, however, they can improve resolution by using an interferometer. This is a system of two or more radio telescopes wired to one receiver (Fig. 5-12). As the name implies, the instrument actually causes interference with radio signals, rearranging their pattern on the recorder. If the radio source is exactly the same distance from each telescope, the waves will arrive at exactly the same time and are said to be "in tune" or in phase. The radio impulse on the recorder is thus increased. When the waves are out of phase, on the other hand, the impulses from the two antennae counteract each and cancel out the effect on both the receiver and recorder.

When the interferometer is in operation, the recorder produces a graph record of incoming radio waves called a *fringe pattern*. Peaks in the graph represent waves in phase, and valleys represent waves out of phase. The fringe pattern of ups and downs can be made finer and sharper by moving the antennae farther apart. If the right distance is

Fig. 5-11. The Arecibo radio telescope, largest stationary radio telescope in the world, is located near Arecibo, Puerto Rico. (Courtesy National Astronomy and Ionosphere Center, Arecibo, P.R.; photograph by Hector E. Roa, Cornell University.)

Fig. 5-12. National Radio Astronomy Observatory interferometer at Green Bank, West Virginia. This system uses three steerable, 85-foot radio telescopes to obtain detail that would normally require a single telescope 5000 feet in diameter. (Courtesy National Radio Astronomy Observatory, Green Bank, W. Va.)

used, only radio sources of small diameters can produce fringe patterns. In the case of a source as big as the sun, for example, one edge would be out of phase while the other edge would be in phase, and no fringe pattern would be formed unless the positions of the antennae were changed. In fact, it is even possible to measure the diameters of radio sources by seeing how far apart the antennae must be moved before the fringe pattern breaks up.

The radio interferometer is a very clever idea borrowed from instruments used for many years to study light. The stellar interferometer, as it is called, consists of a system of mirrors mounted on a girder at the upper end of a telescope. The outermost mirrors are movable. The star image formed by combining light coming over the two paths is crossed by interference fringes that can be made to disappear by adjusting the distance between the two movable mirrors. The exact distance between the two mirrors at the point at which the interference fringes disappears is a measure of the angular diameter of the source. If the distance to the source is known, its real diameter can then be calculated from its angular diameter. The stellar interferometer has been used, for example, to measure the diameter of several giant and supergiant stars such as Betelgeuse and Antares.

SPECTROSCOPY AND PHOTOMETRY

In the middle of the nineteenth century astronomy was revolutionized by the development of two new branches of physics—spectroscopy and photometry. Together they formed a new science called astrophysics, through which virtually all of the important astronomical discoveries of the past 150 years have been made.

The spectrum

In Chapter 1 we noted that Isaac Newton discovered the continuous spectrum of sunlight in 1665. In 1752, Thomas Melvil (1726-1753), a Scottish physicist, found that under laboratory conditions vaporized sodium emitted a bright-line spectrum consisting of narrow, bright-colored lines separated by broad dark gaps. In 1802 William Wolaston (1766-1828) discovered several narrow, dark lines in the solar spectrum but incorrectly attributed them to natural boundaries between the colors. However, in 1815 Joseph Fraunhofer (1787-1826), a German physicist, found about 600 dark lines in the spectrum of the sun and correctly interpreted them as color gaps in the continuous spectrum. Even though Wolaston is credited with discovering the dark-line spectrum, the lines are appropriately called Fraunhofer lines. The stage was then set for the new science of spectroscopy.

It was soon found that each atom had its own characteristic pattern of lines. Thus it became clear that each spectrum contained a clue as to the nature of the atom itself, which in turn explained the different types of spectra. Therefore at this point it will be helpful to take a brief look at the general structure of atoms.

An atom consists of a very small but massive nucleus, usually containing equal numbers of positively charged *protons* and neutral *neutrons*. Orbiting the nucleus are negatively charged *electrons,* usually equal in number to the protons in the nucleus. We can visualize the electrons as orbiting the nucleus at different distances that are called energy levels, or shells. These energy levels are lettered K, L, M, N, and so forth, from the lowest energy level to the highest. Only 2 electrons may occupy the innermost, or K energy level. There may be a maximum of 8 electrons in the L energy level, 16 in the M level, 32 in the N energy level, etc. The oxygen atom, for example, has 8 protons and 8 neutrons in its nucleus and 8 orbiting electrons—2 in the K energy level and 6 in the L energy level.

An atom that possesses its full number of orbital electrons is said to be neutral because the number of negatively charged electrons equals the number of positively charged protons. If an atom absorbs a sufficient quantity of energy, one or more of its orbital electrons may be completely removed from the system. An atom without its full complement of orbital electrons is called an *ion,* and a gas in such a condition is said to be *ionized*. There are, however, degrees of ionization, for an atom may lose all or only some of its electrons.

If an atom is mildly excited by an increase in temperature, 1 or more of its outer electrons will momentarily move to one of the outer empty energy levels. On returning to its normal orbit, the electron loses its excess energy in the form of a photon of light. The emission of this photon of light produces a bright-line spectrum. If, on the other hand, an atom is subjected to a decrease in temperature, one or more of its electrons will temporarily move to a lower energy level. In order to return to its normal orbit, energy is absorbed in the form of a photon of light, and this produces a dark-line spectrum. The atoms of glowing solids or liquids or of hot dense gases under great pressure also emit photons of light at all wavelengths to produce a continuous spectrum. Fig. 5-13 is a simple sketch of the relationship that is described in the following example.

The incandescent light bulb produces a continuous spectrum. However, when the light passes through the cloud of gas at low pressure, the atoms therein absorb some of the photons of light and reemit them at the same wavelengths in all directions. As seen from point *A,* a continuous spec-

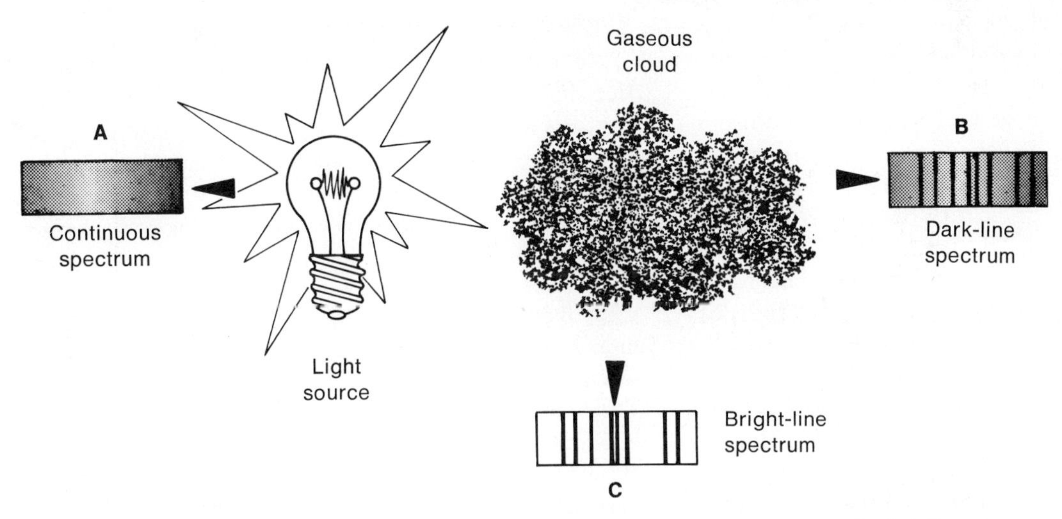

Fig. 5-13. Relationships between the three types of spectra. **A,** Continuous spectrum; **B,** dark-line spectrum; **C,** bright-line spectrum.

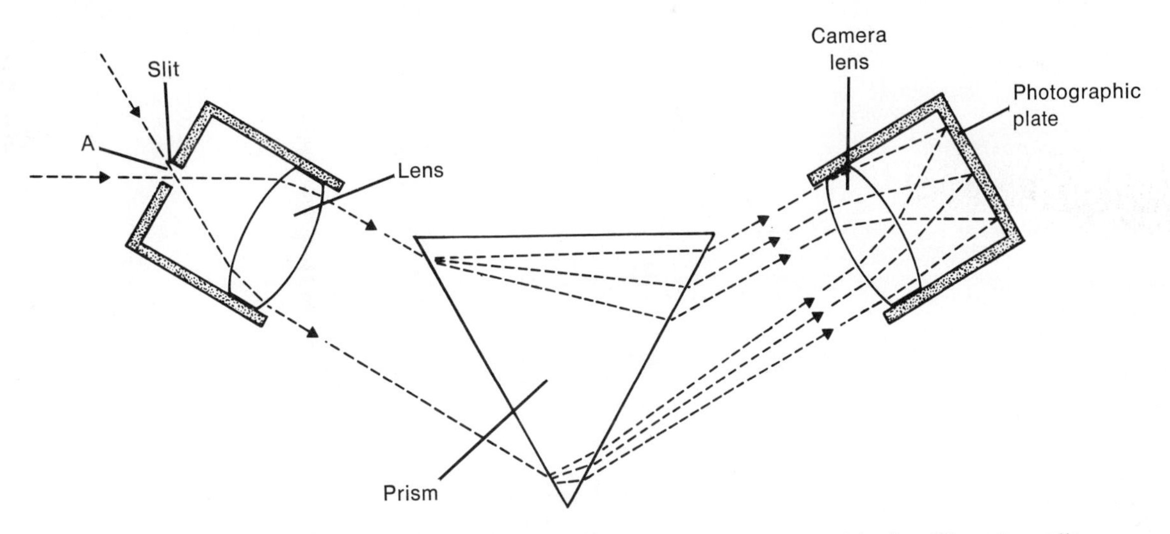

Fig. 5-14. Standard spectroscope. Incoming starlight is focused at point *A* by the objective. The prism splits light into a spectrum that can then be photographed for later study.

trum is formed, but from point *B* a dark-line or absorption spectrum is produced because the continuous spectrum has been depleted of the specific wavelengths of light characteristic of the gas. From point *C*, only the light emitted by the gas cloud is seen, forming an emission or bright-line spectrum.

Spectroscopes

A spectroscope is an instrument that (1) disperses light into its individual wavelengths or colors, (2) brings the resultant spectrum to a sharp focus, and (3) gives the spectrum appropriate size.

The two dispersing media that are commonly used are the prism and the diffraction grating. A prism is a piece of highly homogeneous glass usually cut into the shape of an equilateral triangle, with angles of 60°. Fig. 5-14 illustrates a standard spectroscope containing a 60° prism. A diffraction grating, useful in spectroscopes for observing a wider range of wavelengths, may take several forms. The simplest type consists of a set of fine wires spaced at very close, regular intervals; The most complex type of diffraction grating has at least 15,000 grooves/inch ruled into aluminum deposited on glass.

The spectroscope is mounted on a reflecting telescope

at one of its four basic focal arrangements. The Coudé focus is most important in spectroscopy because the spectroscope remains fixed as the telescope turns. At the other foci, however, the spectroscope is prone to flexure because it hangs on the moving telescope at various angles to the pull of gravity.

Spectroheliographs

The spectroheliograph is a specialized instrument that photographs images of the sun at specific wavelengths of light (Fig. 5-15). A narrow-slit, high-dispersion spectroscope is used, and a second slit is placed at the focus of the instrument exactly at the position of the strongest accessible spectral line, such as hydrogen or calcium. A photographic plate behind the second slit gives a one-wavelength image of only this narrow slit-section of the sun's disc. By moving both the objective lens of the telescope and the photographic plate at the same speed, a one-wavelength image of the entire solar disc is built up.

Photometry, measurement of starlight

In the latter half of the eighteenth century William Herschel devised an approximate method of stellar photometry. His technique was based on the light-gathering power of a telescope, which is directly proportional to the aperture or diameter of the mirror or lens. The effective aperture of a telescope can be changed easily by using a diaphragm. Herschel, for example, noted that the dimmer of two stars seen in the eyepiece of a telescope became barely visible when the instrument was stopped down to a 10-inch aperture. Meanwhile, the brighter star could barely be seen when the telescope was closed down to a 5-inch diameter. The two stars, then, appeared equally bright. Since only one fourth as much light enters the 5-inch telescope as the 10-inch telescope, Herschel reasoned that the brighter star must be sending four times as much light to earth as the dimmer star. In this way, Herschel gauged the brightness of different stars and in effect developed this new branch of astronomy.

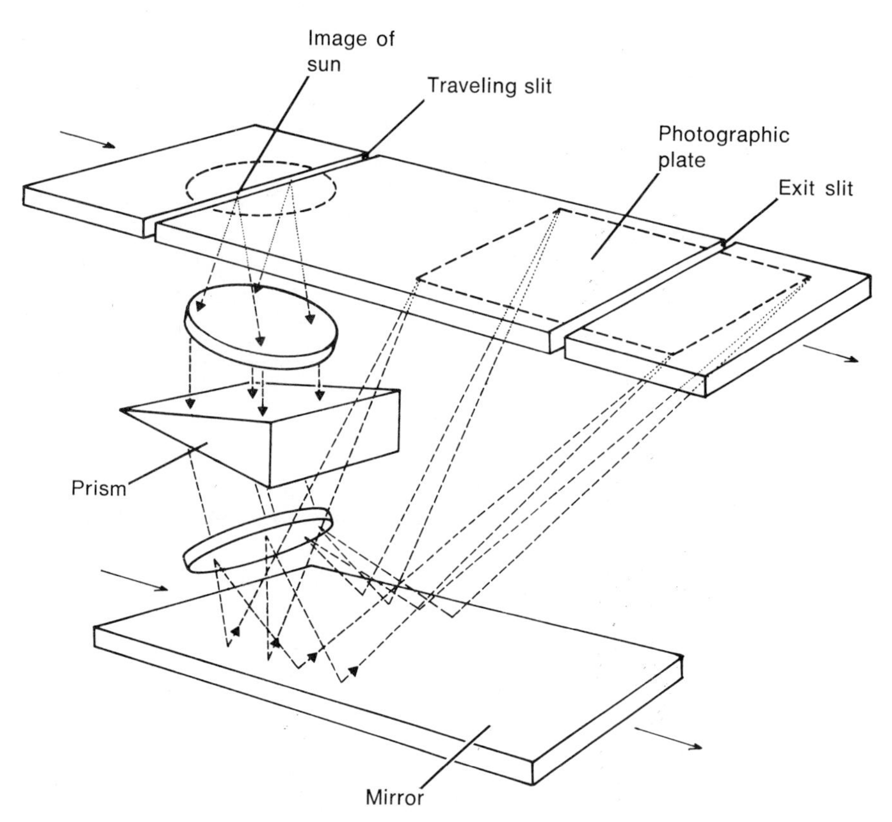

Fig. 5-15. Spectroheliograph. This instrument photographs those emissions of the sun that lie in one particular wavelength. As light from the sun enters a moving slit, it is dispersed to its spectrum of wavelengths either by a diffraction grating or, as shown here, by a prism. Desired wavelength is isolated at the exit slit. As the traveling slit scans the sun to build up an image, the exit slit follows it, leaving a trail on a photographic plate that gradually accumulates into a picture of that particular wavelength.

Photographic photometry

Photographic photometry is a technique that involves comparing the sizes and degrees of blackness of star images on photographic negatives. Time-exposure photography has the same effect as increasing the aperture of the telescope because light can be accumulated on photographic film. Thus by comparing star images and ranking them according to the lengths of the time exposures used, approximate measurements of their brightnesses can be obtained. Photographic photometry is prone to a number of errors, however, that are chiefly the result of atmospheric turbulence. It has for the most part been replaced by more precise techniques.

Photoelectric photometry

More exact measurements of the intensity of starlight are obtained through the use of an instrument called a photoelectric photometer. After coming to a focus at the telescope, the light from a star is passed through a small hole in a metal plate and allowed to strike the surface of a photomultiplier, which consists of photoelectric cells that convert light into an electric current that is then amplified, recorded, and read.

Through the use of photoelectric photometry the problem of comparing star images is reduced to one of measuring and comparing electrical currents—a relatively simple task that can be done accurately. In principle, then, there is only one factor limiting the precision of photoelectrically obtained measurements of starlight intensity; this is the accuracy of statistics. Theoretically, accuracies of better than ten-thousandths of a magnitude are obtainable for bright stars. In practice, however, the earth's atmosphere severely limits routine photoelectric measurements to precisions of about ± 0.01 magnitude. Even with such limitations, it is possible to photoelectrically measure the intensities of stars fainter than those that can be photographed with a given telescope.

ASTRONOMICAL PHOTOGRAPHY

Spectroscopy and photometry alone could never have led to the many discoveries they made possible without the invaluable assistance of photography. Photography provided an ideal means of bringing these two sciences to bear on a huge field of scientific enquiry that would have been futile or impracticable without it. It also enabled astronomers to overcome two major obstacles to their progress—the sheer number of stars and the faintness of the light received from them.

The first successful type of photography, called the daguerreotype, was produced in 1839 by the French artist Louis Jacques Daguerre (1787-1851). Daguerre's invention was based on the less successful work of the French inventor Joseph Niepce (1765-1833). The daguerreotype process consisted of making highly polished, silver-coated copper plates sensitive to light by subjecting them to fumes of iodine. The plates were then exposed to light for a period of time ranging from a few seconds to a few minutes. The image on the plate was then made visible by developing the plates in subdued light with mercury vapor. A quick wash in sodium hyposulfite (hypo for short) and a thorough rinse in water completed the operation.

Aware of the potentials of the new invention, astronomers took an intense interest in photography from the very beginning. In fact, one astronomer presided over its birth while another was present at its christening. Francois Arago, a French astronomer-politician, made the initial public announcement of the invention before a meeting of the Academie des Sciences on January 7, 1839, and gave a complete description of Daguerre's process on August 19 of the same year. John Herschel, the British astronomer and son of William Herschel, coined the term "photography" for the new art.

As strange as it may seem, the first celestial photographs were not made by astronomers. Henry Draper, a New York physician, photographed the moon in 1840. Then Lerebours, a French optician, daguerreotyped the sun in 1842, and George W. Whipple, a professional photographer working under the direction of William Bond at the Harvard Observatory, made the initial photograph of a star in 1850. Finally, photographs of Jupiter and Saturn were made by the Frenchman W. de la Rue in 1857. The first systematic application of photography to extended programs of research in astronomy dates back to the period of 1875 to 1880.

While progress was being made in using photography to study the appearance and positions of the stars, more and more attempts were being made to apply it to the spectral study of their light. For example, as early as 1843 J. W. Draper registered part of the solar spectrum on a photographic plate, and in 1872 Henry Draper succeeded in photographing the spectrum of the star Vega. In 1882, E. C. Pickering of the Harvard Observatory began serious work in stellar spectrography, which culminated in the publication of the *Henry Draper Catalogue* between 1918 and 1924. This work listed the spectral classification of more than 225,000 stars brighter than the eleventh magnitude.

Indeed, photography so revolutionized astronomy that the astronomer can no longer be pictured as sitting for long hours peering into the eyepiece of his telescope. Great telescopes are now used almost exclusively as photographic and photoelectric instruments. Today's astronomer thus spends only a few nights at the observatory performing direct photography, spectral photography, or photoelectric studies. This brief time period is then followed by weeks of laboratory analysis of his data.

CAUTION

An inherent danger exists in the world of instrumentation. Our concept of the universe is apt to be limited or even falsified by the imperfections of our senses and the incompleteness of our knowledge. Even with a realization of these shortcomings, we console ourselves with the belief that the marvels of modern science can make good the deficiencies. Confucius taught that the greatest difficulty involved in hearing was the ear, and the greatest obstacle to seeing was the eye. Plato likened the universe that we are aware of with our senses to shadows on the walls of a dimly lit cave.

Such teachings were an attempt to show that the world of phenomena was only an imperfect and highly misleading manifestation of that which lay beneath. In effect, then, we are only programming ourselves for the expected.

6 *a dwarf among giants*

The sun

Soul of this world, this universe's eye,
No wonder some made thee a deity.
Had I not better known, alas, the same had I.

Contemplations
Anne Bradstreet

The sun is the most underrated and overlooked celestial object visible from earth. Although it faithfully rises above the eastern horizon each morning, its appearance is generally greeted with little more than a yawn. On a clear day most of us give the sun no more thought than whether we will need sunglasses or suntan lotion; on a cloudy day we very often thoughtlessly comment that the sun is not shining at all. Thus we are complacent about the sun's true character and vital importance.

How different, however, was the attitude of many ancient people. They saw the sun as a god to be feared, worshiped, and placated and even offered human sacrifices in their blind allegiance. The early Egyptians believed that their sun god Ra actually sailed across the sky in a boat by day and beneath the earth by night. The ancient Greeks called their sun god Apollo, while the early Slavs named it Bogyarilo (God of Brightness). Later, North American Plains Indian tribes such as the Crow, Arapaho, and Cheyenne performed ceremonial sun dances.

FORMATION AND EVOLUTION OF THE SUN

The sun is our nearest star. Although the daylight now entering your window left the surface of the sun just 8 minutes ago, its energy was created deep in the sun's interior long before human life evolved on earth. Where did the sun come from? How long will it continue to shine? What will eventually happen to it? These are but a few of the many questions that have intrigued people since they first gazed in wonder at the sun.

The sun is believed to have formed some 4.5 to 5 billion years ago through the process of gravitational contraction from a huge mass of tenuous interstellar gas and dust. Why the interstellar material began its contraction still remains a mystery, but that it did occur is virtually beyond refute. As gravity contracted the mass over hundreds of millions of years, the pressure and temperature increased at its center until it finally reached a critical temperature of 20 million degrees F (11 million degrees C). At that moment, the solar furnace was lighted, and a star was born as the sun began to shine. Had the sun been more massive, it would have reached the star stage more quickly and its evolution would have proceeded at a more rapid rate. Once the solar furnace was lighted, a battle immediately began that has continued for billions of years. The foes? Gravity and radiation pressure. *Gravity* is the omnipresent force that attempts to collapse the sun; *radiation pressure* is the force exerted by outward-flowing energy that tends to cause the sun to expand. Although the battle has not yet ended, we can guess at the eventual outcome: gravity is expected to emerge as the final victor.

During its early formative period the sun was probably larger, dimmer, and cooler than the dwarf star it is today. The infant sun was red, in contrast to its present yellow-white appearance. With the continued contraction of its gases, however, the sun decreased in size while increasing both in brightness and temperature (internal and surface). The sun became a normal star when gravity and radiation pressure reached a temporary state of balance. For us on earth this means that the sun is apparently unchanging in size, mass, temperature, energy flow, and brightness. The seemingly static nature of the sun has undoubtedly led to our complacent attitude. Certainly a variation in some aspect of the sun such as size, for example, would attract and hold our attention.

Ironically, the sun is in a constant state of change. It is evolving before our eyes but so slowly that we are unaware of the transformations taking place. It has been estimated that the sun has been in more or less its present condition, increasing slowly in brightness, for at least 4.5 billion years—nearly all of its existence. Before undergoing any obvious alterations, the sun is expected to remain unchanged for another 4.5 billion years.

The sun's long period of relative stability will end when hydrogen, the fuel of the solar furnace, is exhausted. At that time the size of the sun will begin to increase, a process that will continue for millions of years as the solar furnace expands outward in its search for more fuel. In this bloated stage the sun will be known as a *red giant star,* with a mass similar to that it has at present but with an enormously increased volume, lower surface temperature, and lower density. From earth the sun will thus loom across more than 30% of the sky, producing on this planet an environment of evaporating oceans at temperatures that would melt lead. The sun will remain in the red giant stage until most of the hydrogen in its outer portions is consumed.

After the red giant stage, the sun will still be far from dead. During the course of a few hundred million years, gravity should pull the sun's material toward its center, eventually decreasing its size to nearly that of the earth. Under gravitational contraction once again, the sun's brightness is expected to decrease while its surface and interior temperatures increase. The sun in this collapsed form will have reached the *white dwarf stage* of evolution; it should remain in this stage for billions of years.

If we could return to the earth, which would be a charred, lifeless cinder, we would see the sun as only a tiny point of light in a sky almost void of other visible stars. Most of the stars familiar to us, which are more massive than the sun, would have previously ended their normal periods and faded from view. Of those stars less massive than the sun and thus evolving at a slower rate, only a few would be near enough to be seen with the unaided eye.

NATURE OF THE SUN

When projected onto a screen by a telescope, the image of the sun appears deceptively quiet and smooth—almost featureless (Fig. 6-3). Our star, however, is an inferno seething with unimaginable turbulence; it is indeed godlike in its power over the other members of the solar system (Fig. 6-1).

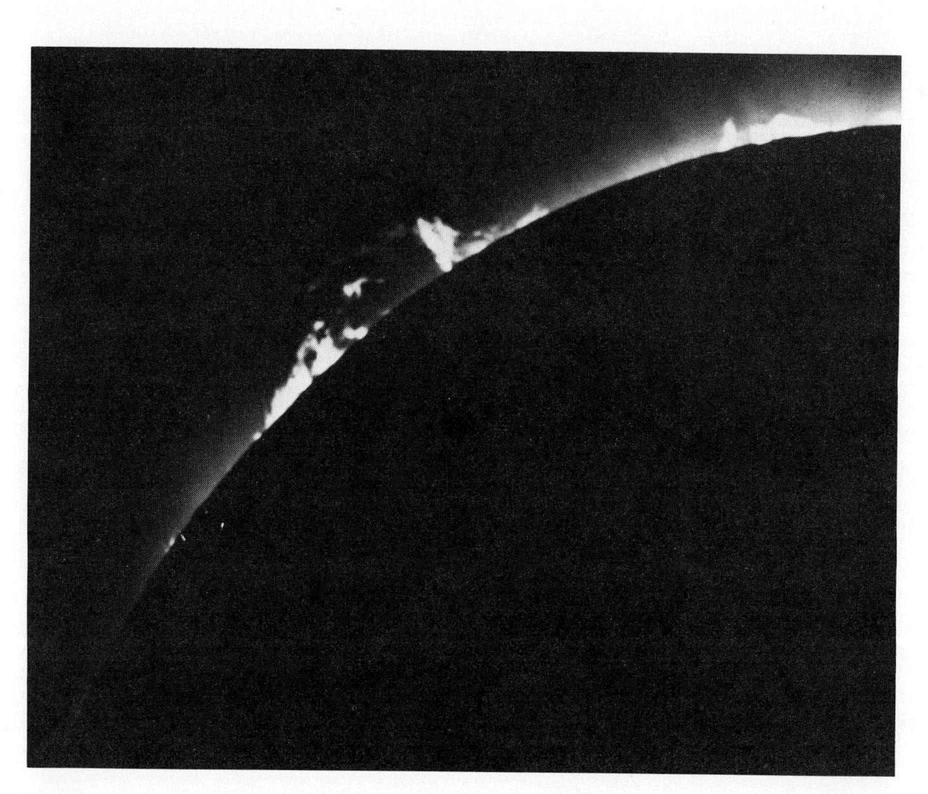

Fig. 6-1. The sun's distance from earth makes its surface appear smooth and conceals the unimaginable amount of activity occurring behind its calm facade. (Courtesy University of California Lick Observatory, Santa Cruz, Calif.)

Distance

From our point of reference here on earth, the sun is very near when compared to the distance of other stars. The sun is 1 AU, or 93 million miles, away from the earth. The next nearest star, Proxima Centauri, is 25 trillion miles away. On a reduced scale in which the sun is represented by a ball 6 inches in diameter, the earth is 18 yards away and Proxima Centauri is more than 2700 miles distant. Is it any wonder, then, that the sun is the only star whose disc and features can be seen even in the largest telescopes? All other stars appear only as points of light.

Size and shape

The average angular diameter of the sun on the celestial sphere has been accurately measured at 32 seconds of arc—almost the same as that of the moon. Combining its average angular diameter with its average distance gives the sun a computed linear diameter of 864,000 miles—108 times the diameter of the earth.

For an object its size, the sun rotates very slowly; as a result, it shows no oblateness. It is therefore almost a perfect sphere.

Mass and density

Mass is the most essential characteristic of a star because it determines to a great extent its other properties and the course and rate of its evolution. For many stars, however, mass is the most difficult quantity to measure. This is not the case for the sun, whose mass may be deduced from its gravitational effects on the earth and other planets of the solar system. It would take more than 330,000 earths to balance the sun's mass on a scale and more than 1 million earths to equal its volume.

The average density of the sun, computed by dividing its mass by its volume, is 1.4 gm/cm^3—approximately one fourth the density of the earth and about the same as that of Jupiter.

Composition and state

The sun is composed of essentially the same chemical elements as the earth (Fig. 6-2), although its percentage composition differs drastically (Table 6). While hydrogen and helium make up 99% of the sun by both weight and volume, 68 other elements have also been definitely identified. The remaining 22 naturally occurring elements have not as yet been found in the sun. Either we have not developed the techniques necessary for their discovery, or these elements are present in such small amounts that the hostile environment of the sun has rendered them undectectable.

The temperature of the sun is so high that all of its constituent elements have been vaporized, including those such as iron that are normally solid on earth. The sun, then, is completely gaseous.

Temperature

When standing in sunlight it becomes immediately obvious that the sun is hot. But why is it hot? How hot is it? Weight pressure that is exerted by the tremendous mass of the sun pressing down on its interior causes the sun's high internal temperature. This temperature, variously estimated at between 20 to 40 million degrees F (11 to 20 million degrees C), produces reactions between constituent elements. These reactions in turn release energy, which then produces radiation pressure from within that balances the weight pressure from without. If the sun's interior were any cooler, it would immediately decrease in size in order

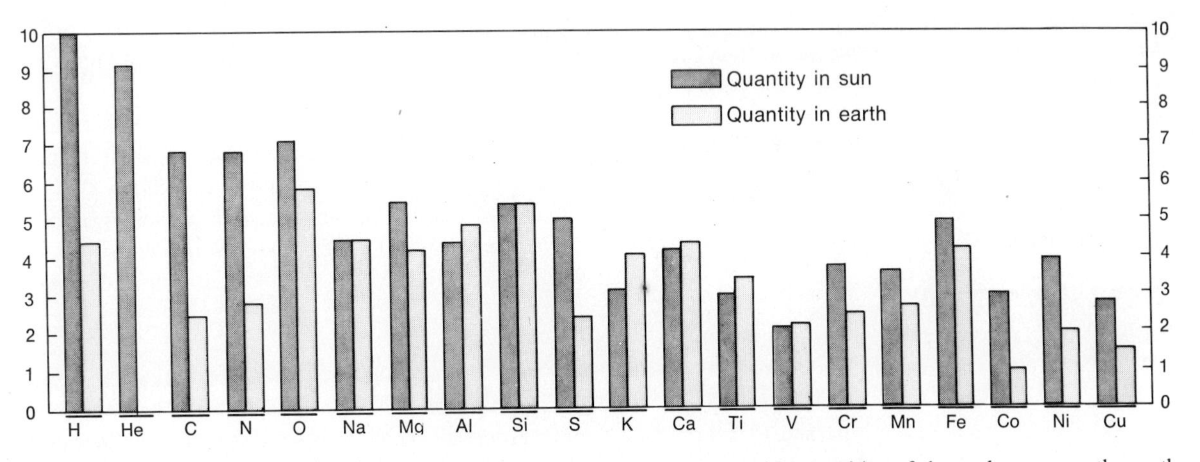

Fig. 6-2. Comparison of quantities of selected elements present in the sun with quantities of these elements on the earth.

Table 6. Elemental composition of total atoms of selected systems expressed as percentages*

Earth's crust†		Moon‡		Meteorite		Human body		Sun§		Universe	
O	46.60	O	40.00	O	32.00	H	63.00	H	92.50000	H	91.000
Si	27.20	Si	19.20	Fe	28.80	O	25.50	He	7.30000	He	9.100
Al	8.13	Fe	14.30	Si	16.30	C	9.50	O	0.06800	O	0.057
Fe	5.00	Ca	8.00	Mg	12.30	N	1.40	C	0.03700	N	0.042
Ca	3.63	Ti	5.90	S	2.12	Ca	0.31	N	0.00930	C	0.021
Na	2.83	Al	5.60	Ni	1.57	P	0.22	Ne	0.00740	Ne	0.003
K	2.59	Mg	4.50	Al	1.38	Cl	0.03	Mg	0.00440	Si	0.003
Mg	2.09	Na > Cr > Mn > S		Ca	1.33	K	0.06	Si	0.00290	Mg	0.002
Ti > H > C						S > Na > Mg		Fe	0.00185	Fe	0.002

*Modified from Selbin, J.: J. Chem. Educ. **50**:306, 1973. These values vary somewhat from source to source, but in general the order of the elements is constant.

†The values for Fe, Mg, S, and Ni would rise considerably if the earth sample were taken down to its core.
‡Mean figures for the Apollo 11 lunar landing site.
§Percent by volume.

to increase pressure to reach an internal temperature high enough to create a state of balance between gravity and radiation pressure. Therefore in order to exist, a gaseous body the size of the sun must be very hot.

The temperature of the sun varies with depth—it is highest on the inside and lowest at the surface. The sun really has no surface in the same sense that a solid object such as the earth has a crust. The outer portion of the sun that we can normally see is called the *photosphere,* and this is the nearest thing to a surface that the sun does have. The photosphere is where solar energy finally escapes into space, and it has the lowest temperature of any portion of the sun—7600° to 11,700° F (4500° to 6800° C). The photosphere has the lowest temperature because it is subject to the lowest weight pressure.

The photosphere of a star is very important, for it acts as a safety valve, reacting to changes in energy flow from within. If the escape of energy at the photosphere of the sun were greater than the flow from the interior, the surface would simply cool. On the other hand, if the loss of energy at the photosphere were less that the outward flow, the surface would heat up until a balance was again reached. A third type of stellar equilibrium is thus introduced, explaining why the surface temperature of a star such as the sun must, at any given time, have a certain value.

Surface gravity and escape velocity

The acceleration due to gravity at the photosphere of the sun is 28 *g*. Disregarding air friction, on earth an object dropped from an altitude of 1000 feet would hit the ground in 8 seconds, traveling at a speed of ½ mile/sec (1800 miles/hr). Dropped from the same height above the solar photosphere, the same object would have penetrated 10 miles below the surface and would be traveling at 30 miles/sec (108,000 miles/hr) after the same elapsed time. In order to leave the sun forever, solar energy or particles must reach a speed greater than 380 miles/sec (1,368,000 miles/hr). This speed is the solar escape velocity.

Rotation and revolution

The sun rotates from west to east on an axis inclined 7° to the plane of the earth's orbit. As a solid body, the earth has a constant angular rotation rate of 15°/hr regardless of geographic location, so that all parts complete one turn in about 24 hours. The earth's rotation speed, on the other hand, varies from 1000 miles/hr at the equator to 0 miles/hr at the poles. As a gaseous sphere, however, the sun does not have a fixed angular rotation rate. At the solar equator, for example, the sun completes one turn in 24 days, 15 hours, while at its poles the rotation period is 35 days. Solar rotation may be confirmed by observing the changing west to east positions of sunspots (Fig. 6-3).

We often tend to forget that the sun also revolves, but around what? The sun is orbiting the nucleus of an immense star system called the Galaxy or Milky Way galaxy (Chapter 8). Its speed relative to other stars in its neighborhood is 12 miles/sec (43,200 miles/hr). All members of our solar system share in the sun's revolution, going wherever the sun goes.

PARTS OF THE SUN

We may think of the sun as having two basic parts: A *body* and an *atmosphere* (Fig. 6-4). Both terms are perhaps misleading, since the word "body" implies a solid struc-

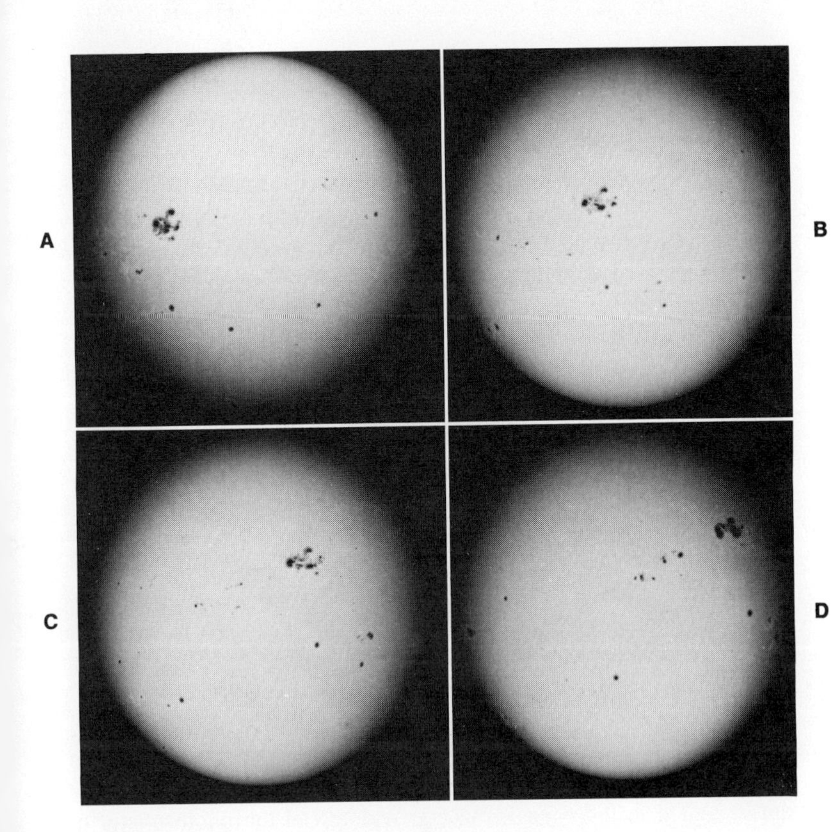

Fig. 6-3. Series of exposures shows solar rotation as plotted by the progression of sunspots across the face of the sun. **A,** August 7. **B,** August 9. **C,** August 11. **D,** August 13. Photographs also show limb darkening, an optical phenomenon caused by the angle at which we view the sun. (Courtesy University of Chicago Yerkes Observatory, Williams Bay, Wisc.)

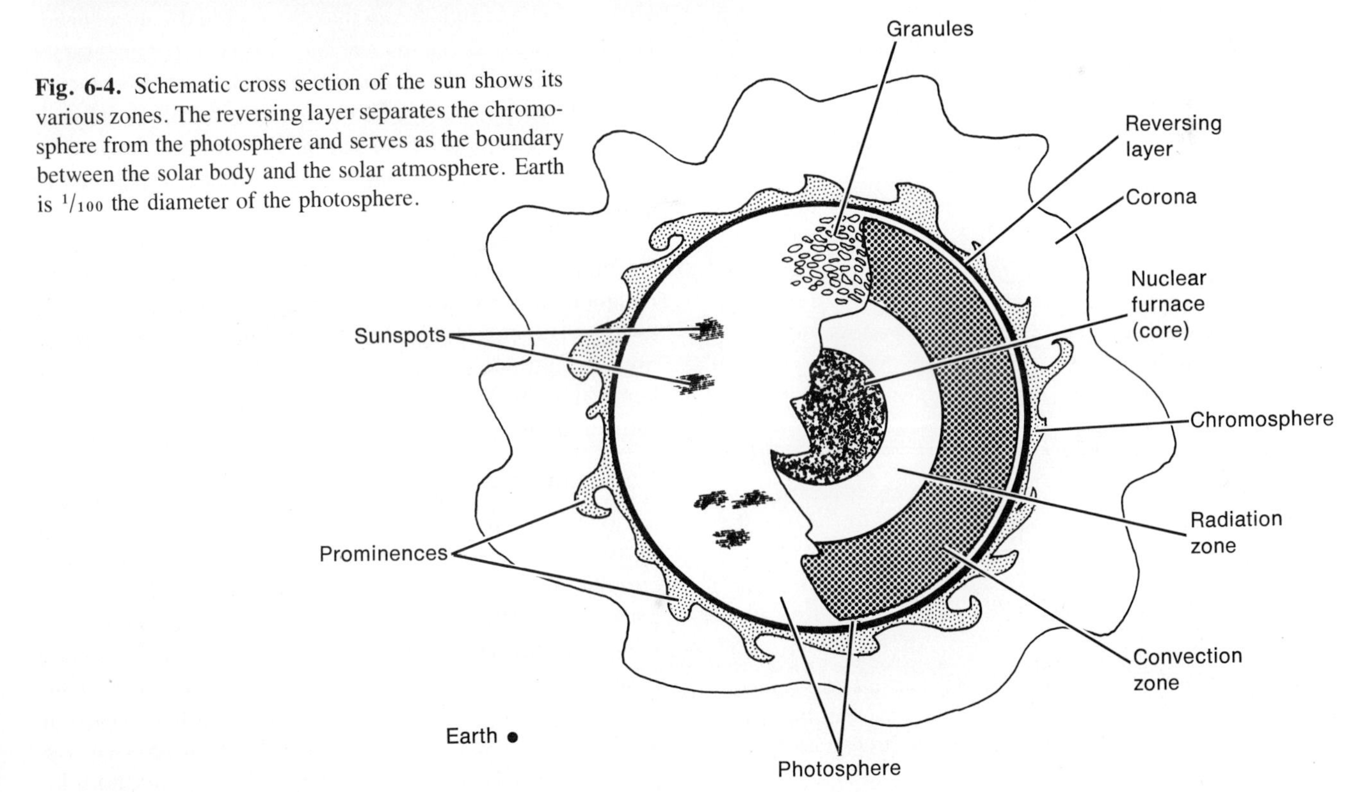

Fig. 6-4. Schematic cross section of the sun shows its various zones. The reversing layer separates the chromosphere from the photosphere and serves as the boundary between the solar body and the solar atmosphere. Earth is $^{1}/_{100}$ the diameter of the photosphere.

ture, and the sun is completely gaseous. When we refer to the body of the sun here, we are discussing the inner portion, the closest thing to a solid portion that our sun has.

The solar furnace

The solar or nuclear furnace is the relatively small heart, or core, of the sun. This is where the sun transforms 564 million tons of hydrogen into 560 million tons of helium each second. The remaining 4 million tons of solar material are converted to energy. The material in the nuclear furnace is called *plasma gas,* sometimes referred to as the fourth state of matter. Plasma gas consists not of the atoms and molecules we are familiar with on earth but of bare atomic nuclei and free electrons rushing wildly through the solar furnace (Fig. 6-5). This condition arises when matter is subjected to pressures and temperatures so extreme that the electrons are completely removed from their normal orbits around the nuclei.

Sun's energy processes

Within the solar furnace, hydrogen fuel is converted into helium by two different means: the *proton-proton process,* commonly known as *hydrogen fusion,* and the *carbon cycle.*

Hydrogen, you may remember, is the most abundant element in the sun. The nucleus of the hydrogen atom is a *proton*. In the plasma gas state of the nuclear furnace the chance collision between two fast-moving protons results in the formation of a nucleus of *deuterium* (heavy hydrogen) and the emission of both a positive electron, or *positron,* and an ultra-small mass called a *neutrino*. The neutrino escapes into space about 2 seconds later, and the positron collides with an electron, resulting in the annihilation of both. However, this collision also releases a parcel (photon) of ultra-high energy called a *gamma ray*. The deuterium nucleus then combines with other protons, growing ultimately into a helium nucleus. The transformations at each step result in the production of gamma radiation (Fig. 6-6).

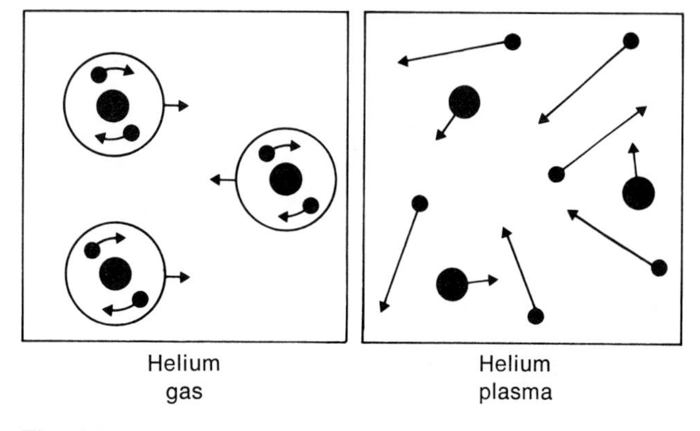

Fig. 6-5. Comparison of helium gas and helium plasma. A molecule of helium gas has two electrons orbiting a nucleus, but in the plasma state, bare atomic nuclei and free electrons rush wildly about.

While the proton-proton process is converting hydrogen into helium, the carbon cycle, a group of reactions involving protons with carbon and nitrogen, is achieving the same end but by different means (Fig. 6-6). It is interesting to note that in the carbon cycle carbon reappears at the end of each cycle. As was the case with the proton-proton process, the neutrinos are lost and the positrons collide with electrons to produce gamma radiation.

In a star like our sun it is believed that hydrogen fusion is the principal energy-producing process and that the carbon cycle is a secondary process. In more massive stars with hotter interiors the carbon cycle prevails, whereas less massive stars with cooler interiors thrive only on the proton-proton process. Regardless of the dominant process, hydrogen in the core of a star is being replaced by stable helium and heavy elements, leading to its eventual doom.

Radiation zone

If the sun were transparent, a photon of energy generated in the solar furnace would exit at the photosphere in about 2 seconds and stream into space virtually unchanged. However, due to the increasing density and temperature toward the sun's center, solar matter obstructs the smooth outward flow of energy. Photons are continually absorbed, reemitted, and reabsorbed along the way, so that an individual photon may require tens of thousands of years to finally reach the photosphere.

Just outside the solar furnace, or core, lies a region encompassing about two thirds of the body of the sun, the radiation zone. Within this zone gamma rays are converted to less energetic, longer-wavelength radiation such as X-rays and ultraviolet rays and are then passed by the turbulent region upward to the photosphere, where the last transformation of energy, mostly into visible light, occurs.

Convection zone

Whenever there are large temperature differences in masses of liquids or gases, vertical flows called *convection currents* are formed in an attempt to equalize the discrepancy. Approximately the outer third of the sun's body consists of a convection zone in which gases are forced to rise in order to pass on the flow of energy from the hotter radiation zone below to the cooler photosphere above. The convection zone also serves another purpose; it removes the products of nuclear transformation (helium) from the solar interior while bringing in fresh fuel (hydrogen). Present calculations suggest that the convection zone must extend to a depth below the photosphere at which the temperature is 3 million degrees F (1.6 million degrees C) and the density is nearly one third that of water. Even though the convection zone accounts for more than half of the sun's volume, it contains only about 2% of the total solar mass.

Photosphere

The photosphere is the outermost portion of the body of the sun, and it is the only part of the entire star that can normally be seen. When we speak of the *disc,* or *surface,* of the sun or say that the sun is rising or setting, we are referring to the photosphere. The photosphere is the only visible part of the sun under normal conditions for at least two reasons. First, the regions of the sun lying outside the photosphere are transparent, while the photosphere itself is opaque. Second, the photosphere is the zone of the sun in which most visible light originates. Our eyes of course can sense only visible light; as a result, we see only the photosphere.

The photosphere is not a surface, it only appears to be. It is actually a relatively thin, gaseous shell no more than 300 miles thick and has no definite boundary. The pres-

$${}_1H^1 + {}_1H^1 \longrightarrow {}_1H^2 + {}_1e^0 \text{ (Positron)} + \text{Neutrino}$$

$${}_1H^2 + {}_1H^1 \longrightarrow {}_2He^3 + \text{Gamma radiation}$$

$${}_2He^3 + {}_2He^3 \longrightarrow {}_2He^4 + {}_1H^1 + {}_1H^1$$

or

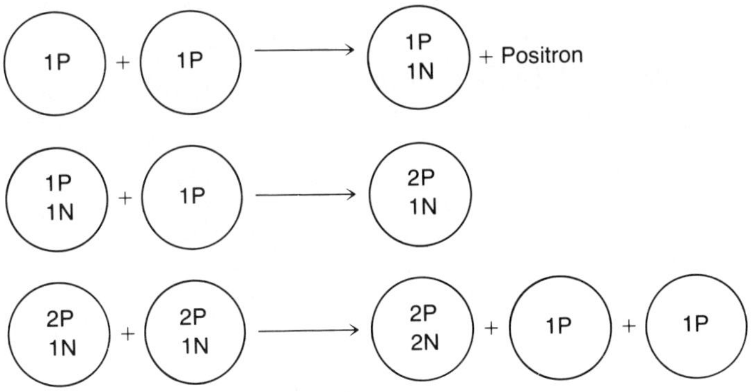

1P + 1P → 1P 1N + Positron

1P 1N + 1P → 2P 1N

2P 1N + 2P 1N → 2P 2N + 1P + 1P

PROTON-PROTON REACTION

$${}_1H^1 + {}_6C^{12} \longrightarrow {}_7N^{13} + \text{Gamma radiation}$$

$${}_7N^{13} \longrightarrow {}_6C^{13} + {}_1e^0 \text{ (Positron)}$$

$${}_1H^1 + {}_6C^{13} \longrightarrow {}_7N^{14} + \text{Gamma radiation}$$

$${}_7N^{14} + {}_1H^1 \longrightarrow {}_8O^{15} + \text{Gamma radiation}$$

$${}_8O^{15} \longrightarrow {}_7N^{15} + {}_1e^0 \text{ (Positron)}$$

$${}_7N^{15} + {}_1H^1 \longrightarrow {}_6C^{12} + {}_2He^4$$

or

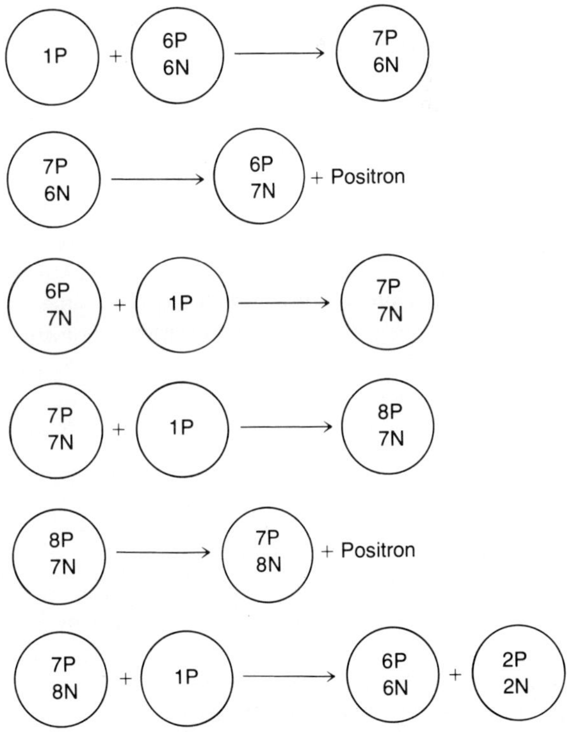

1P + 6P 6N → 7P 6N

7P 6N → 6P 7N + Positron

6P 7N + 1P → 7P 7N

7P 7N + 1P → 8P 7N

8P 7N → 7P 8N + Positron

7P 8N + 1P → 6P 6N + 2P 2N

CARBON-CARBON REACTIONS

Fig. 6-6. Nuclear reactions in the sun.

sure, density, and temperature of the photosphere vary with depth. Over its 300-mile vertical extent, the pressure and density increase by at least a factor of 10, while the temperature varies from 7600° F (4500° C) at the outer zone to 11,700° F (6800° C) inside. A typical point in the photosphere shows a density of about one ten thousandth that of earth's sea level atmospheric density, a pressure of only a few hundredths the atmospheric sea level pressure on earth, and a temperature of 10,340° F (5727° C).

Our knowledge of the solar interior is of necessity based on indirect studies and theoretical considerations. Information about the solar atmosphere must be accumulated during total eclipses of the sun or by the use of special telescopic instruments. Since it is possible to analyze the photosphere directly, it is therefore the best-known part of the sun.

Limb darkening

As we look at the projected image of the sun on the sun screen of a telescope, we can observe that the edge or limb of the photosphere is slightly darker than the central area. This effect is known as limb darkening. It is explained by both the spherical shape of the solar body and the resultant fact that when we view the sun, our line of sight enters the photosphere almost perpendicularly, allowing us to see deeper, hotter regions. Cooler portions of the photosphere appear darker than hotter areas, but this appearance is only relative, for in their own right even the dark regions are very bright (Fig. 6-3).

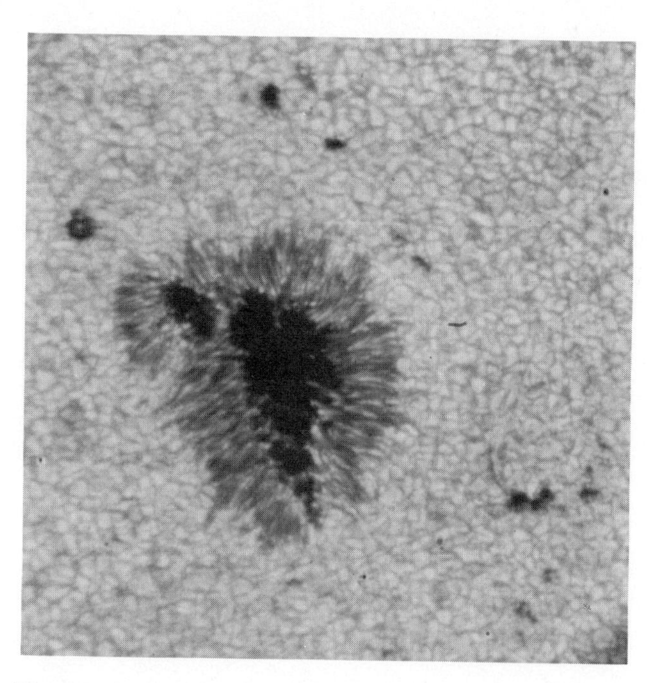

Fig. 6-7. Convective action on the sun's surface causes the beautiful pattern known as granulation, shown here surrounding the large sunspot at center. Columns of hot gases rise from below the photosphere, forming the small, grain-shaped particles. Spreading out and cooling, they sink back toward the surface, taking on the darker colors of the intergranular areas. (Courtesy Princeton University Project Stratoscope [supported by National Science Foundation, Office of Naval Research, and National Aeronautics and Space Administration], Princeton, N.J.)

Granulation

When viewed telescopically or photographically, the photosphere of the sun appears mottled, a phenomenon referred to as granulation (Fig. 6-7). Each individual *granule* resembles a grain of rice and is typically a few hundred miles in diameter. Granules actually are columns of hot gases rising from below the photosphere—the tops of convection currents. When they reach the photosphere the gases spread out, cool, and then sink back below the surface, forming dark intergrannular spaces in the process. Each granule usually lasts only a few minutes before cooling to the temperature of its surroundings.

Sunspots

Sunspots are the first solar feature that were ever noted, for they were observed long before the development of the telescope. However, the fact that they were actually on the sun's surface remained unknown until Galileo's observations in 1610. Sunspots are disturbances that break through the photosphere from below and appear as dark regions on the otherwise bright solar face. The dark appearance of sunspots is due to the fact that they are about 2700° F (1500° C) cooler than the remainder of the photosphere. Sunspots however, are still quite luminous, and if it were possible to remove them from the sun they would shine brightly.

The structure of a well-developed sunspot consists of two distinct parts; a central dark region called the *umbra* and a surrounding striated, less dark zone called the *penumbra* (Fig. 6-8). The motion of gases within a sunspot is complex and still somewhat debatable but appears to consist of both *convective* (vertical) and *advective* (horizontal) flows. Apparently the *penumbra* is a region of rising gases, while the *umbra* is a zone in which gases are sinking. At lower levels the gas seems to flow parallel to the photosphere and away from the center; at upper levels the flow seems to be parallel and toward the center.

Sunspots may range in size from small specks 500 miles across that last for only a few hours to giant regions 100,000 miles in diameter that persist for months and may be seen with the naked eye at sunrise or sunset. Large sunspots are commonly larger than the earth. Sunspots often occur in groups of two to twenty or more. There are a number of sunspots in a group, usually two large ones aligned roughly from west to east, surrounded by the smaller ones. Sunspots may shift their positions slightly on the photosphere relative to other spots, but their primary motion is due to the rotation of the sun.

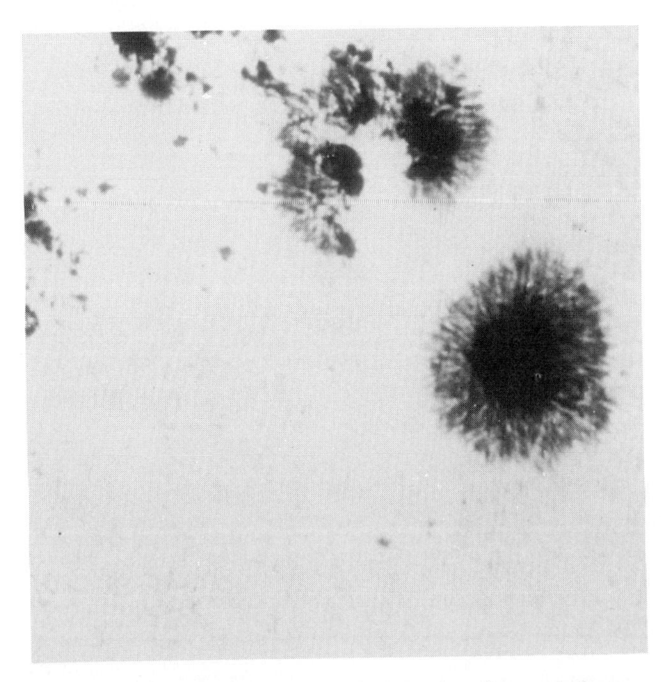

Fig. 6-8. Sunspots. Note the well-defined umbra and the surrounding striated penumbra. (Courtesy University of Chicago Yerkes Observatory, Williams Bay, Wisc.)

The sun has a general magnetic field. This is perhaps not a surprising fact when we consider the immense amount of electrical activity taking place within the sun. Sunspots are the foci of solar magnetic activity, possessing magnetic fields much stronger than the general field. The magnetic fields of sunspots extend from the spots themselves into the surrounding area and often remain after the spot has disappeared. When sunspots occur in pairs one spot has the polarity of a north-seeking magnet and the other has a south-seeking polarity, as though a horseshoe magnet were buried beneath the photosphere under the spots.

Sunspot cycles. Sunspots clearly occur in a cycle that is an average of 11 years in length, although there have been cycles as short as 8 years or as long as 16 years. It should be pointed out here that a single sunspot does not survive throughout the entire cycle. The cycle is based on the observation that short-lived spots are more numerous and are found at different locations at varying times (Fig. 6-9). A cycle begins with the appearance of a few spots or groups of spots near solar latitude 30° N or 30° S. It is rare for a sunspot to occur at latitudes greater than 40°. In time, new spots form at successively lower latitudes and are larger and more numerous. The maximum number of sun-

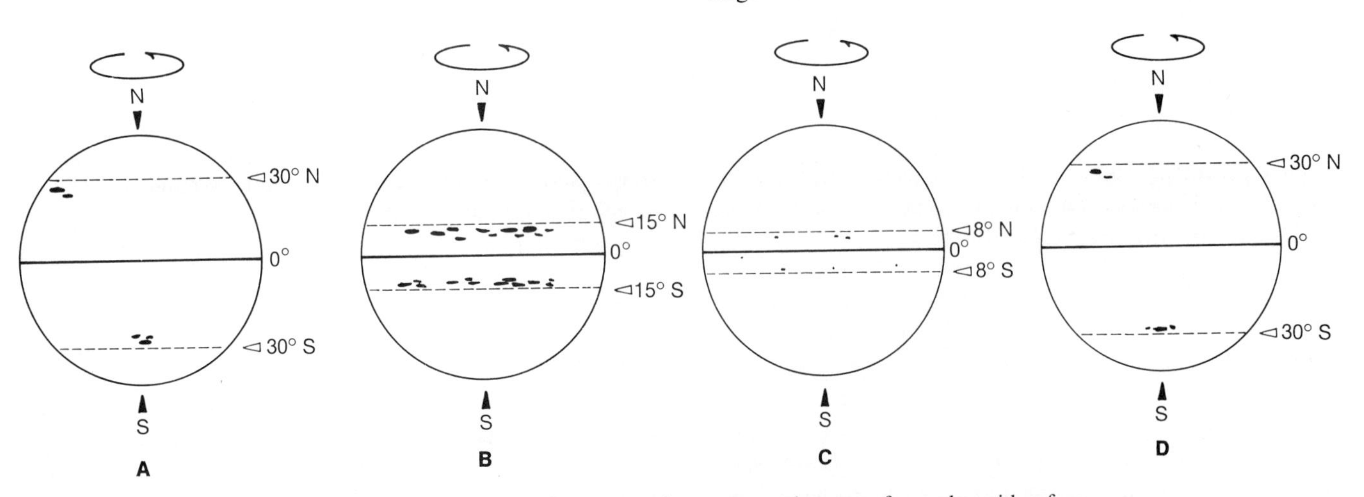

Fig. 6-9. Simplified version of the sunspot cycle. **A,** The sun as it appears at the start of a cycle, with a few spots near 30° N and 30° S. **B,** After 4 years the sun reaches sunspot maximum, when the spots cluster around 15° N and 15° S. **C,** After 11 years, the sunspot band has dropped to a range of only 8° in each direction; these sunspots represent sunspot minimum. **D,** The sun prepares to start a new cycle with the spots again clustering around the 30° N and 30° S bands.

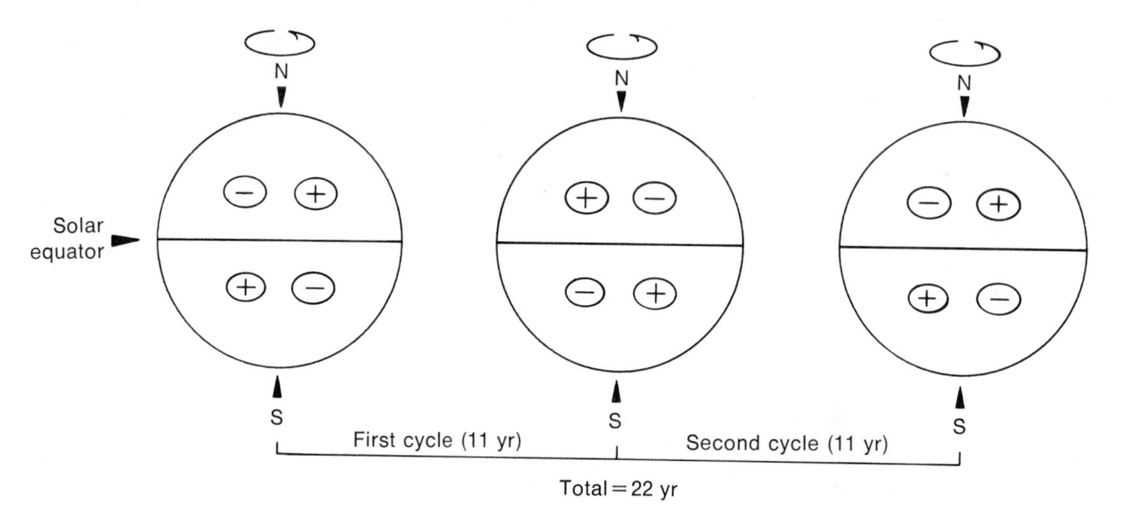

Fig. 6-10. Changing magnetic polarity of sunspots. Note that the pattern north of the sun's equator is always the reverse of the pattern south of the equator.

spots appears 4 years after the appearance of a few spots at high altitudes. *Sunspot maximum* usually takes place near latitudes 15° N and 15° S. After maximum, the number of spots and groups slowly declines during approximately the next 7 years until there are a few small spots near 8° N and 8° S. This period is known as *sunspot minimum*. Usually before the low-latitude spots fade away completely, new spots form near latitudes 30°, marking the start of a new cycle.

In addition to the visual sunspot cycle discussed above, there is also a magnetic sunspot cycle. Many people believe this represents the true cycle of solar activity. It was pointed out earlier that a pair of sunspots show opposite magnetic polarity. A remarkable fact is that the polarity of sunspot pairs in the northern hemisphere of the sun is always opposite the polarity of sunpot pairs in the southern hemisphere (Fig. 6-10). Consider, for example, a pair of sunspots located in the northern hemisphere of the sun. The spot at the right corresponds to the north magnetic pole, and the spot at the left corresponds to the south magnetic pole. At exactly the same time a sunspot pair in the southern hemisphere will have the south magnetic pole at the right and the north magnetic pole at the left. Even more interesting is the reversal of polarities, which takes place with the start of a new 11-year cycle. In the northern hemisphere the spot to the right will then correspond to the south magnetic pole and the spot to the left will correspond to a north magnetic pole. In the southern hemisphere, on the other hand, the spot to the right will have a north-seeking polarity while the spot to the left will have a south magnetic polarity. Two complete 11-year cycles are thus required before magnetic conditions return to their original state. Thus the magnetic sunspot cycle lasts for an average of 22 years.

Reversing layer

The reversing layer is a transparent mass of gases that extends for a few hundred miles above the photosphere. This layer may be considered the dividing line between the solar atmosphere and the solar body, although many scientists include it as part of the chromosphere. The reversing layer is less dense than the photosphere and has a slightly lower temperature. As a result, the gases of this layer selectively absorb certain wavelengths of light coming from below, causing dark lines to appear in the familiar continuous solar spectrum. Just as the moon covers the photosphere during a total solar eclipse, the continuous spectrum of colors is removed, and for an instant the dark lines produced by the reversing layer appear as they really are—bright and colorful. The split-second reversal of the spectrum gives the reversing layer its name.

Chromosphere

The chromosphere is a hot, tenuous, atmospheric shell of transparent gas, extending to a height as great as 10,000 miles above the photosphere. Temperatures increase and density decreases rapidly in the chromosphere. Temperatures range from 10,000° F (5400° C) at the reversing layer to 180,000° F (100,000° C) near the region above. The chromosphere is so named because of its brilliant red color, which is due to the emission of light from hydrogen gas.

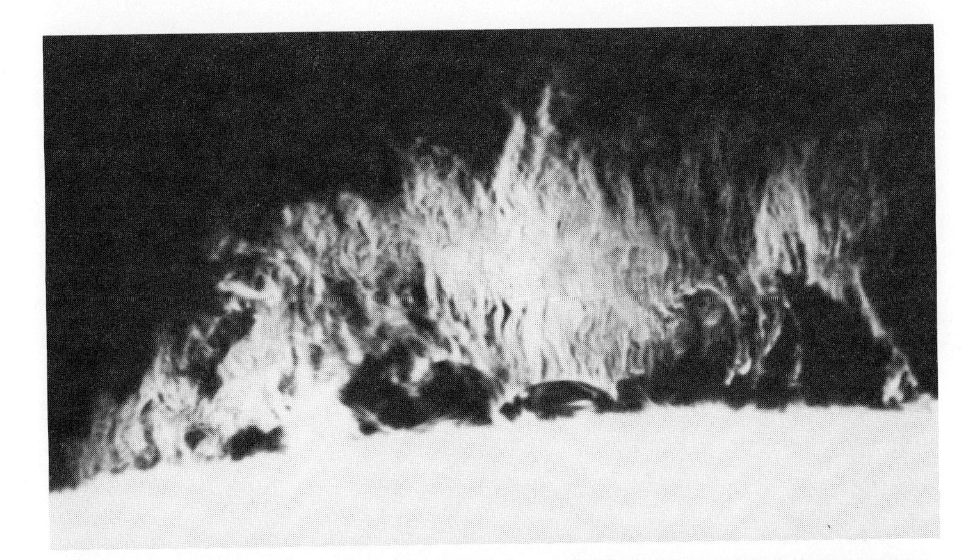

Fig. 6-11. At about 600 miles, the chromosphere separates into relatively cool, high-density regions with higher, pointed areas separating them. Spicules, as the points are called, resemble a prairie fire. (Courtesy Sacramento Peak Observatory, Air Force Cambridge Research Laboratories, Sunspot, N.M.)

Fig. 6-12. Flares reach their maximum brightness in only a few minutes, although they may last several hours before fading entirely. (Courtesy Sacramento Peak Observatory, Air Force Cambridge Research Laboratories, Sunspot, N.M.)

Spicules

Beginning at an altitude of about 600 miles, the chromosphere separates into relatively cool, high-density regions with small, spike-shaped peaks between them (Fig. 6-11). These peaks, called spicules, well up from the lower regions of the chromosphere and either fall back or disappear into the corona at a height of about 20,000 miles. They have been likened to the appearance of a burning prairie. Each spicule, about 500 miles wide, shoots upward at a rate of nearly 20 miles/sec (72,000 miles/hr) and lasts only 5 to 10 minutes. Spicules in the chromosphere seem to be related to granules in the photosphere below.

Plages and faculae

Special photographs of the photosphere that are taken with a spectroheliograph often reveal bright patches called plages in the magnetic fields of sunspots. Light plages that can be seen without the use of special photographic techniques are known as faculae. These clouds of luminous gas are only slightly hotter than their surroundings. Therefore they can best be seen near the limb or edge of the sun where the photosphere is not as bright. The average lifetime of plages and faculae is about 2 weeks; those that persist, however, often herald the birth of a sunspot. Although seen against the photosphere, these phenomena are believed to originate in the chromosphere.

Flares

Occasionally a small region of the chromosphere brightens to unusually high intensity, an event called a flare. Flares reach maximum brightness in only a few minutes, although they may persist for hours before fading out (Fig. 6-12). Enormous quantities of X-rays, ultraviolet light, and radio energy are released during flare activity, as are streams of atomic nuclei. Energy and particles from solar flares affect the magnetic qualitites of the earth, causing compasses to react strangely and disturbing the earth's ionosphere, thus producing radio interference. They also may present a temporary radiation hazard to astronauts traveling in space.

Neither the energy source nor the triggering mechanism of flares has been determined. They are always associated with faculae, and 99% occur near a sunspot. At sunspot maximum they generally form at a rate of five to ten each day. Flares often form at the boundary between the north- and south-seeking polarity regions of a pair or group of sunspots. It is as though vast amounts of energy had been compressed into a relatively small area by the magnetic fields of the spots.

Corona

The third and outermost layer of the solar atmosphere is aptly called the corona, which is Latin for "crown." While the corona is the most extensive portion of the sun, it is less dense and massive than any other region. The temperature of this tenuous layer may be as great as 1.8 million degrees F (1 million degrees C). It is difficult to define the spatial limits of the corona because the part close to the photosphere is visible, whereas the portion at greater distances is not. The visible corona is chiefly illuminated by the light scattered by atomic particles that constantly stream upward from the photosphere. X-rays and visible light, however, are also produced in this region. The visible corona may only be viewed during total eclipses of the sun or by the use of a *coronagraph,* an instrument designed for this purpose (Chapter 5). Although it occasionally extends more than 1 million miles into space, the visible corona actually varies in shape and extent with the sunspot cycle. At sunspot maximum the visible corona is nearly circular and quite extensive. At sunspot minimum, on the other hand, it is elongated in the equatorial latitudes but is not as far-reaching elsewhere. The invisible corona extends at least as far as earth and may reach far beyond Pluto. This expanding envelope consists of rapidly moving swarms of charged particles (electrons and protons), usually referred to as the solar wind.

Prominences

Prominences are the most striking aspect of the visible corona and in many respects of the entire sun itself. They are cool, dense regions of the corona that apparently have formed from the cooling of coronal material. Initially prominences were thought to consist of material thrown upward from the photosphere. However, time-lapse photography has shown that prominences more often than not have a downward motion.

There are three types of prominences: quiescent, loop, and eruptive (Fig. 6-13). *Quiescent* prominences may remain nearly stationary for hours or even days, extending to tens of thousands of miles above the photosphere. *Loop*

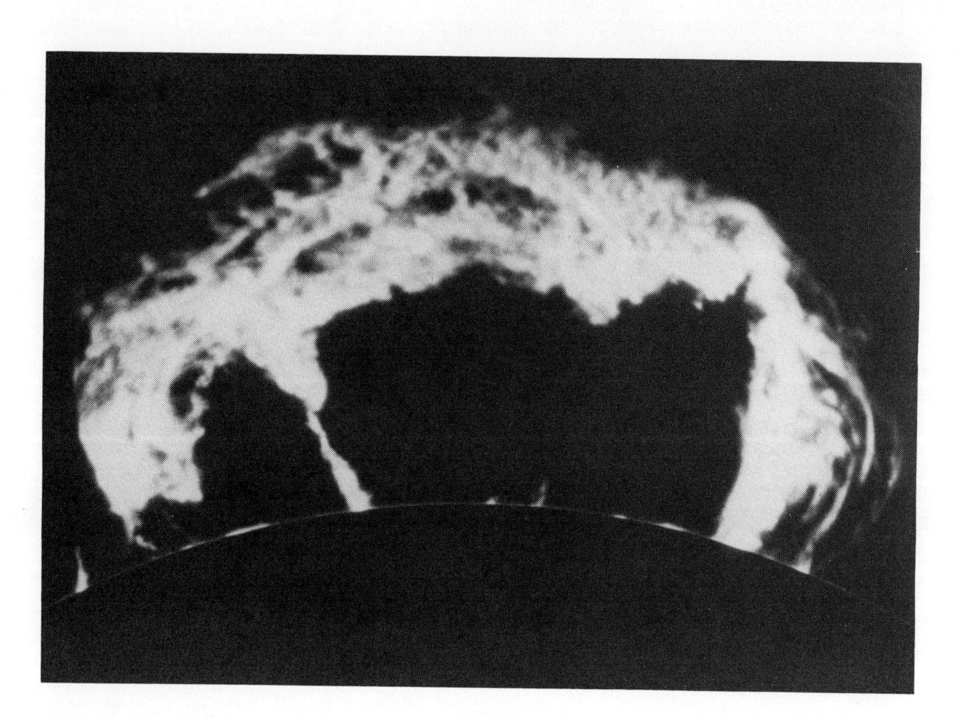

Fig. 6-13. Loop prominences take their shape because of the presence of strong magnetic fields. (Courtesy Sacramento Peak Observatory, Air Force Cambridge Research Laboratories, Sunspot, N.M.)

prominences are more active than the quiescent type and almost always take the shape of a loop or arch whose material exhibits motions suggesting the presence of strong magnetic fields. The most spectacular of the three types is the relatively rare *eruptive* prominence, which may send material upward at speeds as great as 2200 miles/sec (8 million miles/hr). Such high-speed eruptive prominences exceed the escape velocity of the sun, and their matter becomes part of the outward-moving solar wind.

THE SUN'S INFLUENCE ON THE EARTH

That the sun influences the earth is perhaps a truism. Without the sun there probably would not be an earth, for all of the energy we possess on this planet came directly or indirectly from the sun. The sun controls the earth's atmosphere and makes possible life as we know it. Its energy affects long-range radio communications, causes surges in power lines, brings about fluctuations in the earth's magnetic fields, and produces the beautiful auroras of the polar regions—all with a period equal to the 11-year sunspot cycle.

Other effects reportedly related to the sun include:

1. A good correlation between the number of sunspots and the growth rate of trees
2. An apparent relationship between the sunspot cycle and the formation of atmospheric storms on earth
3. A seeming correlation between the number of sunspots and the price of wheat (the smaller the number of spots, the higher the price)
4. A professed relationship between the number of sunspots and the number of furs purchased by the Hudson Bay Company (the larger the number of spots, the more skins are bought)

Knowledge of solar-terrestrial effects is presently in its infancy. What may seem to be a laughing matter today might well become a fact tomorrow. That the sun has a greater overall influence on the earth than any other celestial body, however, is beyond debate.

7 *the lesser lights*

Other stars

Many a night from yonder ivied casement,
Ere I went to rest,
Did I look on great Orion,
Sloping to the west.
Many a night I saw the Pleiads
Rising thro' the mellow shade,
Glitter like a swarm of fireflies
Tangled in a silver braid.

Locksley Hall
Alfred, Lord Tennyson

On a clear, dark night, how many stars can a person with normal vision actually see? Millions? Billions? Surprisingly enough, only about 2000 stars are ever visible at any given time under even the most perfect conditions. Furthermore, a total of only 6000 stars can be seen from both hemispheres of the earth over a period of 1 year. The reason for this is that the human eye is such a poor receptor of the feeble light coming in from distant stars. Any optical aid, however, increases the number of visible stars dramatically. Binoculars, for example, make it possible to observe a total of 50,000 stars, while the use of a small telescope raises that number to 500,000. The largest modern telescopes, especially when used photographically, make visible an almost uncountable number of stars.

STARS: THE HEAVENLY FABRIC

As a suitable beginning to our study of the stars, we must first briefly delineate their place and importance in the universe. The largest known aggregate form of matter is a *galaxy,* which is composed of billions of stars. There is no known star that is not a member of a galactic system; apparently that is an impossibility. All stars visible from earth with the naked eye, binoculars, or a small telescope are members of a single galaxy or star system—the Milky Way (Chapter 8). Although there are millions of other galaxies, the largest telescopes are needed to resolve even those that are closest to us into individual stars. The most numerous celestial bodies and the true fabric of the heavens, then, are indisputably the stars.

Constellations

Observers in ancient times noticed that the stars visible to the naked eye seemed to form patterns in the sky. Imaginations flared and soon legends arose about celestial kings and queens, hunters, and strange animals. Such a group of stars that apparently belongs together to form a figure is known as a constellation. Today, however, the concept of the constellation is somewhat different, for now we use the term to refer to an arbitrarily outlined region of the celestial sphere that may or may not contain a pattern-forming star group (Fig. 7-1). The modern celestial sphere encompasses 88 constellations, most still retaining their ancient names (Appendix A).

Astronomers now know that constellations are only apparent star groups; the individual stars contained in them are at different distances, of different ages, and are traveling in different directions at different speeds—in other words, they are totally unrelated to one another. Constellations take the forms they do because of our vantage point on earth. From the moon or from another planet in our solar system, the star configurations would look much the same. The view from a planet orbiting another star in our galaxy

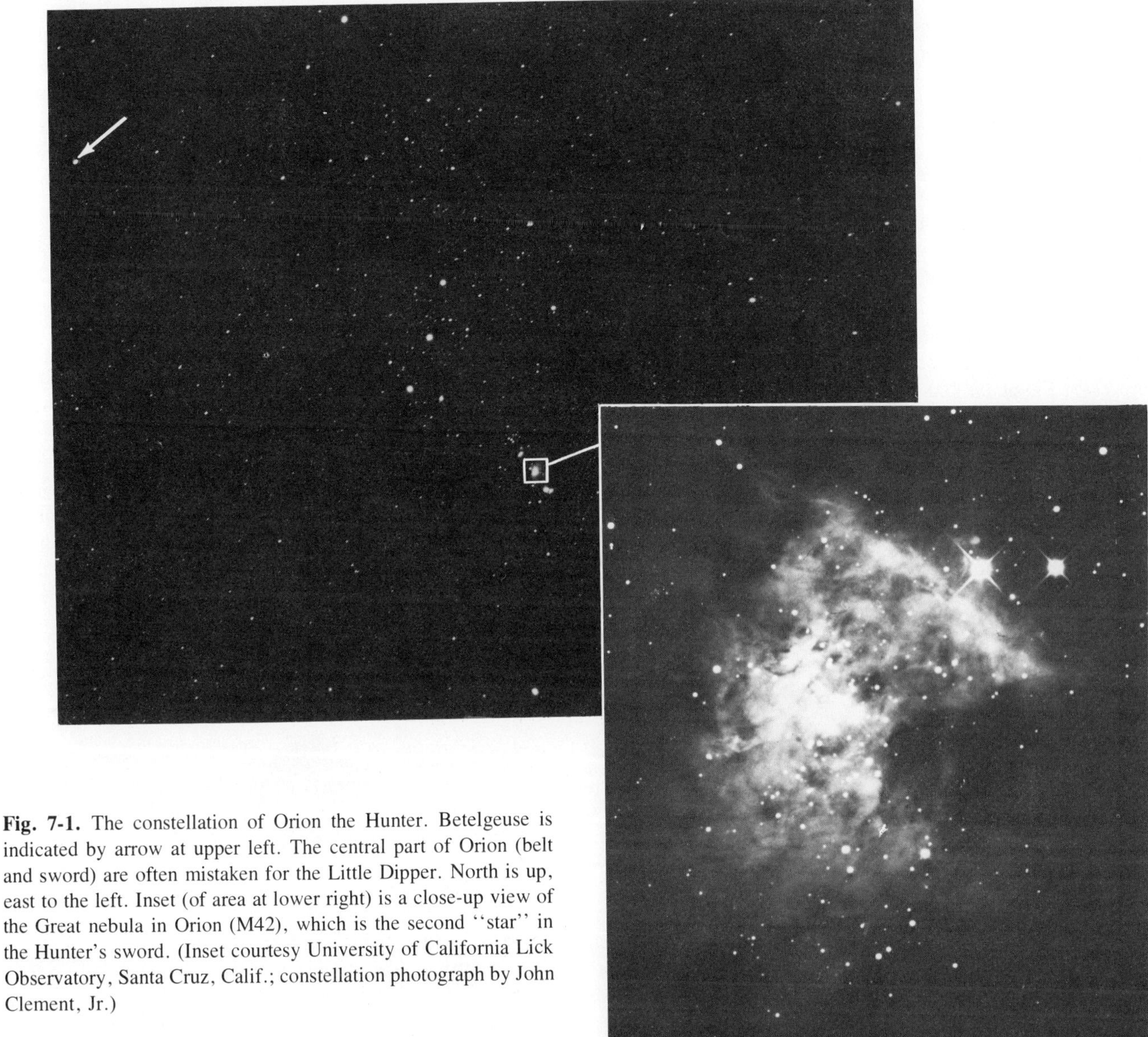

Fig. 7-1. The constellation of Orion the Hunter. Betelgeuse is indicated by arrow at upper left. The central part of Orion (belt and sword) are often mistaken for the Little Dipper. North is up, east to the left. Inset (of area at lower right) is a close-up view of the Great nebula in Orion (M42), which is the second ''star'' in the Hunter's sword. (Inset courtesy University of California Lick Observatory, Santa Cruz, Calif.; constellation photograph by John Clement, Jr.)

would place the stars in a different perspective. New constellations would be formed and our sun, not now a part of a constellation, might become the eye of some strange-looking creature.

Constellations are to an astronomer what a map is to a traveler; they tell him exactly where something is located. We speak of the location of the Beehive open star cluster, therefore, in the constellation of Cancer the Crab, just as the city of Paris is located in France. When an object is said to be ''in'' a constellation, this simply means that the object either has this star group as a background or can be seen through it.

The brightest stars as well as other famous stars within constellations are known by names handed down through generations. While most star names are of Greek or Latin origin, others are derived from the Arabic. In addition, all of the brightest stars in a constellation are given Greek letters, usually in order of decreasing brightness, although exceptions do occur. When several stars in the same constellation are of the same brightness, they are then lettered in order from the head to the foot of the creature. It is now standard astronomical procedure to catalog stars by number and position (right ascension and declination).

Distances to stars

The sun is 93 million miles away from the earth, while the next nearest star is 26 trillion miles away, or 268,144 times more distant. How can such overwhelming distances be measured? The calculation of astronomical distances is probably one of the finest examples of the utility of mathematics. Like the determination of most distances on earth, the distances to many celestial bodies are also computed either directly or indirectly through the concept of triangulation. *Triangulation,* a surveying technique, was known to the early Egyptians but was more fully developed by the Greeks. As the name implies, triangulation involves the construction and mathematical solution of triangles.

A triangle is a geometric figure bounded by three straight lines and having three angles. When the values of any three successive quantities are known, the values of the other three may then be found mathematically. The two possible combinations are either two angles and an included side, or two sides and an included angle. In distance surveying, whether terrestrial or celestial, the usual values of the triangle that are obtained are two angles and the included side, which is known as the *baseline*. To illustrate this concept, first consider the surveying problem of finding the distance to an inaccessible site. To find the distance from *A* to *C* in Fig. 7-2, baseline *AB* is laid out and carefully measured. Angle *A* and angle *B* are then determined. From this information, the distances of *AC* or *BC* can be calculated.

In astronomy we have to deal with "skinny" triangles because the distances to celestial bodies are very great in proportion to the length of the available baseline. The earth's diameter may be used as a baseline to triangulate the distances to nearby objects such as the sun, moon, and planets. The angular position of the object is measured either by two observers on opposite sides of the earth or by the same observer whose measurements are separated by a time interval of 12 hours. Since the baseline, or diameter of the earth, is a known value, and the included angles have been obtained by careful measurement, the distances from earth to the object may thus be calculated.

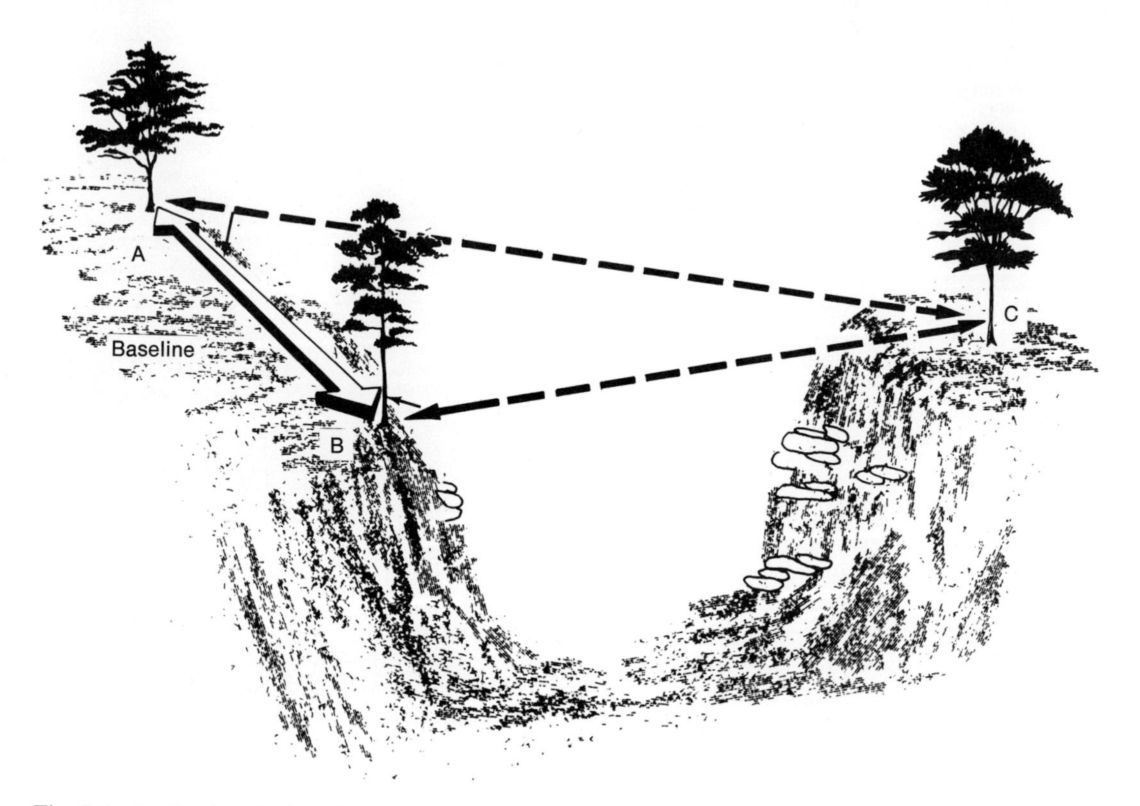

Fig. 7-2. Application of triangulation to determine the distance to an inaccessible point. If angle *B* is 90°, then the Pythagorean theorem can be applied to solve such distances: $(AB)^2 + (BC)^2 = (AC)^2$.

For the stars, however, an even longer baseline is needed. The longest baseline available to an observer on earth is the diameter of the planet's orbit (Fig. 7-3). Measurements of a star's position separated by a time period of 6 months indicates to the observer that he is on opposite ends of a baseline that is 2 AU or 186 million miles long. Even with such an enormous baseline, the angles to be measured are very small, making them not only difficult to obtain but also prone to great error. Direct triangulation as a means of computing distance to the stars has therefore generally proved to be unsatisfactory.

The ancient Greeks, as an argument against the heliocentric (sun-centered) universe, realized that if the earth revolved around the sun, there should be an apparent displacement of the stars due to this motion. The apparent displacement in the direction of a celestial body due to the motion of the observer is known as *parallax*. The early Greeks, unable to detect stellar parallax, became adherents to the geocentric concept of the universe. Even Tycho with his keen, unaided eyes and William Herschel using his telescope could observe no parallax of the stars.

It was not until 1838 that a German astronomer, F. W. Bessel (1784-1846), made the first reliable measurement of stellar parallax. He used the star 61 Cygni, which is just visible to the naked eye in the constellation of Cygnus the Swan. He obtained a parallax of 0.3 second of arc and a resultant distance of 681,000 AU for the star. The calculations Bessel made were so accurate that modern measurements of the distance to 61 Cygni differ from his by no more than 10%.

The distances to the stars are so great that units of measurement used on earth and even those applied to the solar system soon become too cumbersome. Therefore it was necessary to develop new, simple units for astronomical distances that would not become unnecessarily large in computation. Over the years, two such units have evolved —the light-year (ly) and the parsec (pc). The *light-year* is the distance light travels in 1 year at the speed of 186,000 miles/sec. The *parsec* is the distance to a star that has a parallax of 1 second of arc.

A simple relationship exists between parallax (P) and distance:

$$\text{Distance (pc)} = \frac{1}{P}$$

where parallax is expressed in seconds of arc. Using the 0.3 second of arc parallax of 61 Cygni obtained by Bessel, its distance in parsecs would be calculated as follows:

$$D = \frac{1}{P}$$

$$D = \frac{1}{0.3 \text{ second}}$$

$$D = 3.3 \text{ pc}$$

Fig. 7-3. The baseline of the earth's orbit can be used to measure the distance to a star if the angles at E_1 and E_2 can be measured. More accurate is the technique involving parallax. When sighted from E_1, the star is seen at position S_1; when sighted from E_2, it has shifted to position S_2. The angle of shift, P, is the star's parallax.

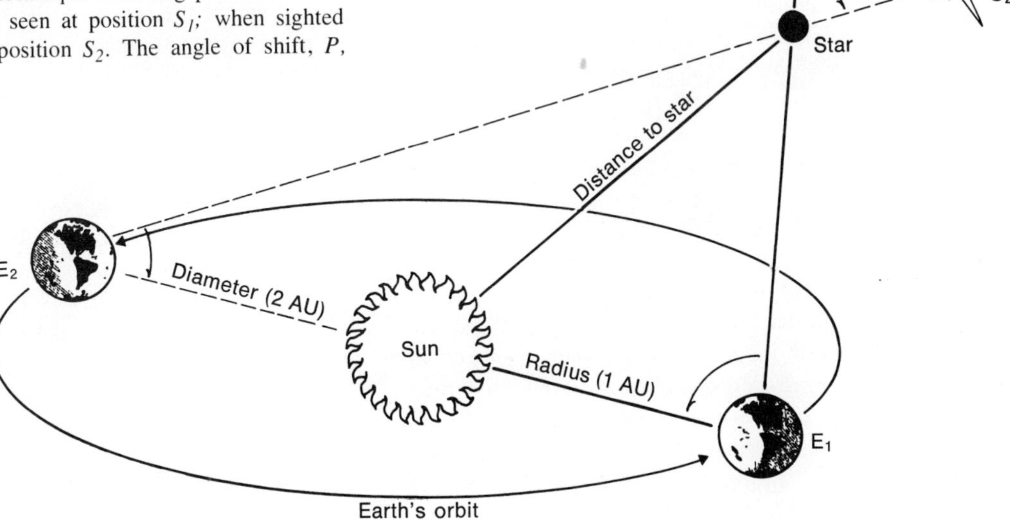

Since 1 parsec is equal to 3.26 light-years, 206,265 AU, or about 20 trillion miles, the resultant distance from earth to 61 Cygni would be 11 light-years, 681,000 AU, or 66 trillion miles.

By 1888, the distances to only about 50 stars had been calculated. The introduction of photography as an astronomical technique greatly speeded up calculations; today several thousand stellar distances have been calculated from their measurable parallax. Unfortunately, the parallax method is limited to the nearest stars, for beyond 100 parsecs the percentage error in computation becomes too high for valid measurements. For more distant stars, indirect means such as those utilizing star brightness must be employed.

Brightness of stars

As was pointed out in Chapter 5, the intensity of light varies inversely with the square of the distance from the source. Therefore at half a given distance a particular light is four times as bright as it actually is, while at twice the distance the light is only one-fourth as bright. A star measured to be one fourth as bright as a similar star of known distance can thus be regarded as being twice as far away.

A glance at the night sky will reveal that all stars are not equally bright. The brightness of stars from their respective distances is called *apparent brightness*. This is not a measure of how bright the star actually is but only of how bright it appears to be.

The early Greek astronomer Hipparchus was the first to classify stars according to their apparent brightness. He cataloged close to 1000 stars, placing them in classes of brightness we now call *magnitudes*. The 20 brightest stars were said to be of the *first magnitude*. *Second-magnitude* stars were somewhat dimmer than those of the first magnitude, and *third-magnitude* stars were dimmer still. The very dimmest stars Hipparchus could see were classed as *sixth-magnitude* stars. However, with the development of the telescope and the later use of sensitive photographic and photoelectric techniques, the magnitude scale has been expanded to include objects brighter than first magnitude and dimmer than sixth magnitude (Table 7).

Table 7. Apparent magnitude scale

Celestial body	Apparent magnitude
Sun	−26.5
Full moon	−12.5
Venus (at brightest)	−4.0
Jupiter and Mars (at brightest)	−2.0
Sirius	−1.4
Vega	0.0
Aldebaran and Altair	+1.0
Polaris	+2.0
Naked-eye limit	+6.5
Binocular limit	+10.0
Limit of 6-inch reflector or 3-inch refractor	+13.0
Visual limit of 200-inch telescope	+20.0
Photographic limit of 200-inch telescope	+23.5

It has been shown photoelectrically that a star of the first magnitude is 2.5 times brighter than a second-magnitude star. A second-magnitude star, then, is $(2.5)^2$ times brighter than a fourth-magnitude star. The difference in apparent brightness between two objects on the magnitude scale may therefore be obtained by raising 2.5 to a power equal to the difference in magnitude between the two objects. The sun, with an apparent brightness of −26.5, is $(2.5)^{50}$ times brighter than the dimmest object seen at the photographic limit of the 200-inch Hale reflecting telescope.

The sun appears to be the brightest star in our sky because of its proximity to earth. Another star, however, may actually be larger and brighter, but it may either appear to be fainter or be altogether invisible because of its greater distance. It is desirable, then, to determine the true or *absolute* brightness (luminosity) of a star for the sake of comparison with other stars. The absolute magnitude is the apparent brightness a star would have if it were at a standard distance of 10 pc from the observer. Stars nearer than 10 pc must be mathematically moved out to that distance, while those beyond the standard distance must be brought in to 10 pc. To illustrate this technique, let us move the sun out to the standard distance using the following steps.

1. The sun is 1 AU from the observer.
2. 1 pc = 206,265 AU
3. 10 pc = 2,000,000 (2×10^6) AU
4. According to the inverse square law of light, the sun would diminish in brightness by a factor of (2×10^6) or 4×10^{12}
5. $4 \times 10^{12} = (2.5)^{31.5}$, or a change toward the dimmer end of the apparent magnitude scale of 31.5 places
6. The sun's apparent magnitude at 1 AU = −26.5; the decrease in brightness = 31.5; therefore absolute magnitude = 5.0

The sun, then, at 10 pc would be a fifth-magnitude star.

Stellar motion

Even though the early Greeks believed that the positions of the stars were fixed, today we know that they are all moving in three-dimensional space. Edmund Halley (1656-1742), the English astronomer who first recognized the return of the great comet now bearing his name, first detected stellar motion in 1718. By comparing early Greek charts with those of his own day, he found that three stars—Sirius, Arcturus, and Aldebaran—had actually changed positions. Although traveling at great speeds, the stars are so distant from earth that many years must pass before any change in their positions is detectable with the naked eye. Thus the constellations one sees today are basically the same as those described and named more than 2000 years ago.

Only two aspects of a star's motion, proper motion and radial velocity, are directly measurable (Fig. 7-4). The other two, true speed and direction of movement, are obtained mathematically. A star's proper motion and radial velocity may be measured with the aid of a telescope. Proper motion (E) is the angular change in a star's position on the celestial sphere. It is obtained by comparing telescopic photographs of a star taken over a period of time from 20 to 50 years. From this knowledge the average annual proper motion can then be computed. This figure is expressed in seconds of arc per year. Bernard's star, the second nearest star to the sun (6 light-years), has the largest known proper motion of any star, 10.25 seconds of arc/yr.

The radial velocity of a star (AC) is the speed with which it moves toward or away from us along our line of sight. It is obtained by measuring the Doppler shift in a photograph of the spectral line placement of a stationary light source such as a carbon arc lamp relative to a photograph of the spectrum of a star. If the spectral lines of the star are shifted toward the red part of the spectrum, recession along our line of sight is indicated. On the other hand, when the lines are shifted toward the violet portion of the spectrum, the star is approaching us along our line of sight. The amount of shift is translated into speed, which is expressed in miles or kilometers per second. Since the sun is also moving with respect to the stars, that portion of the radial velocity due to the solar motion must also be taken into account. Perhaps it is ironic that the study of radial velocities and proper motions of other stars provided us with our first determination of the speed of the sun in its orbit through the Galaxy. The sun is moving toward Vega, a bright star in the constellation of Lyra the Harp, at a speed of 12 miles/sec (43,200 miles/hr).

The tangential velocity and space velocity of a star must be computed mathematically from other known data. The tangential velocity (AD) is that part of a star's motion at right angles to the line of sight. The calculation of the tangential velocity is dependent on both a knowledge of the star's proper motion and its distance (d) from the earth. Once the tangential velocity has been determined, the space velocity (AB), or true stellar motion, may then be obtained mathematically.

Mass and density of stars

In Chapter 6 it was pointed out that the sun's mass could be determined by measuring its gravitational effects on the planets. The same technique may also be used to calculate the mass of a planet if it has a satellite or if it is closely approached by a minor planet. How then can the mass of distant stars be measured when they can be seen only as points of light? A great many stars are members of a *binary system*—two or more stars revolving around a common center of mass. Estimates suggest that from one half to two thirds of all stars are indeed part of such a system. Therefore, when this is the case, the total mass of the stars involved can be computed in much the same way as the mass of a planet is calculated.

However, if a star is not a member of a binary system, its mass may be inferred from its luminosity (absolute magnitude). This is possible because with few exceptions there is a direct relationship between the mass of a star and its inherent brightness. Low-mass stars have low absolute magnitudes while high-mass stars have high absolute magnitudes. There are only a few known exceptions, which will be discussed later in this chapter.

Once the mass and diameter of a star are determined, its density may then be computed. The extremes in star density

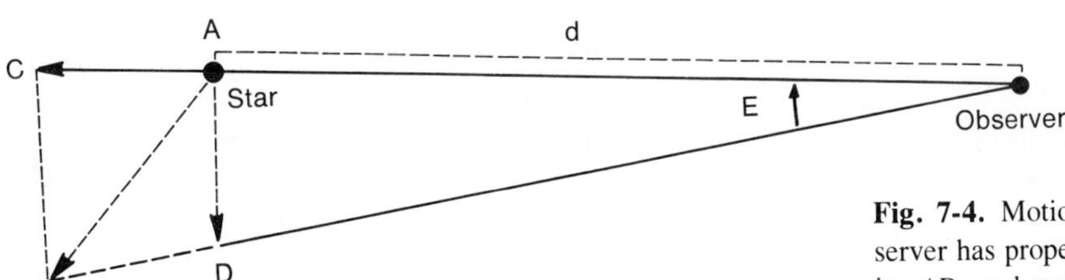

Fig. 7-4. Motions of stars. The star at distance d from the observer has proper motion E, radial velocity AC, tangential velocity AD, and space velocity AB.

are great, ranging from one-ten millionth that of the sun to 10^{14} times the solar density.

Diameters of stars

The linear diameter of the sun is easily obtained from its angular size on the celestial sphere, once its distance is known. Unfortunately, however, the sun is the only star that presents us with a disc large enough to be measured. How, then, can we compute the diameter of a star that appears no larger than a pinpoint of light even in the largest of telescopes?

At least two ingenious methods can be employed to calculate the diameter of stars. For some very large stars like Betelgeuse in the constellation of Orion the Hunter, an instrument called an optical interferometer may be used. The angular diameter of Betelgeuse as measured with this instrument is 0.047 seconds of arc. At a known distance of 300 light-years, Betelgeuse thus has a linear diameter of 250 million miles.

The diameter of stars may also be inferred from knowledge of both their absolute magnitudes and temperatures. A star's absolute magnitude is dependent on the rate at which radiant energy leaves its surface, which in turn is determined by both its temperature and its diameter.

The range in the diameters of stars is perhaps surprising. The largest type of star typically has a diameter of hundreds of millions of miles. Antares, in the constellation of Scorpio the Scorpion, for example, has a diameter 480 times that of the sun. If the sun were at the center of Antares, then Mercury, Venus, Earth, and Mars would all be orbiting within a distance equal to the diameter of this giant star. At the other extreme, some stars are of planetary size, while others are 25 to 50 miles in diameter or smaller (Fig. 7-11, *A*).

Stellar temperatures

Several methods can be used to measure the surface temperatures of stars. The temperature of any radiating body can be calculated from the color of light it emits. We intuitively know that something "white hot" is hotter than something "red hot." The temperature of the body controls the wavelength (color) of most of the intense visible light that radiates from it. Stars are no exception. The sun, for example, has a surface temperature of 10,340° F (5727° C) and a resultant wavelength of greatest intensity in the yellow portion of the visible light spectrum. The spectroscopic analysis of starlight, then, may lead to a determination of a star's surface temperature.

Another technique that may be employed to obtain the surface temperature of a star is its *color index*. It has long been known that obtaining the apparent magnitude of a star, visually and photographically, usually gives different values. This is because photographic film is generally more light sensitive than the human eye. The difference between the photographic and visual apparent magnitude of a star is called its *color index* (Table 8). For example, if a star had a photographic apparent magnitude of +2.8 and a visual apparent magnitude of +2.6, the difference, or color index, would be +0.2. The true color of the star is thus determined; as pointed out earlier, this color is a function of its surface temperature. Today both types of apparent magnitudes are obtained photographically. Yellow-sensitive film records visual apparent magnitude, while blue-sensitive film is used to register photographic apparent magnitude.

Table 8. Color indexes of stars

Typical star	Color of star	Color index	Temperature (°C)
Rigel	Violet to blue	−0.3	21,000
Vega	Blue	0.0	11,000
Procyon	White to blue	+0.4	7000
Capella	Yellow	+0.7	5400
Arcturus	Orange	+1.1	4100
Antares	Red	+1.6	3200

Stellar spectra

Virtually all of the knowledge we have accumulated about stars has come through the study of their spectra. Photographs have been made of the spectra of thousands of stars; almost without exception, these spectra have been found to be similar to the spectrum of the sun—a dark-line spectrum superimposed on a continuous spectrum. This suggests that light originates in the star's photosphere and then passes outward through a stellar atmosphere in which dark lines are formed as atoms absorb specific wavelengths of colors of light.

Stellar data obtained spectroscopically include information on temperature, pressure, composition, radial velocity, rotation, and the presence and strength of magnetic fields.

Temperature, pressure, and composition

It was pointed out previously that the surface temperature of a star may be determined from the wavelength of the most intense light registered in its spectrum. The gas pressure of a star's photosphere, on the other hand, may be inferred from a comparison of the intensities of spectral lines of a particular element under varying degrees of ionization.

The study of stellar spectra reveals a surprising but important fact—the chemical compositions of most stars are almost identical, chiefly hydrogen and helium with smaller amounts of the other elements. Every element is in the gaseous state and emits and absorbs a specific group of wavelengths of light. Thus the identification of the lines in the spectrum of a star indicates exactly what elements are present. On the other hand, a comparison of the intensities of the spectral lines of different elements indicates their relative abundance within the star.

Radial velocity and rotation

The rate at which a star is moving along our line of sight may be measured by an analysis of the shift of its spectral lines relative to those produced by a stationary laboratory source.

Like radial velocity, the rotation of a star may also be measured by a study of the Doppler shift in its spectral lines. The sun has such a large angular size that light coming from its opposite limbs may be analyzed separately. Light from the eastern limb shows spectral lines shifted toward the blue portion of the spectrum, indicating that this part of the sun is approaching us. At the same time, on the other hand, light from the western limb of the sun shows a red shift, meaning recession. The sun is thus found to be rotating at a speed of 4500 miles/hr in its equatorial region. Since all other stars are so far away that they present only a point of light, it may be assumed that the light received from them comes from all parts of their photospheres, including the limbs. Spectral studies of starlight indicate that this indeed is the case, for stars invariably show both a red shift and a blue shift in their spectra.

Magnetic fields

The spectral lines of sunlight, especially those near sunspot regions, are often split. Such splitting of spectral lines, known as the *Zeeman effect,* is thought to be due to magnetic fields. The width of the split is believed to be a measure of the magnetic field strength. Although starlight is so feeble that the splitting cannot be measured, it is possible to measure the resultant broadening of the spectral lines. The broader the lines, the stronger the magnetic fields.

Spectral classes

The ability to classify stars according to spectral characteristics has been a rather recent development. The first known attempt was made by Angelo Secchi in 1863. At that time he classified the stars into four groups based on the arrangement of their spectral lines. The *Secchi system* was later modified and expanded by Henry Draper to include several other types of stars. The present classification system was developed at the Harvard Observatory under the direction of E. C. Pickering. The *Harvard system,* based on the analysis of more than 200,000 spectra, originally consisted of so many stellar types that virtually every letter of the alphabet was used as a class designation. Closer examination, however, indicated that most stars fell into one of seven basic classes. In forming the simpler scheme, the letter designations from the earlier system were retained.

The modern classification system is based on the color-temperature relationships mentioned earlier. The hottest stars are blue in color; the coolest stars are red. Arranged from high to low temperatures, the spectral classes are assigned the letters O, B, A, F, G, K, and M, respectively. Spectral class K has one separate branch lettered R, and class M has two branches designated N and S.* Each of the seven main spectral classes is divided into 10 subclasses

*To remember the letters of all the spectral classes, you can apply the sentence, Oh be-a-fine-girl-(guy-)kiss-me-right-now-sweetheart.

Table 9. Spectral classes

Class	Color	Temperature (°C)	Spectral characteristics	Typical star
O	Violet	30,000	Dark lines of He, O, N, Si; H very weak	λ-Cephei and λ-Orionis
B	Violet to blue	20,000	Dark lines of He, Si, O, Mg; H stronger	Rigel and Spica
A	Blue	10,000	Strongest lines of H, Mg, Si, Fe, Ti, Ca	Sirius and Vega
F	Blue to white	7500	Ca, Fe, Cr, H	Canopus and Procyon
G	White to yellow	5600	Ca strongest line; weak H lines; Fe, Ti, Cyanogen, Hydrogen carbide	Sun and Capella
K	Orange to red	4000	Lines of metals; some of titanium oxide	Arcturus and Aldebaran
M	Red	3000	Titanium oxide predominates	Betelgeuse and Antares

identified by numbers from 0 to 9 to indicate placement within each class. An A_5 star, for example, is halfway between A_0 and F_0. The sun is classified as a G_2 star, which places it 0.2 of the way between G_0 and K_0. The spectral classes and dark-line spectral characteristics and examples are shown in Table 9.

The H-R diagram

The correlation between the spectral class of a star and its absolute magnitude has proved to be one of the most useful tools of the astronomer. Based on the combined work of the Danish astronomer Einar Hertzsprung and the American Henry Russell, the relationship is now shown by a simple chart known as the Hertzsprung-Russell or H-R diagram (Fig. 7-5). It is important to point out here that the H-R diagram does not represent a star's placement in space but only the star's location on the chart of spectral class and absolute magnitude.

The diagram consists of a scale of absolute magnitude arranged along the left-hand edge and notations of spectral class placed along the bottom. When the two values are plotted for stars in the vicinity of the sun, a most interesting fact emerges—there is a close relationship among the mass, absolute magnitude, color, surface temperature, and spectral class of most stars. This is indicated by the placement of the star plots on the diagram. Nearly 89% of the stars within a 50 light-year radius of the sun fall into a narrow diagonal extending from upper left to lower right on the diagram. This narrow zone is called the *main sequence,* or *normal stars*. The stars in the upper left portion of the main sequence are massive and hot with high absolute magnitudes while those to the lower right are not as massive, are cooler, and have lower absolute magnitudes. The sun, at absolute magnitude +5 and spectral class G_2, is on the main sequence.

Stars off the main sequence to the upper right are very large and called *giants,* or *supergiants*. These stars are not normal in that while they are both massive and possess high absolute magnitudes, they also have low surface temperatures. Approximately 1% of the stars in the neighborhood of the sun are classified as giants or supergiants. To the lower left-hand portion of the H-R diagram are other stars that have unusual properties. These are *white dwarfs,* which constitute about 9% of the stars in the sun's vicinity. Since they are low-mass stars with high surface temperatures and low absolute magnitudes, white dwarfs are not "normal."

The H-R diagram has enabled astronomers to make comparisons of stars and even infer their life histories. A star contracting from interstellar gas and dust will reach the main sequence when its gravity and radiation pressure have established dynamic equilibrium. The star's original position on the main sequence is a function of its mass. A massive blue giant star may remain on the main sequence for only a few million years because it consumes its fuel at a fantastic rate. A red dwarf star, on the other hand, may stay on the main sequence for billions of years because it is consuming itself at a much more conservative pace.

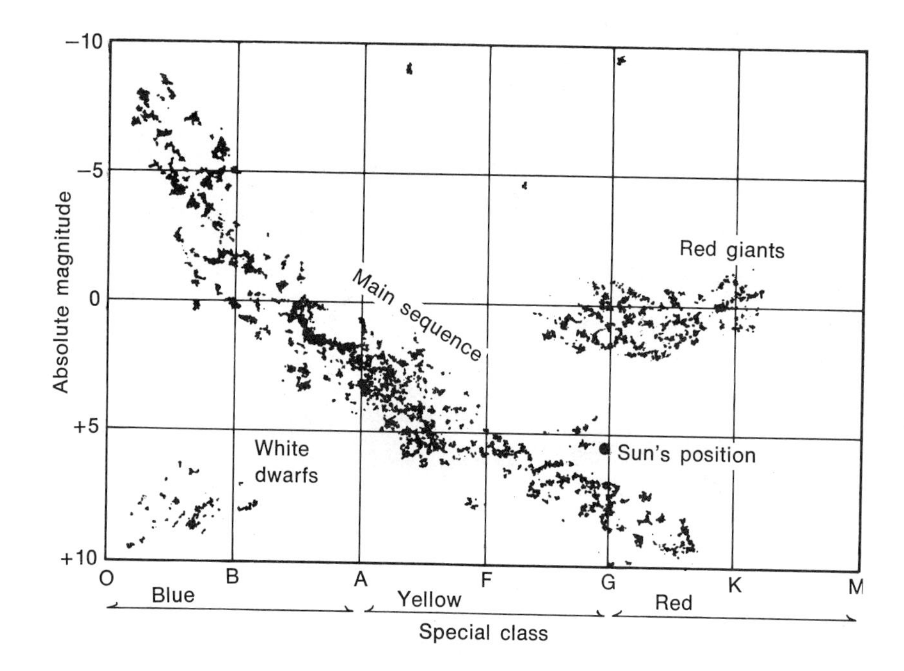

Fig. 7-5. Hertzsprung-Russell diagram. This plot of absolute magnitude versus spectral classes of various stars defines the "main sequence" or "normal stars" as those that fall in the narrow star diagonal.

As the hydrogen fuel is depleted in the nuclear furnace of a main-sequence star, it then increases in size, moving to the upper right-hand part of the diagram to become a giant or supergiant. Eventually, most of the fuel even in its outer layers will be depleted, and the star will collapse to form a white dwarf.

TYPES OF STARS

Stars are classified according to their interactions with other stars. They may form small groups of two, three, or more, or they may "go it alone."

Binary stars

Nearly two thirds of the stars known today fall into the first class, known as binaries because there are two parts to the group, one star and another star or group of smaller stars orbiting it.

The Big Dipper is probably the most familiar star configuration seen from the earth's Northern Hemisphere. The Big Dipper, consisting of seven stars, forms the hindquarters and tail of the Big Bear—the constellation Ursa Major. When you look at one bend in the dipper's handle you can see one second-magnitude star named Mizar. A person with excellent eyesight can barely see another star next to Mizar—this one is called Alcor. The early Arabs named these stars the horse (Mizar) and the rider (Alcor). Although binoculars show the two stars clearly, a small telescope reveals an even more interesting view. Very near to Mizar is still another star called Mizar B, which orbits Mizar just as the earth orbits the sun. Mizar and Mizar B are two stars locked in a mutual celestial dance; they are binary stars.

Mizar B was discovered in 1650 by the Italian astronomer John Riccioli. Since that time it has been found that from half to two thirds of all stars are members of binary or multiple star systems. Binary stars, then, appear to be the normal course of stellar development, while the occurrence of a single star like our sun, seems to be unusual. Some questions are still unresolved. Do planets form instead of a stellar companion? Can planets exist in a binary system? There are no answers as of yet. Since the discovery of Mizar B, a total of seven types of binary stars have been identified. There are optical, visual, astrometric, spectroscopic, spectrum, and eclipsing binaries.

Optical binaries

Optical binary stars are not true binaries but are so classified because to the naked eye they appear to be part of the same star system. Two stars happen to be arranged along our line of sight in such a way that they appear to be side by side. In reality, the two stars are many light-years apart and show different stellar motions, indicating that they are not revolving around each other. Mizar and Alcor are optical binaries.

Visual binaries

Visual binary stars are members of a mutually revolving system. Over a period of time measured in decades they can be seen either telescopically or photographically to orbit each other. Visual binaries are true binaries whose orbits can be calculated, periods determined, and masses measured. Mizar A and B and Epsilon Lyrae, which consists of four stars, are examples of visual binary systems (Fig. 7-6).

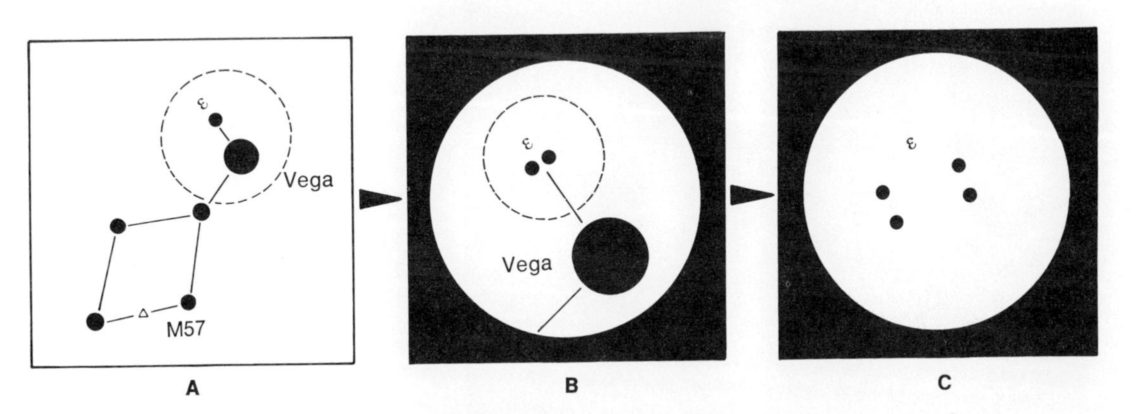

Fig. 7-6. The famous double double star Epsilon Lyrae, ϵ, as seen with the naked eye, **A;** with binoculars, **B;** and with a small telescope at 100×, **C.** What appears to be a single star is actually a four-star group. The location of the ring nebula M57 is also shown in **A.**

Astrometric binaries

An astrometric binary is a star that is believed to be a member of a multiple system because of its unusual oscillatory proper motion. The strange motion of such a star is believed to be produced by the gravitational perturbations of its unseen companion. The astrometric binary shows no other binary characteristics but may prove to be the most important type because it could possess planets rather than another star. If seen from another star, our sun would have a wavy proper motion due to the presence of its planets. It would therefore almost certainly be classified as an astrometric binary star.

When the first expedition leaves earth in search of another planet outside our solar system, a nearby astrometric binary will probably be its destination. In fact, the first target star may have already been selected; the second closest star to the sun, Bernard's star, is an astrometric binary. The computed mass of its unseen companion or companions is 1.5 times that of Jupiter, and the average separation of the two objects is calculated at only 4 AU; both distances are consistent with possible planetary specifications.

A star classified as an astrometric binary sometimes must be reclassified when its unseen companion is identified. The brightest of the fixed stars, Sirius, was first identified as an astrometric binary. Many years later, after the development of new photographic techniques, the companion star of Sirius was first seen (Fig. 7-7). The detection of Sirius B, the first white dwarf star ever discovered, made it necessary to reclassify Sirius itself as a visual binary.

Spectroscopic binaries

As mentioned in our discussion of astrometric binaries, the companion object does not have to be seen directly in order for its presence to be detected. The multiplicity of the system often gives itself away through spectral characteristics. Binary stars move together through space and show the same basic radial velocity. However, because the stars are revolving around each other, the radial velocity of the system varies slightly as the star with the greatest angular momentum first recedes and then approaches along our line of sight. The observed spectral effect is the formation of double lines. If the spectrum of a star shows such an effect, two or more stars must be present, and the primary star is thus classified as a spectroscopic binary. Mizar, Rigel, and Spica are among the more than 1500 known spectroscopic binary stars.

Spectrum binaries

A star whose spectrum shows characteristics of more than one spectral class is called a spectrum binary. Unlike spectroscopic binaries, with which they are often confused, spectrum binaries show no variation in radial velocity but exhibit spectral traits indicative of stars of different temperatures. It is thus believed that spectrum binary stars are those whose orbital planes lie perpendicular to our line of

Fig. 7-7. Tiny Sirius B is the bright dot to the lower right of the six-pointed Sirius A. The two dots to the left are images produced artificially for measurement purposes. (Courtesy United States Naval Observatory, Washington, D.C.; photograph by Irving Lindenblad.)

sight. If this were not true, spectrum binary stars would show characteristics of and be classified as spectroscopic binaries.

Eclipsing binaries

Eclipsing binary stars are those whose orbits are so arranged along our line of sight that the two stars periodically eclipse each other. The interference of light between the two stars causes the visual component of the system to vary in apparent magnitude. This variation may be measured and a light curve drawn, the analysis of which may lead to the determination of the orbital period, shape and inclination of the orbit, and the radius of each star.

As seen from earth, some eclipsing binary stars are involved in total eclipses, while the rest undergo partial eclipses. The inclination of the orbit of the system with respect to our line of sight determines the type of eclipse we see.

Eclipsing binary stars are, of necessity, also spectroscopic binaries and variable stars. This will be discussed in the following section. Algol, in the constellation of Perseus the Hero, was the first eclipsing binary star discovered. A partial eclipsing binary star whose true nature was recognized in 1889, Algol varies between apparent magnitude +2.2 and +3.4 every 69 hours. Today about 3000 eclipsing binary stars have been identified.

Variable stars

There are other stars that also vary in brightness, not because of external factors but because of true internal alterations usually resulting in a change in size. These variable stars are classified as pulsating, eclipsing, or erupting variables.

Pulsating variables

As their name suggests, pulsating variable stars undergo a periodic increase and decrease in size. There are pulsating variable stars among the spectral classes from B to M, but most are giants or supergiants. Three types of pulsating variable stars are known—long-period variables, R.R. Lyrae stars, and Cepheid variables.

Long-period variables. Long-period variables are the most common type of pulsating variable star; most are either red giants or supergiants. Long-period variable stars are also known as Mira type because the classic example is the star Mira in the constellation of Cetus the Whale.

Mira ordinarily has an apparent magnitude of +8, invisible to the naked eye. Every 11 months or so, however, the star brightens to a magnitude of +2.5 and is easily seen with the unaided eye. The diameter of Mira varies from 460 to 550 times the diameter of the sun, with maximum magnitude occurring at minimum size.

R.R. Lyrae stars. The second most abundant type of pulsating variables are the R.R. Lyrae stars, named for a seventh-magnitude star of this kind in the constellation of Lyra the Harp. The R.R. designation indicates a variable star. R.R. Lyrae stars are giants of either spectral class A or F and generally change in brightness by less than one magnitude. These stars have the shortest periods of all pulsating variable stars, typically 7 to 17 hours.

R.R. Lyrae stars are useful distance indicators. They all have mean absolute magnitudes of +0.3. In order to find the approximate distance to such a star, only its average apparent magnitude must be determined from its light curve.

Cepheid variables. The rarest and most valuable pulsating variable stars are the Cepheid variables. These stars are giants belonging either to spectral class F or G, and they are named for the first star of this type that was discovered —Delta Cephei. A fourth-magnitude star in the constellation of Cepheus the King, Delta Cephei varies in magnitude between +4.1 and +5.2 over a period of 5.4 days. Its light curve shows a rapid rise to maximum and then a slower fall to minimum (Fig. 7-8). This star has an average size of 25 times the sun's diameter and, at maximum, increases to 27 times.

Cepheids are the hottest of the pulsating variable stars and as a result are brightest when at their maximum size. There is a faithful relationship between the periods of pulsation and the absolute magnitudes of Cepheid variable stars, making them excellent distance indicators. For example,

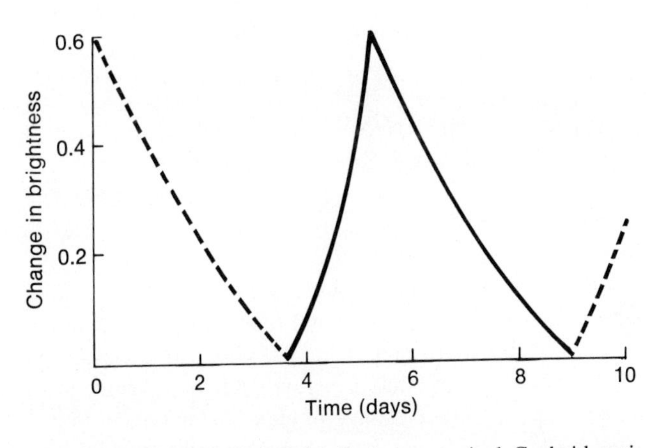

Fig. 7-8. Light curve of Delta Cephei, a typical Cepheid variable star. Note the rapid rise from minimum at about 3½ days to maximum at 5½ days, then the slower descent to minimum again at 9 days.

Fig. 7-9. Two variable stars in the Andromeda galaxy are marked to stand out from the myriad stars in the background. Because of their pulsating brightness, these stars are excellent distance indicators. (Courtesy Hale Observatory, Pasadena, Calif.)

the distances to star clusters within our own galaxy and even to remote galaxies have been determined by using these stars (Fig. 7-9). Since 1950 the study of Cepheid variable stars has forced astronomers to double the size of the known universe.

The brightest, most famous, and most familiar Cepheid variable star is Polaris, the North Star, in the constellation of Ursa Minor the Small Bear. Polaris, also a visual and spectroscopic binary, has a period of light variation of 4 days and a magnitude change of only 0.1, not even noticeable to the naked eye. This star, presently marking our north celestial pole, lies at a distance of 90 pc (293 light-years) and, at maximum light, is 600 times brighter than the sun.

Eclipsing variables

Eclipsing variable stars are not true variables. As is true of eclipsing binaries, they are aligned to our line of vision in such a way that they periodically eclipse each other.

Erupting variables

Erupting variable stars are those that undergo violent structural changes resulting either in the rapid expansion of the star's outer layers or in the explosion of the entire star. The star increases many magnitudes in brightness in only a few days and then decreases in brightness more slowly, often requiring years to completely fade from view. Erupting variable stars take two basic forms, the novae and supernovae.

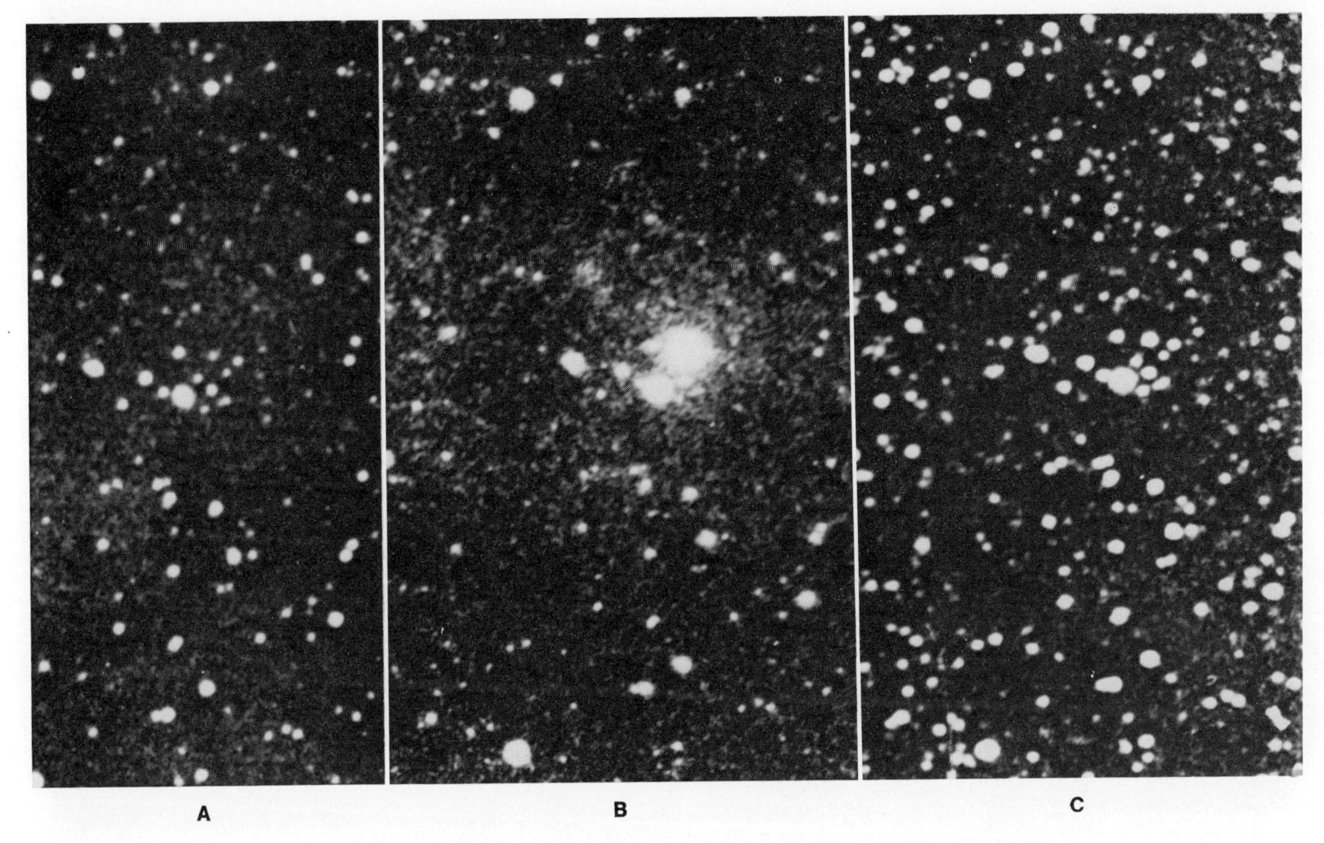

Fig. 7-10. Nova Lacertae of 1910. **A,** August 13, 1907, it appeared as a star of the thirteenth magnitude. **B,** December 31, 1910, it was a star of the seventh magnitude. **C,** September 29, 1911, Nova Lacertae was a star of the eleventh magnitude. (Courtesy University of Chicago Yerkes Observatory, Williams Bay, Wisc.)

Novae. A nova is a star whose outer layers undergo violent expansion. The star does not explode, but as it returns to normal size, an ejected gaseous shell is left behind, resulting in a loss of mass of about 1%. A typical nova reaches maximum light in 3 to 4 days, usually showing a magnitude increase of about 10 over its former apparent brightness. At maximum, a nova may reach an absolute magnitude of −8. The period of fading is generally several months long, and the star usually returns to its prenova apparent magnitude (Fig. 7-10).

The formation of a nova is a rare event—only 20 to 30 occur in the entire Galaxy each year. Still rarer is a nova that reaches a brightness visible to the unaided eye. Most novae are discovered accidentally on telescopic photographs. The same star has been observed to nova more than once, perhaps indicating that this type of star is the longest-period pulsating variable and one of the last stages in variable star development.

Supernovae. The supernova is the explosion of a star, the most spectacular and violent of all stellar events. The occurrence of a supernova is even rarer than that of a nova, taking place on the average only once in 100 years within a galaxy. The increase in apparent brightness may be as much as 20 magnitudes, while absolute magnitude may range from −14 to −30, making it a very luminous object. A supernova reaches maximum brightness in a few days, then often takes years to fade completely from view. After the explosion, an expanding mass of gas called a nebula occupies the original position of the star. The most famous remnant of a supernova is the Crab nebula in the constellation of Taurus the Bull. Recently, a pulsar, the neutron-star core of the supernova, was found near the center of the Crab nebula.

Stars with a mass less than 1.4 times that of the sun probably collapse into white dwarfs as they approach the end of their life cycles. It is now thought that a different fate awaits more massive stars—a supernova. As a massive star collapses, internal pressures and temperatures become

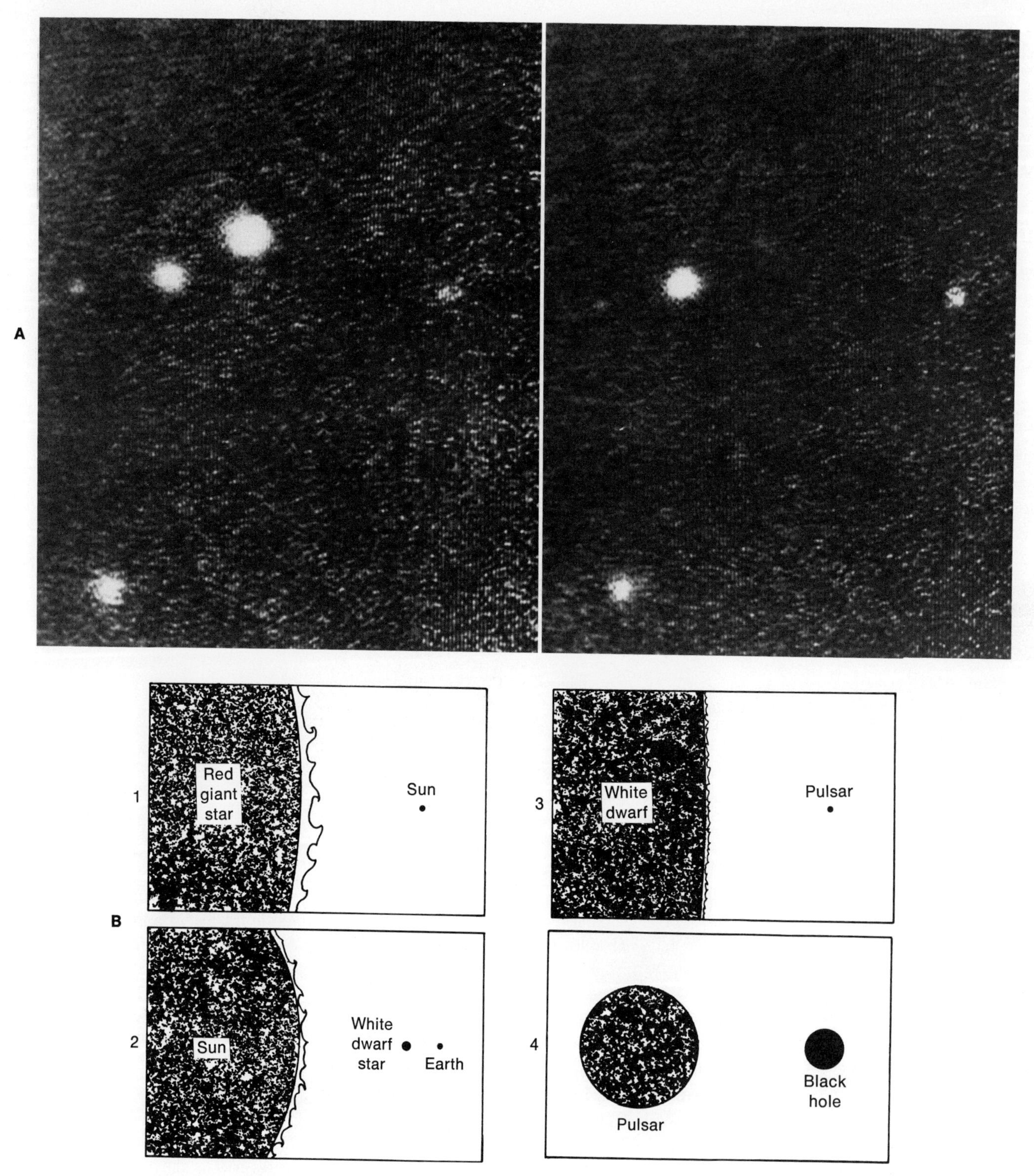

Fig. 7-11. A, Pulsar in the Crab nebula that blinks on and off every 0.033 sec. **B,** Size comparisons between *1,* a red giant star and the sun; *2,* the sun and a white dwarf star (also the earth); *3,* a white dwarf and a pulsar; and *4,* a pulsar and a black hole. (**A** courtesy University of California Lick Observatory, Santa Cruz, Calif.)

so high that every conceivable nuclear reaction comes into play with the formation of heavy elements. The final collapse is a catastrophic event. As central temperatures reach 100 billion degrees F, the star rebounds in a great explosion that disperses into space all of the elements formed during its existence. In this way stellar material is returned to space to be used in the subsequent formation of other stars. The sun possesses some heavy elements that it could not have produced itself. As a result, it is probably at least a second-generation star, containing debris from countless supernova explosions dating back to the early years of our galaxy.

In the 1930s it was proposed that when a star underwent a supernova explosion, a ball of pure neutrons about 20 miles in diameter was left behind, forming a *neutron star*. In 1967 a strange star was discovered in the constellation of Velpecula the Fox, which gives off pulses of radio energy every 1.33728 seconds. This star was appropriately named a *pulsar,* short for pulsating radio star. In 1968 a star was found near the center of the Crab nebula that was not only emitting radio energy every 0.033 seconds but also producing bursts of light and X-rays at the same time interval. To date more than 60 pulsars have been identified, and this type of star is now believed to be the hypothetical neutron star (Fig. 7-11, *A*).

A neutron star, the third fundamental stellar family after "ordinary" stars and white dwarfs, has a mass equal to that of the sun packed into a sphere with a diameter of approximately 20 miles (Fig. 7-11, *B*). This produces a mean density some 10^{14} times that of the sun, a density nearly equal to that found in an atomic nucleus. Neutron stars are the rapidly rotating collapsed cores of stars that exploded—a late stage in stellar evolution. The energy for the radio, optical, and X-ray radiation that pulsars emit comes from their rapid rotation, and the interval between pulses is the star's rotation period. In fact, a pulsar may be thought of as a mechanism for transforming rotational energy into electromagnetic radiation, possibly including cosmic rays that are of unknown origin. As the pulsar spins on its axis, a stream of radiation from its surface sweeps through space like the light from a revolving beacon. If the earth happens to lie in the path of the rotating beam of energy, we receive a burst of radio waves, light, and X-rays once in every turn of the object.

Black holes

With the realization of the connection between neutron stars, pulsars, and supernovae, astronomers now believe that the final pages of stellar evolution have been written. A totally collapsed star, or black hole, the fourth family of stars, has apparently been discovered in association with a binary star system in the constellation of Cygnus. The gravitational field of a black hole is so intense that no particle or radiation of any type can escape one. Passing particles or photons may be caught and pulled into the collapsed star. Thus only through their gravitational fields can these objects be detected.

STAR CLUSTERS

We might say that stars are gregarious. In an astronomical sense this means that they tend to occur in groups and act as a unit in that they are moving in the same direction and at the same speed in space. The largest star groups, to be discussed in the following chapter, are *galaxies,* each consisting of many billions of stars. The smallest groups are binary stars, except for an occasional loner like the sun. In between the two extremes are star clusters, which are made up of thousands of stars. Star clusters may contain binary stars, while all clusters are located in or associated with a galaxy. There are two types of star clusters, the galactic (or open) cluster and the globular cluster.

Galactic clusters

A galactic cluster is a star system containing as few as 20 or as many as 1000 or more stars that are held together loosely by their mutual gravitational attraction. The diameter of a galactic star cluster usually ranges from 15 to 40 light-years, and the system may be as far as 15,000 light-years from the sun. The distances to galactic clusters are determined by a study of Cepheid variable stars. As the name implies, galactic clusters are always found associated with the interstellar medium in the plane of a galaxy. Observers who wish to see a galactic cluster should use either binoculars or a telescope at low power and scan the Milky Way, which is the plane of our galaxy as seen from within.

At present about 900 galactic star clusters have been discovered in our galaxy, and others have been found in relatively close neighboring galaxies. Because of their general lack of white, yellow, and red stars and their close association with interstellar material, galactic star clusters are thought to be young—only a few million years old. The gravitational perturbations caused by the presence of the interstellar medium will cause the stars to separate by greater distances, finally becoming individual members of the general star field.

The two best known galactic star clusters visible without optical aids are both found in the constellation of Taurus; these are the Pleiades and the Hyades. The Pleiades, often referred to as the Seven Sisters and frequently mistaken for the Little Dipper, mark the shoulder of the bull. The Hyades, on the other hand, form a V-shaped star group

Fig. 7-12. A, Double open star cluster in Perseus. **B,** M13 in Hercules is the finest globular cluster visible from the Northern Hemisphere. (**A** courtesy University of California Lick Observatory, Santa Cruz, Calif.; **B** courtesy University of Chicago Yerkes Observatory, Williams Bay, Wisc.)

marking the bull's forehead. Many other galactic clusters can be seen with binoculars. Among the most interesting and beautiful are the Beehive cluster in Cancer and the double cluster in Perseus (Fig. 7-12, *A*).

Globular clusters

Globular clusters, as the name suggests, consist of a group of stars arranged in a rounded configuration. There are nearly 100 known clusters of this type, each composed of tens of thousands of stars. The diameters of globular star clusters range from 75 to several hundred light-years, and their distances from the sun are from 30,000 to 300,000 light-years. R.R. Lyrae variable stars almost always lie within globular clusters, facilitating distance calculation.

Unlike galactic clusters, globular clusters are found outside the galactic plane. They orbit the nucleus of the galaxy and thus are located in any direction on the celestial sphere. Globular star clusters are moving very fast relative to the stars in the galaxy and do not share its rotation. It is often said that a globular cluster is to the galaxy what a comet is to the sun. These clusters must be billions of years old, for blue and white sequence stars are extremely rare and yellow and red stars are abundant.

Among the best-known globular clusters are Omega Centaurus, visible without a telescope although not visible at all from the United States, and M13,* a binocular and telescopic object in the constellation Hercules. Other globular clusters worthy of the attention of the observer are M22 and M55 in Sagittarius and M80 in Scorpio.

*Many celestial objects are labeled "M" followed by a number. The "M" refers to a list of celestial objects cataloged by the French comet hunter, Charles Messier (1730-1817). Messier's original catalog included 103 objects that he knew were not comets. The list (Appendix C) includes star clusters, nebulae, and galaxies. Examples are M31, the Great galaxy in Andromeda, and M42, the Great nebula in Orion.

BETWEEN THE STARS: THE INTERSTELLAR MEDIUM

The space between the stars is not empty. It is permeated by gas and grains collectively called the interstellar medium. The mass of this uncondensed matter represents about 10% of the total mass of a galaxy such as our own. The interstellar material is more dense in some locations than in others but is everywhere less dense than the best vacuum yet produced in the laboratory. Astronomers believe that the uncondensed matter making up the interstellar medium is the birthplace and graveyard of stars.

Interstellar gas

The interstellar gas is composed predominantly of hydrogen atoms, with smaller quantities of helium, together comprising as much as 99% of its total mass. Interstellar gas also consists of atoms of oxygen, sodium, potassium, calcium, titanium, and iron, as well as molecules of water vapor, carbon monoxide, ammonia, and formaldehyde.

The gas is cold, nonluminous, and transparent and thus cannot be seen directly, However, it is detected through the formation of dark lines in the spectra of hot stars that are not attributed to the stars themselves and through emissions of radio energy at different wavelengths such as the 21-cm wavelength of neutral hydrogen.

Interstellar dust or grains

Whereas interstellar gas is invisible, interstellar dust, or at least its visible effects, are apparent. The dust is believed to consist of solid grains with estimated diameters of 10^{-5} cm or smaller. Whether the grains are tiny bits of metal such as iron or small ice crystals of ammonia, methane, and water has yet to be determined.

Unlike the gas, interstellar dust interferes with the paths of photons, resulting in the scattering of light. This causes starlight to appear redder than it really is, similar to the way in which the sun looks redder as it rises or sets through a greater thickness of atmosphere. A comparison of the spectrum of a star with its color index produces a *color excess,* which can then be used to determine something of the properties of the obscuring grains between the star and the observer.

The presence of interstellar dust particles may also be inferred from the photographic counts of stars and galaxies taken in the direction of the Milky Way. It is thus found that there is a deficiency of both stars and galaxies when compared to the theoretical number that should be visible. This indicates the presence of grains that not only redden the light from those objects that are seen but also dims the light, often to the point where some objects are not visible at all.

Nebulae

Interstellar gas and dust almost always occur together. Great, obvious concentrations of these materials are known as nebulae, or clouds. Three types of nebulae have been identified: emission (or bright) nebulae, reflection nebulae, and absorption (or dark) nebulae.

If a cloud of interstellar gas and dust lies near a hot star, the majority of the light coming from the cloud is emitted by the atoms of gas in it. Spectroscopic analysis of this light shows the presence of bright lines, and the cloud is called an emission or *bright nebula.* The Great nebula in Orion, seen by the naked eye as the middle "star" in the sword of the hunter, is a spectacular example.

On the other hand, if a cloud of gas and dust lies near a cooler star, a majority of the light emanating from the mass comes from starlight that has been reflected or scattered by dust particles. The spectrum of such a mass is the same as that of the star with which it is associated—dark lines superimposed on a continuous spectrum. Such a cloud of interstellar gas and dust is called a *reflection nebula.* The brighter stars in the Pleiades open-star cluster are intertwined in reflection nebulae.

Finally, if a mass of interstellar matter is not near any star, it is not illuminated enough to produce either a bright or reflection nebula. In this situation the dust in the cloud simply filters out or blocks starlight coming from stars beyond. The mass is thus seen as a black patch known as a *dark nebula.* The famous Coalsack of the southern Milky Way and the Horse's Head in Orion are excellent examples of dark nebulae (Fig. 7-13).

OUR QUEST FOR THE STARS

At this point in history the stars seem unreachable because of three interrelated quantities: speed, distance, and time. To reach the vicinity of another star a spacecraft must first reach or exceed the 384 miles/sec (1,382,400 miles/hr) escape velocity of the sun. To date, the greatest speed attained by a manned spacecraft has been 18 miles/sec (25,000 miles/hr). Second, the spacecraft must traverse the great distances that separate us from even the nearest of the stars. Proxima Centauri, the sun's closest stellar neighbor, is a full 26 trillion miles away. At a speed of 25,000 miles/hr, it would take a staggering 130,000 years to reach this star, and this assumes a straight-line path, which cannot be achieved in a dynamic universe.

Therefore we must await a breakthrough in velocity

Fig. 7-13. Horse's Head in Orion is an excellent example of a dark nebula. Dust particles in the area are not sufficiently illuminated to glow on their own, but they do manage to block the glow of objects behind them, creating these dark shapes. (Courtesy University of Chicago Yerkes Observatory, Williams Bay, Wisc.)

capability and propulsion systems before we can ever hope to explore the stars. Even with vastly improved technology we must remember that we are human—creatures who need air to breathe, water to drink, and food to eat. New recycled environmental systems must be developed. We also have to deal with the time limitations due to aging. Either suspended animation must become a reality or astronaut families must leave earth on board a spacecraft with the hope that a descendant generations hence may return.

In our contemplation of our future on earth, it would be well for us to consider the realities of space and the folly of a possible exodus of people from an overpopulated, depleted, and polluted planet. Except for an occasional brave explorer, most of us for better or worse will probably live and die on Spaceship Earth.

8 *islands in the sky*

Galaxies

Big whirls have little whirls,
That feed on their velocity,
And little whirls have lesser whirls,
And so on to viscosity.

L. F. Richardson

Many objects visible on the celestial sphere appear as hazy patches of light when viewed with the naked eye, through binoculars, or even with a small telescope. Because of the cloudlike appearance of those objects they became known as *nebulae,* meaning obscure. The famous French comet hunter, Charles Messier, cataloged 103 of these objects that now bear the first letter of his name, "M," followed by a number indicating the order in which the celestial objects were listed (p. 116 and Appendix C). It was not until the advent of large telescopes and the astronomical camera, however, that the true nature of many of these objects was actually discovered. While most were found to be gaseous nebulae and star clusters, the identity of others remained a mystery until early in the twentieth century.

HISTORICAL OVERVIEW

In 1610 Galileo discovered that the Milky Way, that faint band of light extending across portions of the sky, was composed of stars (Fig. 8-1). In 1750, 140 years later, Thomas Wright (1711-1786) published a speculative explanation for the Milky Way in which he stated that the sun formed part of a disc-shaped system of stars and that the Milky Way was our view through the plane of the disc. Shortly thereafter Immanuel Kant (1724-1804), in 1755, recorded his belief that the mysterious nebulae, which telescopes revealed in great numbers, were island universes—other great star systems similar to our own. However, it was William Herschel who in 1785 actually demonstrated the true quantitative nature of our galaxy.

Despite Kant's personal speculation, the Milky Way was generally considered unique, and its approximate center was believed to be the sun. Even early in the twentieth century, astronomers believed that the Galaxy extended only a few thousand light-years from the sun. In 1924, Edwin Hubble (1889-1953), an American astronomer working at the Mt. Wilson Observatory with the newly developed 100-inch reflecting telescope, discovered Cepheid variable stars in M31, now known as the Great galaxy in Andromeda (Fig. 1). Distance calculation to these stars indicated that M31 was 750,000 light-years away—well beyond the limit of the Galaxy. As a result, M31 was found to be a galaxy external to our own. Although Hubble's calculations were later found to be too small by a factor of 3, the error was only in mathematical detail, not in basic philosophy or direction of approach.

The realization that the Galaxy was neither unique nor central in the universe ranks with Copernicus' heliocentric theory as one of the truly great advances in astronomical thought.

TYPES OF GALAXIES

Galaxies take three basic forms—spiral, elliptical, and irregular. *Spiral* galaxies exhibit a spiral structure, with two or more arms arranged symmetrically about a nuclear region. The spirals are further divided into two broad groups—normal spirals and barred spirals. The arms of *normal spirals* originate at the center of the galaxy, while those of *barred spirals* are seen extending outward from the ends of a luminous nuclear *bar*. Within these two groups, spiral galaxies are still further subdivided according to the following three related criteria: (1) the relative size of the nuclear region, (2) the tightness with which the spiral arms are wound, and (3) the degree of resolution of the arms.

Of all bright galaxies, 77% are of the spiral type and

Fig. 8-1. Milky Way in Cygnus. This 3½-hour time exposure shows vast dark clouds of dust blocking the starlight from beyond. Our sun is a "close" neighbor to these stars. (Courtesy University of Chicago Yerkes Observatory, Williams Bay, Wisc.)

four fifths of these are normal spirals. When a galactic measurement is made by volume in space, however, spirals only make up about 30% of the total. Examples of several different spiral forms are shown in Fig. 8-2.

Elliptical galaxies are beautifully symmetric but featureless. They range in shape from almost perfect spheres to highly flattened ellipses. It is the shape that determines the classification of elliptical galaxies. Those that are circular, for example, are known as E_0, while those exhibiting the most highly flattened structure are called E_7 (Fig. 8-3). The largest galactic forms are of the elliptical type. Elliptical galaxies make up 20% of all bright star systems but as many as 60% of the total galaxies in a sample volume of space.

The remaining 3% of all bright star systems are *irregular* galaxies, which lack both symmetry and any recognizable form. In a given volume of space, irregular galaxies make up approximately 10% of the total. Two irregular galaxies are shown in Fig. 8-7.

Attempts have been made to link the galactic types together in an evolutionary sequence. Although generalizations are tempting, the hard facts of mass and angular momentum indicate that the evolution of one galactic type into another is highly unlikely. It therefore appears that the different types of galaxies represent separate species rather than one species seen in different stages of evolution. Within species, however, evolution is still within the realm of possibility, and stellar evolution within any galaxy almost a certainty.

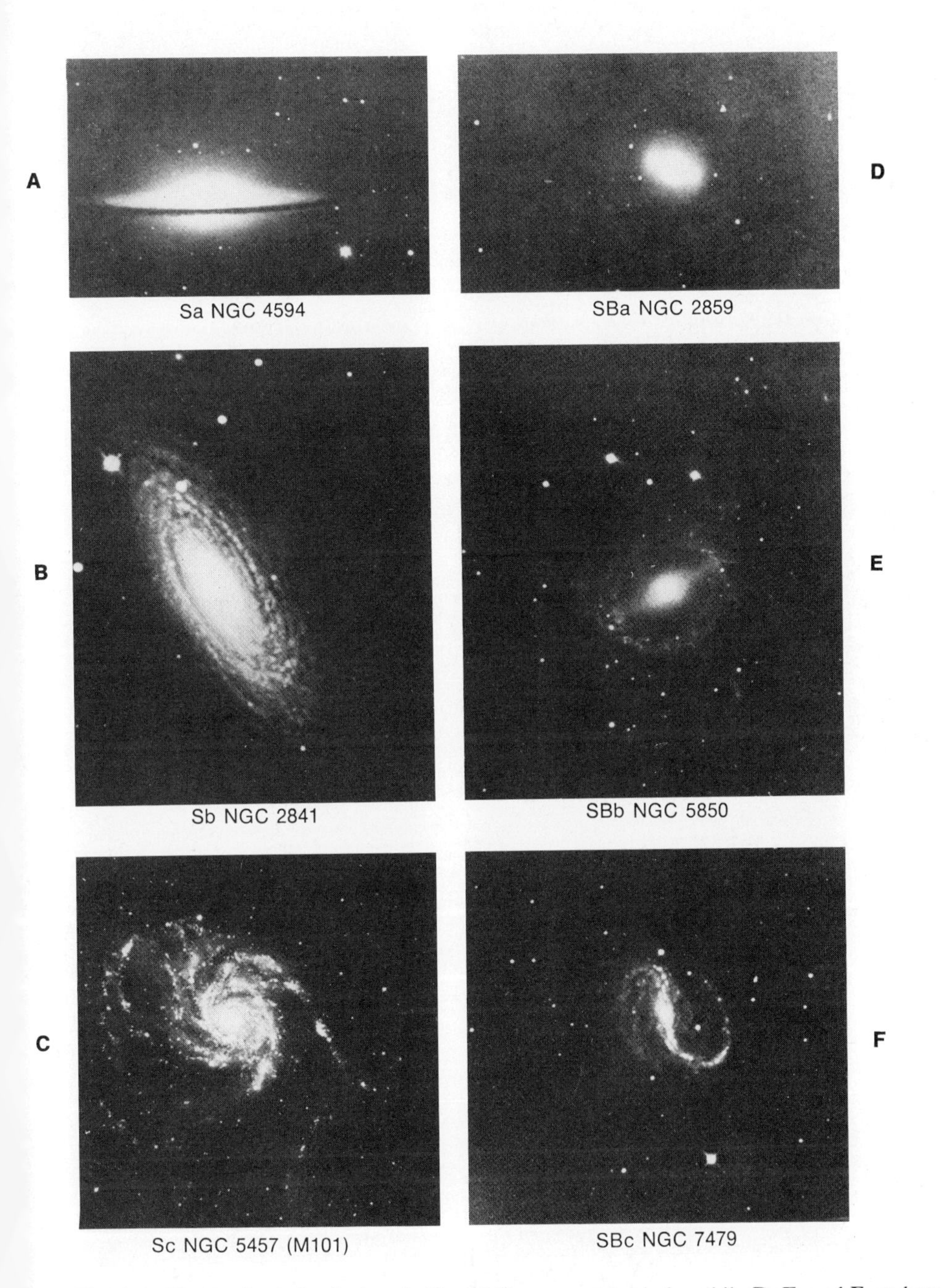

Fig. 8-2. Types of spiral galaxies. **A, B,** and **C** are normal spirals, while **D, E,** and **F** are barred spirals. (Courtesy University of Chicago Yerkes Observatory, Williams Bay, Wisc.)

THE GALAXY

The earth is a small planet orbiting a relatively small, average star, the sun, which in turn orbits the center of a gigantic, rotating star system called the Galaxy or the Milky Way. Taking the shape of a flattened spiral, the Galaxy consists of at least 100 billion stars and huge quantities of interstellar gas and dust. The Galaxy, 100,000 light-years in diameter and 5 to 15,000 light-years thick, has three basic parts: the nucleus, symmetric spiral arms, and the corona, or halo (Fig. 8-4).

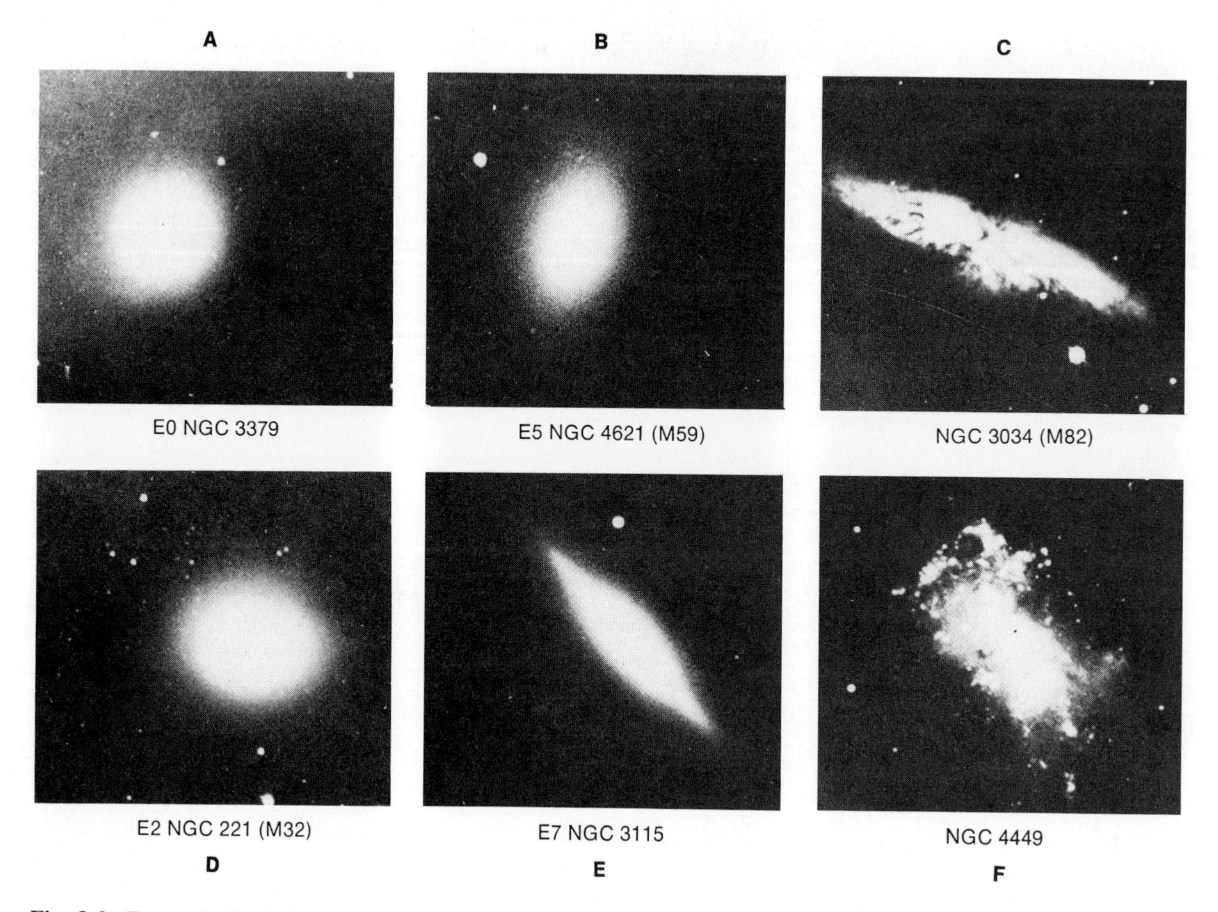

Fig. 8-3. Types of elliptical and irregular galaxies. **A, B, D,** and **E** are ellipticals. **C** and **F** are irregulars. (Courtesy University of Chicago Yerkes Observatory, Williams Bay, Wisc.)

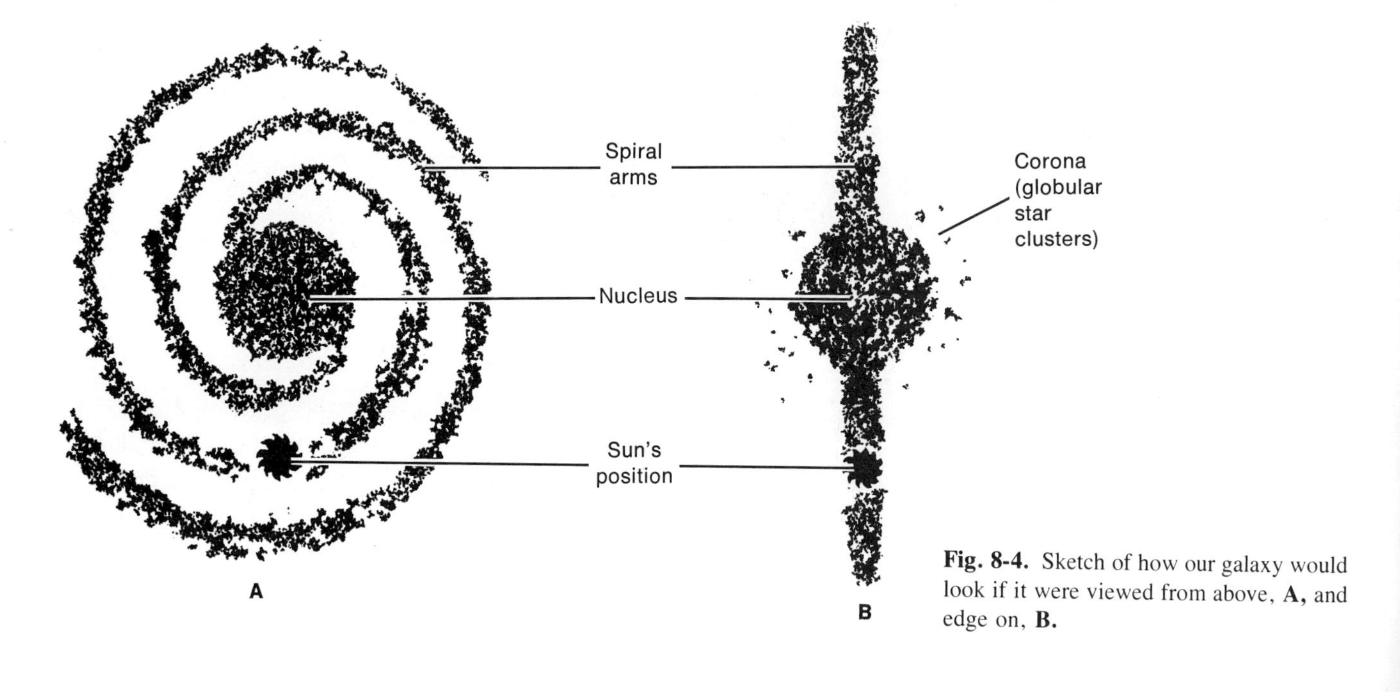

Fig. 8-4. Sketch of how our galaxy would look if it were viewed from above, **A,** and edge on, **B.**

Nucleus

Little is known directly about the nucleus of the Galaxy because extensive clouds of interstellar dust completely obscure our view. A large amount of infrared radiation, however, has been recorded emanating from an oval region in the constellation of Sagittarius. Examination of the region indicates that this oval is the galactic nucleus, a flattened spheroid of stars from 3300 to 4900 light-years in diameter.

Within the nucleus itself is a central core of ordinary stars roughly 3 light-years in diameter. The number of stars per unit volume in the nuclear core is probably about 10 million times what it is in the vicinity of the sun. The stars in the core are thus 200 times closer together than the familiar stars we see. Therefore on a planet orbiting a star in the nuclear core, the stars visible in the night sky would appear some 40,000 times or more than 11 magnitudes brighter than those normally seen from earth. Even at this density, however, the nearest stars would be thousands of times more distant then the planet Pluto is from the sun.

Analyses of the nuclei of other spiral galaxies show that these central regions are composed almost entirely of stars, with no dust and little gas present. The same is also true of the entire structure of elliptical galaxies.

Spiral arms

The most outstanding feature of our own galaxy is its system of spiral arms—gaseous envelopes filled with stars and dust. Features associated with spiral arms that may also be used to trace their location and extent are called *spiral tracers*. Such spiral tracers, including O and B supergiant stars, emission nebulae, young galactic star clusters, and long-period cepheid variable stars, indicate that the Galaxy's arms circle it at least three times (Fig. 8-4). The sun and solar system appear to occupy a position within and close to the inner edge of the second of the three turns, which is roughly 33,000 light-years from the nucleus to one edge of the Galaxy. Once every 200 million years, the sun completes one revolution about the galactic nucleus, a time span referred to as the cosmic year.

The local arm of the Galaxy is called the *Carina-Cygnus* arm because it contains outstanding spiral tracers located between these two constellations. Beyond the local arm lies the *Perseus* arm, 6500 light-years away, which contains the famous Double Cluster (Fig. 7-12). Extending between the nucleus and the local arm is the *Sagittarius* arm, which is approximately 5000 light-years from the sun.

Very little is known about the formation of the spiral arms. An important clue, however, may lie in the motion of interstellar gas within the Galaxy itself. The gas, predominantly neutral atomic hydrogen, forms a thin disc 1200 light-years thick and 100,000 light-years in diameter in the central plane of the Galaxy. The hydrogen in the disc expands outward from the galactic nucleus at speeds ranging from 30 to 100 miles/sec. It is estimated that 200 million solar masses of matter per cosmic year are transported out of the nucleus to other parts of the Galaxy. How then does the nucleus replenish its reservoir of gas? An intriguing possibility might be that new gas enters the nucleus from the galactic halo, a huge sphere of thin gas, stars, and globular star clusters which surrounds the Galaxy.

The spiral arms of the Galaxy were probably formed recently and will be short lived, cosmically speaking. The majority of the stars that presently serve as spiral arm tracers condensed from interstellar material less than 100 million years ago, or half of a cosmic year. The sun and its solar system, on the other hand, have existed for about 25 cosmic years. While the oldest stars in our galaxy are nearly twice the age of the sun, the youngest have ages of less than one hundredth of a cosmic year.

It appears that gravity has doomed the spiral arms to a rather brief astronomical existence. Because of the great gravitational attraction of the massive galactic nucleus, the parts of the arms closer to the nucleus complete their revolutions faster than do those at greater distances. The difference in rotation rates would render the spiral pattern unrecognizable in just a few cosmic years. However, if gravity were the only force controlling the gaseous arms, they probably would never have formed in the first place. Conversely, if the arms came into being through some freakish cosmologic accident, they would surely have dissipated in short order. Therefore it is highly probable that magnetic forces work in conjunction with gravity to control the shape and movement of the gaseous spiral arms. These magnetic forces apparently are able to preserve the spiral pattern more or less intact until the gas in a given arm is exhausted by the condensation and evolution of new stars.

Corona

Surrounding the main body of the Galaxy, the nucleus and the spiral arms, is a nearly spherical system of globular clusters, gas, and old stars called the corona, or halo. Its volume is thought to exceed that of both the nucleus and spiral arms many times over.

The corona was discovered in 1917 by Harlow Shapley (1885-1972) through a study of the locations of and distances to globular star clusters. These clusters, for example, were observed to be strongly concentrated in the direction of the Southern Hemisphere constellations of Scorpius, Ophiuchus, and Sagittarius, almost wholly out of the plane of the Milky Way. Furthermore, they were found to contain R.R. Lyrae and Cepheid variable stars whose distances could be calculated by an analysis of their periods of variations relative to their absolute magnitudes. All globular clusters were thus found to be from 15,000 to 100,000 light-years from the sun, well outside the galactic disc.

The existence of individual stars as far as 6000 light-years outside the galactic plane was confirmed by the discovery of R.R. Lyrae and Mira-type pulsating variable stars not associated with clusters. Although they tend to be concentrated toward the galactic center, noncluster halo population stars are found in both high and low galactic latitudes.

Unlike celestial objects in the spiral arms, the orbits of halo population objects do not conform to the rotation of the Galaxy. The individual stars and globular clusters in the corona move in highly eccentric orbits greatly inclined to the galactic plane. In fact, several long-period, Mira-type pulsating variable stars are moving in nearly hyperbolic orbits that may lead to their eventual escape from the Galaxy.

OTHER GALAXIES

Although it is quite impressive, our galaxy is by no means unique. We now know that galaxies extend outward in space in all directions. In fact, the farther out we look, the more galaxies we see. It was Hubble who estimated that there were approximately 100 million galaxies within reach of the 100-inch telescope. The development of the 200-inch telescope increased this number to a billion. Most likely, billions of other galaxies lie beyond the reach of any present-day instrument.

Galactic distances

Parallax cannot be used to calculate the distance to galaxies because all of these star systems are far too remote. The 186 million mile baseline of the earth's orbit is too short to enable an observer on earth to note any apparent shift in a galaxy's position. With large telescopes, however, Cepheid variable stars can be identified in nearby galaxies. Thus the distances to these stars can be determined, and as a result we can estimate the approximate distances to the galaxies themselves. More distant galaxies are so far away, however, that even the largest telescopes cannot resolve them into stars.

In 1880 the German astronomer H. C. Vogel first demonstrated that the spectra of stars could give information about their motions that could not otherwise be obtained; he applied the Doppler effect to the analysis of starlight. During the latter part of the nineteenth century and the early part of the twentieth century, astronomers at large observatories around the world spent much of their time measuring the velocities of approaching and receding stars in the Galaxy.

Early in the 1920s, V. M. Sepher of the Lowell Observatory made a discovery that was to have a profound impact on our concept of the universe. He noted that the spectra of a number of "nebulae" believed to be in our own galaxy showed red Doppler shifts, an indication that they were receding from us at high speeds. Not long after Sepher's discovery, Hubble established that these nebulae were in fact separate star systems far outside our own galaxy. Even more remarkable, Hubble found that the velocities of recession of galaxies were proportional to their distances. This meant that the more distant galaxies were receding at the fastest speeds. This relationship became known as *Hubble's law*.

As a result of this work, it is now possible to translate a measure of a galaxy's red shift directly into a rough estimation of its distance. One of the most distant known galaxies is located in a faint cluster of galaxies in the constellation of Boötes (Fig. 8-5). The wavelengths of light coming from this galaxy are shifted toward the red end of the spectrum by 45%, indicating a recession speed of nearly half the speed of light. Thus the Boötes cluster is estimated to be about 5 billion light-years away.

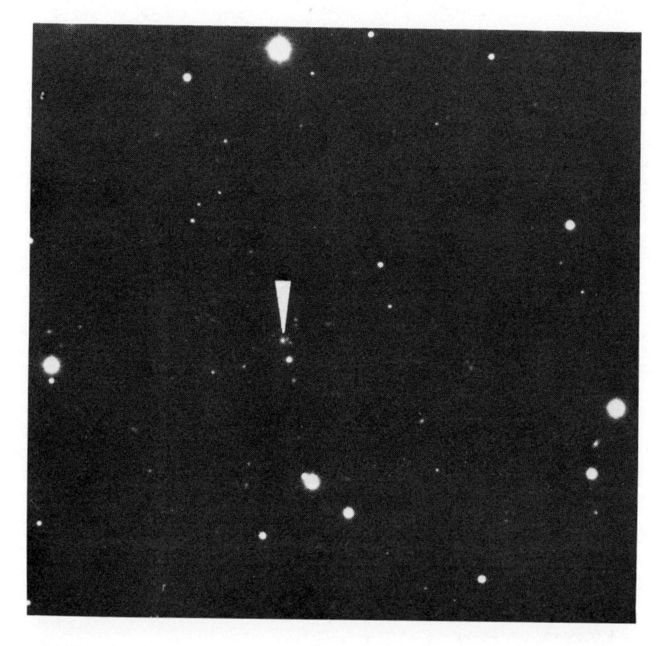

Fig. 8-5. This galaxy, 3C295 in Boötes, is the second most distant galaxy we know. A more distant galaxy was found in 1975 through the use of special computer techniques. The spectral lines of this galaxy show that it is receding from earth at nearly half the speed of light. (Courtesy Hale Observatory, Pasadena, Calif.)

It was the observed red shifts in the spectra of galaxies that finally led to the general acceptance of the concept of an expanding universe and to the development of the steady state and big bang theories of the formation of the universe.

Galactic diameters

Like the diameters of the sun, moon, and planets, the diameter of a galaxy can be determined when both its angular diameter on the celestial sphere and its distance are known. The Great galaxy in Andromeda, for example, has an angular diameter on the celestial sphere of about 4°. At a distance of 2 million light-years, this angular diameter represents a true linear diameter of about 125,000 light-years. The average linear diameter of the 17 nearest known galaxies is approximately 3200 light-years.

Galactic masses

The mass of a galaxy may be determined by utilizing several techniques. Galaxies are often found linked gravitationally in pairs. If the distance between the two galaxies in a pair and their relative velocities are known, Newton's derivation of Kepler's third law can be used to find their total mass.

Another method, used chiefly for spiral galaxies viewed obliquely or edge on, involves first determining the velocity of rotation by measuring the Doppler shift in various parts of the galaxy's disc. As a second step, a rotation curve is plotted to show how the velocity of rotation varies with the distance from the center of the galaxy. Third, the mass is estimated based on the requirement that the outward force produced by rotation must be equal to the inward force of gravity.

While the masses of galaxies are typically 100 billion times that of the sun, they range from 100 million solar masses for some nearby dwarf galaxies to a trillion solar masses for giant elliptical galaxies in more distant parts of the universe.

CLUSTERS OF GALAXIES

Galaxies are not evenly distributed throughout the celestial sphere. There are many small groups of galaxies, and here and there are rich clusters of as many as 9000 members. Clusters of galaxies range from 1 to 10 million light-

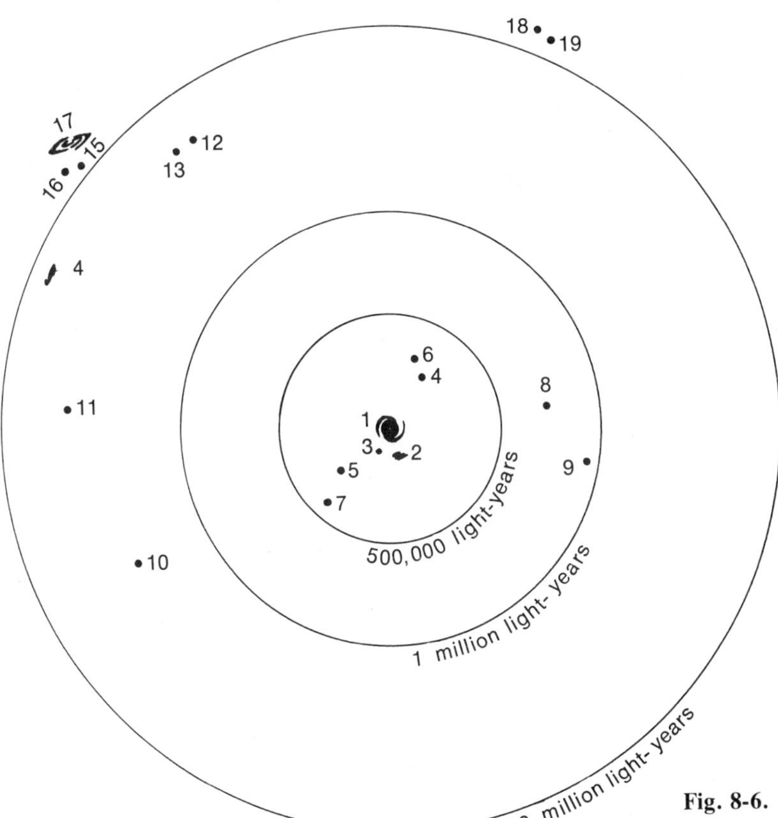

Fig. 8-6. Galaxies of the Local Group. Numbers refer to the first column in Table 10.

years in diameter. Even such huge clusters as these do not seem to be distributed at random in the universe. There is evidence to indicate that there are also super clusters containing perhaps 100 members spread out over 100 million light-years. On scales larger than this, however, the universe appears to be uniform.

The Local Group

Our own galaxy is a member of a cluster of galaxies called the Local Group. The group, approximately 3 million light-years in diameter, consists of 19 confirmed members: 3 spiral galaxies, 12 elliptical galaxies, and 4 irregular galaxies (Fig. 8-6 and Table 10). The brightest members of the Local Group seen from earth are M31, the spiral Great galaxy in Andromeda; the Large and Small Clouds of Magellan, irregular galaxies in the southern hemisphere; and M33, the spiral galaxy in the constellation of Triangulum. At the outer reaches of the Local Group lie Maffei I, a giant elliptical galaxy, and Maffei II, a spiral galaxy, which were discovered in 1968.

M31, by far the largest and most massive of the nearby members of the Local Group, is inclined to our line of sight at an angle of 15°, which means that we see it highly foreshortened. Even so, it covers almost 4° on the celestial sphere and is so bright that sometimes it can even be seen by the naked eye as a faint, hazy patch of light in the constellation of Andromeda. This great galaxy lies at a distance of 2.2 million light-years. With a computed mass of 400 billion times that of the sun, M31 probably contains over 300 billion individual stars.

The nearest external galaxies are the large and small Magellanic clouds discovered by Portuguese seamen in the fifteenth century and named for the famous explorer Magellan. The large Magellanic cloud is at a distance of 160,000 light-years, making it the closest member of the Local Group to our galaxy. The small Magellanic cloud is only slightly more distant. These clouds are irregular galaxies close enough to be regarded by many astronomers as satellites of the Milky Way and bright enough to be visible with the unaided eye, even in full moonlight (Fig. 8-7). The two Magellanic clouds are separated by 22° on the celestial sphere—the large cloud in the constellation of Tucana the Toucan (a tropical bird). The actual separation of the clouds is 80,000 light-years from center to center. Unfortunately, the Magellanic clouds are never visible from the latitude of the United States.

Table 10. Basic data on galaxies of the Local Group

No.*	Galaxy	Type	Right ascension in 1950	Declination in 1950	Apparent magnitude	Distance (1000 ly)	Diameter (1000 ly)	Absolute magnitude	Apparent radial velocity (km/sec)†
1.	Our Galaxy	Sb	—	—	—	—	100	−21	—
2.	Large Magellanic cloud	Irr I	5^h26^m	−69°	0.9	160	30	−17.7	+276
3.	Small Magellanic cloud	Irr I	0^h50^m	−73°	2.5	180	25	−16.5	+168
4.	Ursa Minor system	E_4 (dwarf)	$15^h8.2^m$	+67° 18′	—	220	3	−9	—
5.	Sculptor system	E_3 (dwarf)	$0^h57.5^m$	−33° 58′	8.0	270	7	−11.8	—
6.	Draco system	E_2 (dwarf)	$17^h19.4^m$	+57° 58′	—	330	4.5	−10	—
7.	Fornax system	E_3 (dwarf)	$2^h37.5^m$	−34° 44′	8.3	600	22	−13.3	+39
8.	Leo II system	E_0 (dwarf)	$11^h10.8^m$	+22° 26′	12.04	750	5.2	−10.0	—
9.	Leo I system	E_4 (dwarf)	$10^h5.8^m$	+12° 33′	12.0	900	5	−10.4	—
10.	NGC 6822	Irr I	$19^h42.1^m$	−14° 54′	8.9	1500	9	−14.8	−32
11.	ICI 613	Irr I	$1^h00.6^m$	+1° 41′	9.61	2200	16	−14.7	−238
12.	NGC 147	E_6	$0^h30.4^m$	+48° 13′	9.73	1900	10	−14.5	—
13.	NGC 185	E_2	$0^h36.1^m$	+48° 4′	9.43	1900	8	−14.8	−305
14.	NGC 598 (M33)	Sc	$1^h31.0^m$	+30° 24′	5.79	2300	60	−18.9	−189
15.	NGC 205	E_5	$0^h37.6^m$	+41° 25′	8.17	2200	16	−16.5	−239
16.	NGC 221 (M32)	E_3	$0^h40.0^m$	+40° 36′	8.16	2200	8	−16.5	−214
17.	Andromeda galaxy (NGC 224, M31)	Sb	$0^h40.0^m$	+41° 0′	3.47	2200	130	−21.2	−266
18.	Maffei I	E	2^h32^m	+59° 25′	11.00	3300	—	−19	−165
19.	Maffei II	E or S	2^h38^m	+59° 23′	—	—	—	—	—

*These numbers correlate with numbers in Fig. 8-6.

†Most apparent radial velocity is due to the revolution of the sun around the center of the Galaxy.

Fig. 8-7. Region of the Magellanic clouds. The large cloud (below) and the small cloud are irregular galaxies that many believe are satellites to our own. The oval area to the left of the small cloud is the overexposed image of a star. (Courtesy University of Chicago Yerkes Observatory, Williams Bay, Wisc.)

Other clusters

Clusters of galaxies are numbered in the thousands. The individual clusters themselves may contain as few as four or five members, as in Stephen's Quintet in the constellation of Pegasus the Flying Horse, or as many as 9000, as in the cluster in the constellation of Coma Berenices (Berenice's Hair).

The nearest cluster to the Local Group is in the constellation of Virgo the Virgin, which is 30 million light-years away and contains at least 1000 members. Other well-known clusters include the 100-member Hercules cluster, 350 million light-years away; the Corona Borealis cluster of 400 galaxies, 400 million light-years away; and the 200-member Hydra cluster, 1.1 billion light-years away.

RADIO ASTRONOMY

The first distinct radio source outside our solar system was discovered in 1946 with the use of a radio telescope. The radio energy was found emanating from a spot in the constellation of Cygnus the Swan, and was thus designated Cygnus A. By 1948, radio sources had also been found in the constellations of Taurus, Casseopeia, Centaurus, Virgo, and Hercules. Today, distinct radio sources number in the thousands.

The first identification of a distinct radio source with an optically visible object was made in 1950 by Rudolph Minkowski and Walter Baade of the Mount Wilson and Palomar observatories. Using the 200-inch telescope, they showed that Cygnus A coincided with the position of a galaxy 700 million light-years away. Soon thereafter, Virgo A and Centaurus A were also identified with known galaxies. When a galaxy emits both visible light and continuous radio energy, it is known as a *radio galaxy*. Of the thousands of known radio sources today, it has been shown that more than 100 are of this type. The vast majority of radio sources, however, are not associated with galaxies but with emission nebulae, planetary nebulae, supernova remnants, and quasars.

Exploding galaxies

Circumstantial evidence strongly suggests that enormous explosions have been taking place in the central regions of certain galaxies—perhaps even in our own. When studied optically, radio galaxies, which are usually giant ellipticals, often appear strange in some respect. For example, there are often jets or filaments of high-energy gas extending outward from the galaxy, or frequently very small but intensely bright nuclei are visible. NGC 4651 and M82 are examples of radio galaxies with filaments, while Seyfert galaxies, a type of spiral, exhibit very bright nuclei (Fig. 8-8). Spectroscopic analyses confirm that these galaxies are experiencing violent internal upheavals.

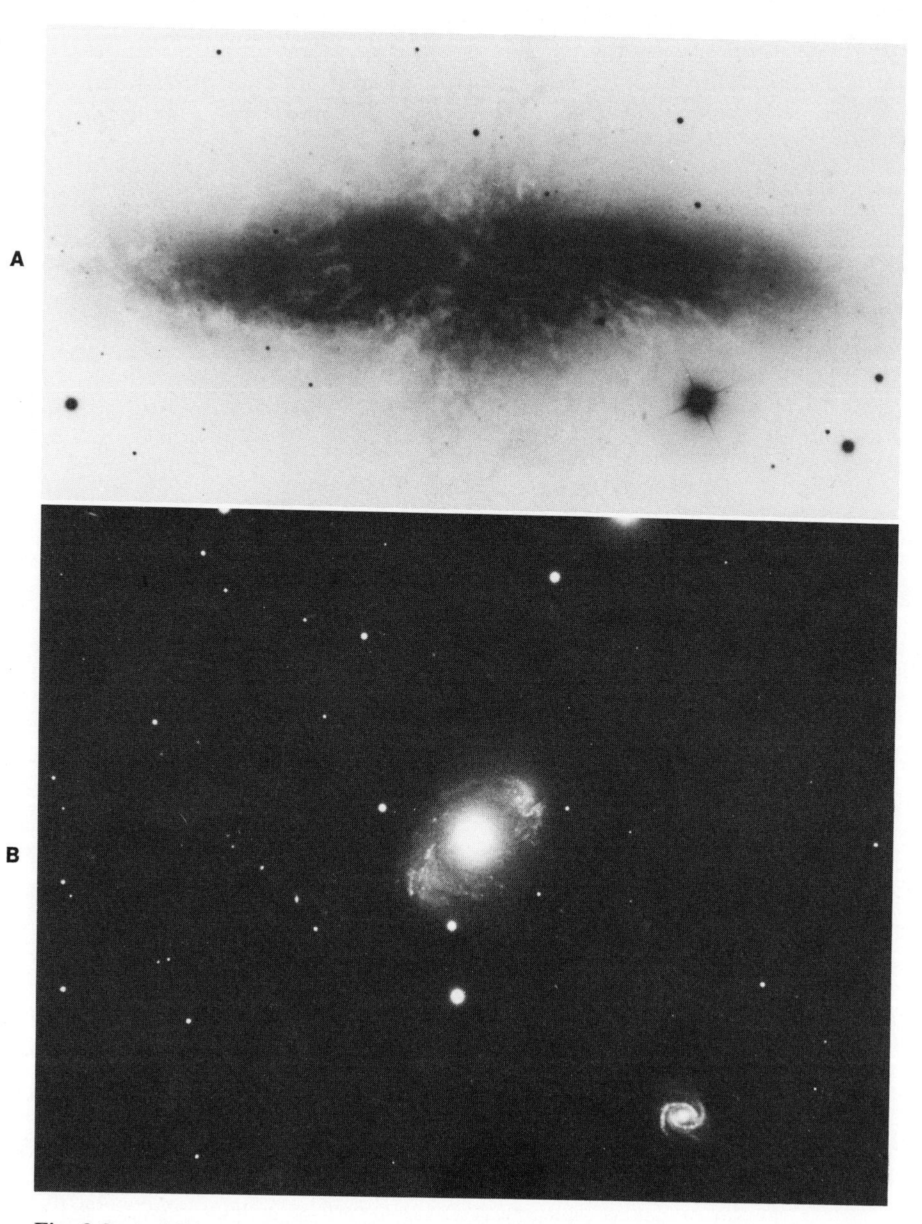

Fig. 8-8. A, M82, an exploding galaxy, shows bands of high-energy gas extending outward from the center. **B,** NGC 4151, a Seyfert-type galaxy, exhibits the bright nucleus typical of radio galaxies. (Courtesy Hale Observatory, Pasadena, Calif.)

Jose Sersic of Cordoba Observatory in Argentina and H. A. Abt of Kitt Peak National Observatory in Arizona have suggested that the explosions of giant elliptical galaxies act as the source of many if not all other galaxies. Sersic writes:

> It therefore seems that in the early stages of the universe, the only galaxies in existence were massive ellipticals. These undoubtedly in some way resulted from earlier cosmological evolution. Instabilities then developed in these high-density systems causing explosions that ejected fragments of various sizes. . . For a brief period during each explosion there was intense radio emission which was now observed in the radio galaxies and quasi-stellar objects.

Although the explanation for exploding galaxies is presently unknown, it is generally believed that these spectacular events are by far the most energetic ever observed by man.

Quasi-stellar objects (quasars)

Alan Sandage of the Mt. Wilson and Palomar Observatories announced in December of 1960 that photographic plates showed a blue star at the precise position assigned to a strong radio source, designated 3C48. Then in 1963 another radio source, 3C273, was identified with a starlike object even brighter than 3C48 in the constellation of Virgo. The more intense light from 3C273 made it possible to photograph and analyze its spectrum. To the amazement of the astronomical world, 3C273 showed a Doppler shift of 16% toward the red part of the spectrum.

The term ''quasi-stellar object'' (quasar) is now used to describe all starlike objects with large red shifts. The number of such objects exceeds 1000 (Fig. 8-9). Their red shifts range from a low of 16% for 3C273 to a high of 223% for OH471. The latter figure corresponds to a recessional ve-

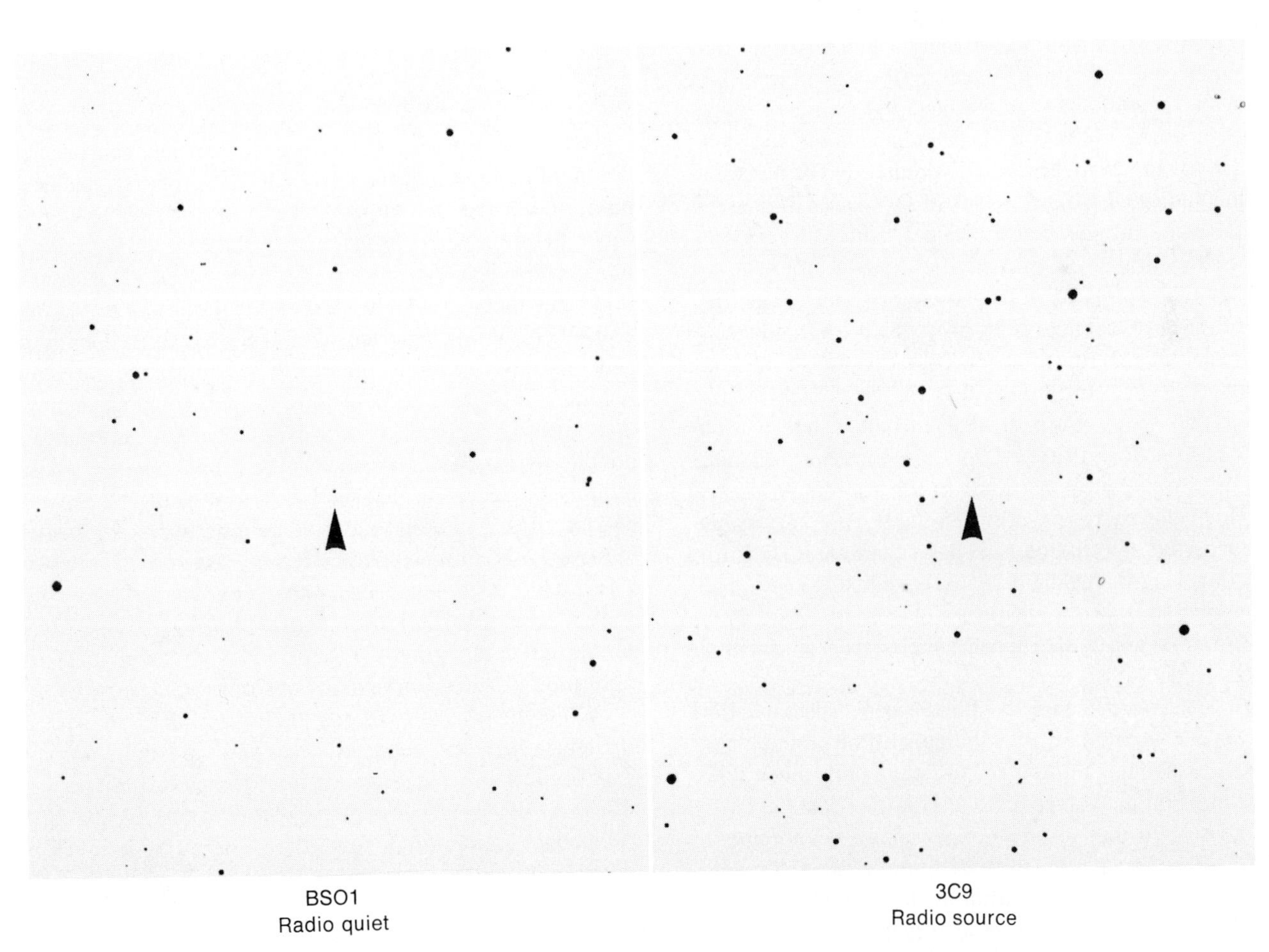

Fig. 8-9. Quasi-stellar objects (quasars) BS01 and 3C9. (Courtesy Hale Observatory, Pasadena, Calif.)

locity of 91% of the speed of light or 169,260 miles/sec and a distance of 12 billion light-years.

With the discovery of the large red shifts of quasars came the realization that there could only be two possible explanations that were compatible with the known laws of physics: either the objects were receding at high speeds or their light was being emitted in gravitational fields far stronger than any previously known. The gravitational field theory fell short on two counts: (1) the duration of the phenomenon would be too short and (2) the object's density would be far in excess of that indicated by spectroscopic evidence. Thus the large red shifts were interpreted as the effects due to rapid recession.

Spectroscopically, quasars exhibit broad, weak emission lines of such elements as hydrogen, helium, carbon, oxygen, nitrogen, neon, magnesium, silicon, argon, and sulfur. Unexpectedly, perhaps, we find these strange objects with familiar compositions not unlike stars and gaseous nebulae in our own galaxy. Some quasars also show spectral absorption lines believed to be produced in a slightly cooler envelope of material, probably a gaseous shell that had been ejected and is now receding from the object. The strength of spectral lines of quasars, like those of stars, provides a crude indication of the temperature and density of the gas in which the lines are produced. The temperatures thus indicated are a few tens of thousands of degrees absolute, while the gas density ranges from 10^4 to nearly 10^7 particles/cm^3.

A significant property of most quasars is their variability in brightness. An example is 3C345, which was studied by Kinman and Goldsmith of the Lick Observatory from June to October, 1965. They found that the brightness of this quasar varied by as much as 40% in only a few weeks. Light variation in quasars is important because it enables astronomers to set an upper limit to the size of the region from which the energy is being emitted. This is possible since the period of variation is fixed by the time it takes for light to travel across the diameter of the radiating body. The best determination of the diameters of quasars range from 3 to 30 light-years, a size much smaller than an average galaxy.

Their small size, coupled with emission spectra indicating gas at low pressure, sets the upper limit to the mass of quasars at about 1 million suns—here again, much less than the mass of an average galaxy. However, their energy output relative to that of an average galaxy is enormous. On the average, a quasar emits 40 times as much energy in the visible part of the spectrum as the brightest galaxy. This is astonishing since a typical bright galaxy contains 100 billion stars.

The true nature of quasars is still being debated. The critical consideration in the study and understanding of their properties is a knowledge of their distances. Although the red shift is generally thought to be due to rapid recession, the matter is still open to question. If, for example, quasars are small, massive bodies at great distances, as most astronomers believe, then they are the most luminous of all celestial objects and virtually require a new type of energy source to explain their brightness. On the other hand, if they are nearby, what is the explanation for their large red shifts? If quasars are on the fringes of the universe, as indicated by their rapid recessional velocities, then they are the most distant and therefore the oldest known celestial bodies. This would mean for us on earth a glimpse of how things appeared at a time near the formation of the universe. If quasars are nearby, then the dimensions and age of the known universe will have to be revised downward for the first time in history.

A solution to the mystery of quasars may lead at last to man's knowledge of how the universe was actually formed—the final cosmology.

A PRIMEVAL FIREBALL

Whereas stars are the fabric of the heavens, galaxies are the building blocks of the universe. The origin of galaxies, then, should be the same as the origin of the universe. Since the discovery of galactic red shifts, two cosmologic models have been developed. These are the steady state and big bang theories. Advocates of the *steady state theory* argue that the universe has always looked as it does now and that new matter is continuously being created out of nothing. New galaxies are thus formed to replace those that have disappeared over the "horizon." By far the most popular hypothesis, however, is the *big bang theory,* whose proponents speculate that the universe began with the explosion of a primordial atom containing all of the matter in the universe. Immediately after the explosion, space was filled with a homogeneous expanding gas. As the gas cooled, local irregularities in density developed and slowly contracted under gravity. The clouds of contracting gas were protogalaxies, or galaxies in formation.

While the discovery of quasars in 1960 did little to cool the debate between the steady state and big bang cosmologists, a further discovery made in 1965 had far greater significance. Robert Wilson and Arno Penzias of the Bell Telephone Laboratories found low-energy cosmic radio radiation filling the universe, bathing the earth from all directions. The existence of such radiation had actually been suggested earlier by Robert Dicke of Princeton University, who believed that if the universe had been formed

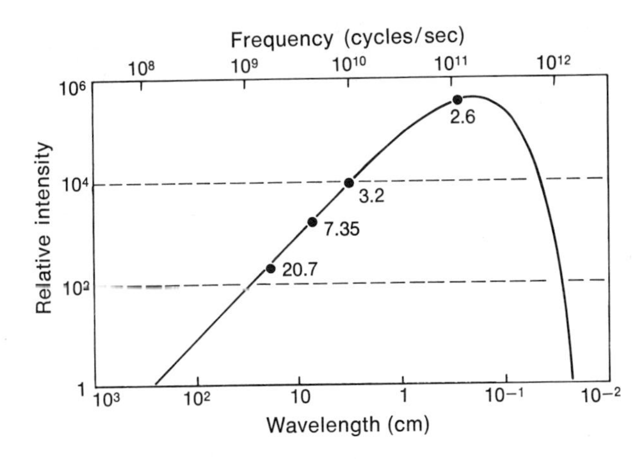

Fig. 8-10. Increase in radio wavelengths with increasing distance, a phenomenon that supports the big bang theory of an expanding universe.

by a titanic explosion, a "primeval fireball" of radiation should survive to this day and be detectable. Dicke was apparently correct: an explosion at least 10 billion years ago started the protogalaxies on their outward journeys. The 1 million-degree flash of radiation has since cooled to 3° A and must be observed at radio wavelengths that increase with increasing distance—further evidence of an expanding universe (Fig. 8-10).

The discovery of background radiation in space made it virtually impossible to seriously consider the steady state theory because the universe according to this theory was never in a compressed state—a prerequisite for a primeval fireball. The big bang was apparently the beginning of an evolving universe that is isotropic—it looks the same to all observers no matter where they are located. Such a universe has no boundary and because of the presence of matter possesses a three-dimensional, uniform curvature.

The surface of an expanding balloon is a good two-dimensional analogy to uniformly curved, three-dimensional space. The galaxies may be represented by inelastic dots attached to the surface of a balloon. As the balloon expands, an ant standing on any dot would see all other dots moving away—the more distant ones moving more rapidly. Our ant should not conclude that he is at the preferred center of the balloon because ants on other dots would see the same effect. The balloon model thus illustrates the general recession of galaxies, Hubble's law, and the lack of a preferred center for the universe.

The fireball radiation may be represented on the same model by a number of evenly distributed ants crawling over the surface of the balloon in all directions. The number of ants in any given area of the balloon decreases as expansion occurs, and regardless of the direction in which the ants crawl they will always move toward dots that are receding from them. In the universe the density of photons in the fireball radiation decreases as expansion occurs. Also, photons from the primeval fireball are forever chasing receding galaxies, undergoing in the process a continuous energy loss that explains the observed increase in wavelengths with distance.

IN RETROSPECT

Although we have come a long way in our search for knowledge of the universe, in a real way we have only begun to unlock its mysteries. We may soon reach a stage of frustration where our mental ability has outstripped our technologic skill. While we are able to peer billions of light-years into the universe with optical and radio telescopes, we are limited to only a few weeks of space travel. This ability will undoubtedly be increased in time, but without major technologic breakthroughs in propulsion and environmental systems, we may reach a point beyond which we are unable to progress—we are limited by our very nature.

First, we require oxygen, food, and water—supplies we must take with us from earth, for no other known sources exist anywhere in the universe. Also, our relatively short life spans are an obvious drawback in space travel that requires hundreds, thousands, or millions of years, even at the speed of light. Finally, our rather fragile mental makeup casts serious doubts on our ability to adjust to long years of separation from the earth.

9 *the blue planet*

Earth

So simple is the earth we tread,
So quick with love and life her frame,
Ten thousand years have dawned and fled,
And still her magic is the same.

The Earth and Man
Stopford Augustus Brooke

Astronomically, there is very little that is outstanding about the earth. It is neither the most massive planet nor the least massive. It is neither the hottest nor the coldest planet. It is not the only planet that has an atmosphere. Other planets also have satellites of their own and rotation periods that are longer or shorter than the earth's. Yet its lack of superlatives may be what makes it unique.

Although massive enough to retain an atmosphere, the earth is not so massive that it crushes living things under its gravitational pull. Its position in the solar system is close enough to the sun to receive just the right amount of radiation to support life as we know it but not so close as to fry its inhabitants. The earth's atmosphere is neither too thick nor too thin, and its composition is such that it acts as a selective filter, allowing only the proper kind and amount of radiation from the sun to reach its surface.

The truly unique aspects of the earth as a planet are the parameters of its satellite, its atmospheric composition, the presence of water on its surface, and its ability to support life. Although the moon is neither the largest satellite nor the satellite closest to its master planet, no other satellite is as large relative to the size of its primary and yet so close. The moon is one-fourth the size of the earth and only 238,000 miles away. This makes the earth and moon virtually a double planet.

The earth's atmosphere is unusual because of its composition. Nitrogen, oxygen, and argon gas, which make up the bulk of the earth's atmosphere, are extremely rare if found at all in the atmospheres of other planets. Carbon dioxide, hydrogen, helium, methane, and ammonia gas, so abundant in other planetary atmospheres, are in short supply in the atmosphere of the earth. Water, probably the most important substance on earth as far as natural environmental processes are concerned, has the ability under normal atmospheric conditions to exist in all three states of matter. Its liquid state, so significant on earth, does not exist to our knowledge on any other planet.

The earth is frequently defined as the planet occupied by man. Perhaps this statement should be altered to read the planet occupied by life, since life of any kind is not known to exist anywhere else in the universe. If present elsewhere, and statistically the odds favor that possibility, life, as on earth, must have developed in response to environmental conditions. Also, as on earth, that life would change in response to alterations in the environment.

EARTH'S COLOR

Light entering the earth's atmosphere is diverted from its normally straight path by molecules of air and water and by dust particles. This effect is known as *scattering*. Short wavelengths of light are more strongly affected than long wavelengths. Consequently, the long-wavelength red light travels more directly through the atmosphere than the short-wavelength blue light. Because the blue light is scattered widely in the atmosphere, we are able to see it coming from all directions.

The blue color of the daytime sky on a clear day, the blue color of bodies of water such as oceans and lakes, and the overwhelming blue color of the earth from space all result largely from this selective scattering of sunlight.

Scattering also explains other color effects that are visible in the atmosphere, such as the reds seen at sunrise or sunset and in dust storms. Near sunrise or sunset, scattering is increased because light must travel on a longer path through the atmosphere (Fig. 27-5). The blue light is largely removed, so that the sun and the sky near it appear orange or red. Dust and smoke produce a similar effect.

EARTH'S SIZE

The earth is not a large planet relative to the others in our solar system. With a diameter through its equator of 7927 miles, a circumference of 24,902 miles, and a mass of 6.6 sextillion tons (6.6 followed by 21 zeros), the earth ranks fifth in size among the nine major planets. Only in specific gravity, the weight of a substance relative to the weight of an equal volume of water, does the earth rank first, with an average specific gravity of 5.5.

The first recorded attempt to measure the size of the earth was made by Erathosthenes (276-195 B.C.), a Greek mathematician. He first measured the angular separation of two locations on earth separated by a known linear distance. He then multiplied the known distance by the fraction of 360° represented by the angular separation (Fig. 9-1). For example, Erathosthenes observed that at Syene, Egypt, on the first day of summer, sunlight struck the bottom of a vertical well at noon. This indicated that the center of the earth, the well, and the sun were in a direct line. At the same time of day and year in Alexandria, Egypt, 5000 stadia (a Greek unit of length) to the north, he observed that sunlight struck the earth's surface at an angle of 7°. In other words, the angular separation of the two locations was 7°, or $^1/_{50}$ of a circle. Since the early Greeks believed that the earth was round, then the distance from Syene to Alexandria was $^1/_{50}$ of the distance around the earth. Multiplying 5000 stadia by 50, Erathosthenes calculated the circumference of the earth as 250,000 stadia. There is some disagreement as to the exact length of the stadia he used because there were several equivalents for the unit in use at that time. If he used a measure equal to $^1/_{10}$ of a mile, he actually came within 1% of the exact circumference as calculated today, which was truly a remarkable achievement.

Once the circumference of the earth is known, its diameter may be obtained by dividing the circumference by 3.14 (π). The radius of the earth is half its diameter. Once the radius is known, the planet's volume may be obtained. By dividing the volume of the earth into its mass, which may be calculated by using a Jolly balance, the specific gravity or density may then be determined. It can now be seen why mathematics is an essential tool, not only in the earth sciences but in all sciences as well.

EARTH'S SHAPE

The ancient Greeks knew not only the size of the earth but also its shape. Aristotle cited two convincing arguments in favor of a round earth. First, he observed that during a lunar eclipse the earth's shadow on the moon was curved (Fig. 3-12). He knew that only a round object could always produce a curved shadow. Secondly, he explained that travelers moving northward or southward from Greece reported seeing stars that were not ordinarily visible, while other stars easily visible from Greece appeared to drop below the horizon. Aristotle pointed out that this effect also was possible only on a round earth.

Today we know that the shape of the earth is a slightly flattened sphere called an *oblate spheroid.* Its *oblateness,* the difference between the polar and equatorial diameters,

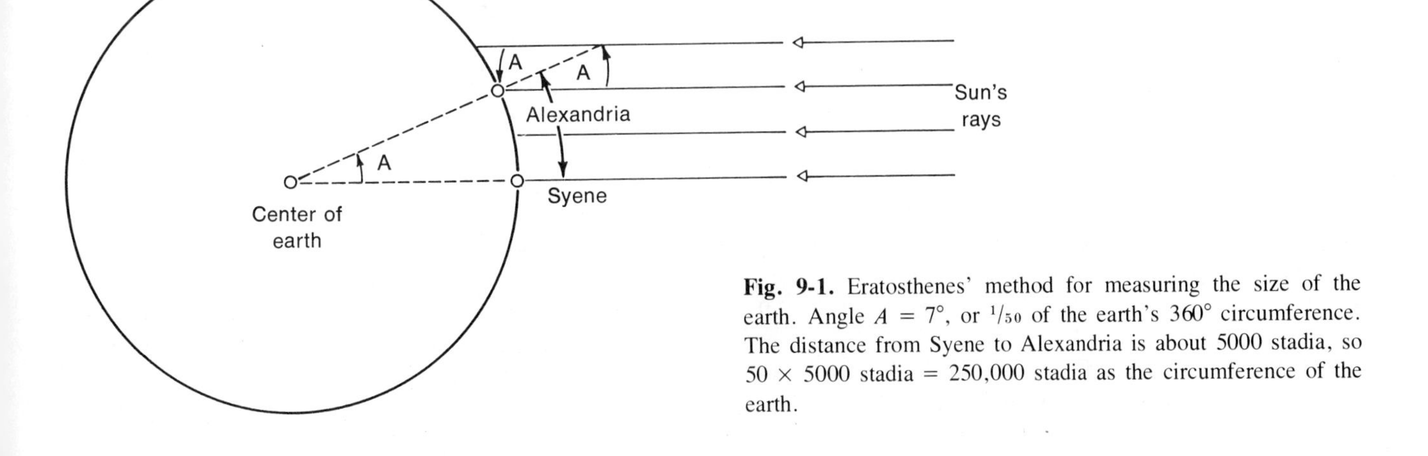

Fig. 9-1. Eratosthenes' method for measuring the size of the earth. Angle A = 7°, or $^1/_{50}$ of the earth's 360° circumference. The distance from Syene to Alexandria is about 5000 stadia, so 50 × 5000 stadia = 250,000 stadia as the circumference of the earth.

is only 27 miles or 1 part in 298. Several other planets also have an oblate shape, so the earth is not unique in this respect. Saturn, for example, has an oblateness of 1 part in 10, the highest in the solar system.

The oblate shape of the earth seems to indicate that it was not always predominantly solid as we find it today. The earth's rotational speed of 1000 miles/hr at the equator, which is cited as the cause of its oblateness, may not be enough to explain its present shape. We can tell that the rotation speed has diminished through the years by comparing the times of modern eclipses with records of early times. The earth is slowing down because of the gravitational braking action of the sun and, to a greater extent, of the moon. From this we can hypothesize that the oblate shape of the earth was set while the planet was in a plastic state early in its history when it was rotating at a greater rate of speed.

MOTIONS OF THE EARTH

Although it is difficult for us to believe and impossible for us to experience, the earth, like all other known celestial bodies, is moving. Because people were unable to feel its motion, they believed that the earth was a fixed body occupying the central position of the universe. Today we know that the earth is just another planet in the solar system and not a spectacular one at that. The many types of motions of the earth, however, deserve closer study at this point because they are responsible for masking the true motions and the appearance of the universe, while playing such a vital role in our affairs.

Rotation

The most dominant motion of the earth and that with effects we can observe is rotation. The planet turns on its own axis, completing one rotation in the natural time unit known as the day. Rotation is not unique to the earth; satellites, stars, galaxies, and other planets also turn on a rotational axis. The rotation of the earth is from west to east at a constant angular speed of 15°/hr, regardless of one's position on the planet. The actual rotation speed in miles per hour, however, does depend on the observer's location. At the equator, rotation speed is 1000 miles/hr, while at the North and South Poles the rotation speed is 0 miles/hr. Halfway from the equator to either pole, the speed of rotation is 500 miles/hr. Regardless of the observer's location, the earth completes one turn each 23 hours, 56 minutes.

The earth's rotational axis is inclined or tilted to the plane of its orbit around the sun at an angle of 23½°. In any one revolution around the sun, the degree and direction of the tilt remains constant, even though the latter is precessing at a rate of 0.01°/yr with respect to the stars because of the gravitational tug of the moon and sun. Again, inclination is not unique to the earth. All other planets in the solar system as well as the moon and the sun are tilted on their rotational axes. No satisfactory explanation of the cause of inclination has ever been found, although it is believed to be somehow related both to the mass distribution of the planet and to the balance of forces that provide a stable orbit.

We do know, however, that the tilt of the earth's axis causes the seasons we experience. Since the rotational axis remains constant in amount and direction as the earth orbits the sun, the net effect is that the sun's rays strike the surface at different angles during the year. Winter is caused by the less direct, less intense rays, while in summer more direct, more intense rays strike the surface. The seasons are not produced, as many people still believe, by the changing distance between the earth and the sun. In reality, the earth is 4,000,000 miles closer to the sun during winter in the Northern Hemisphere than during summer. The revolution of the earth around the sun does cause the changes of the seasons but not the seasons themselves. It should also be pointed out that seasons are not unique to the earth. Since all planets in the solar system possess axial inclinations, they too have seasons of their own.

The earth's rotation from west to east causes the sun, moon, planets, stars, and all celestial bodies to appear to move in the opposite direction, east to west, at a constant rate of 15°/hr. This predictable movement of the sky has enabled astronomers to construct electrical drives that are attached to telescopes properly aligned with the earth's rotational axis. These mechanisms compensate for the rotation of the earth by moving the telescope westward as the planet turns toward the east. This enables an astronomer to track a celestial object visually or photograph it with a minimum of adjustments to the telescope.

Revolution

While the earth rotates on its tilted axis, it is also revolving in an elliptic orbit around the sun. The plane formed by the earth's orbit is called the *ecliptic*. The earth completes one circuit of the sun every 365¼ days, the natural time unit called the *year*. The planet revolves from west to east, the same direction in which it rotates, at an average speed of 66,600 miles/hr (18.5 miles/sec). At *perihelion,* its closest point to the sun, the earth is approximately 91,000,000 miles from its giant master; while at *aphelion,* its greatest distance from the sun, it is about 95,000,000 miles away. The earth's average distance from the sun, known as the astronomical unit (AU), is 93,000,000 miles.

The earth is not the only celestial object that revolves around a dominant gravitational mass. All other planets in our solar system also revolve around the sun. The moon revolves around the earth, while other satellites revolve around their primary planets, the sun and other stars revolve around the nuclei of their galaxies, which in turn revolve around the center of mass of the universe.

The effects of the earth's revolution are not as easily observed as are those of its rotation. In the daytime sky the earth's revolution is evident in the apparent north-south motion of the sun over a period of weeks. As pointed out earlier, rotation causes the sun to appear to rise above the horizon in the east, move westward across the sky, and set below the horizon. At the same time, revolution also causes the sun to appear to drift slowly northward or southward. This effect can be observed by noting the position of sunrise or sunset on the horizon relative to some fixed landmark (Fig. 9-2). The effects of revolution may also be noted by observing the changing length of a shadow cast by a permanent structure such as a utility pole. Like the changing positions of sunrise or sunset, the shadow cast by the utility pole or tree would have to be observed at the same time of day over a period of weeks.

In the night sky the earth's revolution causes a parade of different constellations to come into view over a period of months. Revolution also shows itself in the changing of the seasons. Although the seasons themselves are caused by the inclination of the earth's axis, as pointed out earlier, the progression of the seasons from spring to summer, summer to fall, fall to winter, and winter back to spring, is caused by revolution. If the earth did not revolve, any one location on its surface would experience the same season throughout the year.

Other motions

The earth shares the sun's motion in its orbit through our galaxy, moving toward the bright star Vega at a speed of 12 miles/sec. The earth also is involved in the galaxy's rotation about its own axis. This speed of approximately 150 miles/sec carrys the sun and the earth in the direction of the constellation of Cygnus the Swan. At the same time the earth shares in the Galaxy's motion through the universe at a speed of 50 miles/sec toward a group of galaxies seen in the direction of the constellation Hercules.

We are all therefore armchair astronauts, whether or not we ever leave home, for we are traveling in at least six different directions at a combined speed of more than 800,000 miles/hr.

STRUCTURE OF THE EARTH

The earth consists of three basic parts: the *hydrosphere* (water above, on, or below the surface), the *atmosphere* (the gaseous envelope surrounding the earth), and the *lithosphere* (the body of the planet itself) (Fig. 9-3). Although

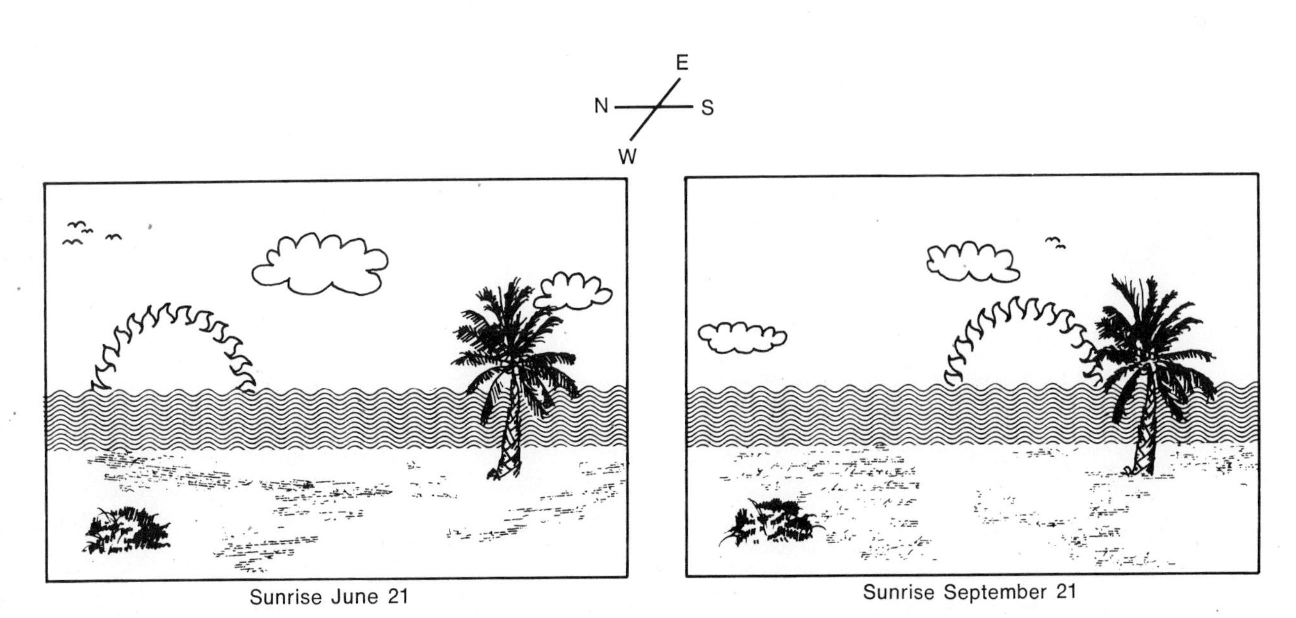

Fig. 9-2. The effects of the sun's revolution can be observed by watching the position of the sun relative to a fixed object such as a tree.

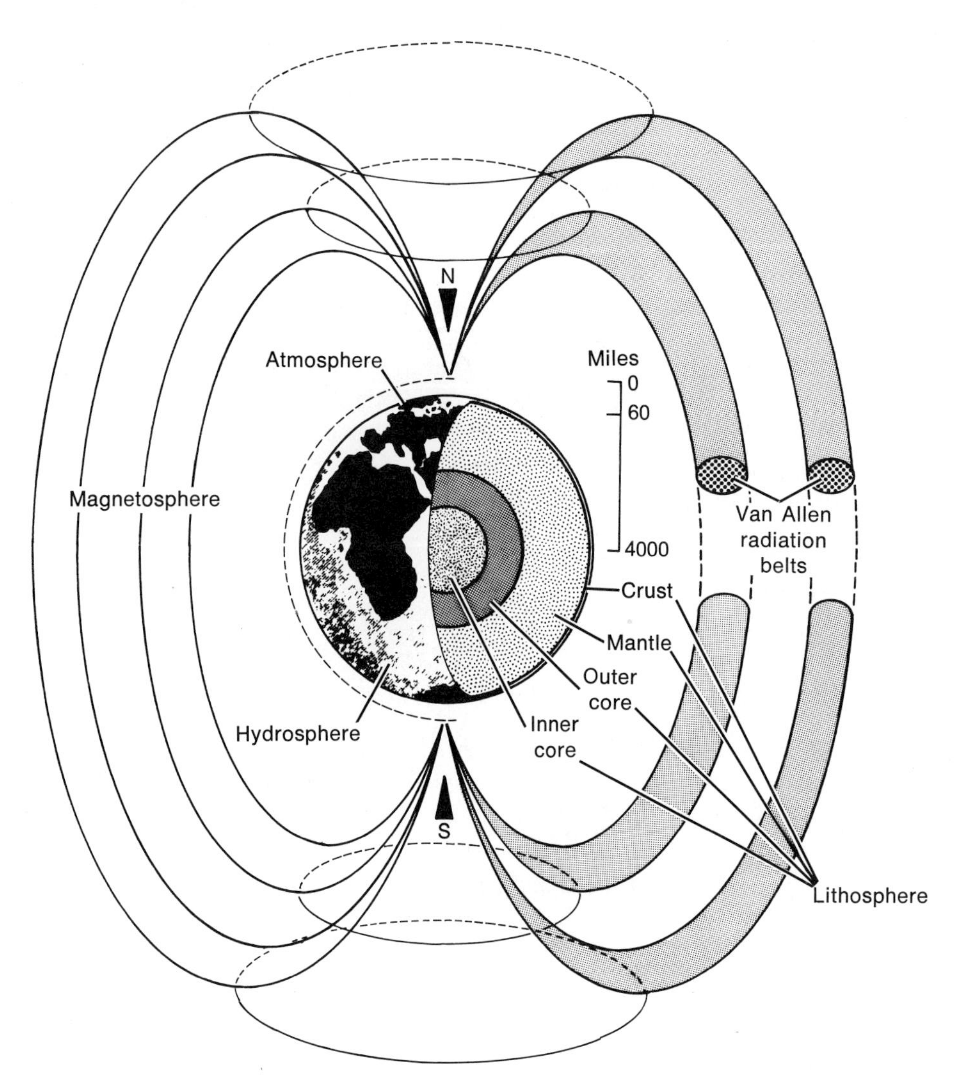

Fig. 9-3. Cross section of the earth and surrounding space. In addition to the more familiar lithosphere, atmosphere, and hydrosphere, we must also consider the magnetosphere, containing the Van Allen radiation belts, as an important part of our environment. These two concentric belts surround the earth like an encircling doughnut.

the earth is the only planet known to have a hydrosphere, all planets in our solar system either definitely do have or are suspected of having atmospheres. Some other planets in our solar system also have a structural body thought to be similar to that of the earth.

Hydrosphere and atmosphere

Since 71% of the earth's surface is covered with salt water, the oceans are our planet's dominant surface feature, Water also exists on the earth's surface as lakes, streams, and swamps. Above the surface, water is also present as water droplets suspended in the air as invisible water vapor. Still another great mass of water is stored on the surface as ice, while below the surface is a great reservoir of groundwater. The water found at these locations is collectively referred to as the hydrosphere (Fig. 9-3). It is believed that most if not all of the water comprising the hydrosphere came originally from the lithosphere as water vapor that later precipitated out of the atmosphere. The process is continuing today, as indicated by studies of gases released during volcanic activity. These gases consist of as much as 95% water vapor. The ocean, constituting the vast majority of the hydrosphere, plays such a vital role in the earth's environment that all of Part III of this book will be devoted to a study of it. Other aspects of the hydrosphere will be discussed later in the text.

The gaseous envelope surrounding the earth is called the atmosphere (Fig. 9-3). The present atmosphere, like the hydrosphere, must have had its origin from within the lithosphere, from which it escaped with the release of volcanic gases. The atmosphere in turn may have formed the hydro sphere through the precipitation of water. Since the atmosphere is invisible except for clouds, we tend to forget its rightful place in the total environment. Part IV of this book is a closer examination of this vitally important part of the earth.

Lithosphere

The body of the earth, known as the lithosphere, makes up more than 99.9% of its total mass (Fig. 9-3). Direct access to the lithosphere is extremely limited. Only in canyons, wells, and mines can we gain a direct glimpse of the earth's subsurface. The deepest canyon on earth measures a little more than 1 mile deep, the deepest mine only 2 miles deep, the deepest well extends only about 5 miles below the surface of the lithosphere. Therefore we must depend on indirect means such as the study of earthquake waves or volcanic activity to gain knowledge of the physically inaccessible lithosphere.

Crust

The study of earthquake waves has shown that the lithosphere of the earth is divided into several concentric layers. From the surface inward to depths ranging from 3 to 40 miles lies the first zone, known as the crust. It is much thinner under the oceans than under the continents, and where the height of the continents is further increased by mountains, the crust is correspondingly thickened. With an overall specific gravity of 2.8, the crust composes less than 1% of the total volume of the earth. The crustal material represents the lightest matter except for gases originally composing the earth; the heavier material sank toward the center of the earth. The difference in the depths of the oceanic and continental crusts can be explained by this separation of heavy and light materials. Oceanic crustal material does indeed have a specific gravity (confirmed by direct measurement) that is slightly greater than the specific gravity of the continental crust.

Mantle

Below the crust and extending a uniform 1800 miles toward the center of the earth is the second zone, known as the mantle. The boundary between the crust and the mantle, where an abrupt change in the nature of the rocks occurs, is called the *Mohorovicic discontinuity,* or *Moho,* in honor of its discoverer, Andrija Mohorovicic (1857-1936), a Yugoslavian geophysicist. A recent attempt to penetrate to the mantle by drilling through the oceanic crust had to be abandoned because of technical and financial problems. The recent discovery that the moon's crust was at least 15 miles thick dampened the spirits of those that had hoped the earth's only natural satellite could perhaps provide data about the earth's mantle.

Indirect evidence, however, has shown that the mantle consists of materials rich in magnesium and iron silicates, a composition probably similar to that of a rock called *periodotite,* which is sometimes exposed at the earth's surface. Like the crust, the mantle appears to be basically solid in structure, but under the stresses of tremendous pressures at depth, it could be plastic. The structure of the mantle is complex and characterized by many sublayers. Its specific gravity ranges from 3.3 near the crust to 5.5 near the zone below, and its volume makes up 84% of the total earth.

Core

Below the mantle lies the core of the planet. The core consists of at least two well-defined parts that together constitute 15% of the volume of the earth. The outer core, extending for a uniform 1400 miles below the mantle, shows plastic characteristics, as indicated by the absence of a type of earthquake wave that fails to pass through substances that are not solid. The specific gravity of the outer core is 9.5 near the mantle and 11.5 near the inner core just below. It is generally believed that both cores are metallic, probably consisting of iron and nickel.

Why should the outer core be plastic while the other zones are solid? The pressure deep within the earth prevents rocks from becoming molten, even though by surface standards the temperature is high enough. In order for rocks to become plastic, basically one of two processes must occur. Either the pressure on the rocks would have to remain constant while the temperature increased, or the temperature would have to remain constant as the pressure decreased.

We now know that the interior of the earth possesses great heat and enormous pressure. Evidence of this can be observed in most crustal rocks, which show signs either of having been melted or subjected to great pressure. Also, the eruption of volcanoes provides direct evidence of subterranean heat and pressure. Furthermore, temperatures recorded in deep mines and wells go as high as 150° F, indicating that temperature increases with increasing depth.

Several theories, none of which are entirely adequate, have been proposed to explain the heat of the lithosphere.

1. The heat is caused by the radioactive decay of elements within the earth.
2. The heat is produced by the tremendous weight of the overlying rocks.
3. The heat is residual heat left over from the formation of the earth.

Although radioactive elements are present in the rocks, it seems that either these elements are there in small quantities or that the heat released from them is not sufficient to produce the magnitude of heat that is observed inside the earth. If only residual heat were involved, it would surely have dissipated during the billions of years since the earth was formed. The pressure exerted by overlying rocks, then, seems to be the best answer, although all of the processes listed above could be involved. The possibility also exists that the lithosphere possesses characteristics and undergoes processes as yet unknown.

If the pressure of overlying rocks produces most of the heat within the earth, then it should follow that the release of this pressure at depth would cause the rocks or other material to become nonrigid. What could cause the pressure in the outer core to be released? This region could be a zone of intense agitation similar in some way to the convective zone of the sun, in which currents of heat are rising toward the surface and then sinking back toward the center. Convection of this sort within the outer core could release the pressure just enough to allow the material in the outer core to become plastic. Such convection currents might also explain the origin of earth movements and volcanic activity.

The inner core extends another 800 miles from the outer core to the center of the earth. With a specific gravity of 14.5 near the outer core and 18 at the innermost point, the pressure at the center of the earth should be approximately 600 million pounds/square inch. The temperature, then, would be about 7200° F, only a little cooler than the surface of the sun. Although we may someday reach the vicinity of the nearest star, it is unlikely that we will approach the center of the earth.

Magnetosphere

Also considered part of the earth are its magnetic fields, which are called the magnetosphere (Fig. 9-3). The invisible lines of magnetic force emerge at the *magnetic poles* of the earth, which are different from the geographic poles about which the planet rotates. The lines loop out into space in a series of great arches encircling the earth like a doughnut from 600 miles to more than 40,000 miles above the surface. The magnetosphere may have originated at least in part in intense convection of the outer core. The inherent magnetic quality of the iron within the earth could also be a factor. Only three other bodies in the solar system have measurable magnetic fields, the sun and the planets Jupiter and Mercury. Both the sun and Jupiter are suspected of exhibiting strong convective agitation.

Electrons and protons from the sun, forming a solar wind, interact with the magnetosphere to produce zones of intense radiation within the magnetic fields. These zones of radiation are known as the *Van Allen radiation belts*. They were discovered by experiments made possible by Explorer I, the United States' first artificial satellite. The belts were named in honor of James Van Allen, who was instrumental in the design of the project. The intensity of the radiation in these belts is directly proportional to the activity level of the sun and influences the other parts of the earth. The auroras of the polar regions, fluctuations in electric currents, and radio fadeouts are but a few of the known effects produced by activity in the magnetosphere.

ABOUT TIME

Although time has been called the fourth dimension, its exact nature is as obscure today as ever. However, its passage can be measured, and people have expended considerable energy throughout recorded history in devising techniques and devices for its measurement.

The basic timepiece for the earth is the earth-moon-sun system itself. The earth's revolution about the sun produces the time unit called the *year*. The moon's revolution around the earth is the basis for the unit called the *month*. The elapsed time between the major phases of the moon form subdivisions of the month called *weeks*. The earth's rotation on its axis forms the unit of time called the *day*, which has further been divided into hours, minutes, and seconds. As our needs have become more sophisticated, we have attempted to divide the day into finer and more accurate subdivisions.

The earliest attempts at keeping time were calendars. Then came efforts to measure smaller units of time by using the changing length of a shadow on a sundial, the burning of knotted ropes or notched candles, the flow of sand through an hourglass or water through a water clock. Later, mechanical devices that were operated by falling weights, swinging weights (pendulums), springs, and finally electricity, crystal vibration, and atomic vibration were introduced. The latter, known as the atomic clock, uses vaporized cesium metal to maintain an accuracy of within 1 second every 300 years.

Zone time

The earth turns on its axis from west to east, completing one turn through 360° each 24 hours. The angular rotation rate is 15°/hr, a rate that is constant everywhere on earth. For time-keeping purposes, the earth has been divided into 24 time zones superimposed on the longitude system (Fig. 9-4). Each time zone includes 15° of longitude, and time in each zone differs from time in the zone adjacent to it by 1 hour. Within each time zone, everyone keeps the time of the central meridian. For each zone not specifically exempted (exemptions are not unusual), all time on earth is computed relative to the time at the prime meridian (Greenwich).

For example, in the time zone called eastern standard time everyone within 7½° east and west of the 75th meridian would keep time with that central meridian. When the time is noon in Greenwich, England, 0° longitude, it is midnight 180° away and the start of a new day. At the same instant at a point five time zones west of Greenwich in the eastern standard time zone, the time is 7:00 P.M. Because of the earth's west-to-east rotation, the new day is continually sweeping away the old day from east to west. Places east of you have times ahead of you and places west of you have times behind yours, 1 hour for each time zone. A jingle that may be used to remember the relationship is "as you go to the east the time doth increase, as you go to the west the time doth grow less." For example, if it is 8:00 P.M. in New York City (eastern standard time), then the time at that exact instant in San Francisco, California (pacific standard time) would be 5:00 P.M., for these two cities are separated by three time zones. Meanwhile, in Greenwich the time would be 1:00 A.M. the next day.

Daylight saving time

Daylight saving time (DST) was first proposed in England in 1907 and adopted in many countries during World War I to conserve fuel and power. In the United States DST was established by Congress during both world wars and subsequently repealed after each conflict.

DST is really nothing more than a clock trick. It was devised only as a means of providing more daylight hours for economy and human activities. In the spring clocks are advanced 1 hour as the number of daylight hours are increasing because the sun rises and sets at a greater angle

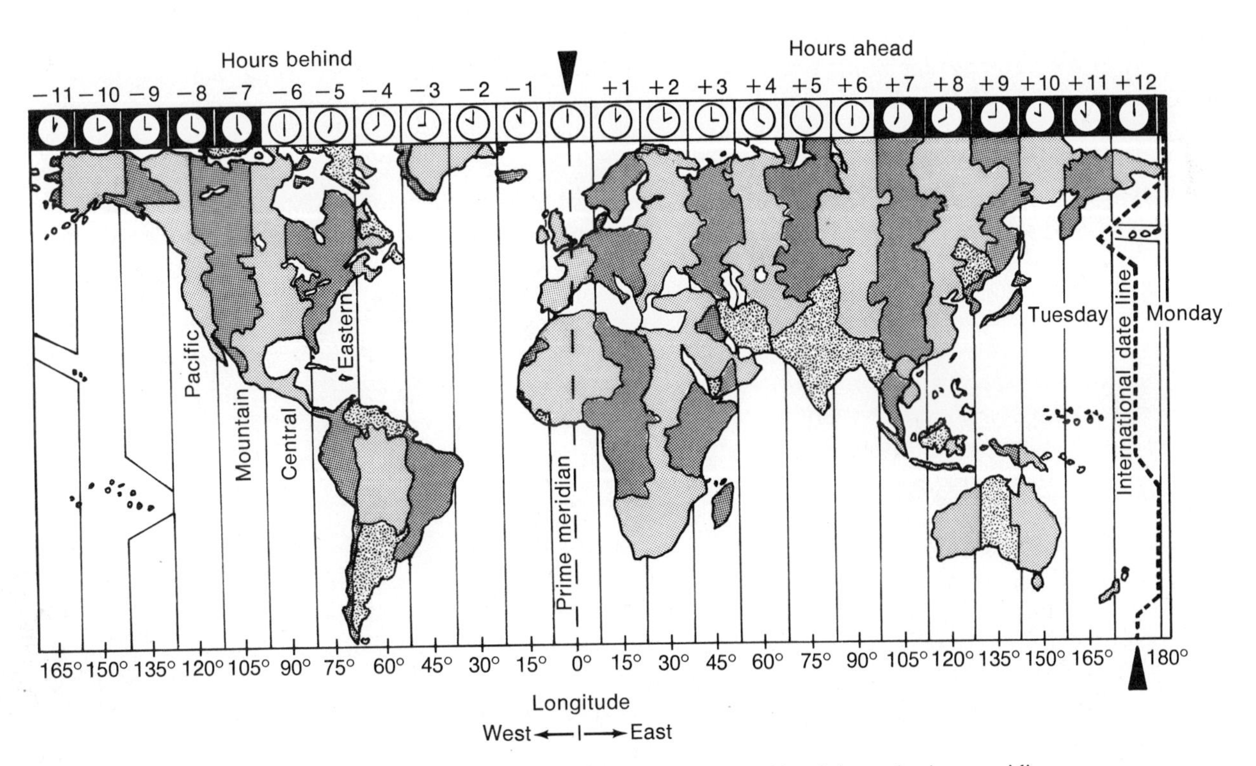

Fig. 9-4. Time relative to Greenwich. A time zone extends 7½° on either side of the main-time meridian. In the United States, there are four time zones: Eastern (75° W), Central (90° W), Mountain (105° W), and Pacific (120° W).

and thus stays above the horizon for a longer time. The clock is turned back 1 hour to standard time in the fall when the number of daylight hours is decreasing as the angle of sunrise or sunset becomes less, causing the sun to stay above the horizon for a shorter time. DST does not really change the number of daylight hours but rather forces changes in schedules to match the clock.

International date line

The most potentially confusing location on earth is the *international date line* (IDL), where the calendar date must, of necessity, change as you cross it (Fig. 9-4). The IDL for the most part follows the 180th meridian, 12 hours from Greenwich. To prevent some confusion, however, the IDL has been constructed to avoid as many populated areas as possible. As a result the IDL zigzags on maps and globes from one side of the 180th meridian to the other.

Except for the exact instant of midnight at Greenwich, two dates exist on earth at all times—today and yesterday. No problem existed with the calendar for a long time, for if you stay at home or travel slowly, tomorrow will come automatically as the rotating earth brings midnight to your locality. Today, however, ships circle the earth in days, aircraft in hours, spacecraft in minutes.

To illustrate the importance of the IDL, consider a passenger on board a hypothetical spaceship in the future taking an around-the-world flight at the speed of light. If he left New York City at 8:00 A.M. Tuesday, flying eastward, he would arrive over London (Greenwich) at 1:00 P.M. Tuesday. As he crossed the IDL, 180° from Greenwich, the time and day would be 1:00 A.M. Wednesday. If our traveler did not correct his calendar at this point, the time and date of his arrival back in New York City would be 8:00 A.M. Wednesday. He was only gone a fraction of a second at the speed at which he was traveling, yet his time would have increased 24 hours.

This very hypothetical example shows the importance of setting the calendar back one day while leaving the watch time the same. Remember that in crossing the IDL the calendar is changed opposite from the way the watch time is changing. Going westward, watch time loses 1 hour for each time zone, so that the calendar must be moved ahead 1 day as the IDL is crossed. Going eastward, watch time gains 1 hour for each time zone, so that when the IDL is crossed the calendar must be moved back 1 day. Our traveler, crossing the IDL from west to east, should have set his calendar back to Tuesday, leaving his watch time at 1:00 A.M. The time and date of his arrival back in New York City would then have been 8:00 A.M. Tuesday.

EARTH PROCESSES

The earth is a dynamic planet, undergoing processes that cause it to constantly change. Like the motions of the earth, these mechanisms are usually so slow relative to human lifetime that they are undetectable. All processes affecting the earth may be classified as either constructional or destructional. The former originate from within the lithosphere, the latter originate outside it.

The processes affecting the earth, and as a result its inhabitants, are of such great importance that they warrant special treatment in a study of the total environment. Now that the physical setting has been established, we shall turn in Part II to an in-depth look at the geologic processes that permeate our existence on our beautiful blue planet—our home.

part two GEOLOGY

When in the dim beginning of the years,
God mixed in man the raptures and the tears
And scattered through his brain the starry stuff,
He said, "Behold! Yet this is not enough,
For I must test his spirit to make sure
That he can dare the vision and endure."

"I will withdraw my face,
Veil me in shadow for a certain space,
And leave behind only a broken clue,
A crevice where the glory shimmers through,
Some whisper from the sky,
Some footprint in the road to track me by."

"I will leave man to make the fateful guess,
Will leave him torn between the no and yes,
Leave him unresting till he rests in me,
Drawn upward by the choice that makes him free—
Leave him in tragic loneliness to choose,
With all in life to win or all to lose."

The Testing
Edwin Markham

Fossils such as this trilobite provide scientists with a means of exploring the earth's history.

10 *cornerstones of the earth*

Minerals

. . . And the least stone that in her warming lap
Our kind nurse Earth doth covetously wrap
Hath some peculiar virtue of its own.

The Glorious Stars of Heaven
Joshua Sylvester

HISTORICAL PERSPECTIVE

The twentieth century has seen a growth in affluence among the nations of the world that is unparalleled in the history of man. This period has also been one of great conflict, as some nations have risen against each other, while still others have been divided by civil wars. The two World Wars of this century alone have resulted in the depletion of more of the earth's human, natural, and economic resources than all previous wars combined. The years since World War II have seen the world in an almost constant state of turmoil, as nations have jockeyed for economic and political advantage.

Most raw materials on which industrial nations depend economically are called *minerals*. Minerals are not only essential to the basic wealth of a nation, but they also provide the vital ingredients for modern civilization. Transportation, communication, construction, heat and light, in fact, every aspect of our technologic societies demands raw materials in the form of minerals. The standard of living of modern industrial nations, which other nations are still striving to attain, has been reached and is being maintained by the development of mineral resources.

Unfortunately, the distribution of mineral wealth on earth has not coincided with the establishment of political and geographic boundaries. Most nations naturally possess some minerals in varying proportions, including at least one of economic importance. No nation, however, possesses sufficient quantities of every mineral resource it needs in order to maintain itself independently.

This uneven distribution of mineral wealth has led to conflict among nations. Although the struggle for the control of mineral supplies is not a new development, it has taken on added importance during the twentieth century as industrialization and technologic advances have spread throughout the world. As the consumption of mineral resources has accelerated, the supply has diminished, further increasing international friction.

Because of their importance, minerals influence national and international politics and economics. This significance is usually masked by superficial developments and may even be classified as secret information in the interest of national security. Throughout recorded history minerals have been important to man. In fact, historical and cultural periods themselves have been named for the minerals that people gradually learned to use (the Bronze Age and Iron Age are good examples). Gold and silver became symbols of wealth and beauty, and the search for these treasures spurred people to engage in exploration, coincidentally extending their horizons. Even today gold and silver are the basis of national and international monetary units and exchange. Certain minerals present in the soil are essential to the growth of plants and to human health. Petroleum, coal, and natural gas, on the other hand, have been and remain the leading source of energy for the entire world.

Many wars have been fought over mineral rights. The Phoenicians (ca 1000 B.C.) and later the Romans (ca 50 A.D.) struggled to gain control of the tin deposits of Cornwall in England. Later battles among the Slavic nations were fought over the lead and zinc mines of Poland, and in the twentieth century the dispute between France and Germany over mineral-rich Alsace-Lorraine can be considered a contributing factor to the start of World War I. Minerals have surely made their mark in history.

Because of the complexity of modern societies, it is difficult to ascertain the role that minerals may have played in recent conflicts. Were the imperialistic tendencies of Germany and Japan during World War II merely spurred on by a desire to control the peoples of the world? Was there also a very real need to furnish their growing industries and populations with raw materials in short supply within their own boundaries? The true motives will undoubtedly remain a mystery, but it seems reasonable to assume that the economic motive of control of mineral wealth played a vital role.

The post–World War II world, involved in the "cold war," has not escaped conflict either. Since 1945 people around the world have engaged in wars of independence for freedom from foreign domination. As a result, the influence of former world powers has diminished and leadership has been assumed by other nations. The United States and the Soviet Union have dominated this period, while the People's Republic of China and the oil-rich countries, particularly those in the Middle East, are emerging as world forces to be reckoned with. Foreign policy has become most involved during this period and with it the role of mineral resources, now strategically buried behind the words "national interest."

A rare glimpse into the continued and increasing significance of minerals can be obtained from a previously secret United States document that was one of the famous "Pentagon Papers" published in 1971. According to this document, President Eisenhower in 1954 approved a policy statement on "United States objectives and courses of action with respect to Southeast Asia." This statement consisted of six items, each discussing the United States' national interests in Southeast Asia. The fifth item was particularly enlightening.

> The loss of Southeast Asia would have serious economic consequences for many nations of the free world and conversely would add significant reserves to the Soviet bloc. Southeast Asia, especially Malaya and Indonesia, is the principal world source of natural rubber and tin, and a producer of petroleum and other strategically important commodities. Furthermore, this area has an important potential as a market for the industrialized countries of the free world.

Whether or not other secret documents of this type exist is mere speculation. Every nation, however, must have policies, written or unwritten, that guide the course of their political and economic action.

Could the continued segregation policies of South Africa revolve around the fact that this African nation supplies three fourths of the gold and most of the diamonds in the free world? Could the Middle East conflict between Israel and the Arab nations be summarized in the following policy statement: The control of the Suez Canal is of vital national interest because it holds the key to the movement of petroleum products out of the oil-rich Middle East? It appears that any effort to stop the flow of oil from this region is bound to result in complex and potentially explosive international developments.

MINERALOGY: A BRIEF HISTORY

The study of minerals, or the science of mineralogy, predates recorded history. Early peoples first used minerals such as quartz, flint, and jade for tools and weapons. Then, with increasing sophistication, people used clay for making bricks and pottery, and oxides of iron and manganese were used in making paints. Garnet, turquoise, amethyst, opal, and other colorful minerals found use as ornaments, while silver, gold, and copper were used to make utensils and ornaments.

Although mineralogy as a formal discipline extends back only to the eighteenth century, the informal or unscientific study of minerals has a much longer history. The Bible mentions six minerals: silver, gold, copper, tin, lead, and iron. The art of *smelting,* the removal of a metallic mineral from its ore,* was developed long before the birth of Christ. The Greek Theophrastus, for example, described a method of extracting mercury from the mineral cinnabar in about 287 B.C., and mercury has been found in Egyptian tombs.

Finally, the development of physics and chemistry provided a rational basis for the description and classification of minerals, while the Industrial Revolution of the eighteenth century furnished increased impetus to the development of mineralogy through greater demand for raw mineral products.

NATURE OF MINERALS

A mineral is a naturally occurring, inorganic, homogeneous substance. Each mineral has a definite internal arrangement of atoms that gives it its unique chemical and physical properties, which vary only within narrow limits for each mineral. Synthetic substances such as plastic or steel could not be considered minerals, nor could organic substances with the possible exception of coal, petroleum, and natural gas. These three, although they are formed through organic processes, are known as fossil fuels or mineral fuels because they are usually of great age and are formed in mineral environments.

*An *ore* is any substance containing valuable metallic minerals for which it is mined and worked.

Since minerals show the same properties throughout their gross structures, any substance not exhibiting this homogeneity therefore cannot be classified as a mineral. In further defining a mineral, the most important consideration is whether the substance shows a definite internal arrangement of its atoms that determines its state and appearance. A mineral must be solid, since only solids fall into this category under normal conditions. While ice, for example, must be considered a mineral, the water from which it is made is not a mineral according to strict definition because it lacks the required orderly atomic arrangement.

Although minor chemical and physical differences may exist among minerals, they are amazingly consistent in the properties they exhibit. It is this consistency, then, that makes minerals relatively easy to identify. The number of minerals identified to date exceeds 2000, and new ones are being discovered at the rate of nearly 25 each year. Of the total, however, only 100 minerals are of great economic importance.

Adding to the problem of mineral identification is the fact that some minerals have several varieties. Quartz, for example, is the second most abundant mineral, and it occurs in no less than eight varieties—each classified as "quartz."

Fig. 10-1. These sparkling uncut diamonds are examples of a mineral composed of only one kind of atom, carbon. (Courtesy Diamond Information Service, New York, N.Y.)

Most minerals are *compounds;* that is, they consist of atoms or groups of atoms chemically combined. However, there are some minerals composed of only one type of atom. Members of this latter group are called *native* minerals. *Feldspar,* the most abundant mineral, has a variety called *orthoclase,* which consists of the compound potassium aluminum silicate ($KAlSi_3O_8$), while the mineral *diamond* is made of only one atom, carbon (C) (Fig. 10-1).

It is a well-known fact that there are 92 naturally occurring elements in the universe. An additional 12 elements have been made in the laboratory, and undoubtedly more will be produced in the future. Minerals, since they are naturally occurring, must consist only of the original 92 elements and are classified on the basis of their chemical composition into eight families. Of the eight families, the silicates are quantitatively the most important because they compose over 90% of the earth's crust and mantle.

Mineral names usually have one of the following origins:

1. Greek or Latin words that give some information about the mineral, such as "barite" from the Greek *barys,* meaning "heavy"
2. Words describing some aspect of the mineral's chemical composition or color, such as "albite" from the Latin *albus,* meaning "white"
3. The names of individuals, such as "powellite" or "millerite"
4. Places in which the minerals were first found, such as "franklinite," from Franklin, New Jersey; "aragonite," from Aragon, Spain; "bauxite," from Baux, France; and "kernite," from Kern County, California

As you can see, mineral names usually though not always end in the suffix "-ite."

Crystals

All minerals, if formed under perfect conditions, may exhibit an external form known as a crystal. Crystals are defined as homogeneous solids bounded by natural flat surfaces called *faces* (Fig. 10-2). The crystal form is an external expression of the three-dimensional internal arrangement of the atoms in the mineral. This arrangement is known as a crystal *lattice.* That minerals are made of atoms arranged in an orderly manner in three dimensions was first proposed in 1669 by the Danish scientist Nicolaus Steno (1638-1686), who also showed that crystals of the same mineral always have identical angles between corresponding faces. The eighteenth-century French mineralogist Hauy first described crystal faces by their positions with respect to three imaginary axes of reference within the crystal. In

1830 the German scientist Hessel showed that all mineral crystals could be classified in one of six crystal systems (Table 11). The existence of crystal lattices was finally confirmed in 1912 by the German physicist Von Laue, who studied the patterns formed as X-rays penetrated mineral crystals.

Perfect crystals are rare, for they need room to grow, and the environments under which they form rarely provide a great deal of space. Where there is limited space, a number of other crystal characteristics may develop. For example, a crystal may be distorted to the point where a crystal of one type may develop characteristics of another. Another example of imperfect crystal development is the type that shows symmetric intergrowths called *twinning* (Fig. 10-2). Very rarely, one mineral is found that has the crystal outline of another mineral, producing what is called a *pseudomorph*.

As a rule, mineral crystals show combinations of crystal faces that are always forms of the classes within any one system. This tendency is so characteristic, in fact, that it is called crystal *habit*. Thus a mineral crystal may be a basic cube with each of its corners cut off to form an eight-sided crystal called an octahedron. Diamond is usually found in this form.

Chemical properties

As pointed out earlier, all minerals are chemical substances, and as such they may be identified by chemical

Table 11. Six crystal systems

System	Appearance	Lengths and angles of crystal faces	Common variations or forms of crystals	Examples
Isometric	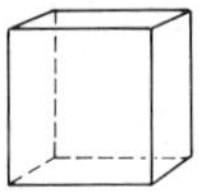	Three equal axes perpendicular to one other	Cube, octahedron, tetrahedron, dodecahedron	Halite (rock salt), galena, diamond, garnet
Tetragonal	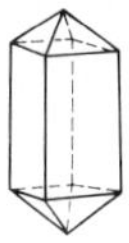	Two equal, horizontal axes perpendicular to a third, which is of a different length	Prism	Zircon, chalcopyrite
Hexagonal	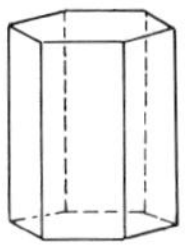	Three equal axes intersecting at angles of 120° running perpendicular to a fourth axis of a different length	Prism	Calcite, quartz
Orthorhombic	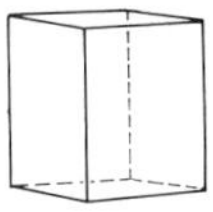	Three unequal axes, all perpendicular to one other	Pyramid, prism	Topaz, sulphur, olivine
Monoclinic	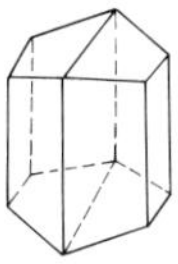	Three unequal axes, two of which are perpendicular to the third	Prism	Gypsum, feldspar, hornblende
Triclinic	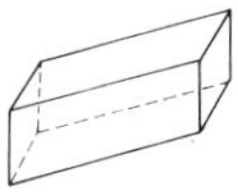	Three unequal axes meeting at any angle other than 90°	Prism	Feldspar

Fig. 10-2. This quartz crystal exhibits an unusual form of crystal growth known as "twinning," which is the symmetric intergrowth of two formerly separate crystals.

means. Since a chemical test usually involves the consumption of part of the sample or an alteration in its appearance, these techniques are used only when a mineral cannot be identified by any other means.

A geologist working in the field may spend most of his time in remote places such as quarries, road cuts, stream beds, canyons, mine shafts, and spoil banks. He must go where the best exposures of rock and mineral environments can be obtained, and since he cannot carry a chemical laboratory with him, he must travel with only small, portable types of equipment. As a result of this limitation, geologists have developed some simple physical tests that can be performed quickly with a minimum of apparatus. Even after returning to the laboratory from the field, they would be very hesitant, except as a last resort, to destroy all or part of the samples that they have so carefully collected. Therefore, although chemical tests are available, with the exception of an occasional acid test for a carbonate mineral, physical properties are most often used as the basis for physical classification.

Physical properties

Minerals can be identified by examining their crystal forms, but since perfect crystals are rare, this technique is also applied only after other means have been exhausted unless the crystal form of a particular mineral is so unique as to be obvious.

The physical properties of minerals usually permit a sample to be recognized on sight or permit the range of choice to be narrowed so that the mineral may then be classified by the application of relatively simple physical tests. It should be pointed out here that most physical properties of minerals are related to their internal atomic structure; several physical properties most often must be examined to identify a sample.

Color

One of the most obvious physical properties of a mineral is its color. The color of a mineral may be the result of nothing more than the inclusion of an impurity or of minute fractures within the sample that hardly alter its overall composition or properties. The natural color of minerals, on the other hand, is due to the types and arrangements of the atoms that make up the sample. In some minerals the lack of color is an important identifying factor. Although coloration alone to the professional eye may be sufficient to either definitely identify a sample or reduce the number of possibilities, this property would have to be considered the most variable and therefore, generally speaking, the least valuable in terms of identification.

Coloration sometimes makes a mineral rare and beautiful and thus prized in the world market as a *gem.** The mineral *corundum,* for example, is relatively abundant and inexpensive. However, when metal oxides tint this mineral red or blue, the resulting varieties of corundum are known as "ruby" and "sapphire," which are collected as gems. Other gems are prized for their size, crystal perfection, or lack of color; all are valued for their rarity. Diamonds and emeralds, the latter of which is a gem variety of the mineral beryl, are other minerals that are collected as rare and precious gems. Other minerals that are not as rare are called semiprecious gems; this group includes amethyst quartz, agate, jade, garnet, and opal.

The rarity of a mineral may actually be a simple matter of controlling the supply. Consider the case of diamonds and gold. Diamonds have been and remain a stable investment during unstable times, and gold is in great demand today worldwide at a time when most national monetary units are under attack by inflation. South Africa, a leading world

*A gem is a mineral cut and polished as an ornament; gems are often called precious stones.

supplier of diamonds and gold, undoubtedly controls the supply of these important commodities on the world market.

Streak

Closely related to mineral color is another physical property called streak, the color of the mineral in powdered form. This is usually more reliable than the color of a gross specimen alone. To examine streak, the mineral sample is rubbed against an abrasive surface such as a piece of unglazed porcelain or a special *streak plate*. Writing with a pencil on paper is one very common method of obtaining a streak sample. The pencil lead, which is actually the mineral graphite, leaves a streak on the paper.

Many hard minerals do not leave streaks, but those minerals that do almost always have a characteristic streak color that may be unrelated to the color of the whole mineral sample. The important iron ore hematite, for example, always leaves a red streak, even though the mineral itself may occur in a variety of colors. Since many different minerals show a streak of the same color, however, this physical property alone cannot be used to identify a mineral sample.

Luster

The surface appearance of a mineral in reflected light is called its luster. This physical property is divided into two basic types—*metallic* and *nonmetallic* luster—just as all minerals themselves are grouped under these two broad headings. Metallic minerals usually look like metal and their colors are usually constant; thus a mineral sample which has this appearance in reflected light is said to show a metallic luster. Although the native metals such as copper, gold, silver, platinum, iron, and mercury obviously exhibit metallic lusters, so do many of the metallic compounds such as galena (Fig. 10-3, *A*), pyrite, chalcopyrite, and magnetite. The mineral pyrite is known as "fool's gold" because of its similarity in color and luster to gold.

The minerals with a nonmetallic luster are much more numerous and variable, and many descriptive terms have been coined to describe them. A mineral may show an *adamantine* or diamond-like luster but yet not be a diamond, whereas *vitreous* and glassy are terms used to describe a mineral that has the look of glass. Other terms employed for nonmetallic lusters are *greasy, silky, pearly, dull,* or *resinous*.

Luster is almost always consistent for the same mineral regardless of its overall color. Although the mineral quartz may be pink, gray, purple, white, or clear, it always has a characteristic vitreous luster. Like the other physical properties mentioned above, many minerals share the same luster, so this cannot be the only criterion used to differentiate them.

Fig. 10-3. A, This specimen of galena shows cleavage. Note how this mineral breaks into smooth, flat pieces. It is this quality that distinguishes cleavage from fracture. **B,** Halite, a mineral that usually shows perfect three-plane cleavage, fractures if broken in a direction not parallel to one of its crystal faces.

Cleavage and fracture

Cleavage and fracture are two physical properties that are very closely related. Both deal with the breakage characteristics of minerals. If, for example, a mineral breaks into relatively smooth, flat, even pieces, usually parallel to its crystal faces, it is said to exhibit cleavage (Fig. 10-3, *A*). The cleavage *plane* represents a plane of weakness within the mineral corresponding to weak attractions or bonds between the atoms in the crystal lattice. Cleavage surfaces may be either perfect or slightly irregular, and more than one cleavage plane may be present. The mineral mica, for example, shows perfect one-plane cleavage to the point where it may be peeled off in very thin sheets. The minerals galena and halite show perfect three-plane cleavage, forming cubes when broken. Minerals that show combinations of crystal faces within any one system may exhibit complex cleavage as well. Diamond, for example, shows perfect octahedral (eight-sided) cleavage.

If a mineral breaks into curved or irregular pieces it is said to exhibit fracture (Fig. 10-3, *B*). Like cleavage, fracture is determined by the patterns of atomic bonding. Bonds that break easily produce cleavage; those that do not produce fracture. The mineral quartz shows breakage only along curved surfaces, which is referred to as *concoidal* fracture. Metals tend to have a jagged appearance known as *hackly* fracture. Other terms used to describe different types of fractures are *uneven* and *splintery*.

Some minerals show only cleavage, while others display only fracture. Many minerals, however, exhibit both properties within the same sample. Diamond, for example, shows perfect octahedral cleavage as well as concoidal fracture. A diamond cutter's job is a most challenging one. Since diamonds of smaller size are in greatest demand and bring a higher price per carat,* larger diamond crystals must of necessity be broken. Great care then must be exercised to determine a direction of cleavage rather than of fracture.

Since breakage patterns are usually constant in the same mineral species, cleavage and fracture are among the most useful physical properties in identifying minerals. Even though many minerals exhibit the same basic breakage pattern, the fact that cleavage and fracture are the external expressions of internal crystal structure makes the directions and numbers of planes and the appearance of breakage surfaces useful in identifying mineral samples.

At first glance, cleavage surfaces may resemble crystal faces. It should be remembered, however, that cleavage and fracture apply only to broken minerals, and minerals do not break into crystal faces but grow into these configurations through the buildup of atoms on the crystal lattice.

Hardness

Hardness refers to the resistance of a mineral sample to scratching by another substance. It should not be confused with resistance to crushing or other properties dealing with the application of stress. Most minerals, especially those that are nonmetallic, are weak in this regard.

A scale of relative hardness that was developed in 1822 by the German mineralogist Friedrich Mohs (1773-1839) is still in use today. At one end of his scale Mohs placed the softest known mineral, talc, calling it No. 1. At the other end he placed the hardest known mineral, diamond, and called it No. 10. He then placed other representative minerals along the scale in order of increasing hardness between the two extremes.

Mohs' hardness scale

1 Talc	6 Feldspar
2 Gypsum	7 Quartz
3 Calcite	8 Topaz
4 Fluorite	9 Corundum
5 Apatite	10 Diamond

This hardness scale is only relative, for the hardness differences between some members of the scale are much greater than between others. Diamond, for example, is at least four times harder than corundum, while corundum is about twice as hard as topaz. All minerals will fit in somewhere on Moh's scale. The mineral pyrite has a hardness of 6.3 and would be located between feldspar and quartz on the scale. To determine the hardness of pyrite, we would take the sample and attempt to scratch various other minerals on Mohs' scale. Realizing that a mineral will not only scratch itself but also every mineral that is softer (with a smaller hardness number), we would soon find that pyrite will scratch feldspar but not quartz. We would thus come to the conclusion that pyrite has a hardness of between 6 and 7. We then would try to determine its hardness number more specifically by attempting to scratch other mineral samples of known hardness between 6 and 7.

Hardness, like most of the other physical properties of a mineral, cannot be used alone for positive identification in most cases because many minerals have the same hardness and many minerals of the same species show different hardnesses in different crystal directions. In such cases as this, a hardness range is given. For example, galena, the well-known lead ore, has a hardness range from 2.5 to 2.7.

Specific gravity

Specific gravity refers to the weight of a mineral sample compared to the weight of an equal volume of water. Water, the most abundant naturally occurring liquid, has been chosen as the standard against which most other substances are compared for measurements of specific gravity. Specific gravity and its close relative, *density,* refer to weight per unit volume. Density, however, is an actual measurement, while specific gravity is a comparison of that measurement with the standard, water. The density of water is 1 gm/cm^3 while its specific gravity is 1.

If we have a regularly shaped object such as a block of wood, we can determine its density easily by first determining its volume and then its weight. To determine its volume, we measure its length, width, and height and multiply these three values together. If the block is 10 cm long, 5 cm wide, and 3 cm high, then its volume would be 10 cm × 5 cm

*A carat is a measure of the weight of precious gems. One carat is 200 milligrams or 0.2 grams. The word is derived from the Arabic *qirat,* a small bean used to measure precious gems. The largest diamond ever found was the Cullinan, which was discovered in South Africa in 1905. It weighed 3106 carats.

$\times$ 3 cm = 150 cm^3. If we then weighed the block on scales and found its weight to be 300 gm, we could determine its density by dividing its volume into its weight: 300 gm $\div$ 150 cm^3 = 2 gm/cm^3. The specific gravity would be simply 2. Volume for volume, then, our sample block of wood would weigh twice as much as water.

Unfortunately, minerals almost never come in regular shapes, and if they do, they usually have so many sides as to make calculation of volume virtually impossible. A rather simple physical principle makes complicated measurements unnecessary. Because two quantities of matter cannot occupy the same space at the same time, as a solid is immersed in a liquid, a volume of the liquid that is displaced by the solid is equal to the volume of the solid. This displacement concept, discovered by the Greek mathematician Archimedes (287-212 B.C.) is known as *Archimedes' principle*. The displacement force is called *buoyancy*.

To use this principle in determining the specific gravity of a mineral, we first weigh our sample on scales. Let us say that our sample weighs 50 gm. We then suspend the mineral from spring scales and weigh it again, this time immersed in water. Now we find that the weight is 20 gm. Our sample has "lost" 30 gm of weight when weighed in water through buoyancy, that water pushed out of the way as the mineral sample was submerged. By dividing the loss of weight in water into the original weight in air, the specific gravity (SG) is obtained.

$$SG = \frac{\text{Weight in air}}{\text{Loss of weight in water}}$$

$$SG = \frac{50 \text{ gm}}{30 \text{ gm}}$$

$$SG = 1.66$$

This mineral, then, weighs 1.66 times more than an equal volume of water.

Specific gravity, when carefully obtained, can be used alone to identify a mineral more accurately than any other physical property. Unless a combination of previously mentioned physical properties fails to identify the sample, the specific gravity, which does require some apparatus, need not be computed through formal calculation. Often the simple technique called *heft,* in which the sample is lifted in the hand and its weight is compared to that of another sample in the other hand, is sufficient.

Special properties

There are several special physical and chemical properties exhibited by some minerals that are not shown by all. Some of these are explained in the following list.

1. *Asterism*. When they are polished, some minerals show a six-rayed "star" shining within the sample as light passes through. Star sapphire is the best-known mineral that shows this property, although others such as rose quartz also display asterism.
2. *Fluorescence*. Several minerals such as fluorite, barite, and celestite convert X-rays or ultraviolet light to visible light.
3. *Thermoluminescence*. Fluorite and certain varieties of calcite give off light when heat is applied to them.
4. *Pyroelectricity and piezoelectricity*. Some minerals, primarily tourmaline and quartz, show the ability to acquire electrical charges as the result of an increase in temperature or the application of pressure.
5. *Effervescence*. This is a chemical property marked by the release of bubbles of carbon dioxide gas from the surface of a mineral as acid is dropped on the sample. This is a positive test for a carbonate mineral such as calcite.
6. *Magnetism*. The mineral magnetite, an important source of iron, is naturally magnetic and may be identified by this property.
7. *Taste*. The mineral halite, natural salt, can be safely identified by its salty taste; however, as a general rule, taste is not a recommended test for minerals, since many of them are poisonous.

MINERAL ENVIRONMENTS

Minerals are formed naturally in three environments—*igneous, metamorphic,* and *sedimentary*. Some minerals can form in only one environment, while others are formed in two or even all three environments. Often minerals are found great distances from the place in which they were formed and in a different environment due to transportation by agents of erosion. Such a location could be called a *secondary environment*. However, environment is so predictable for many minerals that the search for deposits can be restricted to certain primary and secondary locations. Gold, for example, is almost always formed in igneous environments as primary sources but may also be deposited in a secondary sedimentary environment.

It should be pointed out that mineral environments coincide with rock environments in which the three types of rocks are formed. Since rocks are composed of minerals, the discussion of mineral environments means the discussion of rock environments and vice versa. Further discussion of rock environments may be found in Chapters 11, 12, and 14.

AFFLUENCE IN JEOPARDY

People have always considered the earth as an unlimited storehouse for every material they have needed. In the past, when the world population was a mere half billion, this was perhaps true. It was possible to cut a forest, dump wastes in a stream, or strip a mine of some valuable mineral and then move on to a new location.

It is becoming painfully clear today that our exploding populations are fast draining away the mineral resources on which we depend so heavily. At the present rate of consumption, these resources will one day become exhausted. Forests can be replanted, and after several decades they can be cut again. Streams can be cleaned and made useful and beautiful again. Mineral resources, however, are unrenewable; once they are gone they are gone forever. Minerals form so slowly that they cannot be replaced.

We might ask why substitutes cannot be found for our dwindling mineral resources. Some "substitutes" such as plastics have been developed, but we are beginning to realize that general and widespread substitution may not be possible.

A more plausible approach, and one that is virtually inevitable, lies in the recycling process. Throwaway economies can no longer be tolerated in the years of mineral crises that lie ahead. Products of our modern societies must be reclaimed for their mineral value and returned to the economy to be used over and over again. Inherent in the recycling approach is the necessity that the population of the world cut back its present level of consumption to one the earth can more realistically support.

The dawn of the twenty-first century will see a very serious resource problem, which even now is posing a very real and distinct threat. Unless common sense prevails in all aspects of our lives, the length of time it will take to exhaust the earth's known supply of natural mineral resources can be measured in decades, not in thousands or millions of years. The use and loss of all our mineral resources, considered an impossibility less than 100 years ago, is a real possibility during the coming century. Unless steps are taken now to find new supplies of minerals, develop better recycling methods and substitution techniques, and to control populations, the end of our cherished way of life will be at hand and a return to a less affluent standard of living will be necessary.

11 *a fire from heaven*

Igneous activity

How with this rage shall beauty hold a plea
Whose action is no stronger than a flower?

Sonnet 65
William Shakespeare

On August 27, 1883, a volcano on the small, uninhabited island group of Krakatoa in the Sunda Strait between Sumatra and Java exploded after more than 3 months of activity. The final eruption, occurring between 5:30 and 10:52 A.M., blew an estimated 5 cubic miles (8.05 km^3) of rock into the air. The mighty blast destroyed two thirds of the island, which had been as high as 1700 feet (518 meters) above sea level, replacing it with a hole 1000 feet (305 meters) below sea level. The explosion broke windows as far away as 100 miles (161 km) and was heard 2500 miles (4025 km) away in Alice Springs, Australia. A huge ocean wave (tsunami) that reached heights as great as 100 feet (30.5 meters) swept the Pacific–Indian Ocean area, destroying 300 villages and killing 36,380 people. Fine volcanic dust thrown into the air by the force of the eruption circled the earth several times before completely settling out of the atmosphere some 2 years later. Unusually beautiful sunrises and sunsets, caused by the scattering effect of this dust, were reported all over the world for weeks after the eruption. Today, however, a new Krakatoa is being volcanically built on the site of the former mountain (Fig. 11-1).

The explosion of Krakatoa was one of the most powerful and destructive volcanic eruptions ever recorded, exceeded in force only by the eruption of Tomboro, another East Indies volcano, in 1815. There is good reason to believe that still other, more violent events of this type occurred and were unrecorded. The fabled lost continent of Atlantis may have been a volcanic island in the Aegean Sea, that exploded like Krakatoa, resulting in the loss of an entire culture.

Indeed, volcanoes have taken a heavy toll of human life and property throughout recorded history. Probably the most famous such event of all time was the eruption of Mt. Somma, now known as Vesuvius, in 79 A.D. It destroyed the towns of Pompeii and Herculaneum, killing an estimated 10,000 people. Vesuvius has since been active in 1139, 1631, 1779, 1793, 1872, 1906, and 1944. The most violent volcanic eruption in the twentieth century took place on the West Indies island of Martinique on May 8, 1902. The capital city of St. Pierre in the shadow of Mt. Pelée was destroyed as the volcano erupted without warning, sending a fiery cloud of incandescent gas through the city and taking the lives of 30,000 people.

The earth today shows evidence that such violence has occurred throughout geologic time, billions of years before the advent of human life, and that it continues up to the present time. There is also nothing to indicate that the future holds anything different in store. A volcanic eruption, one of nature's most spectacular and feared events, is the surface expression of internal processes at work deep within the earth. Such processes are collectively known as *igneous activity,* which includes the formation, movement, and cooling of liquid rock. Igneous rocks are formed in the process.

ORIGIN OF LIQUID ROCK

As pointed out in Chapter 9, the best evidence to date indicates that the earth's mantle as well as its crust are solid. Nonetheless, large quantities of liquid rock either are blown or flow from the earth's subsurface during volcanic eruptions. Also, when they are exposed at the surface after millions of years of erosion, the structures of rocks formed deep in the earth indicate that the rock was once liquid. This apparent dichotomy can be explained if the origin of liquid rock can be understood. Our knowledge of igneous

Fig. 11-1. Krakatoa, in the Pacific, shows the birth of a new volcanic island at the site of an older one. (From Truby, J. D.: Sea Frontiers **17:**130, 1971; photograph by R. W. Decker. Reprinted by permission of the International Oceanographic Foundation, Miami, Fla.)

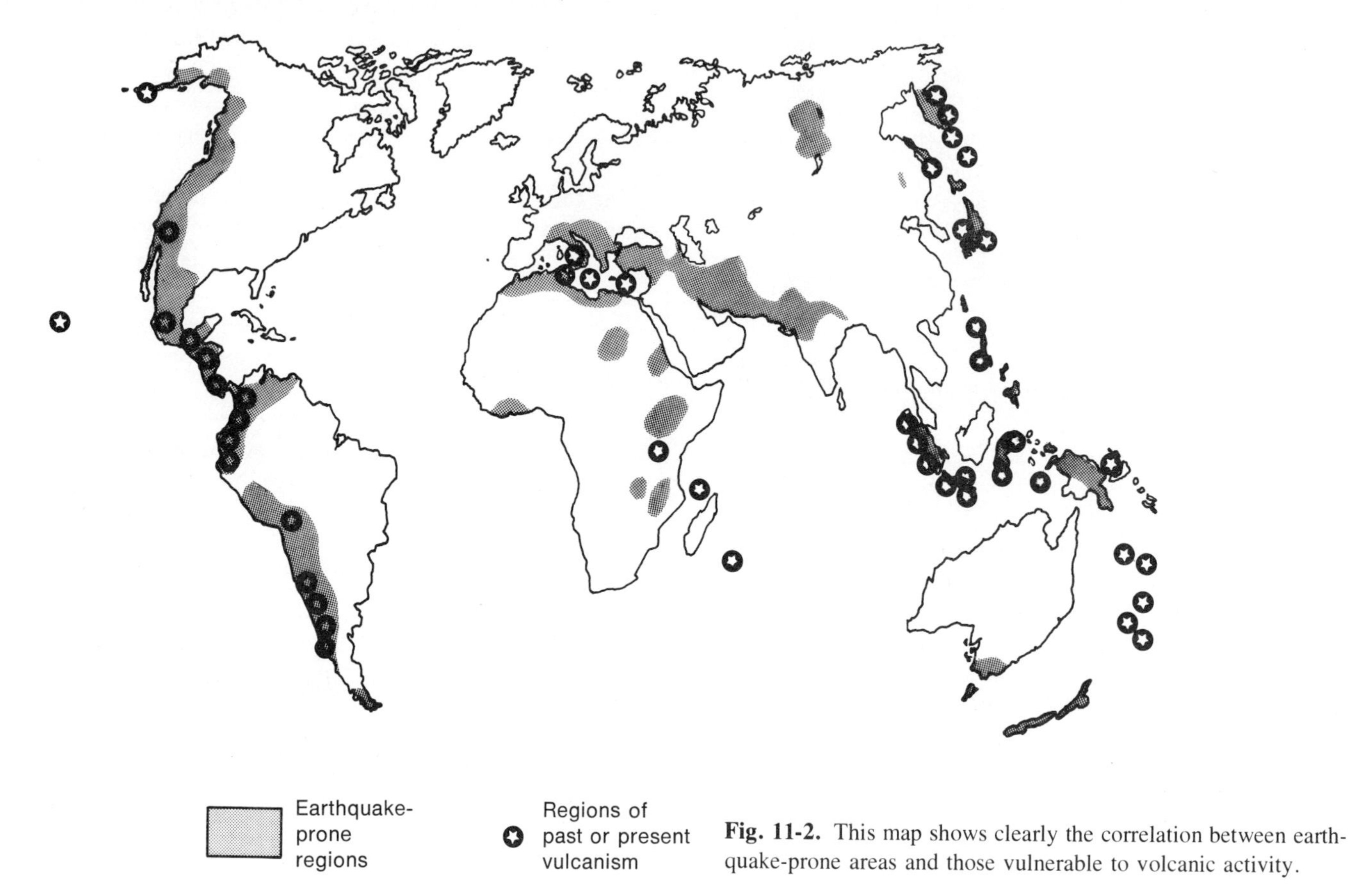

Fig. 11-2. This map shows clearly the correlation between earthquake-prone areas and those vulnerable to volcanic activity.

activity must come for the most part from indirect observations and theoretical considerations.

We know that the earth's interior is hot and that the temperature increases an average of 1.8° F (about 1° C) for each 100 feet of depth, a scale known as the *geothermal gradient*. For example, the temperature in the deep diamond mines of South Africa may climb as high as 140° F (60° C). Even if the geothermal gradient is not constant to the earth's center, the temperature there could rival that of the sun's surface.

The pressure deep in the earth must also be very great. If we assume that pressure plays a key role in the production of heat at the earth's interior, then by surface standards the rocks miles below the surface should be hot enough to melt. However, as the pressure increases, the melting point of the rocks also increases, causing the rocks in the crust and mantle to remain solid. Thus only if the relationship between temperature and pressure is exactly right can the rocks deep in the earth melt. Either the pressure can remain constant while the temperature increases to reach the higher melting point or the temperature can remain constant while the pressure decreases to permit melting at the lower melting point. The pressure reduction theory appears to best fit the observed facts.

For example, there appears to be more than just a casual relationship between igneous activity and earthquakes. When the locations of known volcanoes are compared to the locations of known earth movements, the results are startling, as can be seen in Fig. 11-2. Virtually a one-to-one correlation exists between volcanic activity and earth movements. Earth movements, produced when stresses inside the earth cause rocks to bend or break, could be the triggering mechanism that reduces the pressure on hot rocks deep in the earth, producing liquid rock. Liquid rock that is underground is called *magma,* while the same substance at the earth's surface is known as *lava*.

MOVEMENT AND COOLING OF LIQUID ROCK

Because it is a liquid, magma is less dense than surrounding solid rock. It therefore seeks outlets to regions of lower density, usually upward. If paths of weakness are available in the form of underground fissures or cracks, the magma squeezes through these cracks and may eventually reach the surface. If no fissures are available, the magma may move toward the surface by melting the rocks above it. This process is called *stoping*. Although magma has never been seen extruding (seeping out) at the surface through this process, evidence in rocks indicates that it has moved upward for distances as great as 1000 feet or more.

Almost immediately after it is formed, the magma begins to lose its heat into the surrounding rocks and starts to crystallize. Since magma is liquid rock, it consists of a hot solution saturated with the elements that made up the original rock before it melted. Crystallization into individual minerals begins as the temperature of the magma drops below the melting points of the individual substances. The minerals in the magma with the highest melting points crystallize first, while the others crystallize later. Diamond, gold, cassiterite, chromite, and magnetite are among the important minerals of magmatic origin. Other elements in the magma combine to form gases that later escape at the surface.

THE IGNEOUS ENVIRONMENT

The igneous environment, the site of the formation of igneous minerals and rocks, is divided into two parts—the *intrusive* or *plutonic* environment below the surface, and the *extrusive* or *volcanic* environment at the surface. By a conservative estimate, the intrusive environment makes up 90% of the total, while extrusive igneous activity accounts for the remaining 10%. Extrusive activity has received almost all of the attention of the general public because it affects our environment.

Intrusive activity

Intrusive activity is important not only because it is quantitatively greater than the extrusive environment but also because it is the source of material extruded during volcanic activity.

The main center of activity in the intrusive environment is a *magma chamber,* the area affected by the release of pressure. From this reservoir, magma may melt its way upward or move toward the surface (or in any other direction) through available fissures. The magma eventually solidifies underground to form great masses of rocks known as *plutons,* named for Pluto, the Greek god of the underworld. As a result, rocks that form in the intrusive environment are called *plutonic* igneous rocks.

Igneous bodies

Plutonic igneous bodies may be found in two types of positions—discordant and concordant. A *discordant pluton* is one that clearly does not belong with the surrounding country rock. Since most of the country rock near the surface is in layers, discordant bodies are generally found intruded at an angle through the layers. On the other hand, *concordant plutons* are those that have formed in such a way as to fit into the pattern of the preexisting rock. The intrusion can deform the country rock and still be considered concordant so long as it does not cut through the mar-

Table 12. Plutonic igneous bodies

	Tabular shape	Massive shape	Pipe shape
Discordant position	Dike	Batholith, stock (small)	Volcanic neck
Concordant position	Sill	Laccolith	—

gins of the layers. If the country rock is not layered, determination of position is more difficult.

Igneous bodies may also assume three basic shapes—massive, tabular, and pipe. *Massive* bodies actually have no regular shape but are, as the name implies, an irregular mass of igneous rock. *Tabular* bodies are those with one small dimension and two larger dimensions, similar to the shape of a tabletop. For example, tabular bodies may be long, wide, and thin or thick, long, and narrow. Finally, a *pipe-shaped* igneous body is one that has an oval or rounded shape like a piece of pipe or tubing.

Igneous bodies are classified on the basis of position, shape, and sometimes size (Table 12). For example, a discordant massive igneous body is known as a *batholith* (deep rock) and is a whole magma chamber that has solidified. These bodies are enormous, usually covering thousands of square miles, although smaller batholiths frequently occur. Batholiths are so huge, in fact, that erosion generally exposes only small portions of them at a time. Open-faced quarries are often dug into batholithic structures for the removal of plutonic rock. The best-known natural exposures of batholiths in the United States are the Sierra Nevada batholith in Yosemite National Park in California, the Idaho batholith, and the Boulder batholith in Montana. A batholith with an area of less than 40 square miles is called a *stock,* or *boss;* such a formation is a small upward extension of the larger batholith itself. Stone Mountain, Georgia, just east of Atlanta, is classified as a stock (Fig. 11-3).

A discordant, tabular igneous body is known as a *dike.* It is a fissure into which the magma intruded and in which it later cooled to form solid igneous rock. Dikes vary in thickness from inches to thousands of feet and may extend for many miles. These features are very common and may be seen at many locations where they have been exposed by erosion or excavation (Fig. 11-4).

A concordant, massive plutonic body is called a *laccolith* (rock lake). A laccolith forms when magma, moving upward toward the surface, forces its way between rock layers, lifting the overlying rocks into a dome, which is often later exposed by erosion. Bear Butte in South Dakota is a fine example of a laccolith. Several domes are often formed in the same region to produce domed mountains such as the Henry Mountains of Utah.

Magma that moves through fissures below the surface may intrude between rock layers and solidify there. This

Fig. 11-3. Stone Mountain, Georgia, is an example of a stock (or boss), which is the extending tip of a larger batholith beneath. (Courtesy Stone Mountain Memorial Association, Stone Mountain, Ga.)

Fig. 11-4. Pegmatite dike 10 feet across breaks through to the surface 5 miles east of Burnsville, North Carolina. Note the small, gray intrusion in to the darker rock on the lower left-hand dise of the dike.

forms a concordant, tabular igneous body called a *sill* (Fig. 11-5). Sills, like dikes, may be of different sizes. The Palisades, on the New Jersey side of the Hudson River across from New York City, is the side view of a sill intruded more than 200 million years ago.

The only pipe-shaped igneous body is known as a volcanic *neck*. As magma comes to the surface and exits through a rounded opening called a *vent*, a pile of volcanic debris is deposited around the opening to form the mountain-shaped mass of the volcano. The lava or other volcanic material causes the mountain to get higher over a period of many years as the vent opening at the top extends upward. After the volcano ceases to erupt, the magma in the vent and in the underground feeder conduits from the plutonic environment cools to form solid rock. This hardened magma, more resistant to erosion than the fragmented material that makes up the body of the mountain, remains long after the softer material is eroded away. The volcanic neck, which is the solidified throat of the volcano, shows the typical rounded shape of the original vent. It is discordant in position because the vent maintained itself by cutting upward as volcanic debris was deposited around it. Devil's Tower in Wyoming is an excellent example (Fig. 11-6). Environments associated with volcanic necks have proved to be the most valuable sources of diamonds in the world today. The Kimberly diamond mine in South Africa, the world's deepest mine, has been dug into an ancient volcanic neck. From

Fig. 11-5. This sill on Unalaska in the Aleutian Islands forced its way between layers of hardened lava. The sea cliff exposure is 1200 feet high. The zigzag dike at the lower right is probably a feeder for the sill. (Courtesy United States Department of the Interior, Geological Survey, Reston, Va.; photograph by G. L. Snyder.)

Fig. 11-6. Volcanic neck. Volcanic mountains have pipes at their centers through which magma exits during an eruption. More durable than the mountainside itself, these exit pipes often remain after the mountain is gone. This is the history of the intriguing Devil's Tower in Wyoming. This feature is so unusual that it was preserved as the United States' first national monument. (Courtesy Wyoming Travel Commission, Cheyenne, Wyo.)

this mine diamond crystals are extracted from the plutonic igneous rock periodotite, which is sometimes called "blue earth."

Extrusive activity

As surface dwellers, we are more closely concerned with the extrusive igneous environment than with intrusive igneous activity. When magma is expelled from below the surface to form lava or other volcanic material, it enters the realm of the extrusive environment. Igneous material extrudes or comes out at the surface either through a *vent* or through cracks or *fissures*. Vents often form along a fissure, and fissures are frequently associated with vents. It should be emphasized that extrusive igneous activity involves more than the formation of the well-known volcanic mountain. Any extrusion of igneous material at the surface represents volcanic activity and the rocks thus formed are referred to as *volcanic*.

Lava flows and lava plateaus

If lava extrudes through a fissure, the gases in the hot liquid escape slowly into the atmosphere. The loss of gas reduces the pressure, allowing the lava to ooze from the fissure like blood out of a cut, producing lava flows. Lava flows may extend for many miles, covering extensive areas to depths of hundreds of feet. They often accumulate on top of previously deposited lava flows, producing extremely thick sequences of hardened lava. Such hardened lava flows that cover enormous areas are called lava plateaus. They are generally flat or slightly undulating, for they have either completely or partially buried the underlying landforms previously formed in the vicinity of their outpourings.

One of the best examples of a lava plateau to be found anywhere in the world is the Columbia Lava Plateau of the northwestern United States. This geographic feature exceeds 50,000 square miles in area, covering parts of the states of Washington, Oregon, California, and Idaho. The Snake River, a tributary of the Columbia River, has eroded Hell's Canyon in Idaho to a depth as great as 7900 feet in one location, chiefly through lava flows that were deposited over a period of 70 million years. The Deccan Plateau of India is also a lava plateau, and similar features may be found on every continent.

Volcanoes: sleeping giants

If lava extrudes through a vent, a mountain called a *volcano* is formed. Such mountains are the best-known expressions of igneous activity. There are thousands of volcanoes on earth today, and they are found from pole to pole on the continents, along the continental margins, on islands, or submerged below sea level in the deep ocean basins. Today the areas of greatest volcanic activity are around the Pacific Ocean basin (called the ring of fire) and the Mediterranean Sea, although some volcanic activity has probably occurred at every spot on earth at one time or another. The volcanoes recognizable as such today are all young in the geologic sense because volcanoes are very vulnerable to erosion and cannot last more than a few million years.

Most volcanoes today are classified as *extinct;* that is, they have not erupted within recorded history and are not expected to erupt again. The beautiful chain of volcanic mountain peaks extending northward from California into Canada is classified as extinct. This range, called the Cascade Mountains, includes such famous volcanoes as Mt.

Shasta, Mt. Hood, and Mt. Rainier. Mt. Baker, also in the Cascades, became active in 1975.

Volcanoes are considered *dormant* if they have erupted within recorded time but not recently. Fujiyama in Japan, Mt. Paracutin in Mexico, Mt. Pelée on the island of Martinique, and Mt. Lassen, the southernmost peak of the Cascade Range in California, are all listed as dormant. Some volcanoes whose activity consists only of spewing steam and other volcanic gases on a regular basis are classified as *steaming;* examples of this type are Mt. Vesuvius and Mt. Popocatepetl in Mexico.

Volcanoes that either are presently erupting or that have erupted recently are classified as *active*. Mauna Loa and Kilauea on the island of Hawaii are of this type.

Volcanoes pass easily and often quickly from one stage to another with complete disregard for such classification schemes. We should therefore consider revising activity classifications so that we are not misled into thinking that a volcano classified as "extinct" may never erupt again or that a dormant volcano may "sleep" for 1 million years. Our problem is to recognize the stage a volcano is in at any given time, and we should by all means be pessimistic in our classification to avoid fatal mistakes. The inhabitants of Pompeii and Herculaneum no doubt thought that nearby Vesuvius was extinct or dormant, and the inhabitants of St. Pierre probably thought that Mt. Pelée would never erupt again. In understanding and gaining proper respect for geologic processes such as volcanic activity and earthquakes, the problem is that we often fail to grasp the enormity of geologic time relative to the length of our own history.

Compounding the problem is the fascination that volcanoes hold for us. Their natural beauty, the abundant rainfall that occurs in their vicinity, the extremely rich soil produced by the weathering of volcanic rock, or our own fatalistic attitude make them popular locations for vineyards, plantations, rice paddies, and towns and cities. Naturally, then, when the volcano erupts, human life and property lie in the path of the lava, ash, and gases. Even with knowledge gained from past mistakes, people continue to live on or near these sleeping giants. For example, within a few miles of the steaming volcano Vesuvius lies the modern city of Naples, Italy, with a population of more than 1 million. St. Pierre, rebuilt on the ruins of the city destroyed by Mt. Pelée, has a population of 332,000, almost 10 times the number of people killed in 1902. The island of Sicily, just off the southwestern coast of Italy, has a population of 5 million; the island is dominated by the active volcano Mt. Etna. The islands of Indonesia and Micronesia, the site of Tomboro and Krakatoa, also teem with inhabitants. The islands that make up Japan, the Philippines, and Hawaii are all volcanic. Mt. Rainier, so assuredly classified as extinct, has shown evidence that it could erupt again. If this beautiful 14,410-foot volcano in the state of Washington should erupt, the Seattle-Tacoma area with its population of three quarters of a million people could be placed in jeopardy.

The question, then, may not be whether a volcano will erupt, but when. What can we do to protect ourselves from one of nature's greatest threats? Can we intervene to prevent volcanic eruptions? The answer to this question at the present time is emphatically no. We cannot control forces that have been at work since the formation of the earth. The answer to the first question, then, is that we will have to be aware of the danger in areas of recent volcanic activity, and by recent we mean at least the last 1 million years. In the United States, all areas are relatively safe from volcanic activity except the extreme West Coast, Alaska, and Hawaii. An area of potential volcanic activity of the future would be the Gulf Coast states. Clearly, since we are not about to abandon or relocate cities in volcano-prone areas, whether people want to take the risk of living in such areas becomes a matter for individual decision.

Volcanic by-products. As magma extrudes at the earth's surface, three basic volcanic by-products are produced. These are *lava, pyroclastic* or solid *material,* and *volcanic gases*. Lava is the term applied to magma that extrudes as a liquid from the earth's crust. Lava often reaches a temperature of 1800° F, although this may vary over a rather wide range. Lava, like magma, is a molten silicate; the percentage of silica determines the *viscosity* (thickness, or resistence to flow) of the hot liquid rock. A low silica content produces a thinner or less viscous lava, while a higher silica content forms a thicker, more viscous material. In addition, lava contains dissolved gases; the quantity of gases determines to a great extent the violence with which the lava will break through at the surface.

Although there are three common types of lava, *basaltic* lava, with a silica content of less than 50%, is the most abundant type. This dark-colored lava is associated with eruptions in the ocean basins and sometimes along the continental margins. *Rhyolitic* lava, which is light in color and has a silica content of more than 50%, is associated only with continental volcanoes. *Andesitic* lava, named after the South American mountain range, is intermediate between basaltic and rhyolitic lava in both silica content and color. Andesitic lava is usually associated with volcanic eruptions along the continental margins.

Lavas are classified not only as to silica content but are also categorized according to the combined effects of viscosity and gas content. A highly viscous, slow-moving lava that produces jagged rubble as it cools because of its high gas content is classed as *Aa* lava. A less viscous, faster-moving lava with a lower gas content that cools to form a smooth or ropey mass is known as *Pahoehoe*. Even though both terms are of Hawaiian origin (where the study of volcanoes has naturally been intense), Aa and Pahoehoe lava may be found anywhere in the world.

Lavas contain dissolved gas, just as a carbonated drink contains dissolved carbon dioxide gas under pressure. When the cap is removed from the drink, the gas escapes from the solution because the pressure is reduced. As magma approaches the surface, the pressure on it is also reduced, releasing these volcanic gases. By far the most abundant gas escaping from volcanic eruptions today is water vapor. Oxygen, nitrogen, argon, ammonia, hydrogen sulfide, carbon dioxide, carbon monoxide, and the gases of hydrochloric, hydrofluoric, and sulfuric acids are also present. Because all volcanic gases are extremely hot and many are poisonous, their release contributes greatly to the loss of life during volcanic eruptions.

The role of volcanic gases in the formation and maintenance of the earth's atmosphere is not clear, although they undoubtedly play a major part. As far back as we look in the geologic record, evidence points to the fact that atmospheric erosional agents such as wind, running water, and glacial ice have been at work. From the atmosphere came the products of precipitation that formed the water above, on, and below the earth's surface. The earth, then, has undoubtedly had an atmosphere of some kind since its beginning. Whether the original atmosphere was left over from the formation of the earth, was composed of material that escaped through volcanic activity from the earth's interior, or was a combination of both is impossible to determine at this time. We can hypothesize, though, that the original atmosphere was residual, while the escape of volcanic gases has altered and maintained it through the 4.5 to 5 billion years of geologic time.

If the gas content of the lava is high or if the pressure is reduced suddenly, the liquid rock is thrown out of the volcanic vent together with the escaping gases, just as a carbonated drink may foam over as the pressure inside the bottle is reduced quickly by the removal of the top. As these masses of lava are thrown out of the vent, they begin to cool almost instantly when they come in contact with the much cooler air. As a result, some lava that left the vent in the liquid state will either be solid or well on the way to solidification by the time it strikes the surface a few seconds later. Such material is known as *pyroclastic debris*.

Pyroclastics occur in various sizes. The largest, called *blocks*, are thought to be pieces of solid rock broken away from the vent during an eruption. Blocks commonly weigh several hundred pounds or more; a 2-ton block was reportedly thrown almost 2 miles during an eruption of Mt. Stromboli in Italy. Next in size are *volcanic bombs*, which are usually no more than a few feet in diameter and often have the shape of a football. They get their name from their tendency to whistle like a bomb as they fall. Volcanic bombs are usually twisted on the ends, an indication that they rotate while still in a semi-liquid state in the air. Like blocks, they usually land within a few hundred feet of the vent. Still smaller than volcanic bombs are *lapilli* (or *cinders*), pea-sized particles that are also often twisted. *Volcanic ash* is still smaller material, generally no larger than the head of a pin. *Dust* is the smallest pyroclastic of all, ranging down to almost microscopic size. These finer particles may be carried by the wind hundreds or even thousands of miles away from the eruption. It is virtually impossible to escape from volcanic dust because it is so fine that it can pass through small cracks in the walls of a house, come under doors, etc. Most of the deaths in Herculaneum and Pompeii, and the preservation of minute details of the population's daily lives, were the work of smothering volcanic dust.

Volcanic structure. Although it is difficult to generalize about volcanoes because each one is usually unique in some respect, certain basic characteristics can be noted. For example, all volcanoes may be classified on the basis of structure as either *shield, composite,* or *cinder cone* volcanoes.

The *shield* volcano is characterized by very gentle slopes, generally less than 5°. The mountain itself is made up predominantly of basaltic lava flows, one on top of the other. Pyroclastics hardly ever occur in association with shield volcanoes. The lava, containing less than 50% silica, is nonviscous and has a low gas content. The low viscosity explains the small angle of the slopes, because the lava spreads out for great distances in all directions before cooling. The low gas content of the lava, on the other hand, explains why these volcanoes are considered the least dangerous of the three types and thus are best suited for close study. Shield volcanoes are the largest of all volcanic mountains and are formed almost exclusively in the earth's deep ocean basins. The Hawaiian Islands, Iceland, and the Galapagos Islands off the western coast of South America are composed of shield volcanoes. Mauna Loa, a shield volcano on the island of Hawaii, is considered to be the greatest volcano on earth. From the floor of the ocean to the peak of this mountain is a distance of nearly 30,000 feet, although only the last 13,680 feet are above sea level.

Composite volcanoes are beautifully symmetric mountains with angles of 20° to 30° (Fig. 11-7). These volcanoes are composed of almost equal quantities of hardened lava

Fig. 11-7. Mt. Rainier in the state of Washington is one of the most lovely composite volcanoes in the continental United States. It is now classified as extinct, although recent indications point to an impending eruption. (Courtesy United States Department of the Interior, Geological Survey, Reston, Va.; photograph by B. Willis.)

flows and pyroclastics. The lava extruded from composite volcanoes is more viscous than that extruded from the shield type because it contains a larger percentage of silica. Because they have more gas charged into the lava, composite volcanoes erupt with more violence than shield volcanoes. Although composite volcanoes go through periods in which only lava is extruded, there are also times when only pyroclastics are produced. This intermittent action produces in composite volcanoes a well-layered internal structure that has led to the use of the term "stratovolcano." However, both lava and pyroclastics are often produced simultaneously.

Composite volcanoes are the most common type of volcano found on continents. These mountains may reach elevations of thousands of feet and have basal diameters of many miles. They are often so high, in fact, that their tops are above the snow line, with the result that they are snow capped throughout the year. Some composite volcanoes such as Mt. Rainier actually have glaciers on their sides. The glacial ice is unaffected by the magma below the surface of the mountain because of the insulating blanket of soil and rock. Composite volcanoes are found in all parts of the world but are more often associated with the continents and continental margins. The great peaks of the Cascades such as Mt. Rainier, Mt. Shasta, Mt. Hood, and Mt. Washington; Fujiyama in Japan; Mt. Mayon in the Philippines; Mt. Vesuvius in Italy; and Mt. Kilimanjaro in Africa are all examples of this type of volcano.

The *cinder cone* volcano is the smallest but also the most violent of the three types. Cinder cones are usually found along the continental margins or on island archipelagoes such as the West Indies and the Aleutians. These volcanoes are usually not more than 2000 feet high and have diameters through their bases of less than 1 mile. As the name implies, these volcanoes are composed almost entirely of pyroclastic debris, which is generally deposited in such a way that it produces slopes with angles greater than 30°. Often, however, lava flows emerge from fissures near the base to spread out for great distances away from the cone. The violence of cinder cone volcanic eruptions is explained by the high gas content of the lava. Examples of cinder cone volcanoes are Mt. Paracutin in Mexico, Krakatoa in the East Indies, Mt. Pelée on the island of Martinique, and the San Francisco peaks in Arizona.

Other volcanic features. Volcanic mountains almost always display one or more depressions or holes near their summits. If the depression is produced by the active buildup of pyroclastic debris around the vent, it is known as a *crater*. Crater diameters usually do not exceed more than a few hundred feet. If, on the other hand, the depression is formed primarily by the collapse of volcanic material back into the vent after an eruption, it is known as a *caldera*. The caldera at the top of Mauna Loa on the island of Hawaii, for example, is 3 miles in diameter; the caldera of Mt. Mazama in Oregon is 6 miles across and 4000 feet deep. This caldera contains water to a depth of 2000 feet, forming Crater Lake (Fig. 11-8). This water, however, is not of magmatic origin but is derived from rain and melting snow.

Other features often found in association with volcanic activity include fumaroles, solfataras, paint pots, thermal springs, and geysers. *Fumaroles* are formed wherever hot volcanic gas or steam constantly issues from vents or fissures in the ground. They are produced either through the escape of volcanic gas from magma or by the vaporization of ground water as it comes into contact with or nears cooling magma. If the escaping vapors are sulfur gases, the fea-

A

B

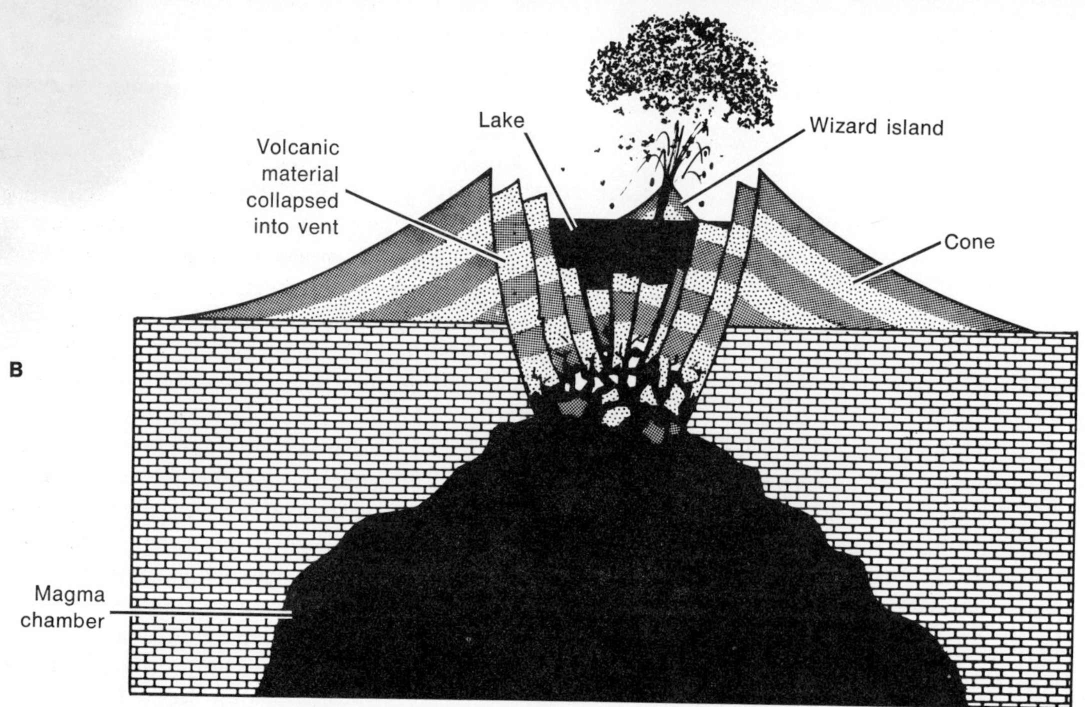

Fig. 11-8. A, Crater Lake, Oregon, is a water-filled caldera formed by the collapse of a volcanic cone after an eruption. Wizard Island, seen in the background, is a small volcanic cone. With a depth of 2000 feet and a diameter of 6 miles, Crater Lake is the second deepest lake in North America. **B,** Diagrammatic representation of the history of Crater Lake. (**A** courtesy Oregon State Highway Department, Salem, Ore.)

ture is called a *solfatara*. Fumaroles and solfataras may be seen in California, Yellowstone National Park in Wyoming, the Valley of the Ten Thousand Smokes in the Mt. Katmai region of Alaska, and in Iceland, Italy, Japan, New Zealand, and Mexico.*

Paint pots, sometimes called *mud volcanoes,* occur as water or gas bubbles upward to the surface through mud, often producing in the process a beautiful array of colors. *Thermal springs,* on the other hand, occur where groundwater is constantly issuing from below the surface at above-normal temperatures. They are usually named on the basis of water temperature, such as boiling springs, hot springs, or warm springs. The warmth of thermal springs results either from the heating effect of contact with cooling magma or from the geothermal gradient due to pressure.

Geysers are intermittent boiling springs that erupt periodically in areas in which groundwater is being heated by cooling magma. The periodicity of geysers is explained not only by a complex underground plumbing system but also by the nature of boiling water under pressure. For example, groundwater fills a network of underground fissures that have only a single opening at the surface. Cooling magma or its escaping gases heat the water in the lower region of the fissures. The weight of the overlying water exerts pressure on the water below, requiring that it be heated well above the normal boiling point of 212° F (100° C), a condition called *superheated*. At the higher temperature the water boils, changes to steam, expands, and pushes the water above it toward the surface. This upward movement reduces the pressure throughout the heated column of water and it also begins to boil. The entire mass of water then becomes unstable and is thrown violently into the air. Water then flows back into the fissures, is heated, and the entire process repeats itself. Old Faithful, the famous geyser in Yellowstone National Park, erupts on an average of every 65 minutes, throwing steam and water as high as 170 feet into the air.

JAMES HUTTON AND THE ROCK CYCLE

The Scottish farmer-chemist James Hutton (1726-1797) is considered the father of geology because of two major contributions he made to the fledgling science. First, he realized that the earth's present features and processes could explain its past. This concept is called *uniformitarianism,* a term coined by the British geologist Charles Lyell (1797-1875).

In a world accustomed to applying the past to the present or future, uniformitarianism is somewhat unusual, for it means that the present is the key to the past. Because geologic time is so immense, the geologist must begin with observable present-day processes and landforms and apply the information gained from these observations to evidence found in the rocks of the earth's crust. For example, geologists observe a volcanic eruption in Hawaii. They record and analyze the conditions surrounding the eruption and the volcanic structures and rocks that are formed. Then, when they find similar structures and rocks 70 million years old at the bottom of Hell's Canyon in Idaho, they can say with certainty that a volcano erupted then as now, even though no man was present with paper and pencil to record the event. Uniformitarianism is the basic concept in geology.

The other main contribution James Hutton made to geology was defining the relationship between rock types known as the rock cycle. In essence, he believed that magma was the original source of all minerals and rocks and that the environments of the earth originated through changes in the original igneous environment.

The modern, idealized rock cycle, shown in Fig. 11-9, assumes that magma was indeed the parent material of all rocks. As the magma or lava cooled, igneous rocks were formed. These rocks, on exposure to the atmosphere, were altered physically and chemically into rock fragments called *sediments*. These sediments were then moved away from their original locations by agents such as streams, wind, glaciers, groundwater, and gravity mass movement and were deposited in water to form sedimentary basins. There, through compaction and natural cementation, the loose sediments became *sedimentary rocks*. As the sediments and sedimentary rocks in the basins became deeper and deeper, the weight of the overlying material created pressure and heat to change some of these rocks and sediments into *metamorphic rocks*. If the heat and pressure became extreme, the metamorphic rocks could melt to form magma, thus completing the cycle.

Like James Hutton, we realize that the rocks we observe today have been through the rock cycle many times. It is doubtful that the rocks of the earth ever went completely through the rock cycle without having the three environments interact. For example, some igneous and sedimentary rocks have been subjected to the metamorphic environment and in the process have become metamorphic rocks. Metamorphic rocks themselves have often been changed into other metamorphic rocks, while all three rock types could have melted to form magma. Likewise, all of them will form sediments as they are exposed to the atmosphere.

The rocks of the earth's crust, then, are "used" in the

*Thermal energy such as this is now being used as a limited source of electric power in California, Italy, New Zealand, Japan, and Iceland. Mexico also expects to put a geothermal power plant into operation soon.

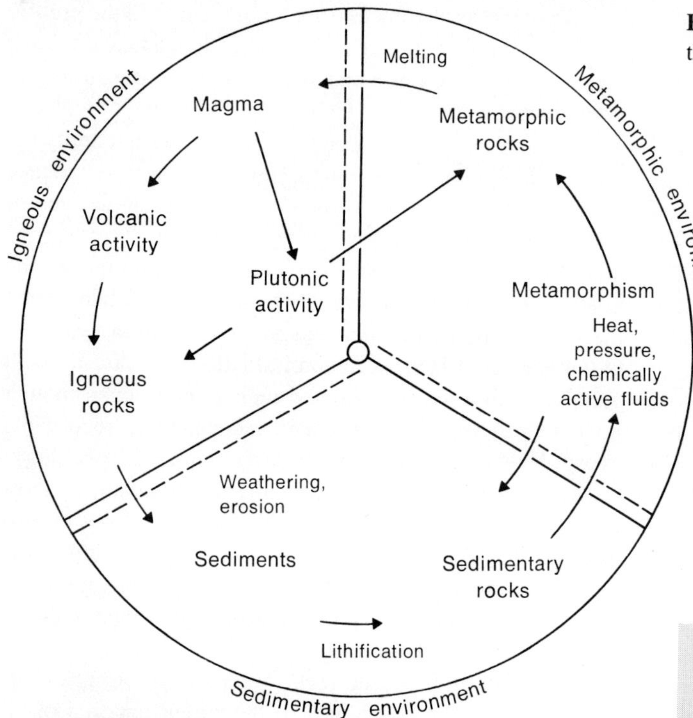

Fig. 11-9. Schematic representation of the rock cycle showing the interrelationships between the three rock types.

sense that they have been through the rock cycle many times. This probably explains why we have not found rocks on earth older than 3.3 billion years, when the age of the earth and the rest of the solar system is estimated at 4.5 to 5 billion years. The moon, still in the igneous state because it lacks an atmosphere and water, has yielded rocks as old as 4.1 billion years. Perhaps only deep below the surface of the earth, beyond our present reach, will the original rocks of the planet be found.

IGNEOUS ROCKS

Thus far we have explored the igneous environment. The final byproducts of the igneous environment are igneous rocks, by far the most abundant type of rock on this planet. Although not as common on the surface as sedimentary rocks nor as abundant in the near-surface environment as metamorphic rocks, we would still have to consider igneous rocks, overall, the most common and important in the geologic sense.

Color

There are many different types of igenous rock, and to establish order out of chaos, attempts have been made to classify them. None of these attempts has been completely successful, for it seems that some igneous rocks, like almost anything one attempts to classify, defy classification.

Fig. 11-10. A, Basalt. The dark color of this fine-grained igneous rock indicates its low silica content. **B,** Granite. The light color of this rock shows its high silica content, and the coarse grain contrasts with the fine-grained basalt to show the variety of textures found in igneous rocks.

Table 13. Igneous rock classification

Texture	Color		Crystals
	Dark	Light	
Glassy	Obsidian	Obsidian	—
Fine grained	Basalt	Felsite, andesite, trachyte, rhyolite	Microscopic or very small
Coarse grained	Gabbro, diorite	Granite, syenite	Large
Porphyritic	Basalt, andesite, trachyte, rhyolite, porphyries	Basalt, andesite, trachyte, rhyolite, porphyries	Large and small
Pegmatitic	—	Pegmatite	Giant
Glassy vesicular	Scoria (large holes)	Pumice (small holes)	—

With this realization, we may use a classification scheme based on silica content (as indicated by overall color of the rock sample) and texture. Such a simplified classification scheme is shown in Table 13. The overall color of an igneous rock is usually but not always indicative of its silica content. A silica-rich igneous rock, for example, will usually be light in color, and a silica-poor igneous rock will normally be dark (Fig. 11-10). Of course, there are intermediate compositions and the ever-present exceptions. All rocks, regardless of type, contain some silica, the combination of the elements silicon and oxygen. This is true because silicon and oxygen are the two most abundant elements that naturally make up the earth's crust, and most rocks, through the rock cycle, have been formed from the reworking of the crust.

Texture

Texture in igneous rocks always indicates the rate at which the magma or lava cooled, for mineral crystals take time to form—the longer the time interval, the larger they grow. Based on the rate of cooling, texture in igneous rocks refers to the size of the mineral crystals that make up the rock. For example, if lava cools very rapidly, perhaps in seconds, crystals will not have an opportunity to form because all the atoms are "frozen" in place without forming orderly arrangements. This produces an igneous rock that contains no crystals and has a texture referred to as "glassy." Obsidian, or volcanic glass, is such a rock. Other glassy igneous rocks may take a different form which is known as *vesicular* rock. Vesicular rocks are formed of lava that has solidified very rapidly around gas bubbles, which create holes in the rock. Pumice and scoria are examples of vesicular, glassy igneous rocks.

A fine-grained texture is produced when magma or lava cools rapidly—perhaps in hours or a few days. Mineral crystals do form but they are too small to be seen with the unaided eye or even with a hand lens. Study of a thin section of such a rock under the microscope will, however, reveal the crystals. Felsite and basalt are good examples of fine-grained igneous rocks (Fig. 11-10).

A coarse-grained rock is formed when the magma cools slowly over months or even years. In such a situation crystals not only form but also grow large enough to be seen with the naked eye. Gabbro and granite are examples of coarse-grained igneous rock. A *pegmatitic* texture, characterized by giant crystals, is formed when magma cools very slowly, probably over a period of thousands of years. The igneous rock pegmatite shows such a texture (Fig. 11-11).

A *porphyritic* texture is produced under variable cooling conditions, when lava begins to cool rapidly and then the cooling slows. Under these conditions, large crystals are found embedded in masses of smaller crystals. Felsite por-

Fig. 11-11. Pegmatite is an igneous rock whose large crystals show that it cooled over a long period of time.

phyry is an example of an igneous rock with a porphyritic texture.

Granite: igneous or metamorphic?

As pointed out earlier, some items always seem to defy classification. In terms of igneous rocks, granite is one of these. In the early days of geology, granite was classified as a metamorphic rock, but over the years it has been reclassified as igneous. Now the trend seems to be toward reclassifying granite as a metamorphic rock. We consider granite an igneous rock here because this is the opinion of most geologists today. However, this does not mean that *granitization,* the term applied to the processes that supposedly convert existing rocks to granite, is not at least sometimes a possibility. Granitization, a metamorphic process, is discussed in detail in Chapter 17.

Uses

Igneous rocks are of great utility in the world today and must be considered, directly or indirectly, the most valuable source of natural materials. Gold, silver, lead, zinc, and other metals and nonmetals are deposited in igneous rocks of all types by the migration of hot solutions expelled from magma. The crystallization of the magma itself also produces mineral deposits of great economic value. For example, the plutonic igneous rock periodotite is the leading source of diamonds as well as a valuable ore of nickel, chromium, and platinum. Pegmatite, the very coarse-grained plutonic rock, is quarried for its mineral content of mica, feldspar, and quartz. The semiprecious gems, rose quartz, smoky quartz, and moonstone (feldspar), come almost exclusively from pegmatite.

Granite, syenite, and gabbro are popular as building, ornamental, and monument stone. They are also used as crushed rock for construction purposes. The volcanic igneous rock pumice is useful as an abrasive in scouring powders, soaps, and dental polish, and it is also used as an insulating material and as a filler for cement and plaster to give these construction materials a lighter weight. The use of fine-grained igneous rocks such as felsite and basalt is limited almost entirely to crushed rock.

IGNEOUS ACTIVITY ELSEWHERE IN THE SOLAR SYSTEM

It is becoming increasingly apparent that igneous activity is not confined to the earth. First, the rocks brought back to earth by the six successful Apollo flights show definite signs that the moon was formed almost entirely in the igneous environment. The few rocks that were not igneous could be classified as metamorphic rocks that may have been formed from the shock of meteorite impact. No sedimentary rocks were discovered on the lunar surface. The moon fundamentally lacks metamorphic rock formed within the moon itself and also sedimentary rock because the normal rock cycle does not occur in the absence of an atmosphere. The presence of hardened lava flows and domes on the moon has been confirmed, however, and volcanic craters have been seen, although none are known to be actively erupting.

Igneous activity has also played the key evolutionary role in the history of the planet Mars. High-resolution photographs returned to earth by Mariner 9 in 1971 and 1972 showed Nix Olympica, a volcano 335 miles in diameter with a caldera 40 miles wide. In addition, a caldera 70 miles wide was found in the vicinity of Nodus Gordii, and other smaller volcanic craters are scattered over the Martian surface (Fig. 4-4). Astronomers who have observed Mars telescopically during its infrequent close encounters (35 million miles) with the earth have reported seeing a red glow in the vicinity of Nix Olympica. This could mean that Mars is still volcanically active. At any rate it would appear that Mars, in its present stage of evolution, is somewhere between the stages of the moon and the earth.

We could hypothesize that the other terrestrial planets in our solar system, those with densities similar to earth's, might also have been the sites of past or present igneous activity. For example, Mercury, Venus, and perhaps even distant Pluto could be geologically more like the earth than anyone would dare to imagine. From the data obtained from Venus and Mercury by Mariner 10, this appears to be the case.

12 *master sculptor of the ages*

Weathering, mass movement, and erosion

The hills are shadows, and they flow
From form to form, and nothing stands;
They melt like mist, the solid lands,
Like clouds they shape themselves and go.

In Memoriam
Alfred, Lord Tennyson

On October 9, 1963, more than 200 million cubic yards of rock and soil slid into the Vaiont reservoir behind a dam high in the mountains of Italy. The landslide, one of the worst in recent history, sent a surge of water more than 300 feet high crashing over the dam. The rushing water was still more than 200 feet high 1 mile down the narrow valley below the dam, and total destruction extended for miles beyond. The catastrophe occurred in 7 minutes and cost the lives of 2117 people. Fig. 12-1 shows the Vaiont reservoir area as it looked before and after the fateful event.

THE SEDIMENTARY ENVIRONMENT

Almost everything found at or near the earth's surface shows the effects of exposure to the atmosphere. Paint peels, automobiles rust, pavement cracks, silver tarnishes, wood rots, skin dries and wrinkles—everywhere the process is persistent as things disintegrate and decompose.

Rocks are not immune to the process, although the changes in them require more time. Solid rocks are gradually broken into smaller pieces and changed in composition to form rock fragments and chemicals called *sediments*. The formation, movement, and deposition of sediments and their subsequent change back into solid rock are the processes of the sedimentary environment, the least severe of the three rock-forming environments.

The tragedy in Italy was the result of the interaction of three often-forgotten geologic processes: weathering, mass movement, and erosion. Weathering loosened the rocks and soil on the slope above the dam, mass movement sent the rocks and soil crashing into the reservoir, and erosion was performed by the water as it rushed down the valley below the dam.

All natural processes that cause rocks to disintegrate and decompose are called *weathering*. There are two types of weathering that attack rocks—*mechanical* and *chemical*. Both mechanical and chemical weathering are largely the result of the earth's atmosphere; without this gaseous envelope, the rocks would undergo very few changes. The moon, for example, is considered to be virtually unchanged since its early history—a celestial fossil preserved by the lack of weathering processes. *Erosion* is the process of moving rocks or soil out of their original environments. *Mass movement* is a rather loose term including agents of both weathering and erosion when they involve large ''masses'' of material; it is not really a separate process in and of itself.

Mechanical weathering

Mechanical weathering includes those processes that cause rocks to disintegrate (larger chunks crumble into smaller pieces) without appreciably changing their chemical composition. Mechanical weathering is at work over the earth's entire land surface but is much more effective at high altitude and high latitude, where the temperature is low and moisture is adequate, and in deserts, where the temperature range is great and moisture is scarce. Mechanical weathering processes include wetting and drying, ice wedging and heaving, exfoliation, organic processes, and thermal changes.

Fig. 12-1. The Piave Valley in Italy before and after the landslide of October 9, 1963, at the Vaiont reservoir.

Wetting and drying

Rocks and soil, like most other substances, are affected by the continual wetting and drying that occur because of exposure to the atmosphere. Rain soaks the rocks, then the sun's heat bakes them. Especially susceptible to disintegration by wetting and drying are rocks and soils rich in clay-forming minerals such as feldspar and mica, which absorb water and expand and then contract when the water is lost by evaporation.

Ice wedging and heaving

Water expands by 9% of its volume when it freezes because the hydrogen and oxygen atoms of which it is made occupy a greater space in the orderly, fixed arrangement of ice than when they are free to move about as a liquid. This explains why ice floats in the water from which it is made, and why 91% of the volume of icebergs is under water. Although rocks generally look nonporous, most contain some pore spaces or cracks into which water may seep. When the water in the rock freezes, the expansion force exerted by the ice causes the rock to break apart. This process is known as ice wedging.

A similar process, called ice heaving, takes place in soil or other unconsolidated material in areas in which the ground is frozen much of the year. As the water freezes, the soil or other material is pushed upward and broken into polygon-shaped sections by the force of expansion. The tundra polygons found in colder climates are formed by this process (Fig. 12-2).

Fig. 12-2. Tundra polygons are formed in cold regions when the land expands on freezing, forcing the top layer up and breaking it apart into these characteristically shaped pieces. (Courtesy United States Department of the Interior, Geological Survey, Denver, Colo; photograph by R. E. Wallace.)

Exfoliation

Several types of igneous rocks that have formed deep underground will, on exposure at the surface, break naturally into thin, curved layers roughly parallel to their own external surface. Such onionlike peeling of rocks is called exfoliation (Fig. 12-3). A large mass of naturally exposed rock that exhibits this property is known as an exfoliation *dome*. Half Dome in Yosemite National Park in California and Stone Mountain in Georgia are good examples of exfoliation domes.

Because exfoliation affects only coarse-grained, intrusive igneous rocks, geologists have concluded that the peeling occurs as a result of the release of confining pressure.

Fig. 12-3. This large slab of granite on the summit of Stone Mountain, Georgia, has started to peel off into layers, or exfoliate. Since the outcrop exhibits this characteristic, it is called an exfoliation dome.

Granite, for example, is a coarse-grained igneous rock that is thought to form deep underground from the crystallization of magma. Over million of years the granite adjusts to the weight of thousands of feet of overlying rocks and sediments. When exposed to the low-pressure surface environment by weathering and erosional processes, the granite readjusts by expanding. This expansion force then causes the rock to exfoliate. In quarrying operations this natural expansion force is used to help remove great blocks of rock from the igneous mass. Once free, the blocks often expand in length by 4 to 6 inches, proving that the mass was indeed under confining pressure.

Organic action

The action of plants and animals also produces mechanical weathering. Plant roots and stems, for example, may grow into the cracks and spaces in rocks, As the plant parts grow, the rocks are forced to split apart. Hoofed animals chip away pieces of rock, and earthworms literally eat their way through soil and rocks. People, however, must be considered the most effective mechanical weathering agents among living things, for with bulldozer and dynamite they can do more in minutes to mechanically weather rocks than nature could do in millions of years.

Thermal changes

Rocks are composed of either intergrown mineral crystals or rock fragments that have been cemented naturally or compacted under pressure. Each of the crystals or rock fragments, like other crystalline solids, expands at a different rate when heated and contracts at a different rate when cooled. This differential expansion and contraction causes the rocks to weaken and crumble over long periods of time. Thermal changes in rocks, however, are so slight that laboratory experiments have failed to record any appreciable change in the rocks under rapid heating and cooling conditions. Perhaps the reason for the lack of laboratory evidence is that other weathering processes are also involved that we have been unable to duplicate under controlled conditions.

However, field evidence shows that thermal changes do

produce mechanical weathering. The evidence ranges from hearsay reports that rocks in deserts may actually be heard cracking and popping at night as they rapidly cool, to photographs of rocks buckled by the heat of the sun and eyewitness accounts of rocks literally exploding during forest fires. As the debate continues, we will assume that thermal changes do play at least a minor role in the disintegration of rocks.

Chemical weathering

Chemical weathering includes those processes that change the chemical composition of rocks and make them decompose as a result. Chemical weathering consists of one or more chemical reactions that usually involve the addition of some atmospheric component to the rocks of the earth's crust. Although we are discussing chemical weathering and mechanical weathering separately, remember that the two types of weathering usually occur simultaneously, even though one form may be dominant. For example, in regions where the temperature range is less extreme and water is abundant, chemical weathering predominates. In fact, the more hot and humid the climate, the more active chemical weathering processes become.

Oxidation

Oxidation is the chemical addition of oxygen to a rock or other substance. Oxygen is the second most abundant gas in the atmosphere today, comprising 21% of it by volume. Oxygen is also the most chemically active constituent of the atmosphere, combining easily with most elements to form oxides. Because of the abundance and chemical activity of oxygen in the atmosphere, oxidation is an important chemical weathering process that acts to weaken the rocks of the earth's crust.

Although oxygen is very active chemically, it combines more readily with some elements than others. Iron, for example, has a strong affinity for oxygen, making this important and abundant metal one of the most susceptible constituents in the rocks to the oxidation process. The oxidation of iron and iron compounds in rocks and soil results in the red or reddish-brown color that is quite noticeable and characteristic of this element. The precious metal silver is another element that has a strong attraction for oxygen. Although silver is rare in rocks, oxidation attacks articles made of this prized native mineral, forming the black tarnish that is so difficult to remove.

Hydration

Hydration is another chemical weathering process affecting the rocks of the earth's crust. This process involves the chemical addition of water to a rock or other substances. Water, the most abundant naturally occurring liquid, is readily available for chemical reactions. After the combination has occurred, the rock is weaker than before and tends to disintegrate more easily.

The process commonly called "rusting" is actually caused by the combined processes of oxidation and hydration. A metallic object rusts much more slowly in a dry climate than in a humid one. For this reason, deserts are chosen for the long-range storage of surplus military equipment such as tanks, planes, and trucks. Unfortunately for the consumer, automobile manufacturers do not follow the example set by the military. By the time a consumer receives his new car, it may have sat exposed to oxidation and hydration for as long as 1 year. In humid climates where groundwater rich in iron compounds is used to water grass, an unsightly reddish-brown stain is left on the exteriors of buildings and on sidewalks and streets. Again, this is the work of hydration and oxidation.

Another interesting result of the combined effects of oxidation and hydration has recently been noted. Along parts of the Colorado River, the rocks are red but the sandbars in the river are white or gray. Since the sand was formed by the weathering of the rocks in the area, it would be expected that the sandbars would also be red. The red color of the rocks in this region is due to the oxidation and hydration of iron compounds. It was discovered that the abrasive action of the sand in the flowing river wears away the superficial red color, leaving the sandbars white, off-white, or gray.

Carbonation

Another important chemical weathering process acting on certain rocks is carbonation, the chemical addition of carbon dioxide. Like hydration, carbonation also involves water. In fact, water plays a vital role in both chemical and mechanical weathering, as it does in virtually all geologic processes.

Rainwater combines with carbon dioxide from the atmosphere to form carbonic acid. This acid reacts with rocks such as limestones and dolomites that are rich in carbonates to form calcium bicarbonate, which easily dissolves in water, while carbon dioxide gas is released. The carbon dioxide then reacts with additional water underground to continue the process. Rocks rich in carbonate minerals are literally dissolved; once they are attacked, the process continues at an ever-accelerating rate.

Another familiar effect caused by the carbonation process is the green tarnish that forms on native copper, copper compounds, and articles made of copper. So characteristic is this green tarnish that deposits of copper in the earth may be recognized on sight.

Soil

An important and vital result of weathering is the formation of soil. Soil is a surface layer of weathered rock that has the physical, chemical, and biologic properties necessary to sustain plant life. Without soil there could be little plant life on land, and without plant life, land animals could not survive.

Soil may either be *residual* (formed in place by the weathering of bedrock) or *transported* in the sense that the soil develops from rock fragments that were moved by one of the agents of erosion. The bedrock or rock fragments that produce the soil is known as *parent material*.

Soil profiles

With time, soils develop a distinct profile composed of layers, or *horizons*. The upper layer is the area of greatest alteration of the parent material, while succeedingly deeper layers show less change. An idealized soil profile is shown in Fig. 12-4.

Horizon A, the top layer of soil, is porous, consisting of very fine particles of rock loosely held together. Soluble minerals are removed from horizon A by downward-moving water. At the same time, partially decayed plant material called *humus* is added at the surface. *Leaching*, or the removal of soluble minerals from horizon A, has led to the use of the phrase *zone of leaching* to describe this layer.

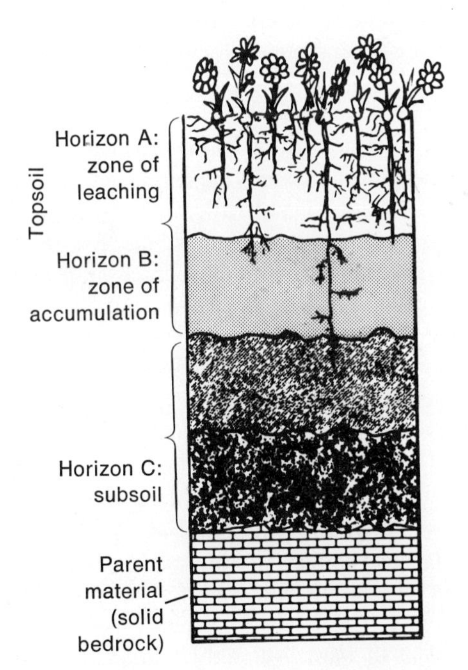

Fig. 12-4. Idealized schematic representation of soil profile.

Below horizon A lies horizon B, a layer also known as the *zone of accumulation* because many of the minerals leached from horizon A are concentrated here. For this reason, horizon B tends to be compact and dense and is often referred to as *hardpan*. Together horizons A and B make up that part of the soil profile known as the *topsoil*.

Below horizon B there is a zone of partially weathered parent material known as horizon C. Since this layer is not fertile and lies below the true soil, this zone is also called the *subsoil*. Below horizon C lies the unaltered parent material itself.

Soil producers

Several interrelated factors are involved in soil formation. The significant soil producers are climate, type of parent material, time, slope of the land, and biologic agents.

Climate and parent material. We might reasonably expect that the major soil producer would be the parent material. However, it has been observed that soils produced from the same type of parent material are different under different climatic conditions. Parent material is an important factor only in young soils; as a soil matures, climate becomes the dominant factor in determining its character. This is true because climate encompasses both moisture and temperature, which are the controlling factors in weathering rates, chemical processes, and biologic processes.

Time. Time is also a factor in soil formation, for it usually takes thousands of years to produce a well-developed soil profile. Because of population pressures, young soils are being forced into premature use with less than encouraging results.

The soils of southern Florida are an excellent example of what can happen to a young soil that is forced into use before its time. Virtually the entire area must be classified as a swamp, which is formed by the overflow of Lake Okeechobee. Originally covered by shallow water, much of the land has been drained for agricultural, commercial, and home development. Canals (artificial streams) have been cut into the ground to drain away some of the groundwater in order to lower the water table enough to produce relatively dry land. The surface material thus exposed is called *muck soil* because it consists predominantly of humus formed under swamp conditions. This crude soil, lacking the profile of older soils, is considered by many to be among the richest in the nation. Its fertility, however, is due only to the high humus content, for otherwise the soil is poor and of short usefulness.

The muck soil is undergoing biologic oxidation in which soil bacteria literally change the humus into carbon dioxide gas, which then escapes into the atmosphere. As a result the soil of the region is disappearing as quickly as 1.2 inches/yr, and estimates indicate that this soil will have to be generally abandoned by the year 2000 (Fig. 12-5). Although fertilizer may impede biologic oxidation, the only final solution to soil subsidence is to place the muck soil back under water.

The example of the changes wrought in south Florida only point out a painful fact: when we tamper with our environment without considering the results, we are the ultimate losers. We have drained, sprayed, fertilized, grown crops, irrigated, and built on the land and have made tests only when problems have become apparent. We have forced soils not suited for the task to support crops, often with staggering yields, only to find later that we must abandon the land because it is worn out.

Fig. 12-5. Soil depth marker in the Everglades shows a loss of 5 feet of soil since 1924. The immature soil turns to carbon dioxide gas and literally disappears at a rate of 1.2 inches/yr. (Courtesy *Miami Herald,* Miami, Fla.)

The dust bowl created by land abuse, the drought in the central and southwestern United States during the 1930s, and the use of soils for only one crop in the southern United States and elsewhere during this century apparently did not teach us a lesson. The full impact of land abuse today is not known, but the experience gained in Florida with drainage and the problems of salt accumulation in the soil through irrigation in the Imperial Valley, California, indicate that the outcome will be less than encouraging.

Slope of the land. Another factor in soil formation is the slope of the land. Steeply sloping land is more susceptible to erosion by running water than flat land, and the continuous removal of the surface layer tends to prevent the development of a soil profile. On more gentle slopes or on flat land the rate of erosion lags well behind the formation of the soil profile. The loss of soil due to erosion can never be completely stopped. The best we can do is to keep the loss to an acceptable minimum.

Biologic agents. The last factor to be considered in soil development is the action of biologic agents. Soil, by definition, must be capable of supporting plant life, but biologic agents are also directly involved in the formation of soil in the first place. For example, lichen are instrumental in beginning the process of soil formation on solid rock. Soil bacteria oxidize plant material that becomes humus in the A horizon, while other bacteria remove nitrogen from the atmosphere to enrich the soil with this vital element. Trees, shrubs, and grass help to retard soil erosion by holding the soil in place with their root systems. The leaves, stems, and roots of these plants also provide humus when they die and begin to decay.

Animals as well as plants contribute their share to soil production. Burrowing animals such as gophers, moles, field mice, and ground squirrels constantly stir the soil by bringing new material to the surface. Earthworms enrich the soil with their body wastes, break up soil particles and small rock fragments by "chewing" and the action of their digestive juices, and aerate the soil and rocks with their constant burrowing activities. Trails underground also allow water to gain quicker entrance into the soil and rocks to enhance alteration of the parent material.

Of all living creatures, only people must be listed as having more of a detrimental than a beneficial effect on the soil. Only we cut and burn the trees, shrubs, and grass. Only we kill small animals directly or indirectly through our constant encroachment on the land to satisfy the pressure of overpopulation.

Mass movement

Mass movement is the natural downslope removal of weathered rock debris by gravity. Rainfall, snowfall, earthquakes, and people must also be included as contributing factors. Mass movement connects weathering processes with those of erosion and overlaps the two, often making it difficult to tell where weathering ends and erosion begins. Mass movement guarantees that a new surface area of rock will be exposed to weathering processes and that erosional agents such as streams, glaciers, wind, and groundwater receive a continuing supply of material to be transported. Wherever the slope of the land exceeds the *angle of repose* (maximum stable angle) of the loose weathered rock debris, mass movement goes to work. Mass movement may be rapid and spectacular, with catastrophic results like the tragedy at the Vaiont reservoir, or it may be imperceptibly slow.

Landslides and avalanches

A landslide is the rapid, downslope movement of large quantities of rock and soil. An *avalanche,* in addition to rock debris, involves large amounts of snow and ice. The triggering mechanism for landslides and avalanches is not always clear but often involves heavy snowfall or rainfall or an earthquake. People may also contribute to the process by underground or surface blasting. Like many other geologic forces, landslides and avalanches must be considered a hazard to people and their property, for they pose a constant threat to populated areas that lie in the neighborhood of steep slopes.

Mudflows

Mudflows are very similar to landslides, with the exception that the moving material has the consistency of mud. Mudflows are often rapid and almost always occur during or shortly after a heavy rainfall in regions that have recently been stripped of vegetation by construction or forest fires. Areas of rapidly melting snow are also susceptible to mudflows. They generally occur in mountains and flow downhill through natural drainage valleys but may occur below any slope. Like landslides, mudflows also create much misery. During drought periods in areas affected by forest fires, caution should be observed by people living below steep slopes.

While mudflows formerly occurred mainly in naturally weathered rock and soil, they now are occurring more frequently in artificial accumulations of debris from mining operations. For example, one of the worst mudflow tragedies of the twentieth century occurred on October 21, 1967, as rain-soaked coal mine residue roared down a mountain to engulf the village of Alberfan, Wales. More than 130 persons were killed, most of them children in a school.

Soil creep

Soil creep is the slow, downhill movement of soil. As in other types of mass movement, gravity and water play key roles in soil creep. Unlike landslides, avalanches, and mudflows, which usually occur rapidly, soil creep is so slow that its movement is not directly visible, although its effects can easily be seen over a period of years. Anything growing or constructed in an area of active soil creep will be affected. Trees, poles, posts, monuments, retaining walls—everything not situated below the zone of creep will tilt downslope (Fig. 12-6). It is imperative that buildings constructed in such an area include a secure foundation below the level of soil creep, preferably into bedrock. Otherwise downslope movement of the structure will occur, resulting at least in cracked walls and floors.

Slump

If mass movement occurs along a curved fissure extending below the surface, it is known as slump (Fig. 12-7). The weight of the debris causes the top of the mass to move downward while the bottom moves upward and outward.

Fig. 12-6. Dislocation of these railroad tracks near Coal Creek, Alaska, is caused by soil creep in the beds underlying them. (Courtesy United States Department of the Interior, Geological Survey, Denver, Colo.; photograph by W. W. Atwood.)

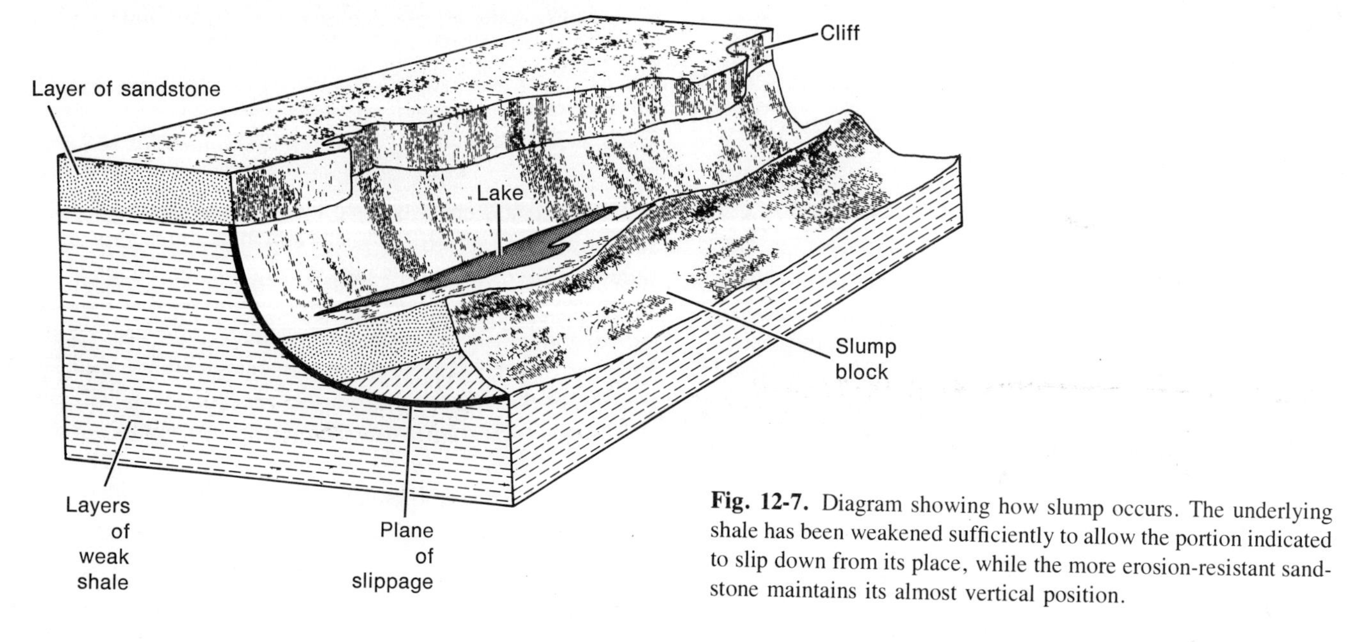

Fig. 12-7. Diagram showing how slump occurs. The underlying shale has been weakened sufficiently to allow the portion indicated to slip down from its place, while the more erosion-resistant sandstone maintains its almost vertical position.

Table 14. Sedimentary rocks

<table>
<tr><th rowspan="3">Materials and diameters</th><th colspan="3">Rocks</th></tr>
<tr><th rowspan="2">Clastic</th><th colspan="2">Nonclastic</th></tr>
<tr><th>Chemical</th><th>Organic</th></tr>
<tr><td>Gravel (> 0.25 inch, 6.25 mm)</td><td>Rounded (conglomerate) or angular (breccia)</td><td></td><td></td></tr>
<tr><td>Sand (0.25-0.01 inch, 6.25-0.25 mm)</td><td>Sandstone (quartz, arkose, graywacke)</td><td></td><td></td></tr>
<tr><td>Silt (0.01-0.001 inch, 0.25-0.025 mm)</td><td>Siltstone or mudstone</td><td></td><td></td></tr>
<tr><td>Clay (< 0.001 inch, 0.025 mm)</td><td>Shale</td><td></td><td></td></tr>
<tr><td rowspan="6">Calcium carbonate (lime)</td><td></td><td colspan="2">Chalk</td></tr>
<tr><td colspan="3">Coquina</td></tr>
<tr><td></td><td colspan="2">Coral</td></tr>
<tr><td colspan="3">Marl</td></tr>
<tr><td colspan="2">Oolite</td><td></td></tr>
<tr><td></td><td>Travertino, tufa</td><td></td></tr>
<tr><td>Magnesium calcium carbonate</td><td></td><td>Dolostone</td><td></td></tr>
<tr><td>Peat</td><td></td><td></td><td>Lignite, bituminous</td></tr>
<tr><td>Sodium chloride</td><td></td><td>Rock salt</td><td></td></tr>
<tr><td>Hydrous calcium sulfate</td><td></td><td>Gypsum</td><td></td></tr>
<tr><td>Hydrous sodium borate</td><td></td><td>Borax</td><td></td></tr>
</table>

The first sign of impending slump is the development of cracks at the top of the mass. Slump may be either fast or slow. Even though it is less known than some of the more spectacular types of mass movement, slump is nevertheless a common occurrence on earth and even occurs on the moon and the planet Mars.

Talus

As rocks weather on the sides of steep cliffs, gravity occasionally pulls a piece of loosened rock debris downslope to collect at the base. These individual rock fragments are called talus. Over many centuries, talus accumulates to produce a slope in the form of a cone. The slope has an angle of repose of about 35° and is known as a talus *cone*. Talus, then, is an example of the rapid movement of small masses of rock debris that take long periods of time to accumulate. Nevertheless, talus is an effective agent of mass movement.

ROCKS THAT FORM AT THE EARTH'S SURFACE

Weathering processes disintegrate and decompose rocks to form rock fragments, organic material, or chemicals called *sediments*. Mass movement carries the sediments downhill. There the agents of erosion transport them to sedimentary basins, where they are deposited in layers. As a result, the rocks formed from these sediments are almost always layered and are known as *sedimentary rocks,* the end product of the sedimentary environment. Because of the close connection between the formation of sediments and weathering and erosional processes, sedimentary rocks are the most abundant rocks at or near the earth's surface.

Sedimentary rocks, then, are made from sediments. The type and size of the sediments determine the classification and naming of the resultant rocks (Table 14).

Clastic sediments and sedimentary rocks

Sediments composed of solid rock or mineral fragments are called clastic sediments. These sediments vary in size from extremely large to extremely small, thus forming an excellent criterion for the subdivision and naming of sedimentary rocks made from them.

Gravel: conglomerate and breccia

Gravel is the general term used for all sediments with diameters greater than 0.25 inch (6.25 mm). A sedimentary rock composed of rounded sediments of gravel size is termed a conglomerate (Fig. 12-8). The rounded shape of the fragments that make up conglomerate indicates that they underwent a long period of intense erosion prior to becoming solid rock. A similar rock composed of angular, gravel-sized fragments is called breccia. Their angularity is indicative of a much shorter period of erosion.

Fig. 12-8. Conglomerate. Notice the coarse, unmixed look of this rock and its large, rounded gravels, which are its distinguishing features.

Sand: sandstone

Sediments ranging from 0.25 to 0.01 inch (6.25 to 0.25 mm) in diameter are called sand, and a sedimentary rock composed of these fragments is known by the general name of sandstone. Sandstones are often subdivided on the basis of mineral composition into quartz sandstone if they consist predominantly of quartz grains, arkose if they are composed of more than 20% feldspar, and graywacke if the sediments are poorly sorted and held together in a matrix composed of the minerals mica, chlorite, and quartz.

Silt and clay: siltstone and shale

Silt is sediment that is 0.01 to 0.001 inch (0.25 to 0.025 mm) in diameter, and clay is any sediment with a diameter of less than 0.001 inch (0.025 mm). Siltstone is a sedimentary rock composed of silt; shale is a sedimentary rock made of clay. Although somewhat similar in appearance, the grittiness of siltstone distinguishes it from shale, which characteristically tends to split apart along roughly parallel planes and has an earthy odor.

Shale that contains a solid organic material called *kerogen* in a proportion that makes it capable of yielding oil when it is heated slowly is known as *oil shale*. Even though fuel has been produced from oil shale in Scotland for over 100 years, in most parts of the world oil shale is an untapped natural resource because oil can be more economically pumped from known underground reservoirs.

In the years of mineral and fuel crisis that lie ahead, oil shale will undoubtedly become a more important rock. The United States and Canada are fortunate to have one of the largest known reserves of oil shale in the world. Except for the ocean, it is often said that the accumulations of oil shale in western Colorado, southwestern Wyoming and eastern Utah are the largest mineral deposits in the world.

Chemical sediments and sedimentary rocks

Chemical sediments are those precipitated from (or dropped from solution in) water. When turned naturally into solid rock, they are classified under the broad headings of limestones, dolomites, and evaporites.

Limestone

A sedimentary rock composed of the mineral *calcite* (calcium carbonate, $CaCO_3$) is called limestone. There are many different types of limestones, but regardless of their variety, they may all be identified by their tendency to bubble vigorously when cold, dilute hydrochloric acid is applied to them. Many limestones not formed by chemical precipitation are directly or indirectly formed through organic processes.

1. The limestone *chalk* is composed of microscopic shells of marine creatures.
2. *Coquina* is a limestone made of lithified seashell fragments (Fig. 12-9).
3. *Coral* is limestone formed by the marine animal coral which extracts calcium from the water around it.
4. *Marl* is an impure, poorly consolidated limestone composed of shell fragments, sand, silt, and clay.

If calcium carbonate is precipitated around extremely small particles of silt, clay, or shell fragments, tiny egg-shaped masses called *oolites* form and drop to the bottom of the ocean. As innumerable oolites become lithified into solid rock, they form *oolitic* limestone.

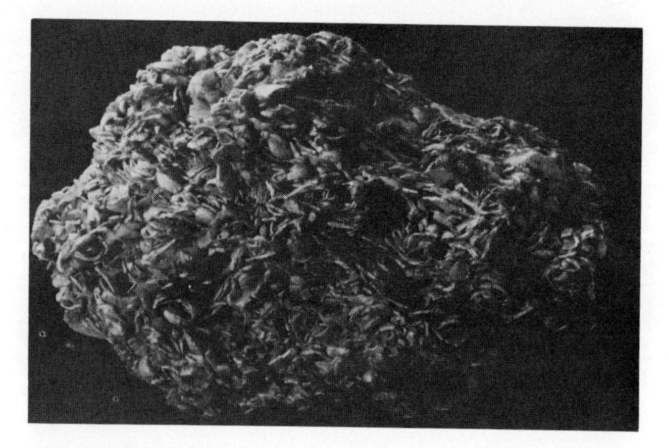

Fig. 12-9. In this sample of coquina, the tiny shells of which it is made can be seen clearly.

A dense, banded limestone called *travertine* is commonly precipitated in caves, caverns, and springs (especially thermal springs), and a spongy, porous, fragile limestone called *tufa* is deposited around shrinking lakes as well as thermal and unheated springs.

Limestones usually lack the strength necessary for building materials, but they are widely used in the chemical industry, as an important component of cement, and as crushed stone.

Dolostone

The sedimentary rock dolostone resembles limestone in appearance but is composed primarily of the mineral dolomite rather than calcite. Unlike calcite, which is a calcium carbonate, dolomite is a calcium magnesium carbonate. It forms either by chemical precipitation or more commonly through the alteration of existing limestones. Dolomite can be identified because it does not effervesce in cold, dilute hydrochloric acid without first being scratched. Dolostone beds of considerable thickness are found in many parts of the world where the rock is used as a building material.

Evaporites

Also classified as chemical sedimentary rocks are *evaporites,* crystalline salts that are deposited as water evaporates from shallow basins such as desert lakes, inland seas, lagoons, and bays. Rock salt, made of the mineral halite, is by far the most common example. In the United States, large quantities of rock salt are found in Michigan, Kansas, and the Gulf Coast area. In the latter location, salt has intruded upward into existing rocks to form salt structures, frequently in the form of domes (Fig. 12-10). Because petroleum and natural gas are often trapped under or against salt structures, geologists search for these and other natural traps as locations for potential fossil fuel reservoirs.

Thick beds of the minerals gypsum and borax are also included as evaporites under the classification of sedimentary rocks. Important deposits of gypsum are found in Utah, Colorado, and Ohio in the United States as well as in Mexico, France, Switzerland, Italy, and Sicily. Borax, useful in a hundred industries, is mined at Searles Lake, California, the world's largest deposit. Borax is also found in the state of Nevada and in Italy and Tibet.

Organic sediments and sedimentary rocks

The only true organic sediment is *peat,* which is partially decayed plant material. As plants die, their remains begin to decay, and if they are buried under other sediments, they may be preserved in the original layers to form coal (Fig. 12-11). By shallow burial, peat is turned into

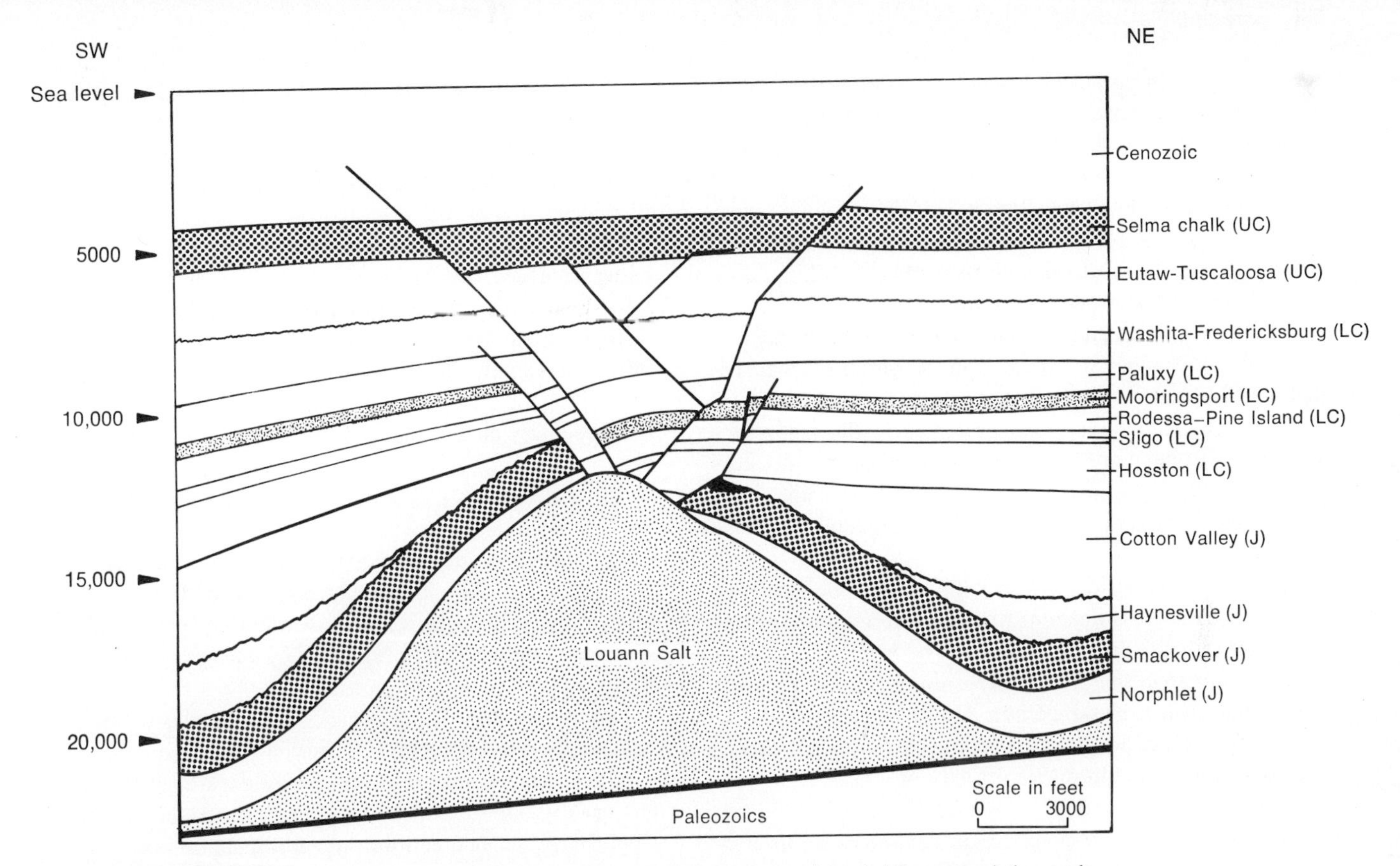

Fig. 12-10. A salt structure on Pool Creek field, Jones County, Mississippi, has intruded into preexisting rocks, forcing them up into a dome. These intrusions often hide oil reserves and hence are widely sought by geologists. *UC* is Upper Cretaceous; *LC* is Lower Cretaceous; *J* is Jurassic.

Fig. 12-11. A seam of coal at Cartwright Hill near Morgantown, West Virginia. The light-colored shale overburden provides the pressure necessary to turn decaying plant remains into coal. (Courtesy West Virginia Geological and Economic Survey, Morgantown, W. Va.; photograph by Richard Hunter.)

low-grade brown coal called *lignite*. Deeper burial, which exerts more pressure and produces higher temperatures, changes lignite into the medium-grade soft coal known as *bituminous*. Still deeper burial produces the high-grade hard coal, *anthracite*. Peat and lignite, and bituminous coals, are almost always found in the sedimentary environment, but anthracite, the most valuable form of coal, is invariably found in the metamorphic environment.

It is a common misconception that natural diamonds are produced by changes resulting from the burial of coal. Industrial diamonds have been made by subjecting pure carbon to high temperatures and pressures in the laboratory and have also been produced in meteorites by the shock of impact. Natural diamonds, however, form from the crystallization of inorganic carbon deep in the igneous environment.

Coal is often considered a rock but at least one difficulty arises in such a classification. Rocks are composed of minerals, and minerals by definition must be inorganic. Coal is obviously organic, but because it forms in layers in the rock environments, it is classified as a rock for there is really no intermediate classification. Coal, together with petroleum and natural gas, should more correctly be classified as fossil fuel.

The thickest known coal beds in the world are in Victoria, Australia, while the most valuable deposits are in Pennsylvania. The coal deposits of Pennsylvania are so important, in fact, that the most significant coal-making period of geologic time, some 200 million years ago, has been named in its honor—the Pennsylvanian period. This period, together with the Mississippian period that preceded it by some 45 million years, are called the Carboniferous period because most of the coal deposits of the world were formed during this 65 million–year span of geologic time.

The major known coal deposits of the world will soon be exhausted. Oil and natural gas have replaced coal as a primary energy source, but even these known fossil fuel reserves could be depleted in the twenty-first century. We will have consumed in a few hundred years what it took the earth hundreds of millions of years to form. It is clear, then, why fossil fuels and minerals are known as unrenewable resources.

SEDIMENTARY FEATURES

A series of features often found in or associated with sedimentary rocks helps geologists to understand the conditions surrounding the formation of this important rock type. The features include stratification, ripple marks, mud cracks, cross-bedding, concretions, nodules, geodes, graded bedding, facies changes, and fossils.

Stratification. One of the most common and characteristic features found in sedimentary rock is stratification, or the formation of nearly horizontal layers, or beds, as more recent sediments are deposited on top of sediments laid down earlier (Fig. 12-12).

The formation of sedimentary rocks always begins at the earth's surface, even though thousands of feet of overlying rocks may have subsequently been formed on top of them. Each layer or stratum of sedimentary rock, wherever it is found, represents what was once the surface of the earth at a time in the geologic past. This surface may have been dry land or the bottom of an ocean, lake, swamp, or stream.

Stratification is so common and basic in geology that a law has been developed to deal with its interpretation. The *law of superposition* states that in an undeformed sequence of rock layers, the oldest layer is at the bottom of the sequence while the most recently formed layer is at the top. The layers between are progressively younger from bottom to top.

Ripple marks. Ripple marks are frequently found in sedimentary rocks. Almost everyone has seen the little waves of sand that develop on sand dunes or at the edges and on the bottoms of streams, lakes, and oceans. These ripples represent the shifting of sediments by rather strong wind or water currents. If these ripples are preserved in solid rock they are known as ripple marks and provide a clue as to the origin of the rocks, even though millions of years have passed since they were last at the surface (Fig. 12-13).

Mud cracks. Mud cracks are found exclusively in sedimentary rocks. They are the familiar features seen on the surface of mud that is drying in the sun. As the water content is lost through evaporation, the mud shrinks and usually breaks into polygon-shaped pieces. If the hardened mud is then turned into solid rock, these cracks will be preserved. Thus when ancient mud cracks are found in solid rock, we can assume that the same alternating wet and dry conditions that produce mud cracks today were occurring when the rock was formed.

Cross-bedding. Sedimentary rocks often show minor angular layers within the overall horizontal bed. This is known as cross-bedding or cross-stratification, and it is produced as wind or water currents change the direction and angle of sediment deposition (Fig. 12-14). If the cross-bedding becomes lithified into solid rock, a hint may later be obtained as to its origin. For example, modern sand dunes as well as stream, lake, and ocean deposits reveal cross-bedding. If rocks thousands or millions of years old show the same feature, we may infer that these rocks were once

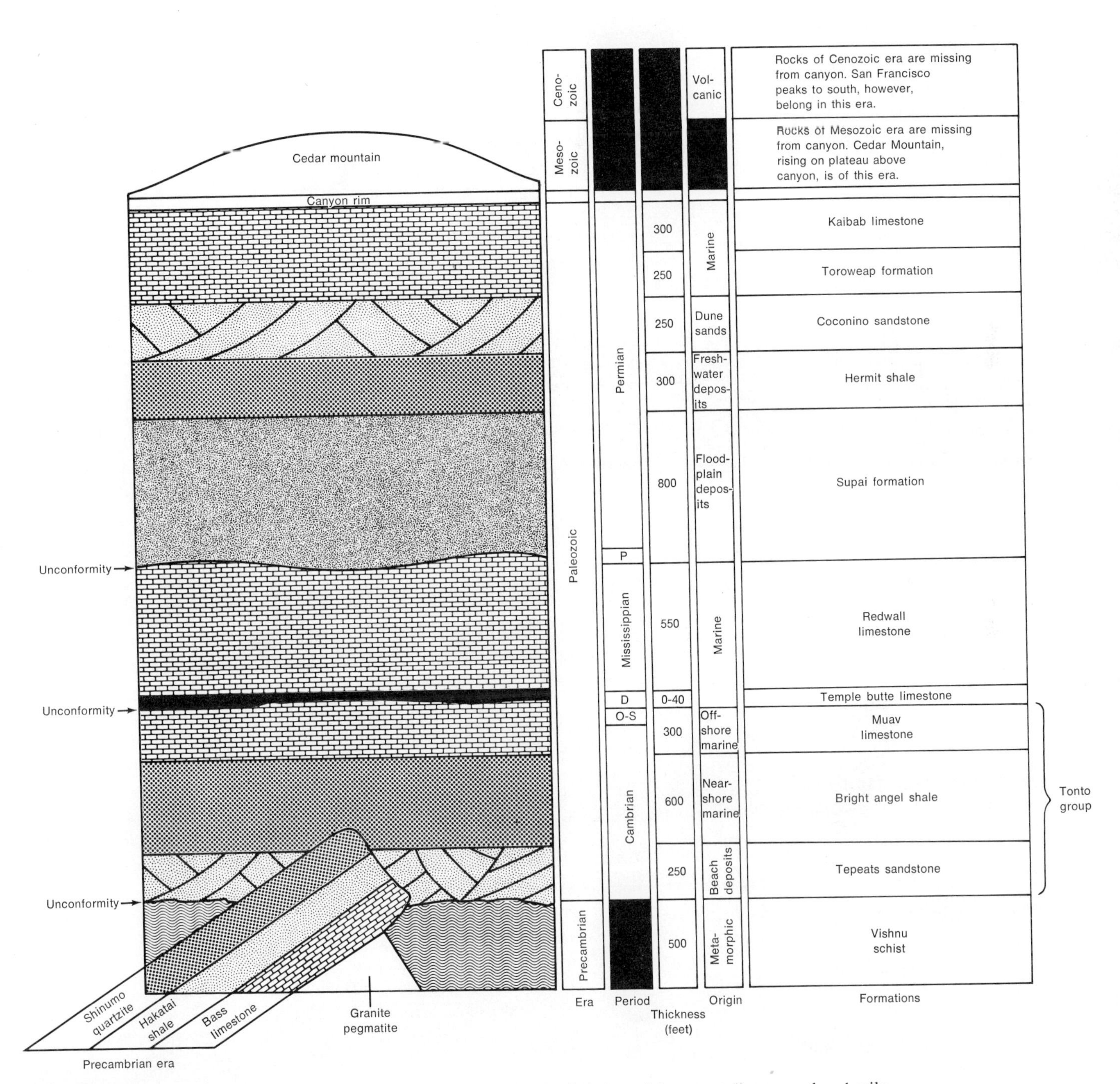

Fig. 12-12. The Grand Canyon. Two billion years of geologic time are written in rock layers totaling more than 1 mile in depth. The canyon, 217 miles long and 8 to 20 miles wide, has been carved by the Colorado River over the past 5 million years.

Fig. 12-13. Ripple marks on the Summerville formation in Colorado show that when this sediment was deposited 140 million years ago the area lay exposed to strong winds or currents.

Fig. 12-14. Cross-bedding in Navajo sandstone, Kane County, Utah. The hardened remnants of an ancient sand dune show us each change in wind direction. (Courtesy United States Department of the Interior, Geological Survey, Denver, Colo.; photograph by H. E. Gregory.)

part of an ancient sand dune, stream bed, lake, or ocean floor.

Concretions, nodules, and geodes. Many sedimentary rocks show features, unlike those already discussed, that were formed after the sediments were deposited. Water moving downward or laterally through sediments not only helps to turn loose deposits into solid rock but also often forms features such as concretions, nodules, and geodes.

Concretions are local, dense concentrations of quartz, calcite, iron oxide, or dolomite. They are usually spherical or disc shaped and are formed where sediments are most easily and more firmly lithified away from a center point or line. The imprint or mold of a fossil is often filled with mineral deposits from ground water to form a concretion cast of the original fossil. Concretions are composed of the same material that cemented the other sediments in the bed.

On the other hand, when a different material forms local, irregular mineral concentrations in the rocks, nodules are formed. For example, in limestones, which are composed almost entirely of calcite, nodules are commonly made of quartz in the form of flint or chert. Quartz, in fact, is by far the most common nodular material. In most cases nodules are formed as quartz replaces the original material, although in other cases it is believed that the quartz may have been deposited simultaneously with the original material.

Geodes are roughly spherical hollow concentrations of minerals that are found in limestones and dolostones but rarely in shale and other sedimentary rocks. Geodes are often composed of an outer shell of chalcedony, a form of quartz, and their hollow centers are lined with inward-projecting crystals of quartz, calcite, dolomite, or occasionally other minerals. Geodes, which weigh less than one would expect because they are hollow, are thought to be formed in water-filled holes in the rocks into which minerals in solution are concentrated.

Graded bedding and facies changes. A feature nearly always found in clastic sedimentary rocks that are formed in deep, quiet water is graded bedding, which is produced by the vertical sorting of sediments according to size as they settle out of the quiet water. Coarser sediments are thus found at the bottom of a bed, while the finer sediments are located at the top. As the sediments are turned into sedimentary rocks, the grading (or sorting) is preserved as an indication of the rock's origin.

Also found in sedimentary rocks is a change in the character of the sediments of which the rock is composed, known as a facies change. Facies changes involve differences in sediment color, composition, or grain size brought about by depositional variations from one place to another. Today in the Mississippi delta in the Gulf of Mexico, for example, facies changes may be traced as coarser sediments are deposited near shore, while finer sediments are carried farther offshore. Facies changes may also be followed in solid rock, where they are a clue as to the changing nature of the sediments and environments in which the rocks were formed. Facies changes are often well exposed in road cuts that have truncated the rocks.

Fossils. There is an important connection between the sedimentary and biologic environments. All living things occupy habitats at or near the earth's surface where sediments are deposited. The natural overlap of the two environments, together with the mild nature of sedimentary processes, makes sedimentary rocks a natural location for the preservation of biologic specimens. The remains or traces of once-living organisms preserved naturally in the crust of the earth are called fossils, and they are extremely rare in any type of rock other than sedimentary. Fossils are a major feature of sedimentary rocks and are of great value as a source of information about the geologic past.

Of all the organisms that have ever lived on this planet over more than 3 billion years of geologic time, most died and decayed without being preserved. However, fossilization has occurred frequently enough to give a record (though it is incomplete) of the development and evolution of life on earth.

In order to have been fossilized, a plant or animal must have been buried quickly, cut off from the agents of weathering and the bacteria that would have produced complete decay. Even under the best conditions of fossilization, most of the specimen decays; only the more resistant parts—the bones, teeth, and shells of animals, for example—are preserved. Often even these hard parts turn into solid rock. *Petrified wood,* for example, really contains no wood at all; the wood itself was converted into solid rock as minerals replaced the wood molecule by molecule. Rare indeed is preservation of soft parts such as skin and muscle. Insects have been preserved intact in amber (Fig. 12-15), fossilized tree resin. Scientists of the Soviet Union have found the entire remains of prehistoric creatures such as mammoths and mastodons frozen in the ice in Siberia.

Fossils also include molds and casts, which are not the actual remains of an animal or plant. The footprint of a dinosaur preserved naturally in the rock is an example of a *mold.* Often the mold is filled naturally by minerals or by scientists with plaster of paris to produce a *cast.* When made naturally, casts are also considered fossils. Sometimes only waste products are preserved, not the organism itself. Fossils, then, include a great variety of types of remains or traces of animal and plant life that are used by

Fig. 12-15. A Caddis fly preserved in amber. (Courtesy American Museum of Natural History, New York, N.Y.)

geologists to study living things of the past, which are preserved almost without exception in the sedimentary environment.

EROSION: LEVELING THE LAND

Weathering and mass movement should not be confused with erosion, although the processes are closely related, since all work together to alter the rocks at the earth's surface. Weathering occurs in place, while mass movement involves the movement of weathered rock debris downslope for short distances. Erosion, on the other hand, picks up and transports the products of weathering for great distances away from their point of origin, thus producing further changes.

This chapter has been a detailed look at weathering and mass movement. The four chapters that follow examine the agents of erosion, thus completing the study of the sedimentary environment. It will help as you proceed to keep in mind the close relationship between weathering, mass movement, and erosion.

13 *arteries of nature*

Stream activity

I hurry amain to reach the plain,
Run the rapid and leap the fall,
Split at the rock and together again
Accept my bed, or narrow or wide.

Song of the Chattahoochee
Sidney Lanier

On May 31, 1889, after many days of heavy rain, an earthen dam that had been built on the Conemaugh River 12 miles east of Johnstown, Pennsylvania, suddenly collapsed. Out of the south fork rushed a torrent of water, devastating the town and drowning more than 2200 people. The Great Johnstown Flood went down in history as the worst flood disaster, as far as deaths were concerned, ever to strike the United States.

As tragic as the Johnstown event was, there have been many disastrous "forgotten floods." One million Chinese lost their lives on the Hwang-Ho River in 1887 and 100,000 more were killed in floods along the Yangtze River in 1911. Floods have undoubtedly been the greatest natural cause of human misery throughout recorded history (Fig. 13-1). Although the exact figures will never be known, loss of life must have been at least 1 billion, while property damage, in terms of modern currency, has amounted to trillions of dollars.

In spite of the problems streams have caused during our brief human history, they have played the key role throughout geologic time as the main leveler of the land in the eternal struggle between constructional and destructional forces.

THE WATER CYCLE

As pointed out in Chapter 9, one of the unique aspects of the earth is its water. Water is an amazing substance in many respects, not the least of which is its ability to exist in all three states of matter—liquid, gas, and solid—under normal atmospheric conditions. On earth, water in its various forms is linked in an eternal cycle, which guarantees that a supply of this vital resource will always be available (Fig. 13-2).

The great reservoirs of water on this planet are the oceans. Energy from the sun causes tremendous quantities of ocean water to change from the liquid state into invisible water vapor by the process of *evaporation*. When the air is cooled sufficiently, clouds form as water vapor changes back into either droplets of water by the process of *condensation* or into ice crystals by *sublimation,* the process of converting a gas directly to a solid without its passing through the liquid state. When clouds contain all of the water or ice they can hold, any excess drops to the surface as *precipitation*.

Precipitation products follow a number of alternate paths back into the atmosphere to complete the water cycle. Some water evaporates directly back into the air when it strikes the surface, or in some cases it returns to a gaseous state before it ever reaches the surface at all. Some water is used by plants and animals in their life processes and then returns to the atmosphere. Some water soaks downward, below the earth's surface to form *groundwater,* while some is temporarily stored as ice. The remainder runs off from the surface to form streams.

STREAM ANATOMY

A stream is a surface flow of water confined to a channel it has made in the earth. Geographers refer to streams by a variety of synonyms such as rivers, creeks, and brooks. Adding to the confusion are such proper names as Congo

Fig. 13-1. In April, 1973, the Mississippi and Missouri Rivers both overflowed their banks, drowning the Midwest in the worst floods in recorded history. West Alton, Missouri, sitting at the confluence of the two rivers, was one of the hardest hit areas. (Courtesy United States Army Corps of Engineers, St. Louis District, St. Louis, Mo.)

River, Kraut Creek, and Adkins Brook. One stream is as important in its own right, geologically speaking, as any other. All streams, regardless of what they are called, share certain basic characteristics.

Source and mouth

A stream has a beginning, called its source (or head) and an end, known as its mouth. The source is where the stream first appears as a surface flow confined to a channel. The stream may begin as the result of glacial melting, as the great streams of Europe such as the Rhine and Rhone Rivers, or as the result of an outflow of groundwater, as does the Withlacoochee River of Florida. Many streams such as the amazing Colorado River have as their source the runoff of water from the land. Some streams begin as overflow from a lake. The St. Lawrence River, which flows from Lake Erie northeastward into the Atlantic Ocean, is an example of this type. However, most great stream systems have multiple sources because they are the by-product of countless and often nameless smaller streams called *tributaries* that combine their flows. The Amazon River of South America (which carries more water than any other stream on earth), the Nile (the longest river in the world), the Congo of equatorial Africa, the Mississippi-Missouri sys-

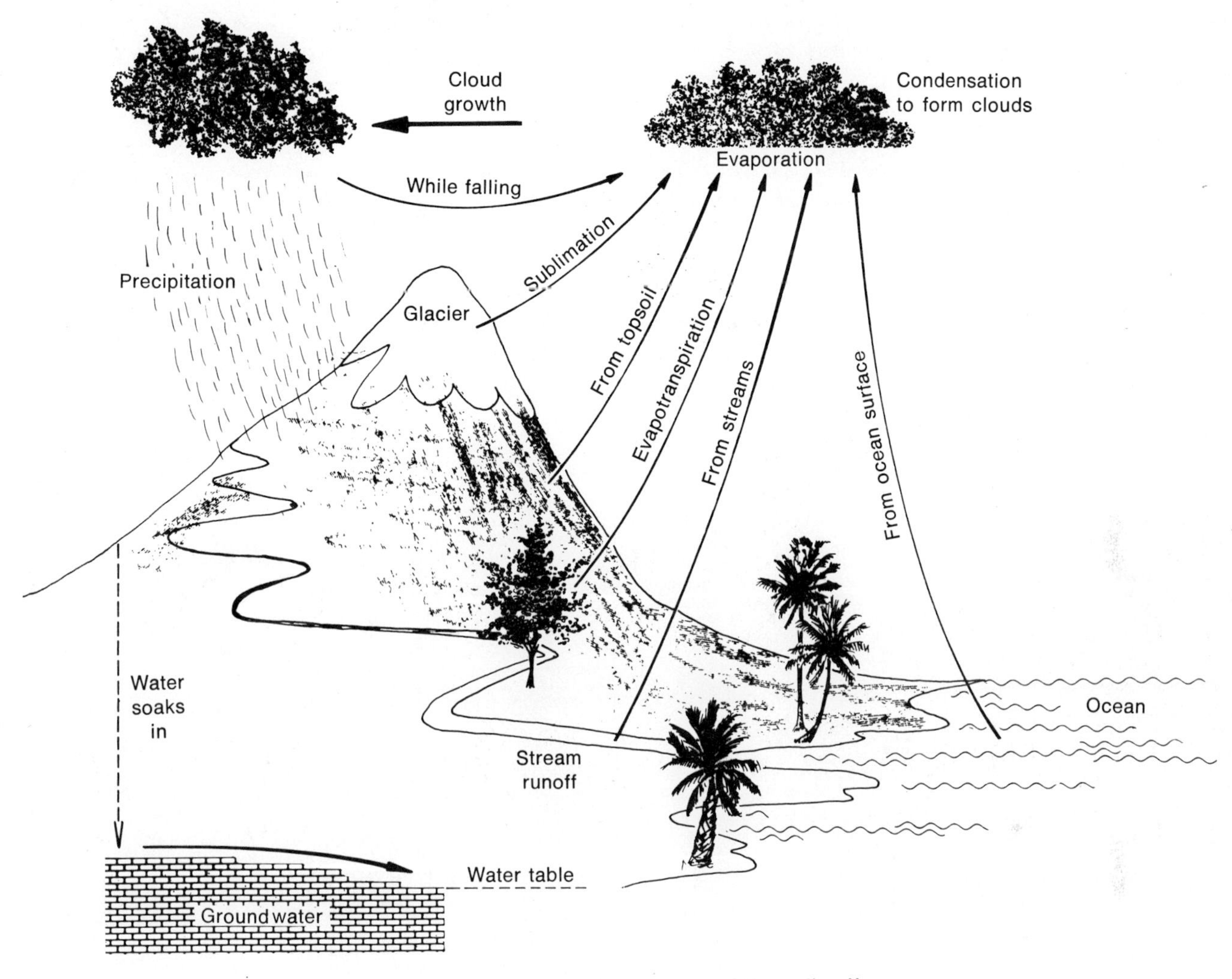

Fig. 13-2. Hydrologic cycle. Nature's eternal process was the first instance of "recycling."

tem of the United States, and the Yangtze River of China are all examples of streams with multiple sources.

The point at which the surface flow of a stream ends is called its mouth. A stream's mouth may be at a variety of locations. For example, it may be found where one stream flows into another of larger size, as the Ohio River flows into the Mississippi. The mouth may be located where a stream flows into a lake, as the Jordan River flows into the Dead Sea. Most great stream systems, like numerous smaller ones, have their mouths in the ocean. The Nile River, whose total length of 4160 miles makes it the longest stream on earth; the Amazon River; the Mississippi-Missouri; the Ob River of Siberia; the Amur, Yangtze, and Hwang-Ho of China; and the Congo are all examples of streams with oceanic mouths.

Channels

Between the source and mouth of a stream is the channel it has formed. The channel represents the path of least-resistant rock in a downhill slope. The direction in which a stream follows is not determined by a particular compass direction but rather by which direction is downhill and where the rocks are least resistant to erosion. Although the Nile is the longest and largest northward-flowing stream, it is not unique in this respect. Two other examples are the Red River of the North, forming the boundary between Minnesota and North Dakota, and the St. John's River of Florida.

A stream channel consists of two parts: the banks and the bed. The *banks* of a stream channel are the part found above the water level, while the *bed* is the part located below the water line. During flood stage, when a stream overflows its channel, the banks are extended upward above average water level by the deposition of sediments to form natural levees. In an attempt to contain flood waters, people have constructed artificial levees using sandbags or in some cases cement (Fig. 13-3).

Channelization

Another increasingly popular technique being applied to streams in an attempt to control water in the channel is called channelization. This method involves the complete modification of channel characteristics by both dredging and constructing "permanent" banks of cement. In the United States, channelization of streams is the responsibility of the Army Corps of Engineers, and has or will involve thousands of miles of natural stream channels.

Flood and erosion control are often given as reasons for channelization, but ironically as often as not the result is just the opposite. For example, channelization of the Black Water River in Johnson County, Missouri, more than 60

Fig. 13-3. Water spilling over this man-made levee on Kaskaskia Island shows an unsuccessful attempt to keep the Mississippi in its channel and out of populated areas. (Courtesy United States Army Corps of Engineers, St. Louis District, St. Louis, Mo.)

years ago had the effect of doubling the slope. This increased the rate of erosion of the river and its tributaries. Since the present channel is much wider and deeper than before, farm land has been lost and bridge repairs have to be made with increasing frequency. Downstream reduction in channel capacity due to the termination of dredging has also caused channel sedimentation and increased flooding.

Channel patterns

The location and characteristics of a stream channel, as pointed out earlier, are largely the result of the slope of the land and the resistance of the underlying rocks to the erosion forces of the running water. Although the slope plays the vital role in some streams and the rocks are more important in others, both are always involved. The earth's land surface is an interlocking mosaic of eroded land, chiefly formed by streams and their tributaries. The land through which a stream flows is referred to as its *drainage basin*.

If a stream channel is primarily the result of land slope, then the stream is known as a *consequent* stream and will develop a type of drainage basin that takes on a corresponding erosional pattern. If, on the other hand, the stream channel is primarily the result of an adjustment to rock structure rather than to slope, a *subsequent* stream is formed.

Effluent and influent stream channels. Even though streams are produced primarily by surface runoff, they still have a relationship with the groundwater environment (this is discussed in detail in Chapter 14). Depending on the climate, streams are either partially fed by the *water table* (the level of water underground) or they themselves feed the groundwater supply (Fig. 13-4). The former type is an *effluent* stream, while the latter type is an *influent* stream. The influent stream channel is usually restricted to arid regions of periodic and generally low rainfall. Effluent stream channels, on the other hand, prevail in humid parts of the world.

Gradient. All streams have a gradient (or slope) that plays the vital role in their development, for streams must flow downhill in response to gravity. Streams actually attempt to flow toward the center of the earth. Although some tourist areas advertise streams that flow uphill, such cases are either optical illusions or hoaxes. The downward flow of a stream is so basic, in fact, that it determines most if not all of its characteristics.

The gradient of a stream changes as the stream flows along; it is steeper at its source and progressively less steep toward its mouth. As a result, when we speak of a stream's gradient we are actually referring to the average slope of the entire stream from source to mouth. Gradient can be computed by using the following formula:

$$\text{Average gradient} = \frac{\text{Altitude of source} - \text{Altitude of mouth}}{\text{Distance from source to mouth}}$$

For example, a stream that has its source at an altitude of 10,000 feet and its mouth at sea level 500 miles away has an average gradient of:

$$G = \frac{10{,}000 \text{ feet} - 0 \text{ feet}}{500 \text{ miles}}$$

$$G = 20 \text{ feet/mile}$$

This means that for each mile the stream flows along the surface, the stream bed (and of course the stream) drop 20 feet in altitude. The actual gradient between two points along the channel could be determined by first carefully measuring the altitude difference between the two points and then calculating their linear separation.

Hydraulic factors

A stream, a fluid in motion, has several interrelated hydraulic characteristics that are vital to its function as a geologic gradational agent. For example, the quantity of water

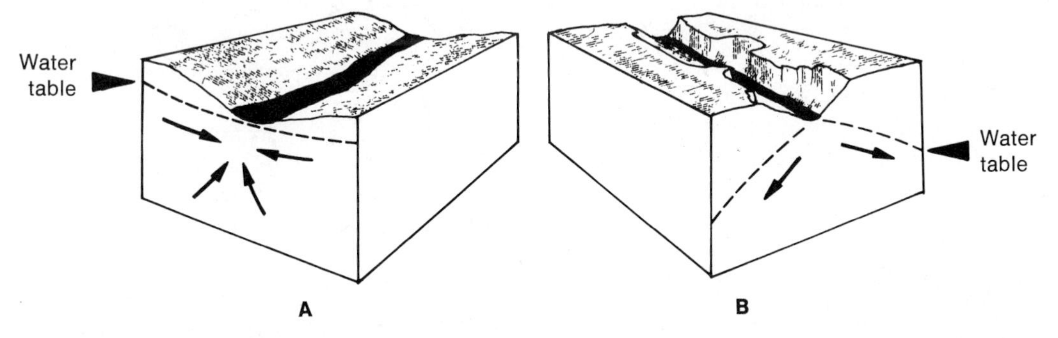

Fig. 13-4. Relationship between the water table and stream channels. **A,** The effluent stream, fed by the underground water supply, is more common in humid climates. **B,** The influent stream, which contributes its share to the water table, is more common in arid climates.

passing a given cross-section of a stream during a specific amount of time is its *discharge*. The discharge rate of the Amazon River, for example, averages 6,600,000 ft^3/sec of water at its mouth. *Volume* is the amount of water contained in a stream, while *capacity,* often confused with volume, refers to the quantity of sediment it is able to transport. The *competence* of a stream, on the other hand, is a measure of the largest sediment particle it can carry. Finally, its *load* is what the stream is actually carrying in the way of sediments. The capacity and load of a stream are exactly analogous to the capacity and load of a truck. We speak, for example, of a truck with a capacity of 2 cubic yards carrying a load of sand. A stream can be discussed in the same terms. All of these factors, as well as others such as gradient, channel roughness, width and depth, and water speed are interdependent and balanced in a very complex manner.

Base level

A critical point for a stream is its base level, the lowest level to which it may erode the land over which it is flowing. Streams can only occupy regions of the surface not already covered with water, and most of the water on the surface of the earth is contained within the ocean. As a result, permanent base level for almost all streams is sea level. The only exception to this would be a stream occupying part of the crust that is below sea level but yet not covered by the ocean. This phenomenon is shown by part of the Jordan River in the Near East because this stream flows in a *graben,* a valley produced when a block of the earth's crust drops down relative to the blocks on either side. If all land on earth were eroded to base level, streams would cease to exist because there would no longer be any gradient.

While permanent base level for most streams is sea level, they may also have *temporary* base levels such as other streams, swamps, and lakes. A temporary base level acts to upset the balanced hydraulic conditions of the stream, and many years are required before readjustments are made. The temporary base acts in such a way as to cut the stream into two parts, while energy is concentrated on the channel to remove the intruder from its path. Sediments are deposited at the temporary base so that the water flowing away from this location, cleansed of sediments, has an increased capacity and as a result a greater erosive power. Thus the land is destroyed at a faster rate than if the temporary base level had not been there.

Natural temporary base levels are of little consequence to a total stream system because they are formed over thousands or millions of years, and the stream adjusts to them gradually. Artificial temporary base levels such as dams, however, upset the balance of a stream quickly and cause immediate problems. The first symptom is rapid sedimentation in the reservoir behind the dam, which necessitates dredging. The result is the same—increased erosion below the dam and eventual increase in erosion upstream.

GEOLOGIC WORK OF STREAMS

Streams, like all other agents of erosion acting to wear the land away to sea level, perform three geologic functions; these are *erosion, transportation,* and *deposition*. Although each of these will be discussed separately, it should be remembered that they interact with each other as well as with other geologic processes such as weathering and mass movement.

Erosion: acquisition of sediments

A stream without sediments is like a piece of sandpaper without sand—it can do very little work. Erosion, the process of acquiring sediments that are used as cutting tools, is a normal function of a stream. Although this process can never be completely stopped so long as the stream exists, people interfere with the natural order by expending considerable energy and expense trying to prevent it. The stream is methodically destroying the one possession that we consider indestructible, namely, our land. All we can do is slow the process down to a level we find acceptable.

Processes of stream erosion

The geologic function of erosion is performed by streams in three ways, through *abrasion, solution,* and *impact*. As stream water containing sediments comes in contact with rocks, the sediments in the water act to wear them away, just as a sheet of sandpaper wears away a piece of wood through friction. The friction between the sediments in the water and the channel is known as *abrasion*. This is the most effective means of stream erosion.

Over a period of many years, stream water dissolves away certain rocks by the erosional process of *solution*. Rocks such as limestone and marble that contain calcium carbonate are very susceptible to this process, as are some other rocks, though to a lesser extent. The interaction between streams and groundwater, except in arid regions, can add considerable solution material as the water underground, which also contains large quantities of dissolved minerals, seeps into the stream channel. In most streams, however, solution is a minor part of the total erosional process.

Sediments that are too large to be suspended in the water of a stream bounce or roll along the bottom of the stream channel. Like colliding billiard balls, the sediments chip and gouge the rocks of the channel and other sediments of similar size in the erosional process known as *impact*. Materials eroded in this way eventually become material for abrasion, which in turn will eventually become products of solution. This gradual change in sediment size brought about by the stream guarantees a continuing supply of tools for erosion.

Erosional effects

Erosion by the processes just mentioned can affect a stream channel in three ways: by downward or bed erosion, lateral or bank erosion, and headward erosion. Although all three effects occur in all streams, they may be more pronounced in one stream than in another or more noticeable in one part of a stream than in another.

Downward erosion (the deepening of the channel) is dependent largely on the stream's gradient; it is more pronounced where the gradient is steep than where the gradient is gentle. This causes a more rapid water flow, thus concentrating turbulence and sediment movement on the bed of the channel. Narrow, deep canyons often result from such concentrated bed erosion. Hell's Canyon of the Snake River in Idaho, for example, attains a depth of 7900 feet.

Lateral erosion (the widening of the channel) is most common where the gradient is less steep and the water speed is slower. The stream follows a sinuous, meandering course that concentrates turbulence and thus erosion on the outside of the curves. Greatly enhancing the effect is the action of mass movement, which causes slumping of the banks and the valley as the stream cuts at their feet. Deepening and widening occurs simultaneously, creating and maintaining the V-shaped valley so characteristic of stream erosion.

Closely related to downward erosion, and actually an extension of it, is *headward* erosion. Not only does a stream erode downward all along its course, concentrating erosion where the gradient is steepest, but it also causes continual erosion of the channel toward the head of the stream. As the stream erodes the land, decreasing the gradient, its energy is always concentrated where the gradient is steepest, which is the area toward the head, or source. Headward erosion, then, can be thought of as the process that lengthens the stream and its tributaries and that constantly increases the size of the drainage basin by the erosion of more and more land.

The process of headward erosion occurs in all streams, so one drainage basin eventually erodes into the land of another stream. When this occurs, the flow of water in one stream is diverted into the channel of the other, producing the phenomenon called *stream piracy*. The stream that does the capturing is known as the *pirate stream* while the one whose water has been diverted is said to be *beheaded*. Where stream piracy has occurred, a gap or valley is formed in the surrounding land. If there is still a stream flowing in this valley, then it is referred to as a *water gap*. If the valley presently contains no stream, it is called a *wind gap* (Fig. 13-5). The Cumberland Gap, famed for the exploits of

Fig. 13-5. This V-shaped valley formed by stream piracy is now known as a wind gap because the streams have disappeared. (Courtesy United States Department of the Interior, Geological Survey, Denver, Colo.; photograph by P. Hayes.)

Daniel Boone, is located where the states of Kentucky, Virginia, and Tennessee meet and is an example of a wind gap. In many parts of the world, water and wind gaps made new exploration possible because they served as doorways through which adventurers could travel onward between high and often otherwise impassable mountains. Even today, highways in the United States tend to follow these "avenues" through the mountains.

Imagine the complications that would arise if a stream, the sole source of water for many towns and farms situated along its banks, should be beheaded, causing it to dry up. With headward erosion, a normal process of all streams, such a development could easily occur.

Transportation: movement of sediments

The second geologic function of a stream is the transportation of the sediments it has acquired through erosion. Streams transport these sediments through the processes of *suspension, solution,* and *traction*. These three actions are closely related to the methods of erosion discussed above.

Abrasive materials are carried in suspension, freely floating in the water. Suspended material usually consists of the smaller-sized sediments such as silt and clay, commonly called mud. At times when the stream is more turbulent, sand may also be transported in suspension. Most people think the worst when they see a stream loaded with mud, unaware that such a stream is a healthy one in the geologic sense. However, this also means that a rapid loss of soil has occurred somewhere in the drainage basin.

When streams transport materials in solution, they are carrying minerals that are actually dissolved in the water. Often called the "invisible load," solution products usually make up only a small portion of the total load of a stream. In some sluggish streams, however, the dissolved load may be considerable. It is estimated that during a 1-year period in the United States alone as much as 275 million tons of rock are removed by solution processes. Throughout geologic time, countless streams have poured countless tons of dissolved minerals into the ocean, the great catch-basin of sediments.

The third method of transportation is *traction,* sometimes called the *bed load*. By this process, sand and small gravels are bounced along the bottom, while larger gravels roll or slide along from time to time. It is difficult to believe that our favorite stepping stone is actually in transit because it may have been sitting for years in the same spot without moving. This type of movement usually occurs during flood stage, when the volume and speed of the water are increased; even then the movement may be very slight. The bed load performs impact erosion and, for the most part, was brought into the stream by the action of mass movement.

Deposition: dropping the load

Deposition, the action of a stream in depositing or dropping the sediments it carries, is the third geologic function of streams. The most critical considerations in stream deposition are a decrease in water speed and a reduction in volume. Sediments are carried because the stream is moving, so naturally the amount of sediment carried is proportionally reduced when the stream speed or volume is reduced. Streams cease to carry sediments when their flow stops.

A decrease in stream speed may occur where a stream encounters a natural obstruction in the channel such as a curve, a large, resistant mass of bed rock, or a boulder that has rolled into the stream. An artificial obstruction such as a bridge piling produces the same effect. In addition, speed is reduced where there is a decrease in the gradient of the stream or as water spreads out over a larger area, as during a flood. Volume, on the other hand, may be reduced whenever water is removed from the stream either by evaporation, seepage into the ground, or by people for irrigation purposes.

Alluvial fans

The gradient of a stream is abruptly reduced as it flows from a mountain onto a plain. A common depositional feature produced at this location, and one that is especially well developed in arid regions of the world, is the alluvial fan (Fig. 13-6). Alluvial fans are composed of coarse, poorly sorted, but layered sediments. Most of the fan is not under water at any one time because it is formed by an intermittent stream that flows only after heavy rains. When a stream does flow over the fan, it is not confined to one channel but flows in many interwoven channels because of the coarse nature of the sediments. When a stream shows such an interwoven channel pattern, it is called a *braided stream*. Sometimes a number of alluvial fans may come together along the base of a mountain to produce a land form known as a *bajada*.

Deltas

When a stream flows into a relatively calm body of water such as a lake, artificial reservoir, another stream, or the ocean, the speed of the water suddenly decreases. At this point, a fan-shaped depositional feature called a delta is formed. Deltas consist of well-sorted, stratified sediments (Fig. 13-7). The delta is the submerged equivalent of the alluvial fan, although it is usually larger and shows much bet-

ter sorting of sediments. Two of the classic deltas of the world are the Mississippi River delta and the Nile delta.

The Mississippi River delta is composed of rock debris that has been stripped from half the surface area of the continental United States over millions of years of geologic time. Furthermore, it is so large that it covers thousands of square miles of ocean bottom beneath the water of the Gulf of Mexico as well as much of the land area of the state of Louisiana. Indications are that the deltaic sediments in some parts of the Gulf of Mexico are as deep as 50,000 feet.

Floodplains

When more water is brought into a channel than the stream can carry, the excess overflows the banks and spreads out onto the surrounding land, producing the condition called a *flood*. Although it may be hard to believe, floodwater is usually more sluggish than the normal stream flow. As water leaves the channel, some coarser sediments are deposited on the stream banks, producing natural levees.

Fig. 13-6. Alluvial fan was formed at the base of a mountain north of Badwater, Death Valley National Monument, Inyo County, California. The stream's gradient decreased at the foot of the mountain, and the stream was forced to drop its load. (Courtesy United States Department of the Interior, Geological Survey, Denver, Colo.; photograph by H. Drewes.)

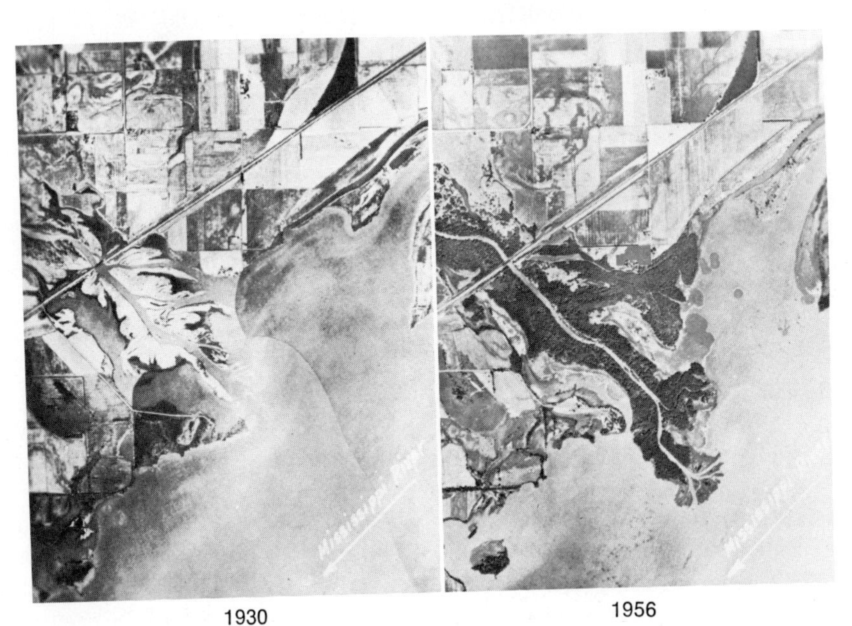

Fig. 13-7. Sediment deposition caused by the drop in speed when a swiftly moving stream meets another stream caused this delta on the Mississippi River at Devil's Creek, Iowa, to grow visibly in only 26 years. (Courtesy United States Department of the Interior, Geological Survey, Denver, Colo; photograph by H. P. Guy.)

As mentioned earlier, most of the sediments carried out of the channel, however, are silt and clay. As the floodwater recedes, these fine sediments are deposited as mud on the land. Through the years this material builds a depositional feature on either side of the channel called a floodplain. The floodplain consists of very fertile topsoil eroded from the entire drainage basin, and floodplain areas are prized bottom land.

Most of the great cities of the world are located on floodplains. These areas were natural locations for the development of population centers because of the rich soil, the abundant water supply, and navigable waters the stream provided. We are paying a terrible price today, as we have throughout human history, for this convenience. Each year the loss of human life and property to floods staggers the imagination. We appear to be making the problem worse by creating artificial conditions that produce excessive runoff, the most common cause of floods. For example, we are replacing natural soil and rock that can absorb water like a sponge with structures of concrete and asphalt that cannot. In addition, we are destroying the forests and grasslands by cutting and burning, thus enhancing runoff rather than absorption of excess water. We must now go to great lengths to control the monster we have created by building costly dams and artificial levees. Streams, however, will continue to flood regardless of what we do to try to stop them because this is a natural part of the geologic function of this important erosional agent.

STREAM VALLEYS

When standing on a mountain peak in one of the world's great mountain ranges such as the incredibly old Appalachians or the relatively new Rockies, it is difficult to believe that streams have carved the valleys they occupy. Although erosion of a stream valley is a slow, almost invisible process that involves the work of several geologic agents, we must realize the enormity of geologic time relative to thousands of human lifetimes.

Streams, then, invariably flow in valleys that they themselves have made, although they have received help from other processes. Except in rare cases, they do not occupy valleys produced previously by some other geologic agent. Two exceptions to this would be streams flowing in grabens produced by earth movement or streams occupying valleys produced by glaciers. Even then, streams work rapidly to modify the valleys into the typical stream-valley mold.

Downward and lateral erosion cause streams to develop valleys shaped like the letter V. In mountainous areas, where the gradient is steep and the stream water is swift, the valley is narrow and the V shape is easy to see. Where the land is low in relief and the valley is very wide, the V shape is still present, even though it takes a highly modified form with the stream occupying the apex of the V.

STREAM CYCLE: FACT OR FICTION?

An interesting but controversial theory concerning stream development is the idea of the stream cycle, which was developed by William Morris Davis (1850-1934). The cycle describes the possible evolution of stream-formed landscapes and divides the life of a stream into stages of youth, maturity, and old age. These stages are roughly analogous to phases of human development, although the cycle has little to do with a stream's chronologic age because all streams are old as far as years are concerned. A stream ages at different rates over different parts of its channel, so that streams in general, and most long streams in particular, have a youthful head, a mature channel, and an old mouth. Moreover, a stream may begin its geologic life in old age, or it may form in maturity. The stream cycle, then, is based not on chronologic age but on its geologic work and hydraulic characteristics.

Youth

A youthful stream, or the youthful part of a stream, has a steep gradient and a very rapid water speed and is located well above base level, usually in the mountains. The headwaters of most streams exhibit youthful characteristics. Also, the rapid flow of water means that the primary type of geologic work will be erosion or, more specifically, downward erosion. In addition, youthful streams have steep V-shaped valleys, straight channels, no floodplains, and numerous waterfalls and rapids, indicative of the resistant obstructions the streams encounter in their channels (Fig. 13-8). Over time, waterfalls are eroded away to form rapids, which are in turn completely destroyed. Finally, flooding in youthful streams can be particularly tragic because the steep, narrow valley and lack of floodplain causes the water to build up higher rather than to spread out. This produces flash floods that may cause entire towns to be quickly inundated or even completely washed away.

Maturity

A mature stream is one that shows a reduced gradient and a correspondingly slower water speed. It either formed from the evolution of a youthful stream or, more probably, originally developed closer to base level. The slower water speed means that lateral erosion is the stream's main type of geologic work. The side cutting widens the valley, leading to the formation of a well-developed floodplain. A mature stream may begin to meander slightly and lacks irregu-

Fig. 13-8. Waterfalls such as Niagara Falls on the New York–Canadian border are indicative of young streams. With age, such features are first eroded into rapids; then they disappear altogether. (Courtesy United States Department of the Interior, Geological Survey, Denver, Colo.; photograph by J. R. Balsley.)

larities such as waterfalls and rapids, which have eroded away before the stream reaches this stage. Finally, the land takes on low, rolling characteristics, although it is still well above base level. The portion of the Mississippi River just south of Cairo, Illinois, typifies a stream in the stage of maturity.

Old age

An old stream exhibits a very gentle gradient and a very sluggish water flow. Its chief geologic work is deposition. The stream has either formed or evolved to very near base level. Thus the land through which it flows is almost flat. An intensely meandering or curving channel is characteristic. In many cases, the stream actually cuts back on itself so that the water follows the shorter course through the merged banks rather than around the old stream curve. This place of breakthrough is called a *cutoff*, and the abandoned meander thus produced is known as an *oxbow lake* (Fig. 13-9). Although such a stream may return temporarily to the old channel during flood stage, the detached channel will first become a swamp and then dry up completely to form a meander scar called an *oxbow*. When flooding occurs in old age streams, the water spreads out on a very extensive floodplain and may cover thousands of square miles of land. However, the water rises gradually, rarely reaching a depth of more than a few feet, and is not swift.

The tendency of an old stream to cut off can have profound national and international repercussions, since streams often form state or national boundaries. Consider the case of the Rio Grande and the boundary dispute between El Paso, Texas, and Ciudad Juarez, Mexico. (Fig. 13-10). Many years ago the Rio Grande cut off at this location, leaving 435 acres of Ciudad Juarez as part of El Paso, for this river forms the boundary between the two cities just as it forms much of the boundary between Mexico and the United States. The Mexicans living in this abandoned loop of the river, not wishing to live in Texas, demanded that the United States return their land. This return was finally accomplished by rerouting the waters of the Rio Grande back into the original channel. Concrete banks had been constructed to guard against future stream cutoff, at least not in the same spot for some time to come. The cost? Millions of dollars for the American taxpayer and ruffled relations between Washington and Mexico City. On August 20, 1970, in an attempt to end further border disputes, President Nixon and President Ordaz of Mexico agreed that the center line of the meandering Rio Grande would serve as the border between the two countries. Although this agreement may temporarily end border disputes, it will not stop the river from meandering. Further cutoffs will necessitate expensive relocation of the people involved as well as payment to the country that loses its land.

Fig. 13-9. These cutoffs, oxbows, and oxbow lakes on the Tallahatchie River in Grenada, Mississippi, are characteristic features of old streams. (Courtesy United States Army Corps of Engineers, Waterways Experiment Station, Vicksburg, Miss.)

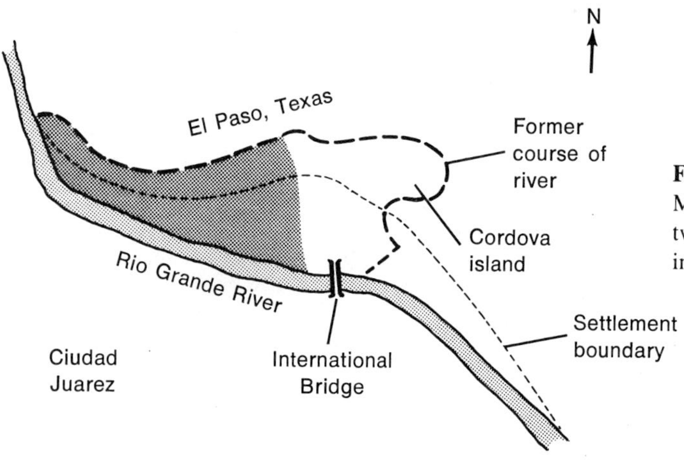

Fig. 13-10. Source of irritation between the United States and Mexico as the Rio Grande, designated as the border between the two countries, wanders merrily across the landscape. Shading indicates disputed territory.

Interruptions in the stream cycle

William Morris Davis realized that interruptions in the stream cycle could and did occur. In fact, *multicycles,* as we can call them, may be the rule rather than the exception. Thus, when we observe stream topography in obvious maturity or old age that nonetheless exhibits many youthful characteristics, the meaning is clear—an interruption in the stream cycle has taken place. The process whereby a stream reverts to youthful traits is called *rejuvenation.*

At least three possible causes of rejuvenation can be given. The most popular explanation for this phenomenon is the uplift of the land by earth movements. When uplift occurs, part or all of the land over which the stream is flowing is raised relative to sea level. This increases the gradient of the stream, which in turn triggers the sequence of events that characterizes a youthful stream. Another equally plausible explanation might be the lowering of sea level, for there have been times during geologic history when sea level has been hundreds of feet higher or lower than it is today. Lowering the sea level also has the effect of increasing the gradient, causing streams to flow more swiftly to the new base level. Apparently even minor changes such as a decrease in stream load or an increase in discharge could have the same result—a deepened channel.

Stream terraces and entrenched meanders are two well-known results of rejuvenation. *Stream terraces,* representing former stands of the stream, are formed primarily as a consequence of lateral erosion. Terraces often occur in distinct pairs on opposite sides of a stream valley, and often there are a series of these features. *Entrenched meanders,* on the other hand, occur where a stream in obvious maturity or old age has cut its channel deeply into the surface to form a canyon while retaining its meandering configuration. Both terracing and entrenchment are beautifully shown in the Grand Canyon of the Colorado River, while a tributary of the Colorado, the San Juan River in Utah, has formed classic entrenched meanders (Fig. 13-11). Obviously, rejuvenation on a truly grand scale has taken place in the breathtakingly beautiful Grand Canyon area.

GRADED STREAMS

Strong evidence indicates that streams are always in a state of equilibrium at any given moment. Such a stream is said to be in *grade* and thus is referred to as a graded stream. All of its physical and hydraulic factors are in a delicate state of balance. However, this does not mean that a graded stream is stable, for these factors are continually shifting, changing, and compensating just in order to maintain equilibrium.

In a graded stream, for example, the water speed adjusts to the gradient of the land, while the load supplied from the drainage basin adjusts to the capacity of the stream. Erosion, transportation, and deposition become adjusted to one another, and the total energy of the stream is nonuniformly distributed over the entire length of the channel.

Any change in these controlling factors acts to temporarily upset the graded condition. Thus compensation will occur in the direction that will tend to absorb the change and

Fig. 13-11. San Juan River in San Juan County, Utah, has created these entrenched meanders. (Courtesy United States Department of the Interior, Geological Survey, Denver, Colo.; photograph by E. C. LaRue.)

return the stream to grade. For example, if the volume of a stream increases during flood stage, more sediments will be picked up and transported, thus maintaining equilibrium at the higher energy level. If, on the other hand, the volume of a stream is reduced, as by drought or irrigation, then sediments will be deposited and the other factors correspondingly adjusted as the stream reaches a state of lower energy.

STREAMS AND THE ECOLOGIC CONSCIENCE

Throughout this chapter the vital role of streams in the total environment of the earth has been stressed. We as part of this environment have been almost totally oblivious to this role and our relationship to it. In fact, we are just now realizing that many of our "successes" are in fact failures. For example, we are discovering that hardly a stream exists from which we can drink or in which we can swim with absolute safety from dangerous pollutants. Many prized fishing and recreation areas have been closed because streams have become open sewers of human and industrial waste.

The Hudson River, so deeply rooted in American history, is a typical example of a dying stream. Lake Tear in the Clouds in the Adirondack Mountains of New York State is the source of the Hudson, and its waters near this location have been described as crystal clear. As the river flows downstate, however, factories and cities pour in huge quantities of filth. Each year New York City alone contributes an estimated 111 million gallons of waste. Long before the river reaches its mouth in the Atlantic, it has become a stinking mass of feces, detergents, dyes and chemicals from countless industrial plants, and assorted garbage of all types. Added to this is the heat from electric power plants and industries that use river water for cooling purposes. This heat produces thermal pollution, which activates the sewage that is already there. It has been conservatively estimated that it will cost the American taxpayers 1 billion dollars to clean up this dying river.

The problem of stream pollution is by no means confined to the United States, for as populations have increased and scientific and technological advancements have become more widespread, so have problems of environmental control. The Tiber River of Italy, for example, receives all of the sewage from the city of Rome; the great Rhine River, flowing through the heart of Europe, has the dubious distinction of being called "the sewer of Europe." The list of polluted streams continues to lengthen while we continue to live as though there were no tomorrow; as a matter of fact, this is becoming more and more a distinct possibility.

14 *the hidden reservoir*

Groundwater

Out of the earth to rest or range
Perpetual in perpetual change—
The unknown passing through the strange.

The Passing Strange
John Masefield

In May of 1965 at Castleberry, Florida, one house was destroyed and three others were heavily damaged as the ground on which they were built collapsed and sank. Luckily, the process took several hours, so that the inhabitants had a chance to escape without injury. The sinkhole produced by the collapse was 20 feet deep and 80 feet in diameter, with cracks extending as far away as 200 feet. This incident was a spectacular reminder of the great unseen reservoir of groundwater that silently performs its geologic work beneath our feet.

ORIGIN OF GROUNDWATER

The best available evidence indicates that most of the water underground came directly or indirectly from atmospheric precipitation. A small portion of water may have originally escaped from magma or lava during periods of igneous activity. Possibly all or part of this magmatic or *juvenile* water was originally atmospheric and simply seeped into the volcanic vents either from existing groundwater supplies or from the ocean. Marine or fresh water trapped in sediments as they were deposited on ocean or lake bottoms is sometimes considered a third source or groundwater. This *connate* water, as it is called, is usually but not always salty, and oil pockets are often found in association with it.

FACTORS AFFECTING GROUNDWATER

Several factors determine the quantity and depth of groundwater. Only rarely is groundwater completely absent. Even in the driest deserts some groundwater can be usually found, although it may not be fit for human consumption.

The first obvious factor in the formation of groundwater in a given region is the quantity and character of the precipitation. Since most groundwater sources are directly linked to local precipitation levels, areas with abundant precipitation will have more plentiful groundwater supplies than areas with less precipitation. A second factor in determining the quantity and depth of groundwater is local temperature. Cold climates are not conducive to the accumulation of large amounts of groundwater for two reasons. First, soil that is frozen for much of the year does not permit much groundwater seepage. Second, such areas receive more snowfall than rain, and since 10 inches of snow are equal in moisture content to 1 inch of rain, there is less water to begin with.

A third factor in determining the groundwater level is the slope, or gradient, of the land surface. As pointed out in Chapter 13, the steeper the land gradient, the more the water tends to run off and form streams rather than soak into the ground. On the other hand, the flatter the land, the better the chances for groundwater accumulation. A basic rule of soil and water conservation in farming provides for contour plowing and terracing on hillsides. This is nothing more than an attempt to reduce runoff, thus slowing the rate of soil loss while promoting the seepage of groundwater.

Still another factor that determines the nature of the hidden reservoir is the type and condition of the rock below the surface. Because of the way in which they are formed, igneous and metamorphic rocks contain few openings through which water can pass, so they are said to be *nonporous*. Sedimentary rocks, on the other hand, are almost always porous since they were made originally from loose sediments. Another important point to remember when considering the rocks underground is their *permeability*. Permeability refers to the ease with which water can pass through the rocks, and it is determined to a great extent by the size of the openings that exist in them. Therefore rock can be porous (with holes) but impermeable (the holes are too small for water to pass through) like the sedimentary rock shale, nonporous but yet permeable like fractured granite, porous and permeable like sandstone and many limestones, or nonporous and impermeable, as is unfractured marble. As a general rule all rocks, regardless of their type, will be permeable to the passage of water to one degree or another. This is because most of the rocks in the groundwater environment are sedimentary, and even rocks not of this type are almost always fractured. The best combination leading to groundwater formation, then, is a rock that is both porous and permeable.

A fifth factor to be considered is the amount of vegetation that covers the ground surface. Plants are gluttons for water, for their root systems are extensive and usually have a mass equal to the part of the plant visible above the surface but with a surface area many times greater. In regions of extensive vegetation, a drop of water has less than an even chance of passing by the root systems without being absorbed. In rain forests and jungles, for example, plants present three lines of defense that guarantee them the water supply essential to their life processes. First is the leaf canopy overhead, which is often so thick that a perpetual twilight exists during the daylight hours on the jungle or forest floor. The canopy breaks the force of the falling raindrops, allowing the water to drip slowly downward and be absorbed by the roots. The second defense is the mat of decaying leaves and plant parts that litters the ground beneath the overhead canopy. This mat also reduces the force of the raindrops, making absorption easier at the roots. Then the extensive root systems themselves act efficiently to absorb the water that reaches them. The net effect is a region with hundreds of inches of rain a year that is nonetheless amazingly deficient in groundwater.

Still another factor in the availability of groundwater is the variable rate of evaporation. It is estimated that of the total precipitation that falls to the earth, less than one third is available for runoff in the form of streams, ice in the form of glaciers or snow fields, or movement below the surface in the form of groundwater. The other two-thirds is either returned directly to the atmosphere by evaporation or indirectly by the evapotranspiration of plants.

Another rarely considered factor is the effect of human activity. The influence of people, seen almost everywhere, is rapidly becoming a major factor in the accumulation of groundwater. We may, for example, be changing the climates on earth by disturbing the vital atmospheric balance when we introduce gaseous and solid pollutants into the atmosphere. This may result in changes in precipitation and temperature levels. We change the gradient of the land at will because with sticks of dynamite and a bulldozer we can and literally do move mountains. Although we have not as yet devised a means of changing the type of rock underground, we affect the porosity and permeability of the surface by replacing natural materials with vast expanses of concrete and asphalt in the form of streets, highways, shopping centers, subdivisions, airports, and the like—all of these are nonporous and impermeable. We destroy or change vegetation at our whim and are consuming the groundwater reserve at an alarming rate.

When all of these factors are favorable, a region has abundant groundwater supplies. The state of Florida, a region whose primary natural resource is groundwater, has (1) an annual precipitation rate averaging 60 inches/yr, 99% in the form of rain; (2) low-lying land with a very gentle gradient; (3) subsurface rock that is chiefly porous and permeable limestone; and (4) a relatively sparse vegetative cover.

ZONES OF GROUNDWATER

When favorable conditions exist, water begins its movement toward the subsurface. Pulled by gravity, the water is drawn downward toward the center of the earth, percolating through the soil and rocks. The depth the water reaches is determined primarily by the point at which it encounters impermeable material—thousands of feet below the surface or perhaps just 1 foot or less. When groundwater reaches such a natural underground dam it begins to accumulate, saturat-

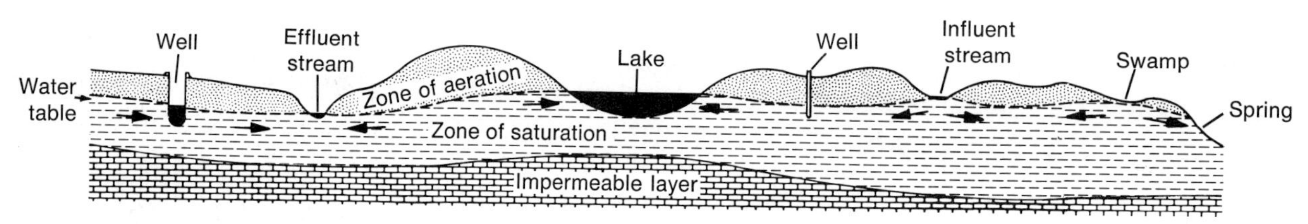

Fig. 14-1. Zones of groundwater and their relationship to the water table.

ing every available pore space and crack. This immediately produces two distinct groundwater zones—the zone of saturation and the zone of aeration (Fig. 14-1).

The *zone of saturation* is the area that extends upward from the top of the impermeable material to the top of the region in which all available spaces and cracks are filled with water. This uppermost level of water in the zone of saturation is called the *water table*.

The area extending from the water table to the surface, in which pore spaces are usually filled with air, is known as the *zone of aeration*. Water in this zone moves downward, while the movement in the zone of saturation is generally lateral, in directions determined by the subsurface slope of the rocks.

Many people mistakenly believe that groundwater rushes through the soil and rocks in much the same way as a stream flows in its channel. Although underground streams do exist, most of the water in the ground flows the way ink does in a blotter, very slowly. Groundwater is often called *fossil water* because hundreds or even thousands of years may have passed since this water was last at the surface.

TOPOGRAPHY–WATER TABLE RELATIONSHIPS

Because of the variable nature of groundwater environments, the water table is not always the same distance underground, even over rather short distances, and it varies in depth with time, since it is thicker during periods of higher rainfall.

Swamps

In areas in which the water table is just under the surface, a geologic groundwater feature called a swamp is produced. Capillary action causes the water to creep upward so that there is no zone of aeration. Almost complete saturation produces this common and significant environment. An estimate today classifies no less than 1 million square miles of the earth's surface as swamp, and swamps were even more prevalent in the geologic past. During the late Paleozoic era some 250 million years ago, for example, most of the present dry land surface was either under water or near sea level. These gradients contributed to poor drainage and high water tables. Swamps flourished, and in this environment newly evolving land plants found a haven. The remains of these Paleozoic land plants served as the raw materials for the greatest coal-making period of all geologic time.

Swamps today are classified as either fresh water or marine. *Marine* swamps are those located near the sea, and they contain *brackish* (salty) water. Marine swamps are very common along the Eastern and Southern coasts of the United States as well as in many other regions of the world. The Great Dismal Swamp of Virginia and North Carolina is an excellent example.

Freshwater swamps, although less common than marine swamps, are still very abundant. They may occur wherever the water table is near the surface or in association with another freshwater source such as a stream or lake. One of the best examples of a freshwater swamp to be found anywhere is in Florida, south of Lake Okeechobee. The overflow from the lake runs predominantly southward through low-lying land with high water tables, producing the famous Everglades (River of Grass). Part of this great swampy area has been preserved in the Everglades National Park. In extreme southern Florida, the freshwater swamp merges with a marine swamp whose water source is the ocean.

Streams

Streams also have a relationship with the water table. In an attempt to dry out swamps, people have cut canals into the land, using these artificial effluent streams to drain off groundwater, thus lowering the water table (Fig. 13-4).

Lakes

Lakes may occur in a variety of sizes, shapes, and forms, but they almost always have a direct relationship with the water table. If not, they will be short-lived in the geologic sense. There are glacial lakes, oxbow lakes produced by streams, and sinkhole lakes. There are also lakes produced by earth movements, volcanic activity, and artificially by people. To be even more precise, it should be said that these are the active origins of the lake basins—groundwater simply fills them.

A lake is usually formed where the water table actually cuts the surface and there is a basin in which the water can accumulate. As in streams, there is some water exchange between lakes and the ground. True groundwater lakes are effluent; that is, the waters flow from the ground into the lake. Other lakes such as reservoirs that are produced behind dams, natural dams formed by landslides or earthquakes, and lakes that develop in craters or calderas of volcanoes are usually influent. Crater Lake in Oregon, for example, is maintained by melting snow and by rainfall, and it feeds rather than draws from the local water table.

The basins of the Great Lakes of the United States and Canada are of particular interest because of the unusual way in which they were formed. Although the basins were originally carved by glacial erosion during the Pleistocene epoch of geologic time, they are no longer fed by melting glacial ice but are instead maintained primarily by groundwater. The lakes at one time drained through the Mississippi River Valley, still later through the Hudson and Mohawk River Valleys, and now drain through the St. Lawrence River.

Another interesting basin is that of Lake Okeechobee, located in south Florida. It was formed on an old ocean bottom, but since the ocean withdrew from the area during the recent Ice Age, the lake, which was probably once brackish, is now fresh and is fed both by local groundwater and the inflow of the Kissimmee River.

The Caspian Sea in Eurasia is the earth's largest lake, with an area of 143,550 square miles. Lake Baikal in Russia is the deepest lake, with a maximum depth of 5315 feet. Lake Titicaca, located between Bolivia and Peru, is the highest lake. The largest salt lake in the world is the Aral Sea in the Union of Soviet Socialist Republics, and the largest salt lake in the United States is the Great Salt Lake in Utah. Many bodies of water that are actually lakes are called seas, while many arms of the ocean might be thought of as lakes when they are not.

Springs

A spring is formed where the water table cuts the land surface but there is no basin into which it can flow. Although most springs occur on hillsides, they can form at any site at which groundwater flows naturally out of the ground. Spring water usually runs at just a trickle, but under special geologic conditions (to be discussed later) the water may emerge with great force and volume.

Spring water, like most groundwater, maintains an amazingly consistent temperature throughout the year. Sometimes, however, because of either its proximity to cooling magma underground or its source very deep within the earth, spring water is unusually warm. Such thermal springs are subdivided into warm springs, hot springs, or boiling springs, depending on their temperature. Springs may also have lower temperatures than average; this type is called a cold spring. When spring waters have a higher than average concentration of dissolved minerals, they are known as mineral springs, sulphur springs, etc.

Something should be said here about the medicinal value of spring water with these unusual properties. There is no medical evidence to indicate that bathing in thermal springs or drinking mineral water has any healing or preventive value, despite the fact that for hundreds of years people have believed in their curative powers.

Wells

When people dig down to tap into the water table the result is known as a well. Such an effort on a large scale produces an artificial lake. There is the classic ordinary well in which a hole is dug downward into the ground until it is below the water table. Groundwater then seeps into the well to form a small reservoir. Modern wells are drilled by driving lengths of pipe into the ground; the water is then pumped out through these pipes. The arrangement is a well regardless of whether a bucket or pump is used.

ARTESIAN FORMATIONS

The groundwater sources just mentioned are all dependent on local precipitation. When it is heavy the water table rises, and when precipitation is light, the water table drops. The precipitation amount, then, causes the level of the water table to fluctuate. This in turn produces varying levels in swamps, lakes, and wells and affects spring flow, for these features are nothing more than surface expressions of the water table. The variable nature of the water table often produces inadequate water supplies in areas in which the rainfall is unevenly distributed during the year.

There is a special groundwater condition called an artesian formation, in which the water source is not local precipitation (Fig. 14-2). In order for an artesian formation to occur, the following conditions are necessary: (1) tilted rock layers with one end exposed at the surface and (2) a permeable layer of rock sandwiched between two impermeable layers.

The rock layer exposure at the surface is called the *collecting* or *recharge* area. It is usually located in a region such as a mountain or swamp that has abundant precipitation. After the water enters the exposed permeable layer in the collecting area, it flows slowly downward, following the tilt of the rocks. Because the permeable layer is located between two impermeable layers and because of the tilt, the water actually becomes pressurized like water in a pipe. This water-bearing layer is called the *aquifer*.

The value of artesian formations is that they are usually quite extensive, often reaching across hundreds or even more than a thousand miles. Thus if it is possible to tap into the water supply of an artesian formation, one is no longer dependent on local precipitation but only on the rainfall in the collecting area of the artesian formation. A local drought would not interrupt a steady supply of water from the aquifer.

Artesian springs and wells

There are two methods of obtaining water from an artesian formation. If a fissure or series of fissures extends from the aquifer to the surface, the water may flow naturally out of the ground, producing an *artesian spring*. Sometimes the water from an artesian spring has enough force to lit-

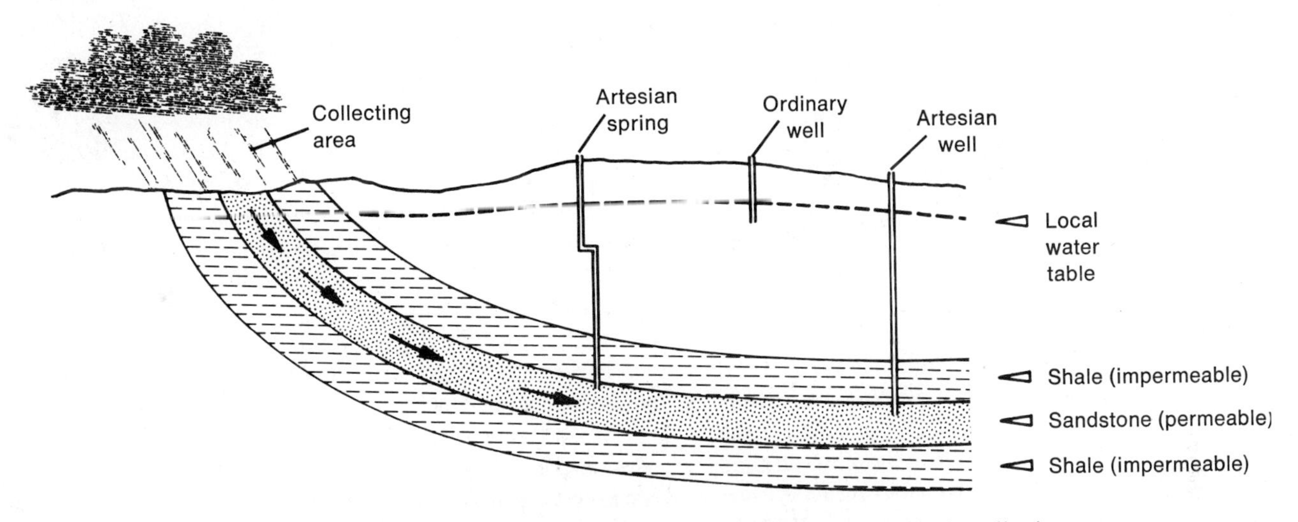

Fig. 14-2. Artesian formation. Water in the system is limited only by the abundance of rainfall in the collecting areas. After collection, the water is channeled through these natural pipes to where it is needed.

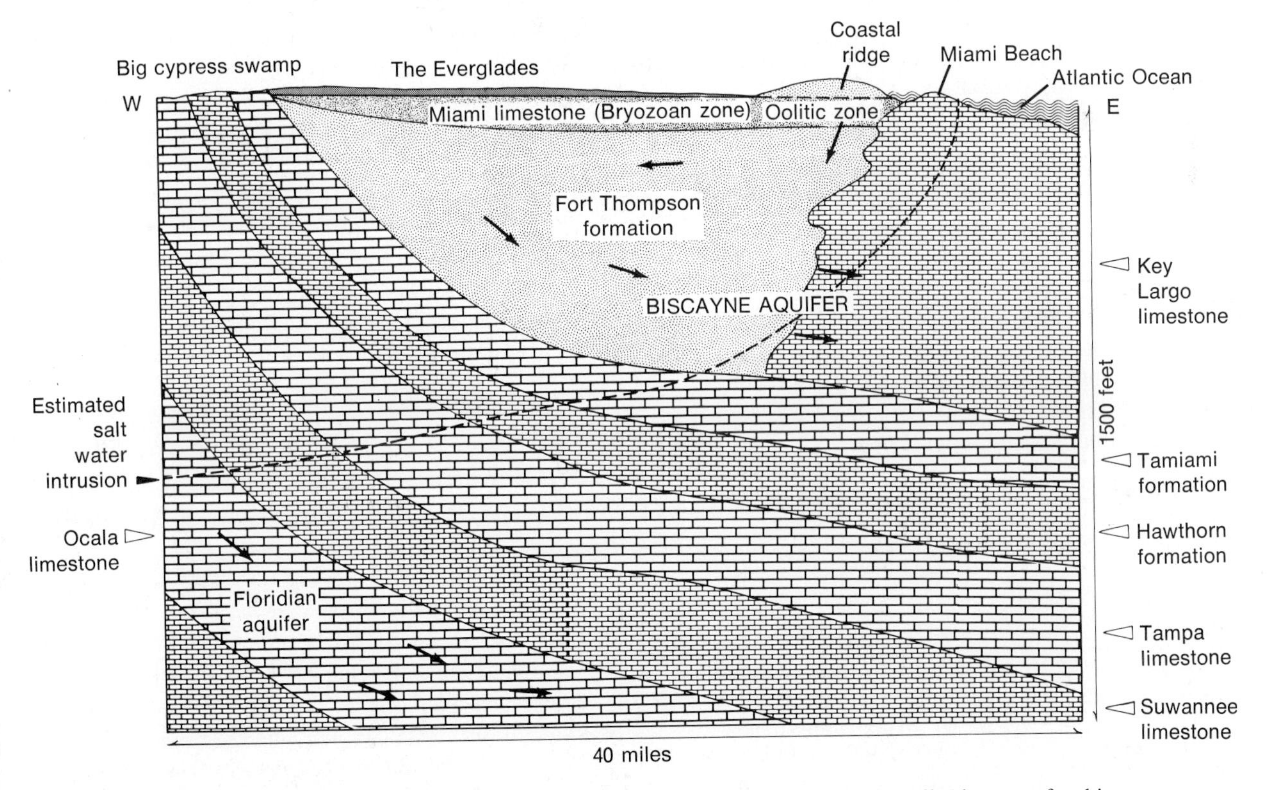

Fig. 14-3. Simplified cross-section of the Biscayne Aquifer. The coastal ridge acts as the collecting area for this formation. The collecting area for the Floridian Aquifer is the Green Swamp, hundreds of miles to the north.

erally jet out of the ground, but more often its flow is a steady bubbling.

Desert oases are explained by the presence of artesian springs that produce a natural underground sprinkling system for the otherwise dry surface. Most of the artesian formations in the Sahara Desert, which covers an area of 3½ million square miles, have their collecting areas in the Atlas Mountains of North Africa. It is ironic that the very mountains that help to produce the desert in the first place should also help provide most of the water the dry region receives.

One of the finest artesian springs in the world is Silver Springs just outside Ocala, Florida. Artesian formations in Florida can hardly be considered typical because the rock layers are so gently tilted. The lack of gradient produces low artesian pressure, which is more than compensated for, however, by water volume, because the limestone rocks underground are made extremely porous by the presence of many solution holes. Of the 75 most prolific artesian springs in the United States, 17 are in Florida. Silver Springs is the largest of these, with an outflow of 500 million gallons/day.

The chief water source in central and eastern Florida is the Floridian Aquifer, whose collecting area is the Green Swamp in Polk County. Occupying limestone rock layers called the Ocala Group, the Floridian Aquifer tilts gently eastward and southward, reaching a depth in south Florida of more than 1000 feet. There is an interesting phenomenon at this location, for there is another aquifer closer to the surface than the Floridian Aquifer. The much smaller but locally important Biscayne Aquifer is even less typical than the Floridian. It furnishes almost all of the potable water to the rapidly growing greater Miami area, for here the deeper Floridian Aquifer is a victim of salt water intrusion and its water is not fit for consumption. The unusual Biscayne Aquifer is shown in Fig. 14-3.

A second method of obtaining water from an artesian formation involves driving lengths of pipe down into the aquifer, as for an ordinary well. Sometimes there is sufficient pressure to bring the water to the surface, but the water must be pumped out more frequently now because increasing numbers of wells are tapping this water source, thus reducing the pressure. Any artificial access to artesian water is called an *artesian* well.

PURITY OF GROUNDWATER

Groundwater is probably our best source of water today from the standpoint of purity, for most surface sources such as streams and lakes are either polluted to the point where their water is unfit for human consumption or they are rapidly reaching this condition. Purity here refers to freedom from disease-causing bacteria and pollutants. The term "best source" should not be construed to mean that groundwater sources are 100% free of these agents, for they are not. It is a common misconception that flowing water in a stream or percolating groundwater purifies itself. People also often believe that because water is cold or clear it is pure. This also is not true. It *is* true that most solid particles are left behind, or filtered out, as groundwater moves through the ground, but most of the bacteria, pesticides, industrial chemicals, and other substances cannot be filtered out in this way. In addition, groundwater carries minerals in solution that cannot be seen but often can be tasted or smelled. Some of these minerals are poisonous to man. Water with an above-average mineral content is called *hard water*. Hard water reacts unfavorably with soap, producing few if any suds.

Groundwater is therefore not necessarily pure. People should make a point of discovering the source of their drinking water. If it comes from a source other than a city water plant, it should be periodically tested for purity. This applies to all groundwater sources, including artesian water.

UNDERGROUND WASTE DISPOSAL

Just as more and more people and communities are turning to groundwater as the source of their drinking water, they are also using underground areas for waste disposal. Human and industrial wastes must be put somewhere other than in our immediate environment, and the idea of pumping these waste products into deep wells is becoming more popular. With proper treatment beforehand, proper research into the nature of the rocks and pattern of water flow of the area, and with proper drilling techniques, this method may have some merit. However, to indiscriminately pump raw and improperly treated wastes into the ground would be another in a long series of mistakes.

THE SILENT INTRUDER

As noted earlier, the Floridian Aquifer in south Florida is so far below the surface that it is intruded by salt water, which makes its waters unfit for consumption. To find salt water underground is not unusual, especially along low-lying coastal regions, for salt water saturates the land beneath the ocean just as fresh water saturates much of the land above sea level. In fact, the saltwater saturation zone actually extends for great distances beneath dry land; the fresh water, which has a lower density than salt water, is actually floating on top of the salt water.

The only factor that prevents salt water from saturating all the land to sea level is the presence of fresh water. The density difference between the two is such that an equilibrium is reached in which there are 38 feet of fresh water below sea level for every foot the water table is above sea level. For example, if the water table is 10 feet above sea level, salt water should be encountered at about 380 feet below sea level. When fresh water is removed at a rate faster than it can be replaced, salt water rises to take its place. Since there is a region of mixing between the two zones, underground water supplies become increasingly salty. The two main producers of saltwater intrusion are drought and human activity; the latter naturally is the greater factor. Our overuse of groundwater and drainage of the land lowers the water table, thus reducing the volume of fresh water and allowing room for the silent intruder.

GEOLOGIC WORK OF GROUNDWATER

Groundwater, like streams, wind, and glaciers, is a destructional, gradational force working to level all land to sea level. All of these agents perform the functions of erosion, transportation, and deposition, even though they may go about their work in different ways.

Erosion

As pointed out earlier, water percolating through the soil and rocks leaves most solid particles behind at or near the surface. As a result, groundwater contains almost no abrasive material. Only dissolved minerals and very rarely some extremely fine clay may be present, although bacteria, chemicals, and the like can penetrate the zone.

Solution, then, must be listed as the only significant erosional function performed by groundwater. Groundwater comes primarily from rainfall, which is slightly acidic because of the reaction between water that has passed through the atmosphere and carbon dioxide. The resultant weak carbonic acid reacts with minerals in the rocks, especially carbonate minerals, changing solid minerals into ionic form in the water. Thus what was solid rock is replaced by a hole where the mineral or minerals were located. As solution holes are produced in the rocks their porosity and permeability are increased, allowing more groundwater to react with more rock. Once started, groundwater erosion therefore becomes more effective with time.

Caves and caverns

A great many erosional features are produced by groundwater. As could be expected from the preceding discussion, most of these features are simply holes of one size or another. Small holes underground are called *solution holes*, while those large enough to admit a person are known as caves. Caves often have smaller solution holes or tubes leading to the surface. They also eventually enlarge and merge to form caverns, or bigger caves.

Some caverns are preserved for future generations as state and national parks or national monuments, while others have been purchased by individuals or groups primarily for tourist attractions. Carlsbad Caverns National Park in New Mexico and Mammoth Cave National Park in Kentucky are excellent examples of government-maintained caverns (Fig. 14-4). Other interesting caverns are Howe Caverns in New York, Luray Caverns in Virginia, Linville Caverns in North Carolina, and Bristol Caverns in Tennessee, to mention but a few.

Fig. 14-4. Stalactites, stalagmites, columns, and pillars as seen in "the painted grotto," one of the many rooms in Carlsbad Cavern, New Mexico. (Courtesy United States Department of the Interior, National Park Service, Carlsbad, N.M.)

Caves and caverns are extremely dangerous unless they are well explored, guided, and lighted. The jumble of rooms and passageways easily leads to disorientation. Slippery floors and continuing solution that is producing more solution holes in rooms and passageways can lead to accidents unless great caution is exercised. Everyone should visit a large cavern to see at first hand the impressive groundwater environment, but extreme care must be the watchword. For those brave souls with a true sense of adventure, there are spelunking societies in which the fine art of cave exploration can be learned from experts and practiced under supervision.

Sinkholes

When the roof of a cave or cavern gives way and collapses, a *sinkhole* is formed. Sinkholes are usually roughly circular and may vary in size from a few feet to a mile or more in diameter (Fig. 14-5). The Devil's Millhopper in Gainesville, Florida, is a fine example of a large sinkhole.

If the water table is broken when a sinkhole forms, the result is a sinkhole lake. Central Florida is dotted with hundreds if not thousands of lakes formed in this way. Exploring underwater caves, caverns, and sinkholes is even more dangerous than exploring caves that are above water level. Sinkhole lakes, complete with cave or cavern rooms and passageways, present a challenge to skin divers and scuba divers, but it should be pointed out that the precautions for cave exploration must be magnified many times for sinkhole exploration.

Natural bridges and tunnels

If part of the roof of a cavern or cave resists groundwater erosion longer than its sides, a natural bridge is produced. The bridge may have at one time been a long tunnel in which an underground stream (actually just concentrated groundwater) flowed. This is one of the few instances in which anything analogous to surface flow occurs underground. Once most of the roof and sides have collapsed, the stream becomes a true stream flowing at the surface, perhaps to return to the ground farther downslope. Natural Bridge, Virginia, is a spectacular example (Fig. 14-6, *A*).

A natural tunnel is considered a stage in the evolution of a natural bridge. About the only difference between the two is the diameter of the opening. After enough time has passed, the tunnel opening becomes enlarged to form a bridge, and eventually the bridge collapses. At Natural Tunnel in Southwestern Virginia, a railroad track has been laid beside an underground stream that runs for about 1000 feet through the mountain (Fig. 14-6, *B*).

Karst topography

Wherever erosional features produced by groundwater but now standing above ground dominate the landscape, the area is said to exhibit Karst topography after the Karst region of northwestern Yugoslavia, where these land forms were first studied in depth. Southern France, parts of Kentucky and Tennessee, central Florida, and southern Indiana are all areas of Karst topography.

Transportation and deposition

Since groundwater contains almost no solid particles, it can transport only dissolved minerals. Therefore the method of groundwater transportation is also called *solution,* for as the water underground moves along, the minerals in solution go with it. The percentage of minerals in solution varies greatly from one locality to another, but it

Fig. 14-5. Series of photographs shows the size of this sinkhole in Butler County, Kansas. Its width is 120 feet, and the limestone walls drop 12 feet to the water level and 25 feet below it. (Courtesy United States Department of the Interior, Geological Survey, Denver, Colo.; photograph by S. W. Lohman.)

Fig. 14-6. A, Natural Bridge, Virginia, is 215 feet high and 90 feet long. It was carved, like the tunnel in **B,** by the combined action of erosion and weathering on carbonate rocks. This type of formation is often found in regions featuring Karst topography. **B,** Natural Tunnel, also in Virginia, is about 1000 feet long, 150 feet wide, and 90 feet high, easily large enough for railroad engineers to allow Mother Nature to give them a hand in cutting through the mountain. (**A** courtesy Virginia Department of Conservation and Economic Development, Richmond, Va.; photograph by Colonial Studios, Richmond, Va.; **B** courtesy Virginia State Travel Service, Richmond, Va.)

can safely be said that all groundwater contains some minerals. When the percentage of dissolved minerals becomes too great for the water to carry, the excess drops out of solution as a *precipitate*.

The processes whereby minerals in solution are precipitated and left behind by groundwater are called deposition. The dissolved groundwater load may precipitate out of solution wherever oversaturation occurs. Oversaturation may be produced by (1) a reduction in temperature, which reduces the capacity of the water to hold minerals in solution; (2) a loss of water by evaporation, which increases the mineral concentration; or (3) an increase in the mineral concentration by the addition of more minerals in solution.

Common locations for groundwater deposition are in solution holes, caves, and caverns above the water table. These features were formed at a time when they were below the water level. When conditions changed and the water table was lowered to a new level, these sites became primarily those of deposition rather than erosion. As groundwater enters the caves and caverns carrying minerals in solution, some evaporation of water occurs; this concentrates the minerals and causes *saturation*. As saturation occurs, the minerals (most often calcite and gypsum) precipitate, building up cave deposits. Since groundwater usually drips slowly downward into these holes above the water table, the cave deposits usually reflect this by forming icicle-like formations. *Stalactites* hang down from cave roofs, while *stalagmites* grow upward from the floor (Fig.

14-4). Often stalactites and stalagmites grow together to form columns and pillars. Great masses of minerals may be deposited in curtainlike shapes on walls or sheetlike masses on floors. The designs and shapes are almost infinite and the features often take on beautiful colors, although whites and grays predominate.

As pointed out in Chapter 12, most sediments are converted to sedimentary rocks through the action of groundwater that deposits minerals between the sediment particles with an assist from pressure. In addition, concretions, geodes, nodules, and vein fillings would not be formed; and the processes of fossilization and petrification could not occur without deposition from groundwater.

15 *the ice age cometh*

Glaciers

With faithful unrelaxing force
Attend it from its primal source.
From change to change and year to year . . .

Matthew Arnold

On May 31, 1970, 15,000 people in the Peruvian town of Yungay were killed when a glacier that had been loosened by an earthquake roared through the Huaylas Valley at speeds estimated at more than 200 miles (120 km)/hr. In less than 2 minutes, ice, mud, and rock had covered the town to a depth of 10 feet (3 meters).

Death by glacier is a rare event, for these geologic erosional agents usually move quite slowly. However, glacial areas are prone to avalanches, or mass movements of ice and rock. The catastrophe in Yungay involved an earthquake, a glacier, and an avalanche so closely intertwined that it was difficult to separate cause and effect. Regardless, "white death" did come to this Peruvian valley as it does to many areas of glaciation each year.

THE GLACIAL ENVIRONMENT

Glaciers are moving masses of snow-covered ice. They form only in areas in which more snow accumulates in the winter than can melt in summer. Such an environment obviously would be a cold climate where precipitation occurs mostly in the form of snow. Many parts of the world receive hundreds of inches of snow each year, but normally the snow melts before a new winter comes. The altitude above which snow remains in the summertime is called the *snow line*. Mt. Rainier (Fig. 11-7) is an example of a mountain that is snowcapped the year round. In snow fields above the snow line, glaciers may form.

Glaciers do not always form, however, in places where there is perpetual snow. There is one more prerequisite for glacier formation, and this is snow depth. Snow and ice must accumulate to a depth at which the mass becomes unstable and begins to move downslope under its own weight. Although the depth of snow and ice necessary to form a glacier varies with the slope or gradient of the land, a minimum of 100 feet could be set for the steepest land. With a decrease in land gradient, an increase in depth becomes necessary.

As already pointed out, glaciers form only above the snow line. An interesting relationship exists between the altitude of the snow line above sea level and geographic location on earth. Glaciers are found in scattered locations from the equator to the poles. At the equator the snow line is at approximately 18,000 feet (5490 meters), while at the poles perpetual snow may be found at sea level (Fig. 15-1). Almost without exception, it can be said that the altitude of the snow line decreases as latitude increases. In the latitudes of the continental United States the snow line varies from 9000 to 13,000 feet (2745 to 3965 meters). With the highest altitude east of the Rocky Mountains (Mt. Mitchell, North Carolina) measured at 6684 feet (2038 meters), perpetual snow does not exist in the central and eastern United States. Active glaciation is occurring, however, in the western United States in parts of the Rockies and on Mt. Rainier and Mt. Shasta, high volcanic peaks of the Cascade range. In the recent geologic past, when world climates were a few degrees colder than today, snow lines were lower and glaciation was much more widespread in the United States and elsewhere in the world.

MOVEMENT OF GLACIERS

Snow accumulates in the snow fields above the snow line, so that each year a new layer forms on top of the old. The weight of the upper layers compresses and consolidates the snow in the lower layers, forming small particles of

Fig. 15-1. Glaciers like this one in Antarctica show that glaciers can form even at sea level at the poles. (Courtesy United States Department of the Interior, Geological Survey, Denver, Colo.)

ice called *firn,* or *nevé*. These particles have a granular texture like that of the remnants of a melting snowbank. As more snow adds its weight to the firn, air spaces are squeezed out and the mass is converted to compact ice. When about 100 feet of ice and snow have accumulated, the great weight forces the lower layers to flow like plastic. Only when the mass begins to move does it become a glacier capable of performing geologic work.

For many years it was thought that glaciers moved simply by slipping over the ground. Although such basal slipping is an important factor in some glaciers, we now know that other processes are involved. The flow of a glacier is due to a combination of (1) internal deforming of ice crystals, (2) melting and refreezing, (3) brittle fracture and faulting in the ice, and (4) basal slippage of the ice over rocks. The weight of the overlying snow and ice produces large ice crystals from small ones. An alignment of the atoms within the crystals takes place approximately at right angles to the applied weight. Movement of the ice occurs as slippage takes place between these layers of atoms. This crystal deformation is not unlike the foliated texture produced in metamorphic rocks (Chapter 17).

Melting and refreezing are thought to be produced by differential weight pressure at the base of the glacial mass. A glacier is almost always thicker upstream than downstream. As a result, glacial ice is constantly melting where the weight pressure is greatest and refreezing where the pressure is reduced. The glacier moves by slipping over this water film. Great stresses build up within a glacier because of the unevenness of the ground surface and uneven distribution of snow and ice. These stresses produce cracks in the brittle ice, planes of weakness along which movement

Fig. 15-2. This valley glacier occupies the entire valley, spreading between and around the mountains of Thule, Greenland.

can occur. It is estimated that as much as 50% of the total motion of glaciers located on steep slopes comes from *basal slippage*—the downslope movement of ice over the surface of the ground due to brute force exerted on the ice from the accumulation source.

The rate of glacial flow depends primarily on the slope of the land, the weight of the accumulated snow and ice, and the geographic location. These three factors are so complexly interrelated that it is difficult to isolate any one factor. However, a few comments can be made about each. Glaciers in temperate (warmer) latitudes are of necessity found in mountainous regions, for only at high altitudes can there be perpetual snow. The slope of the mountains naturally favors basal slippage. Glaciers found at lower latitudes are not as cold as those found in the polar regions, either. As a result of the warmer temperatures, the ice can also move by melting and refreezing. These factors, in addition to the common crystal deformation and brittle fracture, mean that glaciers in the temperate zones move faster than in colder climates. The fastest glacial advance ever recorded was 115 feet (35 meters)/day in the Black Rapids Glacier in Alaska during 1936 and 1937. Most such glaciers, however, flow only a few feet per day. A polar glacier, on the other hand, moves much more slowly, for they flow only by internal plastic deformation of ice crystals and a minor amount of brittle fracture near the surface. The maximum movement of polar glaciers can be measured in inches per day.

TYPES OF GLACIERS

Glaciers may be divided into two main categories: valley glaciers and continental glaciers.

Valley glaciers

Valley glaciers are just that—glaciers that occupy a valley in a mountainous region of the earth (Fig. 15-2). They are not restricted geographically, for mountains are found as frequently at the equator as at the poles. Valley glaciers are the most common type because mountains contain preexisting stream valleys in which snow can accumulate. Most glaciers of this type are small, rarely exceeding 3 miles in length. One notable exception is the Beardmore Glacier in Antarctica, which is 130 miles long and 25 miles wide.

Valley glaciers are found in most high mountain ranges such as the Alps in Europe, the Himalayas in Asia, and the Andes in South America. In North America, valley glaciers occur in parts of the Rocky Mountains and on high volcanic peaks. The mountains of Alaska contain many glaciers of this type; Hubbard Glacier, 80 miles long, is the largest. Valley glaciers are also found in Antarctica and Greenland.

When a valley glacier emerges onto a plain at the foot of a mountain, it spreads over the flat terrain to form what is sometimes called a *piedmont* glacier. Some of these glaciers are fed by two or more valley glaciers. Alaska is the site of many piedmont glaciers, the largest of which is the Malaspina Glacier, which has an area of 1500 square miles. Piedmont glaciers may be considered as a subtype of valley glaciers.

Continental glaciers

A continental glacier, or *ice sheet,* is a body of ice that covers a large part of a land mass. This type of glacier apparently forms from the growth and coalescence of valley glaciers in a cold, humid climate. The combined mass continues to thicken until it covers mountains and valleys alike, with perhaps a few high peaks remaining above the ice.

Continental glaciers today cover most of the Antarctic and Greenland. In Antarctica the ice has an area of about 5 million square miles, 50% larger than the area of the United States, and reaches a maximum depth of about 10,000 feet. The enormous weight of the ice in these locations has caused parts of the earth's crust to subside; in several areas it has sunk well below sea level. If the ice were removed from Antarctica, only a few islands would remain (Fig. 15-3); if it were removed from Greenland, only an inland sea would be left.

Ice sheets are similar to continental glaciers except that they are much smaller and less complex. An ice sheet is a body of ice that rests on a relatively flat land surface, spreading out in all directions under its own weight like pancake batter on a grill. Movement continues as long as more weight is added by accumulating snow. An example is the ice sheet that occupies the high plateau of central Spitsbergen, an island in the Arctic Ocean. As the outward-moving ice reaches the edge of the plateau, it flows to lower elevations as a series of valley glaciers. An ice sheet is nothing more than a small continental glacier.

THE ICE FRONT: RELATIVE MOTION OF GLACIERS

The end of a glacier is called the ice front. Although glaciers move only forward, at the ice front a glacier may appear to advance, retreat, or remain stationary. This relative motion is produced by a struggle between the forces that cause the ice to melt and those that cause the glacier to move. If a glacier moves faster than the ice can melt, then the ice front is said to be *advancing*. On the other hand, if a glacier melts faster than it moves, the ice front is said to be *retreating*. However, if the rate at which the glacier

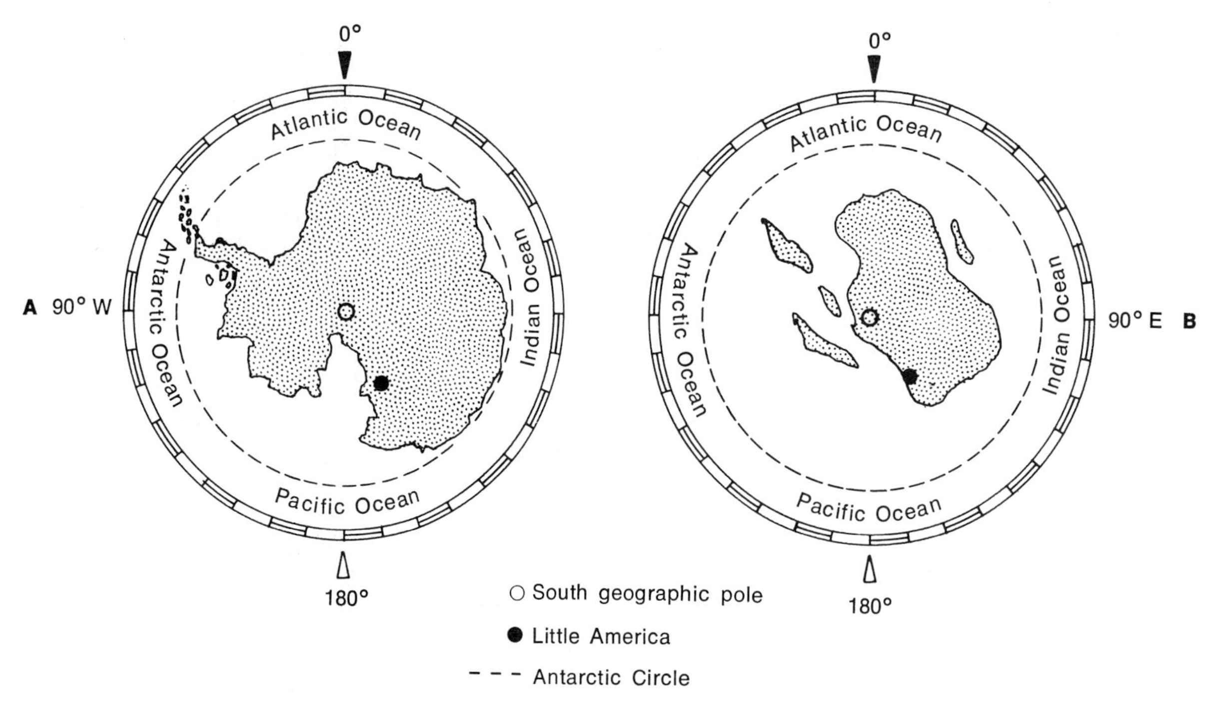

Fig. 15-3. Antarctica. **A,** With ice; **B,** without ice.

moves is exactly balanced by the rate of melting, the ice front is said to be *stationary*.

Glaciers tend to advance during severe winters, retreat during mild summers, and remain stationary during intermediate periods. Relative motion at the ice front of a valley glacier is easy to measure over a period of 1 or 2 years because of its small size. This is done by marking the ice front position with markers attached to bedrock on the sides of the valley. Then, a year later, the position of the ice front can be compared with the position of the markers. Glacial movement and the rate of that movement can be measured by markers placed in a straight line across the glacier relative to fixed markers on bedrock (Fig. 15-4). All parts of a glacier do not flow at the same rate; that part where the ice is deepest moves most rapidly.

Changes in continental glaciers are virtually impossible to detect through direct measurement techniques, even after many years. More subtle techniques, such as measuring the rate of snow accumulation at the centers of accumulation and comparing these to the volume of melt water issuing from the margins of the ice, must be used. If seasonal variations in movement occur in continental glaciers, they are too slight to be measured using methods that are presently available.

Long-range trends in glacial movement are of great interest around the world. Even though glaciers today occupy restricted locations on earth, their advance or retreat would affect all parts of the planet. At any given moment on earth there is a specific quantity of water that is distributed among the ocean, land, and atmosphere. Changes in the percentage of water in one environment will lead automatically to an adjustment in the other two. Today, most of the water is in the oceans. If the average temperature of the earth dropped a few degrees, more water would be stored on land as ice, and this would cause a drop in sea level. If glacial ice were to melt, this would produce the opposite effect. It is estimated, for example, that if all the glacial ice on earth today were melted at the same time, the sea level world-wide would rise by at least 250 feet (76 meters).

What is happening to glacial ice today? Although opinions differ on this subject, it appears that the earth is in a stage in which glaciers are melting, causing the sea level to rise. This is supported by two pieces of evidence. First of all, measurements indicate that more glaciers are retreating than are stationary or advancing. Second, a study of harbor records of mean sea level over long periods of time indicate a slow but steady rise in the level of the ocean.

GLACIERS AS GEOLOGIC TOOLS

Glacial ice today covers about 12% of the earth's total land surface. In the geologic past the ice was much more extensive, but ice has probably never covered more than 25% of the land area. Although limited in extent, glaciers are second only to streams as significant levelers of land. In equal units of time, glaciers can perform more erosion, transportation, and deposition than any other geologic agent.

Erosion

Glaciers perform erosion by two processes, *plucking* and *abrasion*. Bedrock near the earth's surface contains many minute intersecting cracks that hold moisture. In cold climates this water freezes and expands, widening the cracks. Alternate freezing and thawing allows more water to enter the cracks, which become progressively larger until eventually pieces of rock are loosened. When a glacier

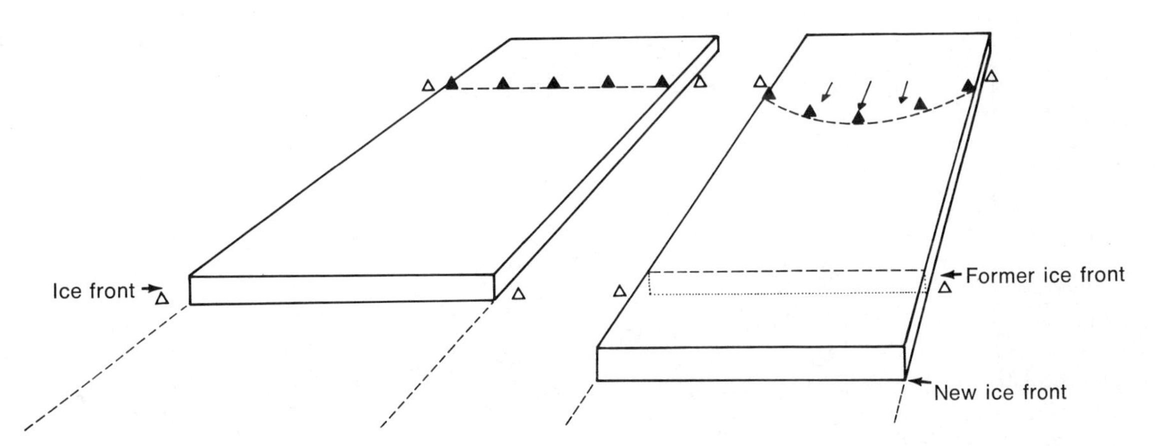

Fig. 15-4. Measurements of glacial movement can be made by setting markers into the valley walls. It was through the use of such markers that scientists learned that the flow is most rapid near the glacier's center.

passes over the loose blocks, they are *plucked* from the ground and incorporated into the ice.

The rock fragments thus frozen into the bottom of a glacier act like a gigantic sheet of sandpaper as the mass of ice moves along. Friction between the fragments of rock and the surface over which the ice is moving produces *abrasion,* which scratches, grooves, polishes, and chips the bedrock lying in its path (Fig. 15-5). Of all the geologic erosional agents, only glaciers can cause such a variety of effects on bedrock.

It has been found that polar glaciers of both the continental and valley types erode and transport less rock debris, volume for volume, than those glaciers located in more temperate climates. For example, the ice at the edge of the Antarctic ice sheet is almost free of rock fragments, a finding that suggests that only slight erosion is in progress in the interior of that continent. The reason for this is that, as pointed out earlier, polar glaciers flow almost entirely by plastic deformation and brittle fracture, while temperate valley glaciers move primarily by basal slipping, melting, and refreezing. Plastic deformation is a process involving ice on ice, while basal slipping, melting, and refreezing bring new ice in constant contact with the ground.

Continental glaciers do, however, perform vast amounts of geologic work, for the quantity of ice contained within glaciers of this type more than offsets the quality of its erosion. Areas of the earth's surface that are now free of glacial ice but that were recently covered by a continental glacier show irrefutable evidence of the erosive power of such massive quantities of ice. In the section that follows, the similarities and differences in erosive effects of the two types of glaciers will be discussed.

Glacial erosional features

Wherever glaciers are found, they leave signs of their passage across the land. The effects of glacial erosion are evident in the largest landforms as well as in small geologic features. The smaller features produced by both continental and valley glaciers include the scratches, grooves, and polish on bedrock that are caused by abrasive material frozen into the bottom and sides of the glacier. Scratches are formed by sand-sized particles or small gravels, grooves

Fig. 15-5. This shiny surface in Tioga Pass, Yosemite National Park, California, was added to the granite bedrock by the polishing action of a glacier's load of stone. The finish has since been weathered off in patches, accounting for the rougher spots.

are made by larger gravels, and polish is accomplished by fine sediment such as silt and clay. Grooves and scratches occur approximately parallel to the direction in which the glacier is moving. This observed fact provides a useful clue in the study of glacial motion in the geologic past.

At the head of a glacial valley, high on a mountainside, ice accumulates in a snow field to a depth at which it begins to move under its own weight. The ice in contact with the surface freezes around rocks of varying sizes that have usually been loosened by the freezing and thawing of water in cracks. As the ice begins to move downward from the snow field to the valley, these rocks are plucked from the ground. Over a period of many years this constant action forms an amphitheater-shaped depression known as a *cirque* at the head of the glacial valley. (Only valley glaciers produce cirques, for the ice in continental glaciers is so deep that most mountain peaks are covered.) Later, when the glacier has disappeared, water may fill the bottom of the cirque, producing a *tarn lake*.

When two valley glaciers occur on opposite sides of a mountain peak, each forms its own cirque. Plucking from the opposing glaciers narrows the rock ledge between the two cirques, forming a sharp ridge called an *arête* (Fig. 15-6). As many as three valley glaciers may form on the sides of a single peak. Cirques cut into a mountain by multiple glaciers form a spectacular three-sided peak called a *horn*, or *matterhorn*, after the famous Matterhorn in the Swiss Alps.

Unlike streams that erode their valleys into the shape of a V, valley glaciers carve valleys in a U shape. The reason for this difference is that streams occupy only the bottoms of their valleys. Weathering and mass movement may

Fig. 15-6. View of the Matterhorn from the Riffelsee, Canton Valais, Switzerland. Glacial cirque formation on three sides of the peak have carved the knifelike ridges (arêtes) on this mountain. (Courtesy Swiss National Tourist Office, New York, N.Y.)

widen the valley, but they maintain the classic V shape. On the other hand, it is not unusual to find a glacial valley half or even entirely filled with ice. The great mass of ice, which contains abrasive material, erodes the sides of the valley as well as the bottom, hollowing out a U shape (Fig. 15-7). In coastal regions, glacial valleys are often invaded by the sea, forming *fjords,* features that may be seen along the coasts of Chile, Norway, British Columbia, Greenland, Scotland, and Alaska (Fig. 15-8).

Several small valley glaciers may combine their masses into one large glacier. As with streams, the smaller glaciers are called *tributaries*. When the ice recedes, however, a very strange topographic pattern remains. Each tributary glacier has cut its own valley as deep as possible. When all the tributary glaciers combine and cut together, however, they can cut much deeper; indeed, the main channel is significantly lower than any of the tributary channels. Thus all along the walls of the main channel one can see smaller channels that enter at higher levels and then fall to the main channel floor in steep cliffs. Such tributary channels that end far above the main channel floor are called *hanging valleys*.

Since streams often flow in valleys vacated by glaciers, it is quite usual for streams to occupy the tributary channels of hanging valleys. When the stream reaches the drop where the tributary glacier joined the main glacier, beautiful

Fig. 15-7. Yosemite Valley as seen from the Wawona Road Tunnel. The U-shaped valleys, hanging valley (right), and hanging valley waterfalls are glacially formed and characteristic of glacial topography. (Courtesy United States Department of the Interior, National Park Service, Washington, D.C.)

waterfalls are formed. One of the better known waterfalls in the United States is Yosemite Falls in Yosemite National Park, California (Fig. 15-7).

Outcroppings of bedrock are reshaped by continental glaciers passing over them. They become gently sloped, smoothed, and polished on the side from which the glacier comes, but are often steep and rough because of plucking on the side toward which the ice moves. Glacially shaped outcrops such as this are called *roche moutonnées* (sheep rock), for at a distance they resemble sheep resting in a meadow. Erosional features common or unique to valley and continental glaciers are summarized in Table 15.

Table 15. Erosional features of continental and valley glaciers

Erosional feature	Continental glacier	Valley glacier
Grooves	Abundant	Abundant
Polishing	Abundant	Abundant
Scratches	Abundant	Abundant
Cirques		Abundant
Arêtes		Abundant
U-shaped valleys	Rare	Abundant
Hanging valleys	Rare	Abundant
Roches moutonnées	Abundant	Rare
Fjords		Abundant

Transportation

The sediments carried by a glacier are its *load*. A load consists of sediments varying in size from boulders the size of a house down to the finest clay. The amount of sediment a glacier can carry is its *capacity,* while the largest rock fragment it can carry determines its *competence*. Glacial ice can carry more and larger sediments than any other erosional agent.

Glaciers transport sediment in basically three ways. Sediments may be frozen into the ice itself, perched on top of the ice, or pushed along in front of the glacier, bulldozer style. Rock fragments frozen into the bottom and sides of the glacier and all of those riding on top of the ice were obtained through the combined efforts of weathering and mass movement. Some of this surface load may grad-

Fig. 15-8. U-shaped valleys formed by glaciers and now invaded by seawater form the unique coastal features known as fjords. This one is in southeastern Alaska. (Courtesy United States Department of the Interior, Geological Survey, Denver, Colo.; photograph by W. W. Atwood.)

ually work its way downward into the ice to become frozen into the body of the glacier. Rock debris pushed along in front of the glacier is the result of basal slippage and is the least important of the three methods of glacial transporation.

Deposition

Glaciers deposit their sediments wherever they begin to melt. Since melting of glacial ice is a constant process that occurs more rapidly at the ice front, so deposition is also a constant process. Glaciers deposit sediments in two ways, as shown in Fig. 15-9. A heterogeneous mixture of rock fragments of all sizes is shown frozen into a block of ice representing a glacier. The block is shown sitting on an incline that represents the sloping surface over which glaciers move. As the block of ice melts, coarser rock debris is slowly lowered to the incline, while finer sediments are carried with the meltwater down the incline to settle out of the water in layers sorted according to size.

All glacial deposits are called *drift,* a name that has persisted from the days when unusual deposits on the earth's surface were thought to have been dropped by the Great Flood of Noah. Drift deposited directly from the melting glacier, like the coarse sediments in the block of ice above, is referred to as *till*. Till consists of a mixture of rock fragments of all sizes—gravel, sand, silt, and clay. Gravels in till almost always show evidence of abrasion, often exhibiting distinct scratches and polish. Sand and silt particles are characteristically angular from crushing in transport.

Drift deposited from meltwater is known as *fluvioglacial deposits,* or *stratified drift*. Unlike till, which consists of a heterogeneous mixture of rock fragment sizes dropped at random, fluvioglacial deposits settle out of water and are therefore sorted according to particle size, producing layers. Stratified drift is deposited wherever meltwater is found—on, within, or below the glacier. The amount of water released from a glacier is so highly variable because of different melting rates that changes in fluvioglacial deposits are numerous and abrupt.

Depositional features made of till

Landforms composed of till are called *moraines*. Accumulations of till deposited along the edges of valley glaciers are known as *lateral moraines*. Where two valley glaciers combine their flow, the inside lateral moraines merge to form a *medial moraine*. Where till has been deposited at the end of a stationary ice front, a *terminal moraine* is formed at the point of the greatest advance of the ice. Long Island, New York, is chiefly a terminal moraine deposited by a continental glacier that melted thousands of years ago. *Recessional moraines* are similar to terminal moraines except that they are deposited closer to the source of ice as the glacier "recedes." They consist of till deposited during stationary periods of an otherwise retreating glacier. *Ground moraine* is the layer of till deposited continuously from a steadily retreating glacier. Morainal types generally merge with one another.

A continental glacier advancing over previously deposited ground moraine may sweep this rock debris into a

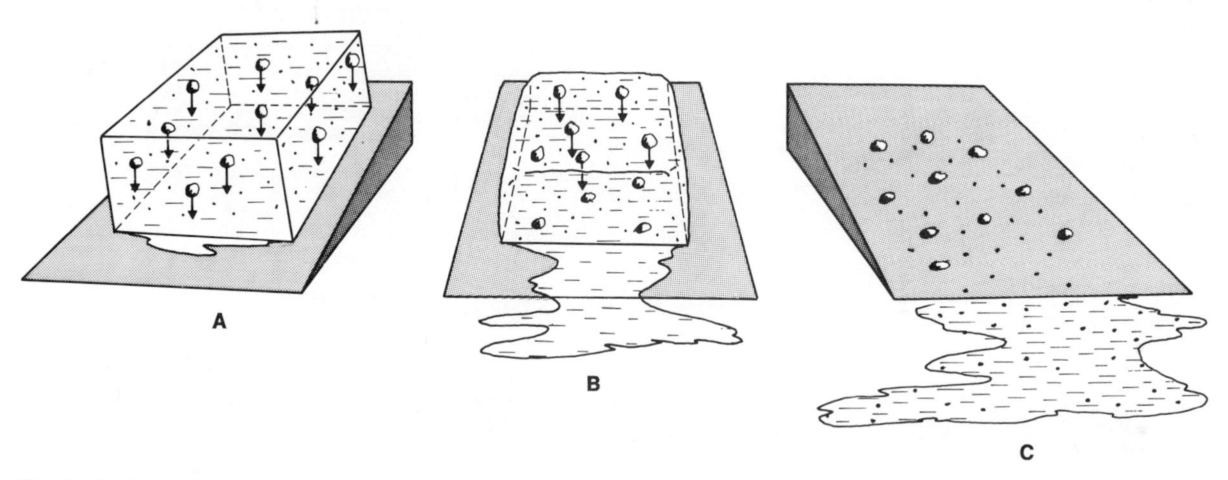

Fig. 15-9. Melting block of ice illustrates how a glacier sorts sediments for deposition. As the ice melts, the bits of sediment frozen inside drop to the floor of the incline. The larger stay exactly where they fall, but the water has the strength to pick up the smaller ones and carry them farther down the incline. As a result, the sediments end up sorted by size, with the larger pieces high on the incline and the smaller ones lower.

series of small, oval-shaped, rounded hills called *drumlins* (Fig. 15-10). Drumlins are usually elongated parallel to the direction of ice flow and may be up to 1 mile long and 150 feet high. Bunker Hill near Boston, Massachusetts, better known for its political than its geologic significance, is a drumlin.

Fluvioglacial depositional features

Tremendous quantities of water are associated with glaciers, for some part of these great ice masses is constantly melting. The water may flow under the glacier in ice tunnels, down into crevasses (deep cracks in the ice), or away from the glacier as streams. Meltwater from these sources often accumulates at the end of the glacier as a *marginal lake*. Smaller particles of rock debris are carried by glacial meltwater and deposited wherever the speed of the water is reduced. Particles are sorted according to size, resulting in the layered characteristics of fluvioglacial deposits.

Water flowing away from the ice front deposits material called *outwash*. Outwash may cover extensive areas to depths as great as several hundred feet, forming an *outwash plain*. Much of the flat land in Ohio, Indiana, Illinois, Wisconsin, and Iowa is actually an outwash plain formed by a melting continental glacier. Streams that flow in tunnels below the ice deposit gravel and sand in their channels. After the glacier disappears, a narrow, winding ridge of sediments called an *esker* marks the former stream course (Fig. 15-11). Eskers may be traced discontinuously for up to 100 miles across the countryside and may reach a maximum height of 100 feet.

As blocks of ice that were buried in till or outwash melt, holes are formed and overlying material slumps down into the void. Such depressions are called *kettles*. Most kettles are only a few feet in diameter, but a few are more than 1 mile across. Depths range from a few feet to as much as 100 feet. Water often fills kettles, forming *kettle lakes*.

In glacial areas fluvioglacial material is often found deposited in low, steep-sided, conical-shaped hills called *kames*. Kames appear to be formed in at least two ways. Sediments may be deposited in an opening in the ice, later slumping down to the surface when the surrounding ice melted. Kames may also be formed as a delta deposit against the ice front of a glacier by a stream that flows from the surface of the ice. As the ice front retreats, the sediments are no longer supported and slump down to form the kame.

The smallest and one of the most interesting fluvioglacial deposits is a glacial *varve*. A varve is a pair of thin sedi-

Fig. 15-10. Two drumlins south of Newark, Wayne County, New York. The egglike shape is characteristic of this type of deposition, as is the parallel orientation. Since drumlins point in the direction of glacial flow, all drumlins in a given area would be parallel. (Courtesy United States Department of the Interior, Geological Survey, Denver, Colo.; photograph by G. K. Gilbert.)

Fig. 15-11. This narrow, winding hill is the Francis Creek esker and it twists through Manitowoc County, Wisconsin, tracing the course of an outwash stream of the glacier that once covered Wisconsin and adjoining states. (Courtesy United States Department of the Interior, Geological Survey, Denver, Colo.; photograph by I. C. Russell.)

Table 16. Depositional features of continental and valley glaciers

Depositional feature	Continental glacier	Valley glacier
Till		
Lateral moraines		Abundant
Medial moraines		Abundant
Terminal moraines	Abundant	Abundant
Recessional moraines	Abundant	Abundant
Ground moraines	Abundant	Abundant
Drumlins	Abundant	
Fluvioglacial		
Outwash plains	Abundant	Abundant
Eskers	Abundant	Abundant
Kettles	Abundant	Abundant
Kames	Abundant	Abundant
Varves	Abundant	Abundant

mentary beds, a fine clay bed on top of a coarser one composed of silt. A series of varves may reach a thickness of many feet. Varves probably are seasonal deposits from marginal glacial lakes, each varve representing 1 year much as the annual rings of a tree represent seasonal growth. The coarse bed would consist of material deposited in the lake during the summer when glacial melting was active and the marginal lake was not frozen over. Wind currents in the lake would keep fine material suspended while the coarser sediment settled out. During the winter when the lake surface was frozen, the fine material would not remain suspended but would settle out of the quiet water to form the fine clay bed. The age of many present and past lakes can be determined by counting varves in the bottom sediments. As a result, the glacial events that formed the lakes can be dated.

Depositional features composed of till and fluvioglacial deposits common or unique to continental and valley glaciers are summarized in Table 16.

EVIDENCE OF THE ICE AGE

Today glacial ice covers about 12% of the total dry land surface of the earth. Of this total, 83% is accounted for by the continental glaciers of Greenland and Antarctica. There is overwhelming evidence that within the last million years of geologic time, glaciers covered an area at least twice as extensive. Less than 1 million years ago, many of today's most populated regions were covered by 1 or more miles of ice. This span of geologic time is known as the Pleistocene epoch, or the Ice Age.

That there was an Ice Age is a relatively new discovery. In 1837 Louis Agassiz (1807-1873), now known as the

Fig. 15-12. Resting on a glaciated surface southeast of Sentinel Dome in Yosemite National Park, California, this boulder was carried from the far north by a glacier, which finally left it here, to remind Californians that their climate was not always the joy it is today. (Courtesy United States Department of the Interior, Geological Survey, Denver, Colo.; photograph by F. C. Calkins.)

father of glaciology, announced to the world that much of England and Northern Europe had been smothered in an ice age. Later he extended his theory to include much of North America. Having been born and raised in Switzerland and educated in Germany, Agassiz had ample opportunity to study the glacial erosional and depositional features found in the Alps.* Finding similar features covering wide areas of the middle latitudes, he developed his glacial theory, which is almost universally accepted today.

In the United States, evidence abounds to support Agassiz's claim. In many locations in the northern tier of states, boulders the size of a house are found resting on rocks of completely different origin (Fig. 15-12). These boulders, called *erratics* by geologists and often referred to by the common name of *haystack boulders,* must have been transported to these locations. The only known erosional agent with sufficient competence to move such a heavy load is a glacier. The only location in the United States where large quantities of native copper are found as ore is in northern Minnesota. However, pieces of this native copper have been found buried in sediments as far south as central Missouri. Streams could not have carried the copper there, for as often as not it has been deposited on high ground between stream courses. The only gradational agents capable of transporting this material over such great distances and depositing it on high ground are glaciers.

Scratched, polished, and grooved bedrock is found along the coast of Maine, in Central Park in New York City, around the Great Lakes, in many locations in the Northern Plains states and Pacific Northwest, and as far south as Grandfather Mountain in western North Carolina. Only one geologic erosional agent can groove, scratch, and polish bedrock all at the same time—glaciers. The Great Lakes themselves are the product of continental glaciation. Erosion by great lobes of ice enlarged and deepened preexisting stream valleys, forming basins in which water accumulated as marginal lakes whose source was melting ice and streams whose drainage was blocked by ice on the north and by terminal moraines to the south. At first the Great Lakes drained through the Mississippi Valley to the Gulf of Mexico. As the continental glacier retreated, lower outlets were uncovered, diverting the water through the Mohawk and Hudson River Valleys to the Atlantic Ocean. Continued retreat of the ice finally exposed the St. Lawrence River Valley, which now drains the five Great Lakes.

That former marginal glacial lakes also existed in the United States and Canada is indicated by outwash deposits, abandoned stream drainage channels, and glacial varves. The largest lake was Lake Agassiz, at its maximum extent larger than all the Great Lakes combined. It existed from about 11,000 to 7000 years ago, covering large areas of what today is Manitoba, Canada, and the state of Minnesota. Formed along the margin of a huge continental ice sheet, Lake Agassiz originally drained southward through the present Missouri-Mississippi stream system to the Gulf of Mexico. After the ice retreated, water from the lake drained northward into Hudson Bay, which was no doubt either carved by glacial erosion or formed when huge accumulations of ice in the area forced the earth's crust down below sea level. Remnants of Lake Agassiz remain as Lake Winnipeg and literally thousands of other small lakes.

Many locations in the Rocky Mountains and Sierra Nevadas show U-shaped valleys. Even though there are no glaciers in these valleys today, we know that they were there at one time. One of the most beautiful U-shaped valleys in the United States is Yosemite Valley in Yosemite National Park in California. Many parts of the northern United States and Canada are littered with drift in the form of ground moraine, terminal moraines, and recessional moraines. There are also abundant eskers, drumlins, kames,

*The Alps show features so characteristic of valley glaciers that other regions on earth exhibiting similar features are said to represent *Alpine topography*.

and kettles. The only agent known to form such features are glaciers.

Evidence of glacial activity is found even in places thousands of miles from the actual location of the glaciers. As discussed earlier, sea level is very closely related to the amount of ice stored on land. With an advance of the ice, sea level would drop. Retreat of the ice causes sea level to rise. In North America, the Ice Age involved at least four stages of glacial advance followed by retreat and long periods of relatively warm climate called *interglacial stages*. The four glacial stages of North America, arranged from oldest to youngest, are the Nebraskan, Kansan, Illinoian, and Wisconsin, each named after a state with abundant evidence of glaciation. The Nebraskan began about 1 million years ago, while the Wisconsin ended less than 10,000 years ago.

The southeastern Atlantic and Gulf Coast states show clear proof that with each advance and retreat of the ice during the Pleistocene epoch there was a corresponding minimum and maximum stand of sea level. Wherever sea level was located for many years, just as along our present coasts, typical coastal features were produced. Abandoned shorelines are obvious in the state of Florida, for on such a peninsula rising ocean water encroaches from three sides. It is believed that these abandoned shorelines are relics of all four glacial stages and the long interglacial stages between them. These changes in sea level also left their mark on other coastal areas around the world.

PEOPLE AND THE ICE

Because areas that nourish modern glaciers are climatically inhospitable, few human works are affected by sedi-

Fig. 15-13. This twin-peaked iceberg broken from the tip of a Greenland glacier is a sample of the most direct contact we normally have with glaciers. (Courtesy United States Coast Guard, Washington, D.C.)

ments carried by moving ice. In fact, it seems that glaciers could have little effect on us and that we could have even less effect on glaciers. However, this may not be the case. Glacial ice is extremely sensitive to climatic changes; it is estimated that a decrease of less than 10° F (5.5° C) in worldwide average temperatures would trigger another ice age. The same magnitude of increase in temperature would cause the ice we have on earth today to melt, flooding low-lying areas around the world.

The key, then, to our effect on glaciers ironically does not lie in the glaciers themselves but in the atmosphere. The sun is the original source of all the earth's energy, but the atmosphere receives most of its heat directly from the earth's surface as visible light from the sun is converted into infrared radiation (heat). This terrestial radiation is then absorbed by water vapor and carbon dioxide in the atmosphere, warming the air. An increase in either the water-vapor or carbon dioxide content of the atmosphere could increase the atmospheric temperature by a few degrees, resulting in the melting of glacial ice and a rise in sea level. Today we are pouring tremendous quantities of both these gases into the atmosphere as combustion products from everything we burn.

Recent findings indicate an even more grave condition arising from the release of particles into the atmosphere. More particles produce more clouds as water vapor condenses on these tiny nuclei. This results in more sunlight being reflected back into space, and this in turn produces a worldwide decrease in temperature. Again, the main culprit is combustion, for substances that burn are almost never completely consumed—the residue escapes as particles into the air.

Since both continental and valley glaciers often reach the sea, great masses may break away from the main body of a glacier and go floating away as icebergs carried by ocean currents (Fig. 15-13). Earlier in this century, icebergs took a terrible toll of human life as they collided with ships. Especially vulnerable were the shipping lanes of the North Atlantic, which are frequented by icebergs from the coast of Greenland. After the *Titanic* disaster of 1912,* the International Ice Patrol was established to keep track of the movement of icebergs in the North Atlantic.

*The *Titanic*, a British ship, struck an iceberg and sank in 2½ minutes with a loss of 1503 lives.

16 *knives of sand*

Wind erosion

Silent and chaste she steals along,
With gentle yet prevailing force,
Intent upon her destined course;
Graceful and useful all she does,
Blessing and blest where'er she goes.

Another
William Cowper

On May 12, 1934, a great storm moved some 300 million tons of soil from the Great Plains of Kansas, Texas, Oklahoma, and Colorado, transporting it eastward where it was deposited in quantities as great as 100 tons/square mile. The effects of this single storm point out vividly why wind is classified as a major mover of earth materials (Fig. 16-1).

Wind is defined as a horizontal movement of air from high to low atmospheric pressure. It is possible to think of a high-pressure area as being like a mountain of air, while a low-pressure area is similar to a valley. Air flows from areas of greater concentration or pressure toward areas where pressure is less. In effect, then, air has a gradient and flows downhill like water, and as a result it can act as a geologic gradational agent. The forces that produce wind are discussed in greater detail in Chapter 29, but at this point we are considering air because of its ability to perform geologic work.

FACTORS OF WIND EROSION

Wind passes over every part of the earth's surface from time to time, so air theoretically has the greatest potential for accomplishing geologic work. However, several conditions must be met in order for wind to be most effective.

First of all, the surface must be dry. Wet sediments tend to stick together, and the moisture also adds weight that makes them too heavy to move. Wind erosion, then, is most effective where the surface material is dry much of the year. Even a temporary drought in an otherwise humid climate will increase the wind's effectiveness as an erosional agent.

A second factor that is involved is whether or not the surface material is loose, or unconsolidated. It should be obvious that loose material is required for movement by any agent of erosion, but especially by wind.

The amount of vegetative cover is the third factor. Plant roots act to hold soil and sediments in place, so wind operates more efficiently when vegetation is scarce.

AREAS SUSCEPTIBLE TO WIND EROSION

Low levels of surface moisture, unconsolidated surface materials, and sparse vegetation are usually encountered in regions with semi-arid or arid climates, especially deserts, the epitome of dry lands. (However, you should remember that other areas also experience the effects of wind erosion, although on a much-reduced scale.)

It is estimated that deserts cover 15% of the earth's land surface. The term "desert" is somewhat misleading, for the only criteria for classification are *precipitation* and *evaporation rates,* not temperature. If an area has an annual precipitation rate of 10 inches/yr or less, or if evaporation exceeds precipitation, no matter what the precipitation rate, the region is called a *desert.* The popular image of a desert as an area of hot, shifting sands does not apply to the polar regions, especially the continent of Antarctica, for they also fit the desert description well. In fact, their precipitation rates are among the lowest anywhere on the earth.

There are at least four types of deserts. Each is produced

Fig. 16-1. Great storms throughout the 1930s moved across the Midwest, dumping tons of loose sand on anything in their way. This photograph, taken in 1936 in Gregory County, South Dakota, shows the depth of the material dumped on unsuspecting areas. (Courtesy United States Department of Agriculture, Soil Conservation Service, Washington, D.C.)

by meteorologic or oceanic conditions that guarantee scarce precipitation or high rates of evaporation. *Tropical deserts* are the first type. They are the largest deserts of the classic description (hot, shifting sands). They lie approximately on the Tropic of Cancer in the Northern Hemisphere and on the Tropic of Capricorn in the Southern Hemisphere. Tropical deserts are formed in these regions by subtropical cells of high pressure in which air sinks and is warmed by compression (Chapter 29). The warming effect increases the capacity of the air to hold moisture and increases evaporation rates; both have a drying effect on the land and inhibit precipitation. The Sahara Desert of northern Africa, the Sonora Desert of the southwestern United States, and the Great Australian Desert are all tropical deserts.

A second type of desert, called *middle-latitude deserts* because they occur between 30° and 50° north and south of the equator, are found in the interiors of Asia and South and North America. Located in the prevailing westerly wind systems, middle-latitude deserts occupy rain shadows produced on leeward sides of coastal mountain systems (Fig. 16-2). As moisture-laden air rises over the mountains it is cooled, releasing its moisture to produce abundant precipitation on the windward side of the mountains. As the air sinks down the leeward sides of the mountains, it is warmed, increasing its capacity and thereby producing drying conditions like those found in the tropical deserts. The Patagonian Desert of Argentina, the Gobi Desert of Mongolia, and the Great Basin Desert of Nevada and Utah in the United States are examples of middle-latitude deserts.

The Atacama Desert of Chile, the deserts of coastal Ecuador and Peru, the Namib Desert of southwest Africa, the desert of lower California, and Mt. Desert Island, Maine

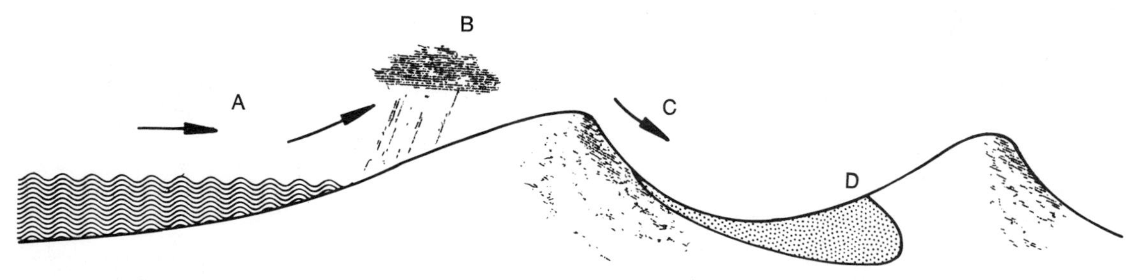

Fig. 16-2. Formation of a midlatitude desert. Moisture-laden air approaches the land from ocean, *A*. Rising to cross the mountains, *B*, it is cooled below its saturation point, dumping its load of water on the ocean side of the mountains. After crossing the mountains, *C*, the air sinks and is warmed, increasing its capacity for water vapor. It then holds all the vapor it has left, instead of letting the rain fall in area *D*. This continued lack of precipitation eventually turns these mountain-protected areas into deserts.

Fig. 16-3. Mt. Desert Island, Maine, is an arid region protected from precipitation by surrounding cold ocean currents. (Courtesy Main Park Service, Augusta, Maine.)

(Fig. 16-3) are all *coastal deserts*. They are formed adjacent to cold ocean currents and upwelling cold ocean water offshore. The cold water cools the air above it, thus reducing the capacity of the air to hold moisture while increasing the rate of evaporation. As this cool, dry air moves onshore, drying effects and reduced precipitation result in desert conditions.

The last desert type, the *polar deserts*, are located north and south of 60° latitude in the coldest regions on earth. In these areas the precipitation rates are low because cold air has a reduced capacity for holding moisture. An anomaly exists, for while the air and soil are moist, other conditions needed for precipitation are retarded. Because evaporation decreases with decreased temperatures, polar deserts are very different from the classic deserts mentioned above. As a result, wind erosion is not common in these areas because glaciation dominates the polar regions.

It is obvious from a study of the geologic past that deserts are not permanent features. Some areas now classified as deserts at one time had more humid conditions, and regions now humid and fertile were deserts at one point in geologic time. These changes may be the result of uplift of coastal mountains, major changes in wind and oceanic circulation, or major variations in the earth's tilt on its axis. An intriguing possibility also exists that the atmospheric and oceanic circulation as well as the earth's axial tilt have remained basically unchanged, while the land masses have changed their geographic positions by continental drift (Chapter 18). Recently discovered evidence indicates that the earth's South Pole, complete with polar ice cap, was located in the Sahara Desert 450 million years ago. Either the Sahara area was once in the present polar region or else a drastic change took place in the rotational plane of the earth. At the same time, coal deposits and fossils indicate that Antarctica at one time possessed lush, tropical vegetation. The final answers are not available, but more and more pieces of this complex puzzle are fitting into place.

LANDFORMS AND WATER FEATURES UNDER WIND DOMINATION

In deserts and arid and semi-arid regions, landforms appear geologically young; that is, they are sharp in outline because water, the main destroyer of rocks, is in short supply. Cliffs and mountains develop rock slopes around their bases called *pediments*. Interior drainage basins in which the source, channel, and mouth of streams lie completely within the area are common, for few streams are able to traverse the region without succumbing to evaporation and effluence into the ground. The few that are able to cross are known as *exotic streams*. They have their sources in a distant humid environment and survive out of sheer volume. Outstanding examples of exotic streams are the Nile of Africa, the Tigris-Euphrates of the Middle East, the Indus of India, the Orange of South Africa, and the Colorado of the southwestern United States.

Around the basins, material eroded by the occasional heavy rains is deposited to form alluvial fans. These are conspicuous in Death Valley, California. If there is adequate water, a salt or alkaline lake may be maintained. Examples are the Great Salt Lake of Utah, the Salton Sea of California, the Aral Sea of central Asia, Lake Eyre in Australia, and the Dead Sea of the Middle East. If such lakes are *intermittent*—present in the wet season and absent in the dry season—which they usually are, they are known as *playas*. Accumulations of salt or other materials are often mined from playa lakes.

Dry stream channels, evidence of the occasional downpours of rain that do occur, are also characteristic of wind-dominated lands. In North America such channels are called *arroyos*. During such a rare but heavy rainfall, flash floods race through the arroyos, making them death traps for animals and any unwary traveler camping there for the night.

GEOLOGIC WORK OF WIND

Wind, like other geologic gradational agents, performs the functions of erosion, transportation, and deposition. The movement and characteristics of wind can sometimes be similar to those of water, and thus many of the processes accomplished by one are also accomplished by the other. However, because of certain basic limitations, some differences also exist.

As was pointed out in Chapter 13, the largest particle that a gradational agent can carry is the measure of its *competence*. It is here that severe limitations are placed on wind as an effective geologic tool. The competence of wind, except under rare conditions, is restricted to particles $^{1}/_{25}$ of an inch (10 mm) or less in diameter. This means that wind can transport sand, silt, and clay. The last two are often referred to as *dust*. Although the ability of moving air to acquire and carry particles of earth materials is much less than that of streams or glaciers, its enormous volume results in large-scale movement under favorable conditions.

Erosion

Without cutting tools, the wind, like water, can do little geologic work in the sense of changing solid rock into sediments. If loose, dry material is available, wind will naturally acquire the necessary tools so there is a one-to-one correlation between the need and the opportunity. With sand and dust in transport, wind erosion is produced in three ways, by *impact, abrasion,* and *deflation*.

Impact and abrasion

Abrasion is the wearing away of rock and other surfaces through the frictional effects of wind-driven sediments, while impact occurs when a sand grain is blown into direct collision with rock, soil, or other materials. It has been found that abrasion and impact are most highly concentrated within the 18 inches (46 cm) above the earth's surface, although sand may occasionally be carried by very strong winds to heights of 3 feet (1 meter) or more. Impact is the more important of the two processes.

The effects of wind abrasion and impact are familiar to those living in dry climates. House foundations, rock cliffs, utility poles, fence posts—indeed, anything projecting out of the ground—often have notches cut into their bases by wind-driven sand. Isolated rocks exposed to these processes exhibit unique characteristics including facets, pits, and a high-gloss polish. All rocks that show these characteristics are known as *ventifacts* (Fig. 16-4). In addition, those rocks with well-developed facets are referred to as *dreikanter* (three-sided). It is virtually impossible to keep surfaces painted in regions of active wind erosion, for impact and

Fig. 16-4. Ventifacts, or wind-polished rocks. Note the high gloss on certain stones and the lines cut into others. (Courtesy United States Department of the Interior, Geological Survey, Denver, Colo.; photograph by M. Q. Campbell.)

abrasion quickly wear away the paint. The paint on automobiles, for example, is often stripped off to bare metal, and window glass can become so scratched from impact as to be opaque.

Wind erosion, however, is most effective in areas of low moisture, so that rusting is of little consequence. Desert air is such an excellent preservative, in fact, that aircraft and other pieces of handware are often stored for long periods in these areas with little detrimental effect. Tanks and armored vehicles used in desert battles in World War II have been found in the Sahara Desert almost exactly as they were left more than 30 years ago. In most parts of the world, rust would have reduced them to rubble.

Deflation

The third erosive process of wind is deflation, the actual en masse movement of unconsolidated sediments. While abrasion and impact are rather slow processes, deflation is often swift and massive, as in the great sandstorm of 1934.

Wind, like most agents of erosion, attacks on a *differential* basis; that is, it erodes the weakest materials first. Differential erosion by deflation produces depressions in the ground called *blowouts,* which may be a few feet in diameter and a few inches deep or a mile or more in diameter with a depth of many feet. Deflation also removes finer material from large expanses of desert areas, leaving behind only coarse rock called *desert pavement*.

Differential erosion by wind also affects isolated masses of solid bedrock, especially those that consist of a resistant layer underlain by a more susceptible layer. As time goes by, the nonresistant material beneath is eroded faster than the resistant layer, producing strange forms often referred to by such names as table rock, pedestal rock, or balance rock, or by names suggested by their shape, such as Camel Rock in New Mexico. Bedrock may become honeycombed or develop caves and natural bridges carved by the wind. The longest natural bridge in the world, Landscape Arch, was formed in this way. Located in Arches National Monument in Utah (Fig. 16-5), Landscape Arch is a natural sandstone bridge with a span of 291 feet 100 feet above the canyon floor. In one place, wind erosion has narrowed its thickness to only 6 feet.

Differential wind erosion, whether by abrasion, impact, or deflation, receives an important assist from weathering process and infrequent but locally heavy rainfall.

Transportation

The second geologic function of wind is that of transportation—the movement of sediments acquired through erosion. Wind-driven materials are moved in three ways. Heavier material, primarily coarse sand, is transported by a process called *saltation,* in which the grains roll or bounce along the earth's surface, rarely exceeding a height of 18 inches (45 cm) above the ground. Finer sand driven by strong winds may be lifted to much greater heights or even become suspended by air currents for short periods.

Suspension is the second method of wind transportation. It has been shown that turbulent air is more effective at this process than air that flows smoothly. Suspended sediments may reach an altitude of thousands of feet and travel across linear distances of hundreds of miles.

A third method by which the wind moves fine sediments is known as *surface creep*. This is the slow, downwind movement of grains of sand that are too heavy to be transported by saltation or suspension. Surface creep should not be confused with soil creep (Chapter 12), which is the slow

Fig. 16-5. Landscape Arch in Arches National Monument, Utah, was produced by differential wind erosion. The wind carried off the softer underlayers, leaving behind the tougher top layers. (Courtesy Utah Travel Council, Salt Lake City, Utah.)

downhill movement of soil by gravity. Surface creep is the result of impacting sand grains. The kinetic energy of an average-sized sand grain $^1/_{250}$ to $^1/_{25}$ inch) is such that on impact it can move a stationary surface grain six times as large as its own diameter or 200 times its own weight. Saltating or suspended grains of sand strike larger grains on the surface, causing them to move downwind, producing surface creep.

The havoc produced by sandstorms and duststorms is well known, but the value of wind-transported materials is often overlooked. One important area in which dust plays a key role is the weather. In order for water droplets and ice crystals to form into clouds, tiny particles are needed on which water droplets and ice crystals can form. These small, suspended particles are known as *condensation* or *sublimation nuclei* and may consist of salt crystals from the ocean, meteor dust, volcanic dust, pollen or seeds, or more important over most land areas, dust from soil and rocks.

Deposition

Deposition is the dropping of sediments that are being transported by an erosional agent. In Chapter 13 we discussed the ways in which streams drop material when their speed or volume is reduced. Since wind can operate like a fluid such as water in many respects, it is not surprising that many of the hydraulic forces at work in streams are also operative in terms of wind movement and action.

The largest particle the wind can carry (its competence) is proportional to the wind's speed. The highest wind speeds occur in tornadoes, and in these storms the competence of the wind is virtually unlimited, as demonstrated by the incredible damage they cause. Although tornadoes are of little value geologically because they are relatively rare, small storms of a local nature, they do vividly illustrate the lifting power of the wind and how objects are dropped as wind speed diminishes.

Speed therefore is the critical factor in wind deposition, just as it is in wind erosion and transportation. Whenever the speed of the wind is reduced, air-transported sediments are dropped. Wind speed can be diminished in a number of ways. Reduction can occur naturally, for wind is never constant in speed and direction for extended periods of time. Some of the material being transported by a wind of 40 miles/hr will be deposited as the speed drops to 20 miles/hr. Obstructions produce deposition as well, for the speed of the wind must decrease as the obstruction either blocks the wind altogether or slows it as the wind must pass over or around. As the wind's speed is reduced, its capacity to carry sediments is also reduced, and some material is dropped as the wind adjusts to the lower energy state. Any irregularity in the surface over which wind is passing qualifies as an obstruction. It may be a house or a rock, a street curb or the wheel of an automobile, a hill, a mountain, or a fence—the list could go on and on.

In regions where wind is the major gradational agent, the landscape is characterized by wind-formed erosional and depositional features. Wind deposits take the form of *sheets, drifts,* or *dunes*. These three can occur together or separately, but where sand and dust are present in almost unlimited quantities, drifts and sheets are most common.

Sheets

Dust deposits over a large area are called sheets. Sheets vary in thickness, but they usually thin out downwind from the source of the dust. These deposits cover much greater areas than do either drifts or dunes. The dust that composes these sheets may come from drought-striken lands, desert areas, volcanic eruptions, or glacial areas. The great storm of 1934 deposited a sheet of wind-blown material over extensive areas of the Central United States. In 1912 the eruption of Mt. Katmai, a volcano located on the Alaskan Peninsula, produced dust that was deposited in sheets extending as far as 100 miles (166 km) from the eruption. Even at that distance, the dust deposit was 10 inches (25 cm) thick.

Classic examples of wind-deposited dust sheets are the loess accumulations of the central United States, northern China, and Europe. *Loess* is a yellowish, unconsolidated, unstratified, wind-deposited dust of great age that is composed of small, angular mineral fragments. It forms thick deposits, sometimes as deep as 50 feet (15 meters) or more. These deposits are capable of standing in almost vertical cliffs in stream banks and road cuts. Most of the loess deposits of the United States and Europe are believed to be derived from Pleistocene (Ice Age) glacial deposits to the north. The widespread, fertile loess accumulations of China consist chiefly of dust blown from the Gobi Desert. This curious type of deposit is also found in the Sudan, downwind from the Sahara Desert.

Drifts

Drifts are an accumulation of sand that forms on the lee (protected) side of an obstruction such as a bush or rock. Drifts form because the obstruction produces a wind shadow, an area of relative calm, on its downwind side. This area of calm traps sand grains, gradually producing an elliptic, smoothly rounded pile of sand pointed in the wind direction. Unlike sand dunes, drifts do not migrate but are fixed in place. The picture that comes to mind when most people hear the word desert—mile after mile of hot, burning sand—consists mainly of drifts, not dunes.

Dunes

A sand dune is a mound of wind-deposited sand that is usually bare but may be covered by vegetation. Dunes can be formed wherever an obstruction lies in the path of wind-driven material. They always have a gently sloping windward side and a steeply sloping leeward side, differing from drifts primarily because of their tendency to migrate downwind. Sand dunes are by no means confined to deserts, past or present, for they may also be found along shorelines of oceans and lakes and on stream floodplains in arid or semi-arid regions.

Dunes are usually composed of sand-sized particles of rocks or minerals. Quartz grains are by far the most abundant dune constituent, but in some localities other minerals or rock debris predominate. The dunes of the White Sands National Monument in New Mexico, for example, are composed almost entirely of the mineral gypsum. Dunes in Hawaii, Italy, and the Azores are composed of black sands derived from volcanoes. The beautiful pink beaches and dunes of the Bahamas are made from eroded coral rock. In the polar regions, dunes are composed of sand-sized particles of ice.

The height of dunes ranges from a few feet to nearly 1000 (1 to 300 meters), while their length may vary from 20 to 30 feet (6 to 9 meters) to many miles. Dunes come in a variety of shapes that are determined by wind speed, wind direction, and the availability of sand. Additional factors include the rate at which sand is supplied from the source and the amount and distribution of vegetation, the criteria used to classify sand dunes.

Dune types. Where sand is scarce and the wind direction is more or less fixed, a linear, nearly symmetric dune forms parallel to the wind direction. This type of dune is called a *longitudinal* dune. Where the wind direction is more or less at right angles to the sand supply, a *transverse* dune will be produced. Common on shorelines of oceans and lakes, this type of dune is often very long but rarely higher than 15 feet (5 meters). Like longitudinal dunes, transverse dunes are linear, but they are perpendicular to the wind.

One of the most beautiful, abundant, and well-known types of dune is the *barchan* or *crescent-shaped* type (Fig. 16-6). Formed where the wind direction is constant and the wind speed is moderate, barchans usually form in groups and may reach heights of 100 feet (30.5 meters) or more. These crescent-shaped dunes generally form in areas whose surface is free from irregularities. *Seif* dunes are similar to barchans except that one of the leeward projecting points of the crescent is poorly formed or missing. This curious type of dune is produced where occasional wind direction shifts occur without a corresponding shift in the source of sand supply. Seif dunes are the largest of all the dune types, often attaining a height of nearly 1000 feet (305 meters) and a length as great as 60 miles (100 km). They are so large, in fact, that they can be plainly seen from satellites orbiting the earth.

A *parabolic* or *blowout* dune is produced where there is a thick vegetative cover, part of which covers the sand. Where there is no vegetation, sand will be removed by deflation, forming a blowout. More and more sand is removed from the blowout and carried to the lee side, while

Fig. 16-6. Barchan dunes tell geologists that wind conditions were constant and moderate during their formation. (Courtesy United States Department of the Interior, Geological Survey, Denver, Colo.; photograph by G. K. Gilbert.)

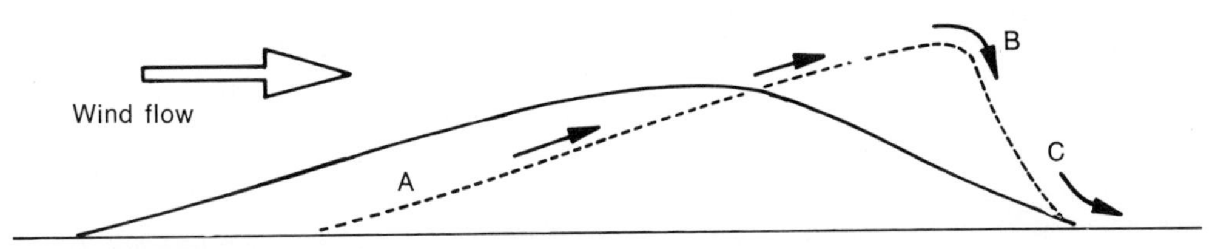

Fig. 16-7. Solid line shows dune's original position. Wind picks up sand from the low end of the dune *(A)* and drops it on top *(B)*, where it falls down the side *(C)*. When all the sand on the hind side of the dune is moved to the front side, the dune has shifted.

the vegetation limits sand movement on either side. The result is a parabolic dune whose wings or points project windward instead of leeward, as is the case of a barchan dune.

Dune migration. Dunes tend to migrate downwind, often moving several inches a day or as much as 100 feet (30 meters) each year. An understanding of the nature of this movement may be obtained by studying Fig. 16-7. Sand moves up the gently sloping windward side of the dune by the transportation process of *saltation*. When the particles reach the crest of the dune, they slump down into the wind shadow formed on the leeward side, thus creating a steep slope, usually 35°. Sand, then, is almost continually moving from the windward side to the leeward side of the dune and as a result, the entire land form migrates downwind until it is blocked by some topographic feature or covered by vegetation that serves to hold the sand in place.

Migrating dunes may bury forests, fields, roads, and even towns. Sand dunes moving over highways are persistent nuisances and require that men and equipment be on duty constantly during the peak migration season unless the dunes can be stabilized. Where sand blows due to natural conditions, people must deal with it as best they can. Where the potential for dune migration exists, but the source has been stabilized naturally by vegetation, care must be taken to ensure that the natural cover is not altered or removed. It is incumbent on city planners, highway departments, subdivision developers, farmers, and the public in general to be aware of the geologic conditions that permit the movement of fine sediments by the wind.

PEOPLE IN DRY LANDS

As populations have expanded, people have moved into the arid and semi-arid regions of the earth, a trend that will no doubt continue. Cities such as Phoenix, Tucson, and Yuma in Arizona, for example, are precariously perched in such areas—sustained not by rainfall, but by irrigation from springs, wells, and streams. At present, 2% of the arid land on earth is irrigated, and as the population increases, so will the percentages of irrigated land. When people intervene with nature, problems arise. Irrigation, an apparent blessing, may eventually create conditions that will lead to wind erosion at a faster rate than before.

This apparent anomaly exists because of the very nature of dry lands. As was pointed out earlier, deserts and other dry regions have extremely high evaporation rates. The introduction of additional water other than rainfall concentrates mineral deposits, loosely termed salts, in the soil. The effects of salt buildup is slow, leading us to believe we are successful in turning arid lands into viable land. Imperial Valley, California, a rich truck-farming region reclaimed from the desert through irrigation early in this century, is a good example. Conservative estimates place the remaining useful life of the Imperial Valley at about 20 years before it will be necessary to allow the land to return to the desert from which it came.

We are enjoying a degree of success in our battle to retard dune migration and soil loss due to the action of the wind. Success is due primarily to an understanding of the processes involved and the lessons learned from past mistakes. We construct dune fences to slow down the migration of dunes, in effect forming drifts. We plant tree lines around our property and beside our highways to reduce the force of the wind. We plant grass and other vegetation on dunes, drifts, and sheets in an attempt to hold them in place. In the short run we are usually successful, but in the long run we will have to deal with mixed results, for wind will continue to perform its geologic tasks with complete disregard for us and our property.

17 *changes from the depths*

Metamorphism

Yea, all fair forms, and sounds, and lights,
And warmths, and mysteries, and mights
Of Nature's utmost depths and heights . . .

Sidney Lanier

In the eastern United States the Blue Ridge Mountains extend from southern Pennsylvania to northern Georgia. These mountains are part of the 230 million-year-old mountain system called the Appalachians. The Blue Ridge is actually a series of ridges trending northeast to southwest. Just east of the Blue Ridge, extending from southern New York to Alabama, is the Piedmont, a low, rolling hill country that is also part of the Appalachians. Much of the Blue Ridge and Piedmont region is composed of rocks that were once buried as deeply as 10 miles below the surface. More than 200 million years of uplift and erosion have brought them into view.

The rocks that make up this region are neither the flat-lying sedimentary rocks nor igneous rock. Except for some igneous intrusions, the area is composed mainly of a third type of rock known as *metamorphic* rock. These rocks probably formed in a geologic structure called a *geosyncline,* which is a subsiding, sediment-filled trough in the earth's crust. As thousands of feet of sediments accumulated in such a trough, the weight of the load increases the pressure and therefore the temperature of the deeply buried sediments and sedimentary rocks, changing their physical and chemical characteristics to form new types of rocks.

METAMORPHIC AGENTS

All processes that change existing rocks into entirely new rocks are known as *metamorphism.* These processes include pressure, heat, and chemically active fluids.

There are at least two distinct kinds of pressure involved in metamorphic changes. The first type could be called *static,* or *weight, pressure,* for as sediments accumulate, they exert enormous pressure due to their weight alone. Weight pressure is most effective in a very deep structure such as a geosyncline, but it is undoubtedly a factor in almost all parts of the earth's crust.

A second form of metamorphic pressure is called *stress,* or *dynamic, pressure.* This involves the application of a directed force that results in the folding, crumpling, and crushing of rocks. Rocks that have undergone stress pressure are almost always found in mountainous areas.

Metamorphic rocks show evidence of formation at temperatures between 300° and 1500° F (148° to 815° C). The heat that brings about metamorphic alteration may be produced by both weight and stress pressure or by contact with magma. Because of their close relationship, it is difficult to say with certainty whether pressure or heat came first, but many geologists believe that heat plays the key role in metamorphism. Regardless, both are almost always involved in metamorphic processes.

Chemical fluids may also produce metamorphic alterations. These fluids are usually water and gases that have escaped from magmas, but they may also be residual fluids already present in the pores or spaces in the existing rock. These hydrothermal solutions transport ions into the country rock to produce chemical reactions that form new minerals or alter or enlarge existing minerals by the process called *metasomatism.*

METAMORPHIC TYPES

Each of the agents of metamorphism acts in a distinct metamorphic environment. These environments, which can overlap, produce *regional, dynamic,* and *contact* types of metamorphism.

Regional metamorphism

Regional metamorphism occurs over an extensive area in the earth's crust, often involving thousands of square miles of land to depths of thousands of feet. This type of metamorphism is probably produced through the deep burial of sediments and rocks in geosynclines; regionally metamorphosed rocks may represent the roots of ancient mountain ranges. The Piedmont of the eastern United States was probably formed by regional metamorphism and could represent the core of the Appalachian geosyncline. Under a weight pressure as high as 45,000 pounds/square inch (psi) and a temperature as great as 1500° F, the preexisting rock would react by developing many minerals not usually found there in other circumstances, although there would be little change in bulk composition. Unique metamorphic minerals produced in such an environment include chlorite, epidote, staurolite, and the minerals of the garnet group. In addition, most but not all of the metamorphic rocks formed in this environment show a layered or banded alignment of the constituent mineral crystals. This layering is called *foliation* (Fig. 17-1).

Dynamic metamorphism

Dynamic metamorphism results primarily from stress pressure exerted during folding and faulting of the earth's crust. Like rocks of regional metamorphic origin, rocks of dynamic origin are almost always found in mountain areas of geosynclinal formation; the two types of rocks are different merely because they are formed at different stages in the development of the geosyncline. These rocks are twisted, bent, and wrinkled although, like the rocks produced through regional metamorphism, they almost always show foliation of their mineral crystals.

Contact metamorphism

When magma intrudes into country rock, the vast quantities of heat, water, and gases released may alter the existing rock, creating a new rock of metamorphic origin. The magmatic heat bakes the country rock, while the hydrothermal solutions bring about metasomatic changes. Since rocks are generally poor conductors of heat, the metamorphic alterations brought about by baking will usually not extend more than a few feet from the contact zone. Hydrothermal fluids, on the other hand, may extend far beyond the area of magmatic contact. The zone of metamorphism surrounding intrusive igneous bodies, especially batholiths and stocks, is called an *aureole,* meaning ''halo.''

METAMORPHIC ENVIRONMENTS

The three types of metamorphism, regional, dynamic, and contact, may be thought of as three distinct but often overlapping metamorphic environments. In these environments, country rock may undergo changes in structure, composition, texture, or any combination of the three. For example, sometimes only one change, such as structural alteration, is apparent. This is the case with the low-grade metamorphic rock slate, which has developed such a fine crystal alignment that it shows a unique breakage characteristic called rock cleavage (Fig. 17-2). In other cases, two changes are visible, as in the structural and textural alterations that occur when limestones and dolostones are con-

Fig. 17-1. Gneiss, a common metamorphic rock, shows foliated texture. Note the different-colored bands; these indicate different crystals grouping together to form each band.

Fig. 17-2. When a metamorphic rock breaks off in flat, even, bands it exhibits cleavage. This sample is slate, the metamorphosed form of the sedimentary rock shale.

verted to marble. Sometimes a complete change in composition, structure, and texture occurs, as in the rock schist, which is completely different from the original rock.

Sedimentary rocks are more frequently metamorphosed than igneous or older metamorphic rocks, simply because igneous and metamorphic rocks originally were formed in and became adjusted to the high temperatures and pressures that exist underground. Furthermore, neither igneous nor metamorphic rocks lie in layers or contain pore spaces, as do sedimentary rocks. Such planes and pore spaces serve as pathways for hydrothermal solutions. The great crystalline masses in which igneous and metamorphic rocks naturally occur are not readily affected by dynamic forces. As a result, these rocks are not as susceptible to metamorphic changes. Sedimentary rocks, on the other hand, were formed and became stable in surface or near-surface environments of relatively low pressure and temperature. When these rocks find themselves in the rigorous conditions of the metamorphic environment, alterations of otherwise stable constituents are inevitable. Due to the varying intensities in metamorphic environments, however, changes are not always complete. A metamorphic rock that retains recognizable sedimentary characteristics is termed a *metasediment*.

While metamorphism produces some minerals common to all rock types, it can also form entirely new minerals that are considered to be almost exclusively metamorphic. The minerals of the garnet group and talc, graphite, serpentine, kyanite, and staurolite, to mention but a few, are found almost exclusively in the metamorphic environment.

Many of the minerals and most of the structures found in metamorphic rocks indicate the grade or intensity of metamorphism that has occurred. Since the metamorphic environment encompasses a great range of conditions between the sedimentary and igneous environments, a bewildering variation is possible. This makes metamorphic rocks in many ways the most complex of the three rock types. To show the grade sequence, we can trace the possible development of the sedimentary rock shale, which is often found in the metamorphic environment. Under moderate pressure and temperature conditions in the regional or dynamic zones, the constituent clay minerals of the shale begin to recrystalize to form microscopic crystals of mica and chlorite. These align in layers perpendicular to the applied pressure. This process forms the low-grade metamorphic rock *slate,* which exhibits the best foliation and rock cleavage of all metamorphic rocks. Under conditions of deeper burial or greater dynamic stress, the slate may change into the medium-grade metamorphic rock phyllite, in which the crystals are larger and more distorted, resulting in less pronounced rock cleavage. If the intensity of metamorphic conditions increases even further, crystals continue to enlarge and new minerals start to form. When the crystals are large enough to be seen individually with the unaided eye, the high-grade metamorphic rock *schist* is produced.

Although the metamorphic environment has a less direct effect on people than the igneous and sedimentary environments, it nevertheless plays a key role in the development of the earth. Without it, the rock cycle would have been short circuited, leaving only sedimentary and igneous rocks and their associated minerals. Without metamorphism many useful minerals would not exist. Among these are graphite, used in pencil lead, lubricants, and crucibles; corundum, including the gem varities ruby and sapphire; scheelite, an important ore of tungsten; pyrophyllite, used as a carrier for insecticide dust; talc, used in talcum powder and sink and table tops; serpentine, used as flame-proofing material; mica, invaluable as an electric insulator, a lubricant, and as Christmas tree ''snow''; the garnets used as gems and abrasives; and andalusite and kyanite, useful in sparkplugs and other porcelains requiring high levels of heat resistance.

METAMORPHIC ROCKS

As we have seen, metamorphic rocks are formed by alterations or changes in existing solid rocks. (If these rocks had melted altogether they would have cooled to become igneous rocks.) Metamorphic rocks are classified primarily on the basis of texture and composition of the country rock and the type and degree of metamorphism.

As pointed out earlier, igneous rocks are classified on the basis of composition and texture, while sedimentary rocks are categorized on the basis of texture and origin. Texture is also perhaps the best criterion for the classification of metamorphic rocks. It will be recalled that texture in igneous rocks refers to the size of the constituent mineral crystals. In sedimentary rocks, texture refers to the size of the sediment particles that made up the rock. In metamorphic rocks, however, *texture* refers to the alignment of mineral crystals.

Metamorphic rocks either show an alignment of their mineral crystals, obvious in gross form or thin section, or they do not. Those rocks in which crystals are aligned are said to exhibit a *foliated texture,* while those lacking this property are said to have an *unfoliated* or *massive* texture. Foliated texture is by far the more abundant of the two types. Some basic metamorphic rock characteristics are summarized in Table 17.

Table 17. Metamorphic rock characteristics

Rock	Environment of formation	Texture	Composition	Rock derived from
Slate	Dynamic or regional	Foliated	Clay minerals	Shale
Phyllite	Dynamic or regional	Foliated	Mica	Shale, fine-grained igneous rocks, or slate
Schist	Dynamic, regional, or contact	Foliated	One mineral, usually mica	Shale, fine-grained igneous rocks, slate, or phyllite
Gneiss	Regional, dynamic, or contact	Foliated	Quartz, feldspar, or mica	Coarse-grained sedimentary and igneous rocks
Quartzite	Regional	Massive	Quartz	Sandstone, conglomerate, or breccia
Marble	Regional or contact	Massive	Calcite or dolomite	Limestone or dolostone
Hornfels	Contact	Massive	Shale, limestone, sandstone, or basalt	Shale, limestone, sandstone, or basalt

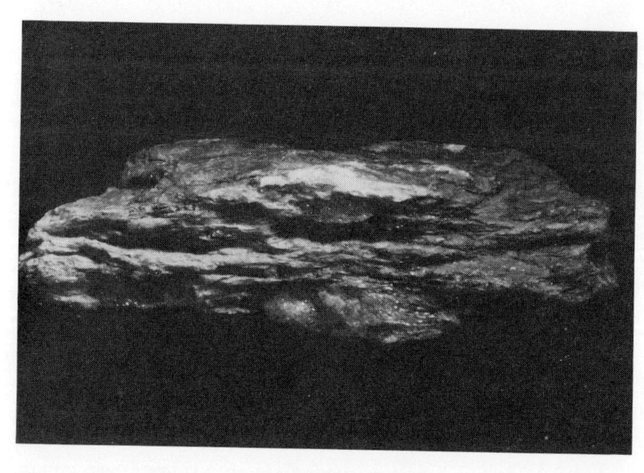

Fig. 17-3. Schist is a high-grade, foliated metamorphic rock. This side view shows the crumpling of the crystals in the intense heat and pressure conditions of the metamorphic environment.

Foliated texture

Slate is a foliated metamorphic rock that has been transformed in the dynamic or regional environment from the sedimentary rock shale. The individual mineral crystals in slate are microscopic and aligned in distinct layers, producing almost perfect rock cleavage. Mica and chlorite are the most abundant minerals in this rock, which may occur in a variety of colors from blue-gray to green, red, or brown. Slate has found use as a roofing and flooring material, in blackboards, as flagstones for walkways, and in ornamental work. Also the best and most expensive billiard tables have slate beds that are guaranteed to withstand variations in temperature and humidity without warping. However, slate is being replaced for most of these purposes by less expensive, often artificial materials.

Phyllite is a foliated metamorphic rock formed in the dynamic and regional environments from the metamorphism of shale, slate, or lava. It contains more and larger mica crystals, which give this rock its characteristic sheen. Phyllite often has a wavy surface appearance indicative of the intense pressure that has been exerted on it during formation. It is very similar to slate in color but generally lacks well-developed cleavage.

Schist is a foliated metamorphic rock (Fig. 17-3). It represents the final product of the heat and pressure alteration of fine-grained sedimentary and igneous rocks or the further metamorphism of existing metamorphic rocks such as slate or phyllite. Schist, which forms commonly in the dynamic and regional and more rarely in the contact environment, is made up chiefly of one mineral, usually mica. However, other minerals may be dominant, producing several varieties of schist, each named after the predominant mineral. In addition to mica schist, there are talc schist, graphite schist, chlorite schist, and hornblende schist. Sometimes several mineral cyrstals share dominance, producing such rocks as garnet-mica schist or staurolite-kyanite schist. When the rock is produced from the intrusion of igneous material it is called *migmatite*. Schist is foliated, but the intensity of its formative environment has commonly produced rock that is severely crumpled and bent.

The metamorphic rock *gneiss* forms mainly in the regional environment, but sometimes occurs in the dynamic and contact environments as well. Gneiss is produced by the alteration of coarse-grained sedimentary and igneous rocks and shows a foliated texture in the form of banding. The crystals are so large and interlocked that this rock shows no cleavage. Its ancestors may be sandy shales, shaly

Fig. 17-4. Marble is the product of the exposure of limestone or dolostone to the metamorphic environment. Its lovely crystalline structure makes it a prized building stone.

sandstones, granites, and other coarse-grained rocks. Subtypes of gneiss are named for their most conspicuous mineral or origin; muscovite gneiss, biotite gneiss, hornblende gneiss, granite gneiss, and syenite gneiss are examples. An *injection gneiss* is formed in the contact zone from the permeation of igneous material and, like the schist mentioned above, this gneiss is also known as *migmatite*.

Unfoliated or massive texture

The unfoliated metamorphic rock *hornfels* forms in the contact metamorphic aureole surrounding igneous intrusions. This compact, fine-grained black rock is produced at temperatures in excess of 1500° F (815° C), well above temperatures found in the dynamic or regional environment. Hornfels is made from the alteration of shale, limestones, sandstones, or basalt and looks very much like basalt.

Marble is a massive metamorphic rock formed in the regional or contact environment from the alteration of limestones or dolostones (Fig. 17-4). It differs from the original rock in that it has larger and more densely packed crystals. Marbles are white in pure form but may be tinted by iron oxide, carbon, serpentine, or other impurities in the original sedimentary rock to form attractive shades of yellow, pink, brown, black, or green. Especially prized are the black marbles of Belgium and the white marbles of Italy. In the United States, fine marble deposits are found in Vermont, Tennessee, and Georgia. Even though marble is more dense and thus more resistant to weathering than its ancestral limestones and dolostones, it will not hold up well in some warm, moist climates. Marble has been used for centuries by builders and sculptors.

When sandstone is regionally metamorphosed, the massive metamorphic rock *quartzite* is formed. Since quartz is very stable even in the metamorphic environment, the only change that occurs in quartzite is that the quartz crystals become larger and more firmly welded together as the original lithifying material recrystalizes. The result is a very hard, resistant rock that stands up well to weathering in all climates. Pure quartzite is white, but impurities in the original country rock may produce varieties that are gray, reddish, yellow, or brown. Quartzite, like marble, is an excellent building material.

GRANITIZATION

As pointed out in Chapter 11, it is possible for granite, which is almost always classified as an igneous rock, to be formed in the metamorphic environment. In the process, which is called granitization, it is believed that ions are carried into the country rock by hydrothermal solutions, gradually building up elements such as potassium and sodium that are characteristic of granite while removing elements such as calcium, magnesium, and iron.

Certain sedimentary formations tend to support this theory about granite formation. In these localities, sedimentary rocks that were originally deposited in continuous layers grade into schists and migmatites, which in turn grade into rocks that have the composition of granite but that exhibit a crude type of foliation. These rocks finally grade into pure granite. Supporters of granitization say that the grading represents intermediate steps in the metasomatic alteration of existing rocks to granite.

Just so that you will not think that the issue of granitization is settled, a few opposing points are given for the sake of argument.

1. Supporters of the concept of granitization say that batholiths and other large intrusive plutonic bodies do not show the displacement of country rock necessary to prove that granites are intrusive and thus igneous. One could counter with the argument that the intrusive process of *stoping* (melting and consuming the country rock) does not imply large-scale displacement or, indeed, any displacement at all.
2. Supporters say that granites often merge into rocks that are definitely of metamorphic origin, which in turn often grade into unaltered sedimentary rocks. It could also be pointed out that the enormous heat from the cooling magma could have changed the country rock through contact metamorphism and metasomatism and that grading represents varying degrees of metamorphism rather than degrees of granitization.

Fig. 17-5. Xenoliths like the small black band in this photograph are often found in granites and add spice to the metamorphic-igneous debate regarding the origin of granite. Inset is a close-up view of the xenolith. (Courtesy North Carolina Granite Corporation, Mt. Airy, N.C.)

3. Supporters say that small granitic masses are often formed apart from any known magma source. One could argue that there is no reason to believe that granitic masses have to be large and that the small mass could have been its own magma source.
4. Supporters say that the lack of *xenoliths** gives definite proof that granite is metamorphic (Fig. 17-5). Where xenoliths are found, supporters say that all of the original country rock was altered to granite except these small pieces. One could disagree and say that the lack of these fragments represents complete melting of the original rock and nothing more, and that when they are found they represent unmelted rock in the ancestral magma.

The presence of xenoliths (inset in Fig. 17-5) represents one of the best arguments in favor of the magmatic origin of granite. So convincing are the arguments of the magmatists that a check today would probably find nine out of every ten geologists in agreement with the concept of the predominant magmatic origin of granite. These nine would probably say that granitization is possible in perhaps 15% of all granites. The granitizationists, on the other hand, would say that as much as 85% of all granite is of metamorphic origin. Until definite proof is found one way or the other, the battle will rage unabated, and classification of granite will be a matter of individual preference.

*Xenoliths are small fragments of unmelted metamorphic or other rock types that are found embedded in some granites.

18 *the restless earth*

Diastrophism

Moving of the earth brings harms and fears,
Men reckon what it did, and meant . . .

A Valediction Forbidding Mourning
John Donne

Good Friday, 1964, was a memorable day for residents of Alaska and the Pacific coastal regions, for at 5:36 P.M. the most severe earthquake ever recorded in North America rocked the area with a force estimated to be equal to that of 12,000 Hiroshima-sized atomic bombs (Fig. 18-1).

This disturbance, which registered 8.5 on the Richter scale of earthquake magnitude (Table 18), was felt over an area of 500,000 square miles. It left 115 people dead, 4500 homeless, and caused $750 million in property damage—more than 100 times what it cost to buy Alaska from Russia in 1867. Ernest Gruening, United States Senator from Alaska, declared that his state had suffered a disaster surpassing in magnitude any that had ever occurred in the entire nation's history. Few would disagree with his statement.

THE NATURE OF DIASTROPHISM

The earth is often referred to as terra firma (firm earth), a term that implies an unchanging, static planet. Nothing could be further from the truth, for the earth is not only undergoing changes from without, but there is much evidence to indicate that the earth has experienced and is still undergoing changes from within. Earthquakes, for example, are a result rather than a cause of these changes. They are indications of the vibrant nature of the earth. Although earthquakes cause great devastation, they are actually a constructional rather than a destructional force in the geologic sense. They are only part of a much larger system of events called *diatrophism,* the movement of the earth's solid rock crust.

That diastrophism occurs is beyond refute, but its exact cause is presently unknown. It is obvious that great internal forces are at work in the earth, causing rocks to bend, break, melt, flow, and drift. These great forces are often so violent that rocks change their very substance, forming completely new ones. Not only can diastrophism produce metamorphism, it can also produce the conditions that allow solid rocks to melt deep in the earth, bringing into existence the igneous environment.

Before any final conclusions can be reached about the exact nature of diatrophism, one basic question must be answered. How did the earth form? Did it begin in a hot, gaseous state and then cool to a liquid and then to the basic solid we know today? Did it begin as a cold body through the accumulation of solid particles swept out of space and then increase in temperature? These may not be the only two possibilities, but they certainly seem to be the two extremes. All we can say today is that most evidence seems to favor the ''hot earth'' theory. The bulk of the planet is composed of igneous rocks, those that cooled from liquid form. There can be no question that the earth is hot inside and that this heat is escaping from the interior. These great heat or convection currents moving upward through the earth's viscous mantle could be the cause of crustal movements.

TYPES OF DIASTROPHISM

Historically, two types of diastrophism have been identified, epeirogeny and orogeny. *Epeirogeny* is the gentle upward or downward movement of large areas of the crust. These movements may affect an entire continent at one time. *Orogeny,* on the other hand, is much more localized, affecting long, narrow zones of the crust. In such zones rocks are tortuously bent, broken, and deformed in con-

Fig. 18-1. Cleanup crew clears debris from the Los Angeles Veterans' Hospital after the 1971 earthquake that shook the Pacific Coast. (Courtesy United States Department of the Interior, Geological Survey, Denver, Colo.)

Table 18. Richter scale of earthquake magnitude

Magnitude	General classification	Expected results in populated areas	Average number of annual occurrences
8 and above	National disaster	Destruction virtually complete; people dazed	1
7 to 8	Major earthquake	Major damage to all buildings; cracks in ground	10
6 to 7	Destructive earthquake	Some buildings collapse	100
5 to 6	Damaging earthquake	Heavy furniture moves; some damage to poorly constructed buildings	1000
4 to 5	Strongly felt earthquake	Sleep disturbed; fragile objects (glass, plaster, trees) broken	10,000
3 to 4	Small earthquake	Some breakage of fragile objects (glass, plaster)	100,000
0 to 3	Tremor	Just able to be felt; may be some tinkling of glass	1,000,000

trast to the action in epeirogenic areas, in which there is virtually no change in the rocks themselves.

Today it is increasingly apparent that there is a third type of diastrophism that is worldwide in extent and may itself be the cause, at least in part, of epeirogenic and orogenic movements. For lack of a better name, we can call this third type *continental drift,* with the full realization that it, like earthquakes, is a result and not a cause. How continental drift is related to the other two types of diastrophism is not clear at this time. However, assuming that epeirogeny and orogeny are related to continental drift, a discussion of this larger concept would seem to be in order before any further attempts are made to understand the smaller movements of the earth's crust.

The continents are adrift

In 1912 the German geophysicist Alfred L. Wegener (1880-1930) proposed that the continents had moved across the surface of the earth. His rationale was based more on geography than geology. It seemed to him, as it does to anyone who studies a globe, that the continents should fit together like a gigantic puzzle. The idea of continental drift was not held in very high regard until 1968. Today the theory of continental drift is generally accepted. This is the shortest turn-around period ever experienced by a major theory in the history of science.

The amazing reversal of opinion came about primarily as a result of seven events.

1. The midAtlantic ridge, which proved to be a location where new ocean basin material was coming to the surface, was discovered and studied. Deep drilling in the area also showed that rocks were progressively older with increasing distance on either side of the ridge, thus proving that the crust was pulling apart at this location.
2. Findings in the field of paleomagnetism (the study of the earth's ancient magnet fields as indicated by the alignment of magnetic minerals within the rocks) provided evidence that the continents had changed their positions with respect to the magnetic pole.
3. Similarities among fossils found on widely separated continents indicated that the continents were once much closer together.
4. Intercontinental relationships between ancient rock structures showed a close link in their formation and age.
5. Findings in paleoclimatology (the study of ancient climates) gave evidence that widespread climatic changes such as ice ages had occurred that could not be explained by purely meteorologic processes.
6. Deep drilling indicated that some ocean basins (the Atlantic, for example) were much younger than had been previously believed.
7. Computer reconstruction of the continents at the 600-foot (1830-meter) continental slope level gave an almost perfect fit.

In the 1920s Wegener proposed that all of the present continents were once joined in a single supercontinent, now called *Pangaea*. The concept of Pangaea is not universally accepted, for some scientists prefer a model based on two separate supercontinents, one in the Northern Hemisphere called *Laurasia* and one in the Southern Hemisphere known as *Gondwana*. The most widely accepted theory now stipulates that Pangaea broke apart to produce Laurasia and Gondwana.

The mechanism of continental drift may be explained in theories regarding *plate tectonics* and *sea-floor spreading*. The theory of plate tectonics suggests that the earth's crust has been broken up into 10 major parts, called *plates,* and many smaller plates, perhaps by convection currents rising from the interior. *Sea-floor spreading* could explain how these plates move about once they have been formed. It is believed that crustal plates are colder and more dense (and therefore heavier) on one end than on the other. The heavier end would sink downward into the mantle along certain zones that are represented by deep sea trenches,

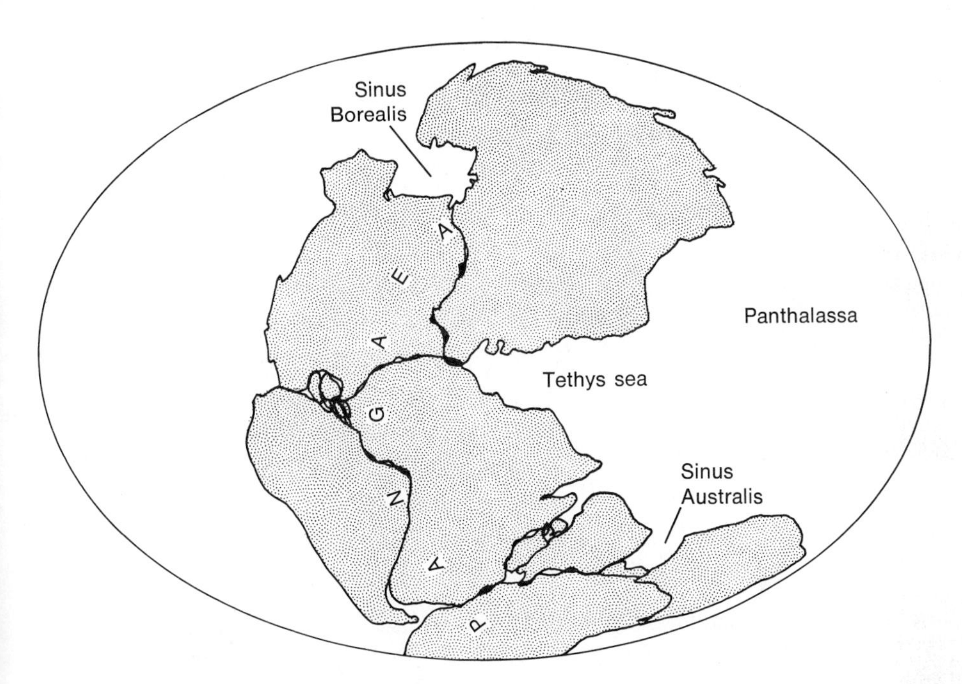

Fig. 18-2. Scientists believe that Pangaea, a universal land mass from which all present continents are derived, would have looked something like this about 200 million years ago. The fit is based on corresponding features at a depth of about 6000 feet.

which are especially common around the edges of the Pacific Ocean. This subsidence would cause the plates to break along great rift zones into which viscous mantle material and molten basalt would flow from below. The upwelling material would then form a new ocean basin as the rift widened.

In a recent popular reconstruction of Pangaea, the supercontinent was an irregularly shaped land mass covering about 40% of the earth's surface (Fig. 18-2). Pangaea was surrounded by the universal ocean known as *Panthalassa,* later to become the Pacific. The great land mass was located approximately where the present day Atlantic Ocean is found. The supercontinent contained several indentations, or arms of Panthalassa, just as our present continents are marked by gulfs, bays and seas. One of these to the east was the *Tethys Sea,* partially separating Eurasia from Africa, today probably represented by the Mediterranean Sea. To the north was *Sinus Borealis,* later to become the Arctic Ocean. To the southeast was *Sinus Australis,* a bay of the Tethys Sea partially separating Australia from India. The Gulf of Mexico could be a remnant of an eastward projecting arm of the universal ocean that is called *Sinus Occidentalis*.

Two extensive rifts were allegedly in Pangaea about 200 million years ago during the Triassic period of geologic time. By the end of this period some 180 million years ago, the ancestral Atlantic and Indian Oceans had formed as this rift widened by seafloor spreading. The rift to the north split Pangaea from west to east, forming Lourasia (composed of North America, Europe, and Asia) and separating it from Gondwana (composed of South America, Africa, India, Australia, and Antarctica). Soon after the formation of the northward rift, a second rift to the south split South America and Africa away from the remainder of Gondwana. India was soon thereafter split away from Antarctica and began its rapid migration to the north.

During the Jurassic period, from 180 to 135 million years ago, the Atlantic and Indian Oceans continued to open, forming the Labrador Straits and the Bay of Biscay, while the Tethys Sea began to close on its eastward end. At the end of the Jurassic period, South America began to break away from Africa to form the South Atlantic. The process was complete by the end of the Cretaceous period 65 million years ago. A rift developed on the eastward side of what would become Greenland and began to break in the direction of the ancestral Arctic Ocean. Africa drifted northward about 10° and rotated counterclockwise. Eurasia rotated clockwise, continuing the movement it shared with North America during the Triassic.

During the Cretaceous period, North and South America

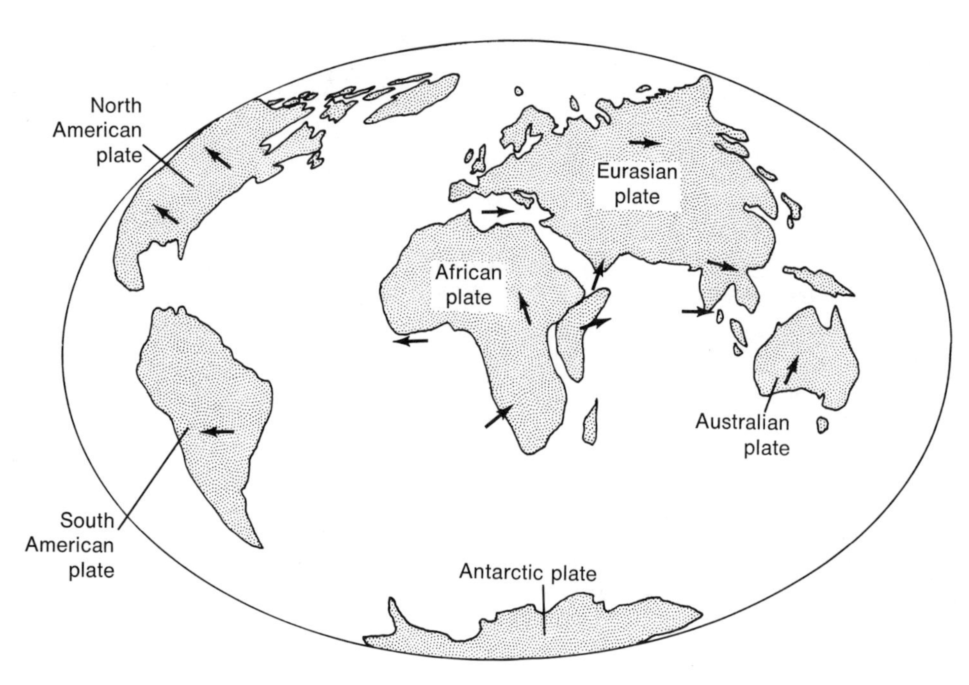

Fig. 18-3. Scientists' speculation of our world's appearance 50 million years in the future. Note the break between North and South America and the wider space between Africa and Europe.

became joined by a narrow land bridge that was formed by volcanic activity. Australia became detached from Antarctica and drifted slowly northward to its present location. India completed the longest journey of any continent by colliding with Eurasia after a drift of more than 6000 miles. As the Indian Plate collided with the Eurasian Plate, one overrode the other, producing the Himalaya Mountains. The extraordinary height of these mountains is thus explained by the fact that they are two continental plates thick.

During the last 70 million years, or the Cenozoic era, final adjustments were made as the continents took the positions they occupy today. Greenland completed its split away from Europe as did the British Isles, while the Caribbean Sea partially closed. The Atlantic continued to widen, halting and reversing the northwestward drift of Eurasia. As the Mediterranean Sea continued to close, the forces exerted by the interaction of the Eurasian and African plates formed the mountain range today known as the Alps.

Looking back over the 200 million years since the breakup of Pangaea, we find that Antarctica moved less than any other continent, remaining near its original location. North America drifted more than 5000 miles (8333 km) toward the northwest. Miami, Florida (26° N, 80° W), for example, had it been in existence 200 million years ago, would have been located in the South Atlantic near Ascension Island (8° S, 14° W). South America drifted slightly northward but to a much greater extent westward. Australia, Africa, and India all drifted northward. Eurasia also showed a slight drift to the north, but clockwise rotation was its predominant motion.

Have the continents completed their drifting tendencies? There is no reason to believe that the continents, like magic, started to drift 200 million years ago, or that they will not continue to drift in the future (Fig. 18-3). Perhaps the continents will drift until they all collide to form a single, irregularly shaped continent and then start to drift again as forces are applied from below. We are only at a very early stage in our understanding of the drift mechanism and how events of the geologic past are related to it. The movement of the continents must represent diastrophism on the grandest of all scales.

EVIDENCE FOR DIASTROPHISM

Diastrophism, like most geologic processes, is usually a very slow process. It is estimated, for example, that North America is moving westward at a rate of one body length in a lifetime. Millions of years are required for the uplift of mountains. Sections of earth occasionally move a few feet in less than a minute along rift zones, producing strong earthquakes, although most movements are slow. Even so, much evidence proves the existence of diastrophism.

The very existence of land above sea level today is evidence of diastrophism. It can be shown that enough time has passed for all land on earth to have been worn down to sea level by the destructional forces of weathering, mass movement, and erosion. Only the work of constructional forces such as diastrophism and igneous activity could offset the ravages of geologic time.

Epeirogenic evidence

Epeirogeny is the deformation or changes in the earth's crust that produce general features of relief or surface characteristics. For example, flat-lying sedimentary rocks containing marine fossils are found high above sea level in many parts of the world. Clearly, the creatures represented by these fossils once lived in the ocean, died, and were buried and fossilized in the sediments of a long-lost sea. Then, over millions of years, the entire area was slowly uplifted by forces from within the earth. The Grand Canyon of the Colorado River is a classic example of a region affected by epeirogenic diastrophism.

Geologists in most parts of the world have undertaken to accurately measure the altitude of the land above sea level. In the United States this is the responsibility of the United States Coast and Geodetic Survey. Wherever such measurements are made, a brass disc called a bench mark is set into the ground; on it is recorded the original measured altitude. Later measurements taken at the same bench mark will show any change in the altitude of the land. Such studies are quite revealing, for measurements indicate an increase in altitude while others show a decrease. In the Los Angeles, California area, for example, the San Fernando Valley is sinking at the rate of ½ inch (1 cm)/yr, while the San Gabriel Mountains are rising at a rate of ¼ inch (50 mm)/yr. Parts of the United States Gulf Coast are sinking at rates as great as ¼ inch (50 mm)/yr. In the Northern Baltic Sea region, uplift is occurring as rapidly as 3 to 4 feet (1 meter)/100 yr. The changing altitude of bench marks around the world indicates that epeirogenic movements are a continuous process.

Orogenic evidence

Orogeny is the folding and warping of the earth's crust, especially in the creation of mountains. Again, there is also much evidence to indicate that such great forces within the earth can deform rock structures over extensive areas. This evidence includes the presence of *folds, fractures,* and *faults,* which are almost always found in association with past or present mountain ranges.

Folds

Solid rocks in the earth's crust are often found bent but not broken. Such wrinkles, called folds, are the result of horizontal pressure that shortens the crust. Folds may vary in length from a few inches to several miles. It is estimated, for example, that if the folded rocks of the Appalachian Mountains could be straightened out, the width of the range would increase by at least 40 miles. In the Himalayas and Alps, the increase in width would amount to several hundred miles.

Upfolds in the rock are called *anticlines,* while corresponding downfolds are known as *synclines* (Fig. 18-4). Where rock strata are bent in only one direction, upward or downward, a *monocline* is formed. When both sides or limbs of a fold slant in the same direction, an *overturned* fold is produced. This type of formation indicates great pressure from only one direction. Upfolds and downfolds that are about as long as they are wide are called *domes* and *basins,* respectively.

Fractures

A fracture, or *joint,* is a break in the rock along which neither part of the rock has moved. Fractures are one of the most common features found in all rocks. Most fractures, however, are not produced by diastrophism. Those not produced in this way are called *primary* fractures, for they form simultaneously with the rocks themselves as a result of contraction (shrinkage). For example, shallow igneous intrusions such as dikes, sills, volcanic necks, and lava flows often exhibit a type of fracturing called columnar jointing. As the molten rock cools, it contracts slightly, breaking the mass into vertical columns. The Palisades Sill of the Hudson River, the Giant's Causeway of Northern Ireland, the Devil's Postpile in California, and Devil's Tower, Wyoming, show columnar jointing (Fig. 11-6). Primary fractures are also common in sedimentary rocks, for as water-saturated sediments dry out to become solid rock, some shrinkage occurs.

Secondary fractures are those that were produced after the rocks were formed. The presence of such fractures in the crust indicates that the rocks have been subjected to forces great enough to break them without producing relative motion. Fractures associated with diastrophism may be formed as a result of *torsion* (twisting), *tension* (stretching), or *compression* (squeezing) and are almost always restricted to orogenic belts.

Faults

A fault is a break in the rocks along which the rocks on one or both sides have moved. Only relative movement can be measured in most cases, for faults involve great blocks of rock that may extend underground for many miles. It is virtually impossible to tell whether one block remained stationary while the other moved or whether both blocks moved at the same time.

The zone of displacement is called the *fault plane.* The plane is composed of two directions: the *horizontal,* referred to as the *strike* and the vertical, known as the *dip,* which is always measured at right angles to the strike. The amount

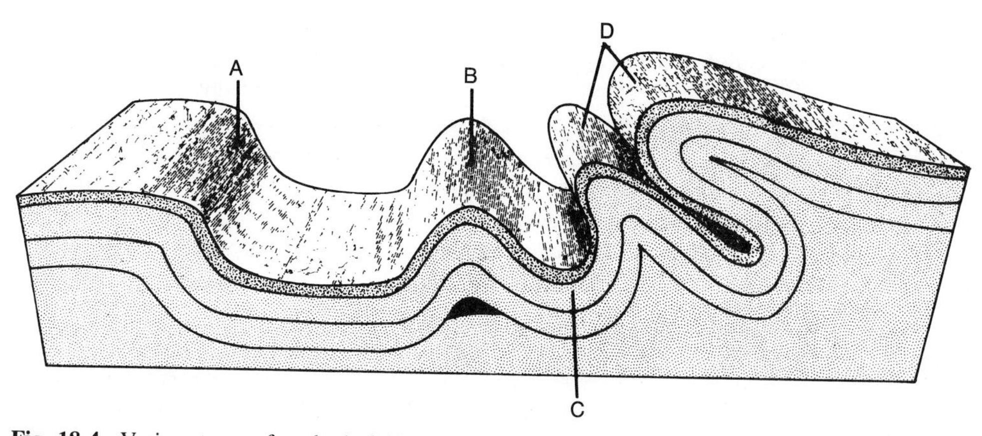

Fig. 18-4. Various types of geologic folds. A monocline *(A)* is bent in one direction only (in this case, down). An anticline *(B)* is a fold in which both sides are bent and the surface is forced up. These often act as traps for oil and natural gas. A fold that is pulled down is a syncline *(C),* and more intense pressure produces overturned folds *(D).*

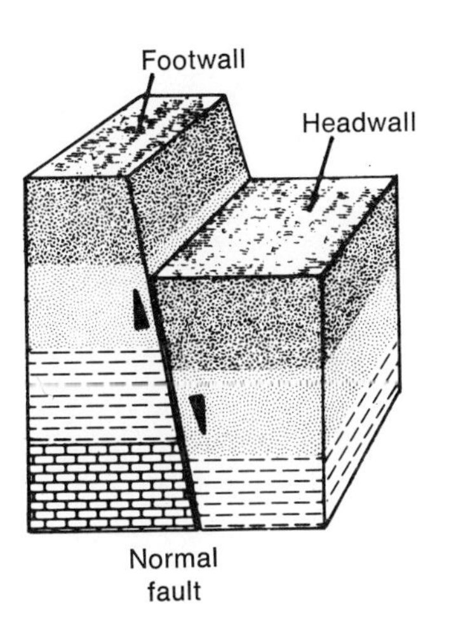

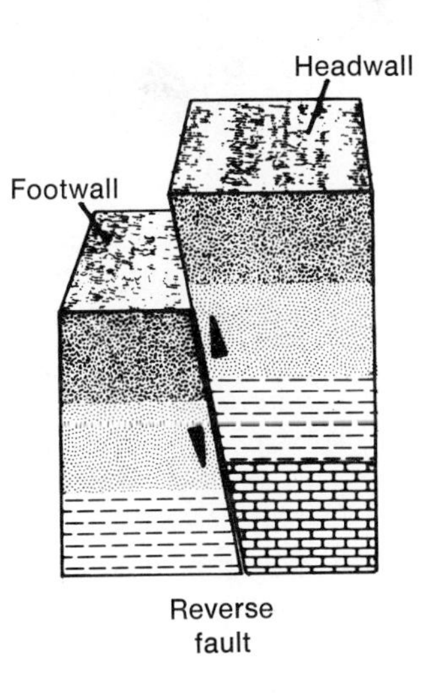

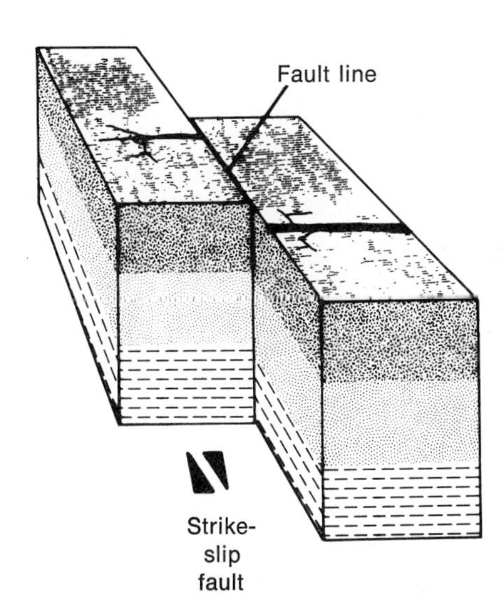

Fig. 18-5. Relationships between the two walls of the three types of faults, also showing the directions meant by strike and dip.

Fig. 18-6. The San Andreas fault in California is the reason for the recurrent rumors that the Golden State will someday break off and slip into the sea! (Courtesy United States Department of the Interior, Geological Survey, Denver, Colo.; photograph by J. R. Balsley.)

of movement along the fault plane is the *displacement* of the fault.

Much of the fault terminology in use today comes from mining operations. Faults are frequently the sites of valuable mineral deposits, for these breaks are paths of weakness into which mineral-rich magma or groundwater may have flowed. Consequently, many mine shafts have been dug downward along the natural slopes of faults. Miners walking up and down the sloping tunnel would have one fault block over their heads, while the other block would be below their feet. This experience led to the application of the term *headwall* to the fault block above the break, while the block below the break became known as the *footwall*.

Faults that show movement confined to the dip (only vertical movement) may be classified on the basis of relative motion of the headwall. For example, if the headwall moves down relative to the footwall, a *normal* fault is formed. If the headwall has moved up relative to the footwall, the result is a *reverse* fault. Faults produced by movement only along the strike are known as *strike-slip* faults (Fig. 18-5). The fault plane of such a fault often intersects the surface, where it is called a *fault line*.

San Andreas fault line. The most famous and closely studied strike-slip fault in the world is the San Andreas fault line in California (Fig. 18-6). This famous fault line extends for more than 500 miles from the Gulf of California to Point Reyes, north of San Francisco, and then along the coast and under the ocean for another 100 miles to the vicinity of Cape Mendocino. Accurate surveys show that the crust on either side of the fault line is still bending slowly, with relative movement amounting to as much as 2 inches (5 cm)/yr in some places. The total horizontal displacement since the end of the Jurassic period, 135 million years ago, has been more than 350 miles (583 km).

Movement along the fault line in 1906 caused the disastrous earthquake and resultant fire that destroyed much of San Francisco. During this earthquake, land west of the fault moved northward by as much as 21 feet (7 meters) relative to the land on the other side. This northward motion was traced by measuring the displacement of highways, fences, and other structures that crossed the fault line. Similarly, the total movement along the fault since its creation was measured by noting the displacement of large-scale features such as mountain ranges.

RESULTS OF DIASTROPHISM

Continental drift and epeirogenic and orogenic movements of the crust have left clear evidence of their work. Now we can examine two other results of diastrophism, mountains and earthquakes. They may also be listed as evidence for diastrophism.

Mountains

A geographer defines a mountain as any part of the land that projects above its surroundings. Geologists, on the other hand, are more interested in how a mountain came into being, and would define it as a region of disturbed or deformed rocks. They therefore see mountains as part of geologic activity. Geologists classify mountains in three broad groups: (1) *volcanic* mountains, (2) *erosional* mountains, and (3) *diastrophic* mountains.

Volcanic mountains

Volcanic mountains are produced in the extrusive igneous environment by the extrusion of lava through a rounded opening called a vent and are composed of layers of volcanic rock and debris. (Their origin and structure are discussed in detail in Chapter 11.) Although volcanic mountains are not directly a result of diatrophism, many geologists believe that the original molten rock (magma) that extrudes to form the volcano was produced by movements within the crust.

Erosional mountains

Erosional mountains are not of major concern to us now because they are formed in ways not directly related to diastrophism. As their name implies, erosional mountains are areas of high relief simply because agents of erosion have lowered the surrounding, less resistant material. The Catskill Mountains of New York are of this type, as are the buttes and mesas found in arid climates.

Diastrophic mountains

There are three types of mountains that result directly from diastrophism: (1) fault-block mountains, (2) domed mountains, and (3) folded mountains. *Fault-block* mountains are those produced along fault zones as segments of the crust were forced upward relative to adjacent blocks that moved downward. The Sierra Nevada Mountains in California comprise a single fault block more than 400 miles (667 meters) long. Its eastern edge rises abruptly along the zone of faulting and tilts gently westward beneath the relatively young sedimentary rocks of the San Joaquin Valley. The Grand Teton Mountains of Wyoming are a series of deeply eroded fault blocks (Fig. 18-7).

Domed mountains are formed by the equal compression of rock from all sides, producing an upfold of equal length and width. The compression force is often the intrusion of magma between rock layers below the surface. Domed mountains are always small and rounded and do not extend unbroken for miles. The Zuni Mountains of New Mexico, the Uinta and Henry Mountains of Utah, parts of the Adirondack Mountains of New York, and the Black Hills of South Dakota are all examples of domed mountains.

Folded mountains, the most common type, are much longer than they are wide, often extending unbroken for hundreds and even a thousand miles or more. The rocks in such mountain ranges were deposited as thick layers of sediments in a structure called a *geosyncline,* a subsiding, sediment-filled trough in the earth's crust. The geosyncline continually subsided as sediments accumulated. Eventually, the rocks in the structure were subjected to strong horizontal forces of compression that crumpled them into anticlines, synclines, monoclines, and overturned folds. The folding was often accompanied by faulting, metamorphism, and the intrusion of magma. Most great mountain ranges are of this type. The Appalachians and Rockies of North America, the Andes of South America, the Alps of Europe, and the Himalayas of Asia are all *folded* or *geosynclinal* mountains.

Geosynclines and isostasy. The concepts of the geosyncline and isostasy are two of the most basic theories in geology. Many geologists consider attacks on them almost irreverent. However, these theories are being questioned in light of recent findings on continental drift. Geologists are investigating to see whether folded mountain ranges and other geologic structures could have been formed by forces created during the collision of drifting crustal plates.

The *geosynclinal theory* was developed in 1857 by

Fig. 18-7. The Teton Mountains of Wyoming are classified as fault-block mountains. These impressive peaks are the eroded remnants of great blocks of tilted rocks. (Courtesy Wyoming Travel Council, Cheyenne, Wyo.)

James Hall (1811-1898), a New York paleontologist, who based it on his studies of the Appalachian Mountains. The term ''geosynclinal'' was coined by J. D. Dana (1813-1895) of Yale University several years later. The theory suggests that certain areas of the earth's crust that are weaker than others have a tendency to sink and naturally accumulate sediments eroded from the land. These areas are called geosynclines. The weight of the sediments further increases the rate of subsidence, or sinking. After thousands of feet of sediments have accumulated, their weight exceeds the load limit of the crust. This produces a vertical readjustment of the crust, which slowly thrusts the geosyncline upward into folded mountain ranges.

The theory of *isostasy* is often used to explain this and other vertical movements of the earth's crust. Much of the early work on the theory was done by J. H. Pratt (1826-1893), a British mathematician. The word ''isostasy'' itself was suggested by C. E. Dutton (1841-1912), an American geologist, in 1889. The theory of isostasy suggests that parts of the earth's crust are in a state of equilibrium with adjacent parts. Imbalances in one part bring about readjustments in the adjoining areas. According to this theory, the continental crust is lighter than the more dense oceanic crust. In order for the continental rocks to balance the rocks of the ocean basins, more continental material is needed. This difference could explain why continents are naturally higher than ocean basins. Once the balance is reached, any extra weight added to the continents (such as great masses of ice) would cause them to subside, while weight removed (by melting ice or erosion, for example) would produce uplift. Deposition of sediments in a geosyncline would add extra weight, upsetting the balance and forcing the whole system to readjust. Geologists hypothesize that this ''readjustment'' involves a flowing of lower crustal material to the area of the geosyncline to help support the extra weight. Since there is already rock in the region this extra matter is flowing into, the additional material must force the overlying rocks up to make room for itself. Subsequently some adjustment occurs in the area from which the additional material came; this area, with some of its rock missing, subsides.

There is much evidence to support both the geosynclinal and isostasy theories. The presence of marine sedimentary rocks and fossils in all folded mountains points to an origin below sea level, which supports the geosynclinal theory. Isostasy is supported by the observed fact that regions recently covered with tremendous loads of ice are rising, while areas covered with great masses of ice today are often below sea level due to the weight of the ice. The continued existence of mountain ranges such as the Appalachians after more than 200 million years of erosion also supports isostasy, for if these areas were not rising they would have long since been leveled.

The role that continental drift plays in the uplift of folded mountains is not clear at this time. It appears that plate collision has played a vital role in the formation of the Alps and Himalayas, but what of the other folded mountains such as the Rockies, Andes, and Appalachians? Could geosynclines be subduction zones where heavier crustal plates are sinking into the mantle? Have plates collided in the geologic past to form other folded mountains? Is isostatic uplift a byproduct of plate collision? At this point there are more questions than answers. The coming years will be exciting as some of these questions and others not yet asked are resolved.

Importance of mountains

Mountains have played a major part in the history of civilization. For example, they have been barriers to travel and to national growth and expansions. On the other hand, mountains have often acted as fortresses against invasion. During World Wars I and II, the Alps helped to isolate Switzerland from the conflicts that engulfed the rest of Europe.

Mountains also have an important effect on the weather. They force moisture-laden air to rise, resulting in the release of the moisture as rain or snow as the air cools. The economic importance of mountains is manifold. Their forests provide most of the wood used in lumber, paper, and myriad other products. Mountain streams provide much of the water power that generates electricity. Mountains are the greatest source of minerals such as gold, tin, copper, and lead because the formation of such rich mineral deposits is related to the processes of mountain building. Mountains are also rich in natural beauty, making them popular retreats for vacations and leisure activities.

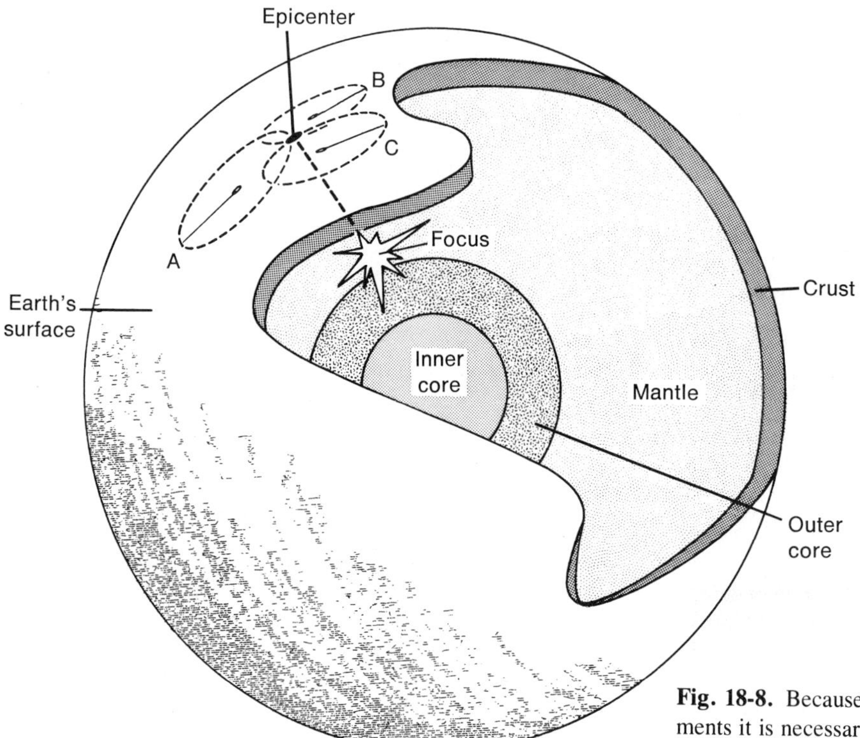

Fig. 18-8. Because of the indirect nature of earthquake measurements it is necessary to measure an earthquake from at least three seismic stations to locate the epicenter. From that measurement the focus, or initial break, can be calculated.

Earthquake!

The most spectacular result of diastrophism and one of the geologic processes we fear most is the earthquake. This is a violent shaking of the ground that is produced as rocks break or move along previously formed breaks. The location of the initial break or movement is called the *focus* of the earthquake (Fig. 18-8). The focus is usually deep within the crust or mantle of the earth. The point on the surface immediately above the focus is called the *epicenter*. It is here that the strongest shocks occur.

At the instant that a break or movement occurs, two types of shock or seismic waves are produced at the focus. These two waves, called *body waves* because they travel within the earth itself, follow the same path but move at different speeds due to differences in wave motion. Although the speeds of the two waves are different, the speed ratio of one to the other remains constant. The faster wave, traveling at 5.5 miles (9.2 km)/sec, is called the *primary* or *P wave*. Its wave motion is a back-and-forth vibration in the direction toward which the wave itself is moving. The slower body wave, known as the *secondary* or *S wave*, travels at a speed of 3 miles (5 km)/sec with a horizontal wave motion at right angles to the direction of propagation.

A third type of seismic wave, known as a *surface wave* because it forms and moves only at the surface, is also produced. This wave is called a *long* or *L wave* because of its tendency to follow the longest possible path. L waves are thought to be modified S waves, for their motion is also at right angles to the direction in which they travel. Long waves have a forward speed of 2.3 miles (4 km)/sec. There is also a second type of surface wave, called a Rayleigh wave after its discoverer Lord Rayleigh (1842-1918). It moves in vertical vibration in the overall wave direction at a speed of 1.6 miles (2.7 km)/sec.

All seismic waves increase their speeds slightly in dense rock and reduce their speeds in less dense rock. S waves do not move through liquids at all. (It was this observation that led to the theory that the outer core of the earth is not solid.)

The destructive effects of earthquakes are the result of the actual displacement of atoms that takes place within rocks and other substances as seismic waves pass through them. A well-known form of wave motion can be used as an example to help demonstrate the displacement process. A water wave has a motion similar to a Rayleigh wave. An energy source, usually the wind, produces a vibration that moves through the water as a wave. The water itself does not move forward, only the wave. An investigation of water waves shows that an individual water particle would transcribe an elliptic, reverse orbit as the wave passed (Fig. 23-2).

The amplitude of a seismic wave is the most critical factor, for it is this wave characteristic that produces the destructive displacement. An amplitude of ½ inch (1 cm) is destructive, and an amplitude of 1 inch (2.5 cm) is catastrophic. L waves, which travel through our surface environment, have the greatest amplitude and do most of the damage.

Study and detection of seismic waves

The branch of geology that deals with the study and detection of earthquake waves is called *seismology*. The instrument used to detect earthquake waves is the *seismograph* (Fig. 18-9, *A*). Seismic detecting stations have been set up around the world to determine when earthquakes occur and to collect data on such factors as their duration, location, and severity. Unfortunately, a seismograph cannot predict when or where a shock will occur, for by the time a seismic record is available the earthquake has already taken place. Earthquake waves take time to move from the focus to a recording station. The only means we have of preventing severe earthquake damage is to avoid areas of the earth that are known to be seismically active.

A single seismic station can record whether an earthquake occurred, when the shock took place, and how strong it was, but it cannot identify the epicenter. Seismographs continuously record the constant small shocks called microseisms. This record is called a *seismogram*. An abrupt change in the amplitude and frequency of the trace heralds the arrival of earthquake waves (Fig. 18-9, *B*). The time can be read directly from the seismogram. P waves, traveling faster than the S waves, arrive first at the seismic station. The slower S waves are recorded a short time later. P and S waves are formed simultaneously at the focus and follow the same path through the body of the earth at a constant speed ratio. Because of this, the distance to the epicenter may be obtained by subtracting the arrival times of the two waves and substituting this time in a distance formula. Once the distance is known, the magnitude of the earthquake may be computed.

Seismographs are not accurately directional. To determine the exact location of the epicenter, at least 3 seismic stations must pool their distance findings. Circles are drawn on maps using as a radius the distance as measured from each station. The point where the three circles intersect on the map is the location of the epicenter (Fig. 18-8).

Richter scale of earthquake magnitude

The *intensity* of an earthquake is the measured amplitude of the seismic trace as recorded at a station somewhere in the world, and represents the earthquake's effect at that particular place only. Intensity is a variable, for the ampli-

A

B

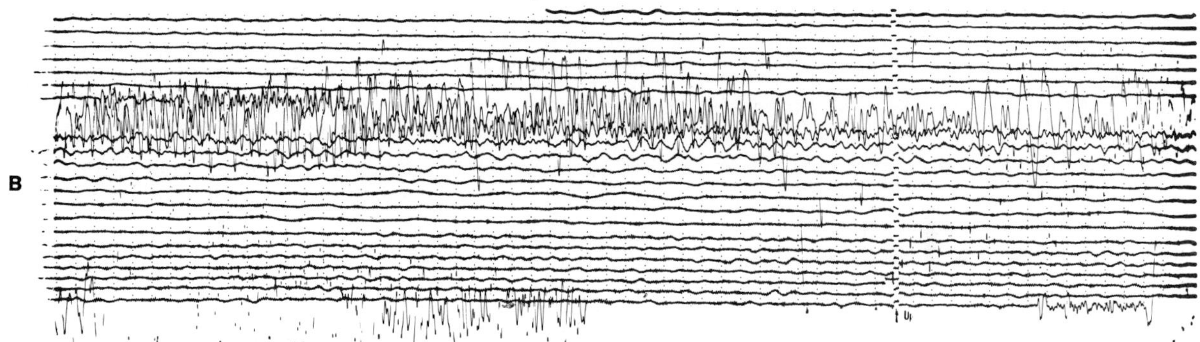

Fig. 18-9. A, Horizontal seismograph. The heavy weight to the left resists movement as the earth, building, and seismograph frame move. An ink trace of this movement is made on the rotating drum just in front of the weight. **B,** Seismogram (or seismograph's reading) of the earthquake in San Fernando, California, on February 9, 1971, which registered 6.6 on the Richter scale. The abrupt change in the pattern of the lines, in this case a short distance in on line 7, marks the beginning of an earthquake. (Courtesy Lamont-Doherty Geological Observatory of Columbia University, Palisades, N.Y.)

tude of shock waves decreases with increasing distance from the focus. A much more useful concept is that of magnitude, a single measure of the earthquake's force that does not vary from place to place.

Magnitude is roughly proportional to the total energy released by an earthquake at its focus. It is defined in terms of the amplitude of the trace that would be made by a standard seismograph located 60 miles (100 km) from the epicenter. The magnitude scale in common use today was devised by the American seismologist C. F. Richter in 1935. Sixty miles was chosen as the standard distance because Richter's work was done at the California Institute of Technology, 60 miles from the San Andreas fault line.

The Richter scale is nonlinear. A rating of 7.9, for example, represents a very large release of energy; it is 10 times larger in terms of the energy involved than a rating of 6.9 and 103 times larger than a rating of 4.9. The Richter scale is presented in Table 18.

The San Francisco earthquake of 1906 had a magnitude of 8.3, the Good Friday earthquake of 1964 registered 8.5, and the first atomic bomb had a magnitude of 5. The greatest earthquake ever recorded occurred in Ecuador in 1906; its magnitude was 8.9.*

Tsunami

When the focus of an earthquake is under or near an ocean, seismic ocean waves are sometimes produced (Fig. 18-10). The international name used for such waves is *tsunami,* a Japanese word meaning "great wave."

A tsunami is a series of traveling ocean waves of long wavelength and frequency. In deep water their wavelengths may be as great as 100 miles or more while their amplitude measures only a few feet. In fact, they cannot be seen from the air or felt by ships at sea. They may reach speeds of 600 miles (1000 km)/hr. As the waves approach shallow water, their speed decreases while their amplitude increases. Wave heights of 100 feet (30 meters) or more have been recorded. The arrival of a tsunami is often but not always preceded by a withdrawal of water from coastal

*Observations of this earthquake were translated into the Richter scale. This is done for earthquakes that occurred prior to 1935, when the Richter scale was developed.

Fig. 18-10. Damage from a tsunami. The tsunamis produced by an earthquake in Alaska on March 27, 1964, washed these vessels into the heart of Kodiak Island. (Courtesy United States Department of the Interior, Geological Survey, Denver, Colo.; photograph by W. R. Hansen.)

areas. The withdrawal represents water being fed to an incoming wave that is beginning to break. (A similar effect may be observed on a greatly reduced scale by observing normal water waves breaking along coasts all over the world.) Rarely is the first wave the most destructive. It is believed that resonance* in the water of the focus area builds a subsequent wave to maximum.

In 1948, a seismic sea-wave warning system (SSWWS) was put into operation in the Pacific Ocean by the United States Coast and Geodetic Survey, with headquarters at Honolulu, Hawaii. Since that time, no tsunami has struck the Pacific area without warning.

Earthquake damage and effects

Earthquakes on land can have tragic effects. As the shock waves pass through populated areas, buildings, streets, and bridges collapse. Gas pipes and power lines are severed, compounding the problem with the threat of fire. Water supplies can be contaminated, food spoils, and bodies decay, leading to the dangers of famine and disease. Dams may be broken, causing floods. Damage and death are always proportional to the density of the population.

The Good Friday earthquake of 1964, the most thoroughly studied and best documented shock in history, affected water levels in wells, rivers, and aquifers all over the world. The quake triggered some 50 avalanches, the largest involving 13 million cubic yards of rock. The land level was altered in an area of about 70,000 square miles in south-central Alaska; 23,000 to 35,000 square miles were elevated, in places by as much as 33 feet. The submarine extent of land movement is not precisely known, but elevations of as much as 49 feet have been measured.

As tragic as the Good Friday earthquake of 1964 was, there have been other earthquakes that caused a much greater loss of life. In Shensi Province, China, in 1536, 830,000 people died as a result of an earthquake, and 300,000 perished at Calcutta, India, in 1737. Chiefly as the result of a tsunami, 30,000 people were killed at Lisbon, Portugal, in 1755. The list of tragic earthquakes is much longer.

During the course of any single year there are about 1000 shocks that do some damage, 100,000 that may be felt by humans, and 500,000 that are detected by seismographs. It has been estimated that since the year 1 A.D., 6 million people have lost their lives as a result of earthquakes.

*Resonance is a vibration produced in one body as a result of waves from another vibrating body. For example, a vibrating tuning fork will cause another tuning fork of the same pitch to vibrate.

19 *a search for the past*

Geologic history

But the old man turns the pages
of the rock—illuminated ages,
Tracing from earth's mystic missal
the antiquity of man.

Through the Ages
William Canton

The geologic history of the earth encompasses a time span so vast that it staggers the imagination. We can comprehend the immensity of geologic time somewhat better if we reduce its scale.

Suppose that approximately 1 million years of earth history are represented by a distance of 1 foot on a linear time scale. The formation of the earth would be located at a distance of about 5000 feet (just less than 1 mile) from the reader. The oldest rocks ever found on earth would be at 3300 feet. The oldest fossils, bacteria and algae, would be found at about 3000 feet. The beginning of the Paleozoic era of geologic time, represented by the oldest rocks that contain abundant fossil evidence of life, would be found at 525 feet. The beginning of the Mesozoic era, the Age of Dinosaurs, would be located at 180 feet. The beginning of the Cenozoic era, in which we are now living, would be at 70 feet. The oldest apelike ancestor of man would be at 20 feet. The beginning of the Pleistocene epoch, the Great Ice Age, would be found at 12 inches. The advent of Homo sapiens (modern man) would be located at ½ inch. On the same scale the fall of the Roman Empire would be at $^{1}/_{64}$ inch, the signing of the United States Declaration of Independence would be at $^{1}/_{512}$ inch, and an average human life time would span less than $^{1}/_{1000}$ inch. Human life, then, has existed on earth for less than one ten-thousandth of geologic time.

Prior to the eighteenth century, prevailing thought concerning the history of the earth was a literal interpretation of the Book of Genesis in the Bible. The heaven and earth were created in 6 days, and since 1 day with God was as 1000 years, it naturally followed that the earth was 6000 years old. Because of the incredible slowness of geologic processes, people would see little change in the earth during this short interval of time. It is perhaps easy to understand why people for so long believed that the earth was formed in 4000 B.C. In their minds, 6000 years represented an extremely long period of time.

Our present understanding of the earth's history, still incomplete, is an accumulation of knowledge acquired through years of observation, thought, and study. As has often been the case, however, it remained for economic motivation to stimulate studies that led to an explosion of knowledge concerning the history of the earth. It was the search for mineral deposits in the early years of the Industrial Revolution that provided the major impetus for the development of geology in general and historical geology in particular. As more mineral-rich areas were found and knowledge of them increased, local rock relationships were recognized by miners. Gradually fossils became valued as data that could be obtained from a large portion of the rocks of the crust and interpreted to ascertain time relationships. Rocks and fossils became synonomous with the passage of geologic time, and it was soon learned that the story of the formation and evolution of the earth and its life lay in the rocks.

UNIFORMITY AND EVOLUTION

Two of the most controversial theories ever proposed were uniformitarianism and organic evolution. Although the uniformity of natural processes had been investigated as early as 600 B.C. by the Greek scholar Xenophanes of Colophon, it remained for James Hutton in 1795 to state the basic principle of uniformitarianism beautifully and simply: "In examining things present we have data from which to reason with regard to what has been; and from what has actually been, we have data for concluding with regard to that which is to happen hereafter."

If this conclusion, that an understanding of nature's past operations could be obtained from observations of present natural geologic processes, was shocking to the eighteenth-century mind, another statement by Hutton must have shattered it: "The result, therefore, of our present enquiry is that we find no vestige of a beginning, no prospect of an end." With this statement, the immensity of geologic time was first envisioned, and *uniformitarianism* became the heart of geology.

By the close of the eighteenth century, all of the ingredients were at hand for the appearance of another theory that was to have even more far-reaching effects. The concepts of organic variation, environmental change, and natural selection—supported by a fossil record of extinct organisms—awaited combination. Charles Darwin (1809-1882), an English naturalist, stated in 1859 in *On the Origin of Species* that all organisms had evolved over a long period of time through gradual changes from common ancestors. It was noted that although the fossil record was incomplete, it revealed many examples of organic variation and modification from the past. Fossils showed evidence that some species had become extinct while others had survived.

If uniformitarianism had raised questions as to the accuracy of the 4000 B.C. birthdate for Adam, organic evolution seriously challenged his creation from dust at all. More than 100 years after the publication of *On the Origin of Species* and nearly 200 years after the work of Hutton, the two theories are still being debated.

AGE OF ROCKS AND FOSSILS

In reconstructing the history of the earth, it is necessary to know the sequence of geologic events that have occurred in a region and the approximate time of their occurrence. Fossils, an important tool in historical geology, must be correlated with other fossils and their ages must be determined. Two techniques are used to determine the age of rocks and fossils.

Relative age

It is not always possible or necessary to determine the exact chronologic age of a rock or fossil in order to recon-

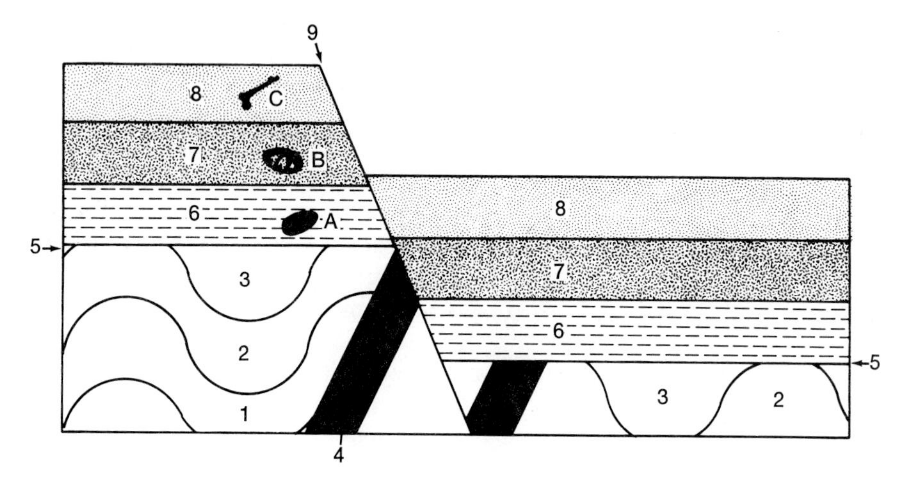

Fig. 19-1. Law of superposition enables us to understand the sequence of events that produced the rock formation schematically shown here. First, layers *1, 2,* and *3* were deposited. Next, they were folded. Folding had to have occurred before the intrusion of dike *4;* otherwise the dike would also be folded. After dike *4* intruded, erosion produced the unconformity, *5.* Then layers *6, 7,* and *8* (including fossils *A, B,* and *C*) were deposited. We know that these layers were not deposited before because they are not folded. Layers *6, 7,* and *8* must have been deposited after the intrusion of the dike *(4)* because it does not cut through them. Some time after layers *6, 7,* and *8* were deposited, the normal fault, *9,* was formed. The fault was formed at that point because the layers are identical on both sides of it; if the fault had been there first, the layers would droop over it rather than be split by it.

struct the sequence of geologic events that have taken place. In fact, much important work in this area was completed before exact dating techniques were developed by determining relative age—whether certain rocks and fossils are older or younger than other rocks and fossils.

One useful technique for determining relative age is the application of the *law of superposition*, developed by Nicolaus Steno (1638-1687) in 1669. The law of superposition states in effect that in any sequence of undeformed sedimentary rock layers or other surface-deposited materials such as lava the lower layers must be older than the upper. In an exposure of sedimentary rock strata, as in the Grand Canyon in Arizona, we can be sure that the lowest layer is the oldest and the top layer is the youngest of the sequence (Fig. 19-1). Although Steno's law of superposition is almost self-evident in its simplicity, it has been one of the most basic and useful tools of geology.

In 1799 William Smith (1769-1839), an English surveyor, developed the *law of faunal succession*, which proved to be such a significant technique for measuring relative age that Smith is known as the father of historical geology. The law of faunal succession, stated in the light of modern knowledge, means that in any sequence of sedimentary rock layers the fossils contained in the lower layers are older than those in the upper layers (Fig. 19-1) and that particular layers can be recognized over a broad area by their unique fossil groups. William Smith had provided geologists with a key by which the doors to past geologic time could be unlocked.

Two other important concepts regarding relative age are the *law of intrusion* and the *law of original horizontality*. The law of intrusion states that an igneous body such as a batholith or dike is younger than the rocks in which it is found, for they had to be there before the igneous body could intrude into them (Fig. 19-1). The law of original horizontality states that any deformation of the rocks (such as fractures, faults, or folds) is younger than the rocks themselves because the rocks must be present before they can be broken, displaced, or bent (Fig. 19-1).

Absolute age

Techniques that are now available make it possible to determine the absolute chronologic age of rocks and fossils. A rock can thus be dated as being 200 million years old rather than just older or younger than the rocks around it. Although the name implies that absolute age determination is exact, geologists expect these dating techniques to be within 10% of the actual age of a specimen.

The most reliable techniques for determining absolute age are based on the change that takes place in the atoms of radioactive elements in rocks and fossils of the earth's crust. Radioactive methods are based on the fact that unstable elements lose electrons and particles of their nuclei and that through this process they change gradually into new elements. When a stable end product or element is made, the substance ceases to be radioactive. The age of such a sample is measured by performing a chemical analysis of the ratio of the amount of radioactive element to its stable end product. This system can be relied on because the rate at which radioactive change or decay occurs is constant. Radioactive elements and their end products that have been used in absolute age dating are: (1) uranium (→ thorium → lead), (2) rubidium (→ strontium), and (3) potassium (→ argon). The radioactive clock of each of these elements was set at the time of its formation—almost always from the crystalization of magma. As a result, the use of radioactive dating methods generally has been confined to igneous rocks. Attempting to date metamorphic or sedimentary rocks using radioactive techniques would give the time that had passed since the rock was molten, not the time at which it was metamorphosed or deposited as a sediment. The oldest known rock on earth has been radioactively dated at 3.3 billion years, while the estimated age of the earth is about 5 billion years, a figure based primarily on the age of meteorites.

A useful method for determining the age of sedimentary rocks containing fossils no older than 40,000 years is through the use of carbon 14 (^{14}C), a radioactive form of carbon, which is an element contained in all living things and their remains. Cosmic rays from space coming through the atmosphere collide with nitrogen atoms to form ^{14}C. The ^{14}C combines with oxygen to form radioactive carbon dioxide, which mixes with the normal nonradioactive carbon dioxide in the air. This carbon dioxide is absorbed by plants, which are in turn eaten by animals, and the radioactive substance is absorbed into their flesh. Thus all plants and animals contain both normal carbon (^{12}C) and radioactive ^{14}C. As long as an organism is alive, the rate of absorption of ^{14}C balances the rate at which it disintegrates into ^{12}C, its stable end product. At death, the absorption of ^{14}C ceases, so that by analyzing the ratio of ^{14}C to ^{12}C, the age of a fossil can be determined (Fig. 19-2). The half-life of ^{14}C is only 5600 years, so its usefulness is limited to recent fossils and archeologic artifacts. In order to date the oldest fossils, geologists do not use the ^{14}C method. Instead they use radioactive dating techniques to determine the age of the rocks in which the fossils were found or the ages of other rocks in the vicinity. Techniques for determining relative age are then applied. The oldest known fossils are estimated to be about 3 billion years old.

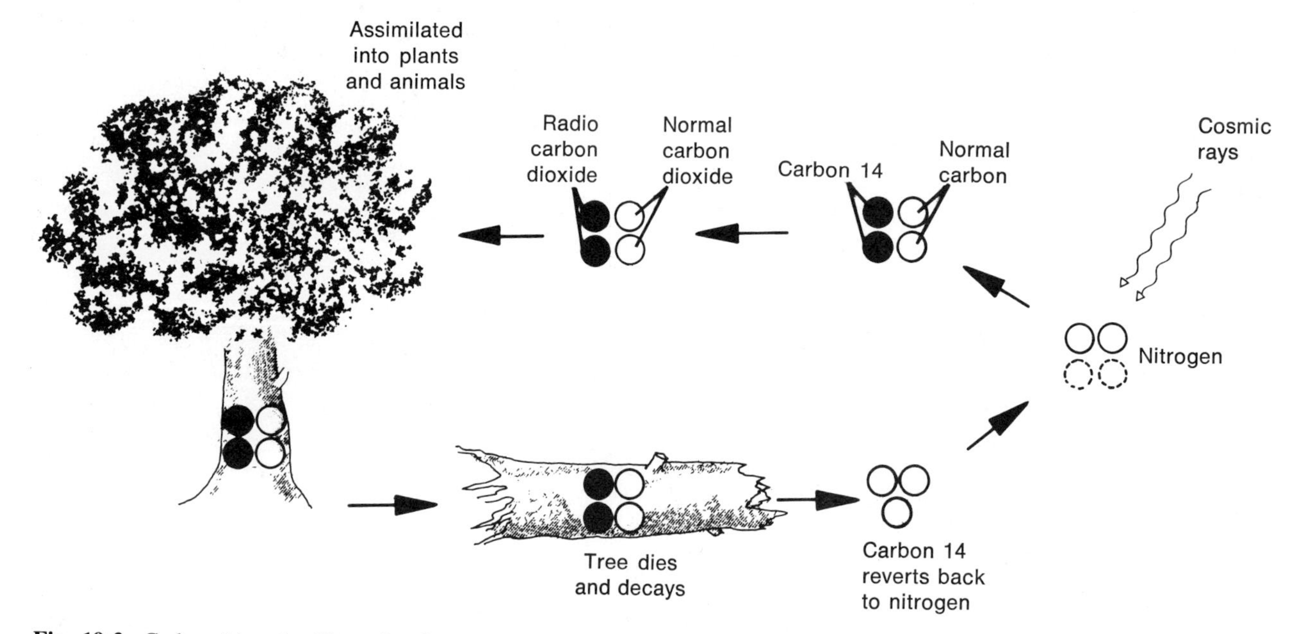

Fig. 19-2. Carbon-14 cycle. The ratio of normal carbon to carbon 14 allows scientists to determine the age of certain fossils.

PREHISTORIC TIME SCALE

The many years of studying the rocks and fossils of the earth's crust and painstakingly applying the techniques for determining relative and absolute age have culminated in an incomplete but ever-growing knowledge of the history of the earth. One outgrowth of this knowledge has been the development of a colossal calendar known as the geologic time scale. The parts of the scale are based almost entirely on the law of faunal succession, which was developed through the study of fossil groupings and their worldwide correlation.

In the Book of Genesis we read that the heaven and earth were created in 6 days. The original Hebrew word *yom* had two meanings—day and age. If *yom* was intended to mean "age" rather than "day," then there is not such a significant conflict between geologic knowledge and the religious beliefs of some individuals, for the original meaning could have been the longer time intervals geologists today call *eras*. The breakdown of geologic time into six primary divisions based on stages of fossil development is perhaps more significant than previously believed.

The era we are now living in is known as the Cenozoic (recent life) era. Prior to the Cenozoic was the Mesozoic (middle life) era. Before the Mesozoic was the Paleozoic (ancient life) era. Even earlier were the Proterozoic, Archeozoic, and Azoic eras. For simplicity, the last three are often collectively called the Precambrian era.

The fundamental working units of the geologic time scale are the *periods,* which refer to a specific interval of time. Most of the periods are named for a location where rocks of this age are found well exposed. Much of the early work in geologic history was done in Europe, and therefore most of the geologic periods are named for locations in England, France, Germany, and Russia. It should be remembered that the period names are applied to actual rocks containing specific groups of fossils. The Precambrian era contains no periods; the Paleozoic era has seven, the Mesozoic era has three, and the Cenozoic era has two. The two periods of the Cenozoic era have been further subdivided into epochs.

The geologic time scale is shown in Table 19. Included is the approximate absolute age for the beginning and ending of each period or epoch, and the approximate duration of each. Refer back to Table 19 as we proceed with a very brief summary of the major geologic events, climatic conditions, and the evolution of life as told by the fossil record in the rocks. Special emphasis will be placed on North America in general and on the United States in particular.

Precambrian era

The oldest rocks found anywhere on earth are those of the Precambrian era. They may be any one of the three rock types, but metamorphic and igneous rocks predominate. The most abundant exposures of Precambrian rocks are found in more or less centrally located areas of each continent called *shields*. Continental shields are the deeply

Table 19. Geologic time scale

Era	Approximate duration (years)	Period	Approximate duration (years)	Epoch	Approximate duration (years)
Cenozoic	63,000,000	Quaternary	2,000,000	Recent	50,000
				Pleistocene	1,000,000
		Tertiary	61,000,000	Pliocene	12,000,000
				Miocene	12,000,000
				Oligocene	11,000,000
				Eocene	12,000,000
				Paleocene	14,000,000
Mesozoic	177,000,000	Cretaceous	72,000,000	Mesozoic era not divided into epochs	
		Jurassic	55,000,000		
		Triassic	50,000,000		
Paleozoic	370,000,000	Permian	50,000,000	Paleozoic era not divided into epochs	
		Pennsylvanian	30,000,000		
		Mississippian	35,000,000		
		Devonian	60,000,000		
		Silurian	20,000,000		
		Ordovician	75,000,000		
		Cambrian	100,000,000		
Precambrian	4,000,000,000	Precambrian era not divided into periods or epochs			

eroded, deformed roots (or bases) of mountains that have long since disappeared.

Unfortunately Precambrian rocks have not yielded the secrets of the formation of the earth or the beginnings of life, even though these events must have occurred during Precambrian times. The reason for this is that Precambrian rocks have undergone perhaps as much as 4.5 billion years of geologic change. The original rocks and any fossils they may have contained have been destroyed in the continuing processes of the rock cycle. Even if the original rocks had not been destroyed, early life was probably frail and almost incapable of being fossilized due to a lack of hard parts such as teeth, bones, and shells. As a result, Precambrian rocks contain a meager fossil record.

However, algae and fungi have been found preserved in Precambrian rocks in Canada. The imprint of a jellyfish-like animal has been found in Australia, and other fossils have been found in the Grand Canyon. Other types of primitive plant and animal fossils have reportedly been discovered in some Precambrian rocks in other parts of the world, but in most cases the validity of these reports is questioned. Little is known, then, about the formation of the earth and its early life forms.

Most of the Precambrian rocks of the United States, as in many parts of the world, are complex and badly eroded. As a result, it is difficult to decipher them. However, it is believed that extensive mountain building and volcanic eruptions occurred and that seas covered most of the continent. The most significant land mass was the Canadian Shield, a large mass of rock that today underlies much of central Canada. The extremely rich and valuable iron ore deposits around Lake Superior were formed during Precambrian times. By the end of the Precambrian era the climate of the world had changed and glaciers were widespread.

The great unconformity

An *unconformity* is a buried erosion surface found between older and younger masses of rock. An unconformity represents a gap in the geologic record and shows a change from an environment of deposition to an environment of erosion and then back to an environment of deposition. As a result of such a change, which can be caused by uplift of the land or a decrease in sea level, an erosional surface develops. When sediments are again deposited in the area, the erosional surface is buried to become part of the geologic record. The unconformity, representing what used to be the surface, is of value as a record of earth movements. Unconformities range in extent from small, localized features to buried surfaces that can be traced over thousands of square

Fig. 19-3. One of the best-known unconformities in the world lies in this section of the Grand Canyon (arrow), where dark-colored Precambrian rocks yield to lighter Paleozoic. The abrupt break indicates that for some time deposition stopped and erosion occurred; eventually, deposition resumed. This break is the only record we have of the intermediate years. (Courtesy United States Department of the Interior, Geological Survey, Denver, Colo.; photograph by L. F. Noble.)

miles. Missing time intervals represented by unconformities may encompass a few years or millions of years.

The greatest unconformity ever found in the geologic record occurs between Precambrian rocks and the younger rocks found on top of them (Fig. 19-3). This unconformity, representing the longest "missing" interval of geologic time, is virtually worldwide in extent. It is an indication that late Precambrian time was marked by a widespread withdrawal of the sea from the continents, exposing much land to attack by erosional agents. The lost time interval represented by the great unconformity has been estimated at up to 100 million years.

Paleozoic era

Below the great unconformity lie Precambrian rocks, which contain few if any fossil signs of life. Above the buried erosion surface, on the other hand, are rocks of the Paleozoic era, which contain evidence of a veritable explosion of life that was remarkably varied and well established. This leads many geologists to conclude that the rocks destroyed by late Precambrian erosion contained valuable evolutionary records in the form of fossils that will never be found. Regardless, the oldest rocks of the earth's crust that contain abundant fossils are said to represent the Cambrian period, named after the Cambridge region of England where they were first studied. (It should now be clear why rocks older than this are called Precambrian.)

Early Paleozoic era

The early Paleozoic era, which is subdivided into the Cambrian, Ordovician, and Silurian periods (Table 19), is known as the Age of Invertebrates because creatures with-

Fig. 19-4. Reconstruction of an early Paleozoic (Middle Cambrian) seafloor in western North America showing free-swimming trilobites, the dominant form of early Paleozoic life. (Courtesy Field Museum of Natural History, Chicago, Ill.)

out backbones dominated the seas during this interval of geologic time (Fig. 19-4). Evolution had not yet produced the first land plants or animals. Marine organisms, on the other hand, were numerous. Plant life consisted of a form of algae somewhat similar to today's seaweed. The animals were more varied—many organisms were the ancestors of our present-day sponges, corals, jellyfish, bivalves, sea slugs, crabs, lobsters, and shrimp.

A prominent life form during the Cambrian period was a creature known as a *trilobite,* a distant relative of today's king crab. Although trilobites became extinct in the late Paleozoic era, they evolved rapidly during the early Paleozoic into a variety of forms that spread over the entire world. These two factors—its rapid development and spread and its worldwide disappearance at a set time—make the trilobite, like other organisms that have similar histories, useful as *index* or *guide fossils*. These allow for worldwide dating of the rocks in which they are found. The trilobite *Olenellus,* for example, is found only in Cambrian rocks. Important index fossils of the Ordovician period were organisms called *graptolites,* an ancestor of today's corals and jellyfish. Significant index fossils of the Silurian period were *eurypterids,* somewhat similar to present-day horseshoe crabs, and *nautiloids,* whose only remaining living relative is the pearly nautilus. The Silurian period also marks the appearance of the first land plants. These plants were basically seaweed that adapted gradually to the air environment. Over the next 400 million years these plants were to evolve into more than 350,000 species, less than half of which are marine.

The early Paleozoic era set a geologic pattern in North America that was to continue for more than 300 million

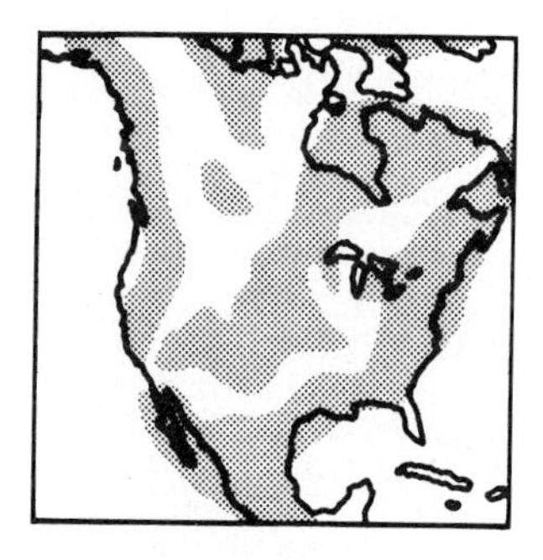

Fig. 19-5. Paleozoic land area in North America. Shaded areas were land, light areas were water during the Ordovician period about 450 million years ago.

years. The dominant geologic structures were three geosynclines that were located along the edge of low-lying land extending from Mexico northward through the Canadian Shield to Greenland (Fig. 19-5). To the east was the Appalachian geosyncline, while to the south was the Ouachita geosyncline. To the west, the Cordilleran geosyncline was developing. Even though the three geosynclines must have been formed during the Precambrian period, the first mountain building to occur in the early Paleozoic era was during the Ordovician period, when the northern part of the Applachian geosyncline was uplifted. One of the best remnants of this uplift is the Taconic Mountains of eastern New York State.

The early Paleozoic era was a time of rapid changes in sea level. The peripheral geosynclines (except where uplift occurred) and much of the existing continent lay under shallow seas. The early Paleozoic climate was generally mild, with extensive areas experiencing arid conditions. Even though the climate was mild in many parts of the world, we could infer that the rapid changes in sea level that occurred during this interval of the earth's history were somehow linked to glacial ice in the polar regions, for a similar condition exists today. The Silurian period in the United States is especially noted for the combination of arid climate and widespread shallow inland seas. During this period extensive beds of marine limestones and dolostones were deposited, including the Niagara limestone cap that produces Niagara Falls. Vast salt deposits were also formed from the Great Lakes to Virginia.

Middle Paleozoic era

The middle Paleozoic era consisted of the Devonian and Mississippian periods. The geologic pattern set in North America during the early Paleozoic continued into the middle Paleozoic. Sediments that were eroded from the land continued to accumulate in the geosynclines that had formed earlier. A strong uplift of another section of the northern Appalachian geosyncline occurred during the Devonian period, affecting large areas of New England and eastern Canada. This mountain-building episode, called the Acadian orogeny, formed the granite and marble masses of New England. The continental regions continued to be periodically inundated by shallow seas that covered much of North America. These widespread seas, in combination with the warm, moist climate prevalent during the Mississippian period and the time that followed, set the stage for one of the most interesting and important developments of geologic time.

The plants that had taken hold on the land during the Silurian period found a haven in the warm swamps of the Mississippian and Pennsylvanian periods that followed (Fig. 19-6). Many land plants grew to impressive size, often reaching heights of 40 feet (12 meters) or more, even though most were as biologically simple as the ferns, cattails, and rushes that are their present-day relatives. Dead plants accumulated in the stagnant swamp water, and sediments rapidly covered them. This organic debris, compressed under hundreds or thousands of feet of subsequent sediment deposits, eventually turned into coal. Although coal-making began during the Mississippian period, the process reached its zenith during the Pennsylvanian period. While plants dominated the land, animals in the form of fish dominated the seas. For this reason, the Devonian and Mississippian periods are collectively called the Age of Fish. Although invertebrates continued to flourish in the seas, they would never again be dominant.

Late Paleozoic era

The late Paleozoic era consists of the Pennsylvanian and Permian periods. While the Mississippian period was a time of widespread deposition and orogenic quiet in the United States, the Pennsylvanian period saw some uplift of parts of the southern geosyncline to form the Wichita Mountains of Oklahoma and Texas. One of the most remarkable aspects of the Pennsylvanian period was the large number of cyclic deposits involving organic debris that would later form coal. The process was so widespread and complete, in fact, that the Pennsylvanian is known as the Great Coal Age. In Europe, the Mississippian and Pennsylvanian periods are collectively named the Carboniferous period.

Almost overshadowed by the coal-making events was another critical evolutionary development. In the swampy lowlands during the Pennsylvanian and Permian periods a life form called the *amphibians* had evolved. For this reason, the late Paleozoic era is known as the Age of Amphib-

Fig. 19-6. Restoration of late Paleozoic (Pennsylvanian) swamp life. It was this lush plant life that died and formed the coal beds for which the Pennsylvania period was named. (Courtesy Field Museum of Natural History, Chicago, Ill.)

ians. The story of the amphibians probably began in the Devonian period, as fish trapped in evaporating pools during this arid time gradually learned to move across the land on their fins looking for more water and food. Even today, several species of fish live out of water for extended periods. For the first time, animals had reached the dry land and had become firmly entrenched there.

The Permian period ended the Paleozoic era. This period saw violent and geologically rapid mountain building, including the uplift of the remainder of the Eastern geosyncline to form the majestic Appalachian Mountains. The remainder of the Southern geosyncline was also uplifted to form the Ouachita Mountains of Arkansas and Missouri. The present-day Appalachians and the Ouachita and Ozark Mountains are but erosional remnants of these once lofty peaks. With the uplift of the Eastern and Southern geosynclines, the seas withdrew from the interior of eastern North America for the last time.

The violent changes that took place during the Permian period also affected global climates. In contrast to the warm, moist conditions of the Pennsylvanian period were the glacial climates of the Permian. Many organisms became extinct during this period; surprisingly, the marine invertebrates seem to have suffered most. The reason for this was probably the rapid retreat of the seas from the continents. This retreat, caused by the upheaval of the land and the storage of water on land as ice, created shifts in the marine environments toward deeper water, and many shallow-water invertebrates were unable to adjust.

Mesozoic era

The 180 million years after the close of the Paleozoic era are known as the Mesozoic era. The Mesozoic saw a shift of major geologic events in North America to the western part of the continent as the east underwent rapid erosion. Sediments continued to accumulate in the Cordilleran geosyncline until it too was finally uplifted to form the Rocky Mountains—an event that, along with similar activity elsewhere in the world, brought the Mesozoic era to a close.

Early Mesozoic era

The early Mesozoic era is known as the Triassic period. During the Triassic period much of the area west of the Plains states was covered by shallow seas. Except for some volcanic islands, the west coast of the United States was almost completely covered by water. In most of the world the climate was warm and moist, but arid conditions prevailed in what is today Wyoming, southern Montana, and northern Colorado. For the most part, the Triassic period in North America was not an active mountain-building period. With the coming of the middle Mesozoic era, however, such activity would increase and culminate in late Mesozoic

Fig. 19-7. Restoration of middle Mesozoic (Jurassic) seafloor showing free-swimming belemnoids and oyster-like pelecypods. The marine reptile *Ichthyosaurus* is shown in the background. (Courtesy Field Museum of Natural History, Chicago, Ill.)

times with the *Laramide Revolution,* the uplift of the Rocky Mountains.

Although the early Mesozoic era found most of the world geologically quiet, the evolution of life was proceeding actively. Between the Paleozoic and Mesozoic eras, striking changes in life forms occurred. The reason for this was the wave of extinction that took place in the late Paleozoic era. As many types of organisms died, this left ecologic niches free for occupation by the hardier stock that became the life forms of the Mesozoic era. In the seas the dominant invertebrate animals were the *mollusks,* distant ancestors of today's snail, clam, oyster, squid, octopus, and chambered nautilus. Especially important as index fossils of the early Mesozoic era were groups of now-extinct mollusks known as ammonoids and belemnoids (Fig. 19-7). On the land evolution was occurring uninterrupted at a rapid rate. The amphibians were on the decline and a new form of life, the reptiles, were developing rapidly. The reptiles, which first appeared in the Pennsylvanian period of the Paleozoic era, were the first vertebrates to adapt to the land, air, and water environments. The adaptation was so complete that the Mesozoic era is known as the Age of Reptiles.

As already pointed out, ammonoids ruled the Triassic seas. From the Permian period to the end of the Mesozoic era, ammonoids experienced three crises. Like many other marine invertebrates, they nearly died out at the end of the Paleozoic era. Only one stock was able to survive into the Mesozoic era; from this group evolved an even wider, more diverse ammonoid population in the Triassic period. Late in the Triassic, however, the numbers of ammonoids dropped once again, and only one type survived into the middle Mesozoic. As before, this stock flourished on a grand scale. Late in the Mesozoic era, however, the ammonoids finally became extinct as the group vanished from the earth.

The ammonoids warrant such close attention because their near extinctions are so clearly seen in the geologic record that the Mesozoic era has the best-established chronology of the entire geologic time scale. The boundary between the Paleozoic and Mesozoic eras is marked by their first extinction. The boundary between the early Mesozoic (Triassic) and middle Mesozoic (Jurassic) periods is marked by the second ammonoid extinction. The third extinction marks the boundary between the late Mesozoic (Cretaceous) and the Cenozoic eras.

On land during the Triassic period a common reptile group called *thecodonts* developed. Thecodonts are significant because they gave rise to the dinosaurs as well as to flying reptiles, crocodilians, and birds. Thecodonts were small creatures, only about 3 feet in length, with strong hindlegs, small hand-like forelegs, and long tails used for balancing on two legs. Before they became extinct late in the Triassic period, thecodonts had set the stage for the most unusual of nature's experiments—the dinosaurs. From the thecodont stock, two types of dinosaurs evolved. These dinosaurs are classified primarily on the basis of the arrangement of their hip bones. One group had typical repitle hips, while the second group had hip bones similar to those of birds, which had not yet evolved. *Coelophysis,* the first true dinosaur, appeared late in the Triassic period.

Middle Mesozoic era

The middle Mesozoic era is known as the Jurassic period. Like the Triassic period, the Jurassic period in North America was dominated by erosion in the East and deposition in the West (Fig. 19-8). The Jurassic period in the West saw the encroachment of a shallow sea over the Cordilleran geosyncline (Fig. 19-9). This shallow arm of the ancestral Arctic Ocean, named the Sundance Sea, was the site of the most active deposition in Western North America during the Mesozoic era. Later, in the Jurassic period, sedimentary rocks in eastern and western Nevada were faulted and folded to form the ancestral Sierra Nevada Mountains. Although the present-day Sierra Nevadas were formed later, the diastrophism of Jurassic times was accompanied by the intrusion of granite batholiths, exposures of which can be seen today in Yosemite National Park, California.

The Jurassic period saw the rapid evolution of the dinosaurs into unusual forms, often of massive proportions. Especially prevalent during this period were members of the reptile-hipped group of dinosaurs. Worthy of mention was *Ceratosaurus,* a meat-eating dinosaur that attained a maximum adult length of 24 feet (7 meters). Also common was *Struthiomimus,* often called the "ostrich dinosaur," whose diet probably consisted of fruit, insects, and dinosaur eggs,

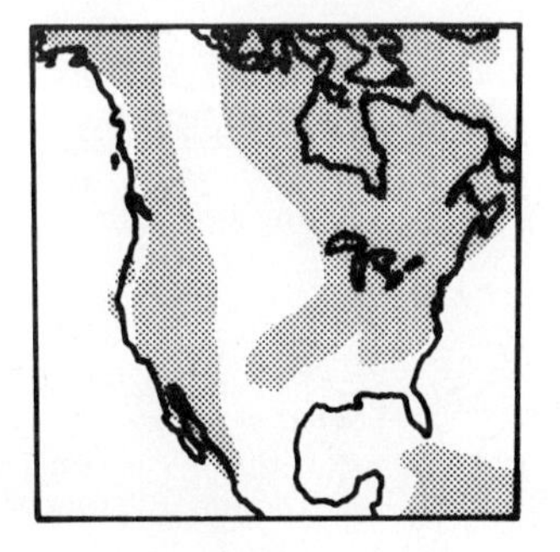

Fig. 19-8. Paleogeographic map of late Mesozoic (Cretaceous) time. About 100 million years ago a shallow sea extended through much of the present-day Rocky Mountain states.

Fig. 19-9. Restoration of late Mesozoic (Cretaceous) seafloor in central United States, showing many long, straight ammonoids and one coiled variety as well as pelecypods and gastropods. (Courtesy Field Museum of Natural History, Chicago, Ill.)

Fig. 19-10. The dinosaur *Stegosaurus*. (Courtesy American Museum of Natural History, New York, N.Y.)

for it lacked teeth in its beaklike jaws. Although the bird-hipped vegetarian dinosaurs were more common during the late Mesozoic era, one well-known member of this group that lived only during the Jurassic period was the armored beast *Stegosaurus* (Fig. 19-10).

Late Mesozoic era

The late Mesozoic era is known as the Cretaceous period. In North America, as in much of the world, the Cretaceous was a period of widespread encroachment of the seas over the continents and much mountain building. More than half of the present area of North America was drowned under shallow Cretaceous seas, the last time such a widespread invasion occurred. This period saw the beginning of the development of the coastal plain, which extended from New Jersey southward through Florida and westward through the Gulf States to Mexico. Seas covered this region and also extended from the present Gulf of Mexico northwestward through the present Rocky Mountain and Plains States and across western Canada to the Arctic Ocean.

The Cretaceous period ended with the Laramide Revolution, the greatest mountain-building event since Precambrian time. The great Cordilleran geosyncline, which had been filling with sediments since the Precambrian era, was uplifted to form the Rocky Mountains of western North America. At approximately the same time a similar geosyncline in western South America was uplifted to form the Andes. The Laramide Revolution brought to a close the Mesozoic era. The warm, moist climates of the Mesozoic era were to be replaced by the cold climates of the subsequent Cenozoic era. This change in climate would have widespread effects on the life forms of the Mesozoic era, as these organisms attempted to adapt to the Cenozoic world.

One of the life forms that (perhaps luckily) did not survive into the Cenozoic era was the dinosaurs. These reached their evolutionary zenith during the Cretaceous period and became extinct at its end. Dominant during the Cretaceous were the reptile-hipped dinosaurs such as *Brontosaurus, Brachiosaurus,* and *Tyrannosaurus*. *Brontosaurus* was a giant vegetarian dinosaur that reached a length of 80 feet (24 meters) and a weight of many tons, while *Brachiosaurus,* also a vegetarian, was even larger. *Tyrannosaurus,* generally considered the most fearsome beast ever to walk the earth, was a flesh-eating dinosaur that, unlike his two vegetarian brothers, walked on two legs (Fig. 19-11). *Tyrannosaurus* was often 50 feet (15 meters) long, 20 feet (6 meters) high, and weighed as much as 10 tons.

The evolution of the bird-hipped dinosaurs also continued into the Cretaceous period. In many respects, these animals were more highly evolved than the reptile-hipped forms. One main difference between them was the wide range of ecologic adaptation of the bird-hipped dinosaurs. On the land were *Iguanodin,* a 30-foot (9 meter) dinosaur that walked on two legs, and *Trachodon,* a duckbilled dinosaur with webbed feet that must have spent much time in the water. The last major group of dinosaurs to evolve were the horned type. *Protoceratops* was a small horned dinosaur only about 6 feet (2 meters) long, while *Triceratops,* the best-known of the horned dinosaurs, reached a length as great as 30 feet (9 meters) and had a huge head with three large horns (Fig. 19-11).

While dinosaurs were evolving on land, other lines of reptiles had adapted to the water and air environments. Aquatic reptiles had begun to evolve as far back as the Permian period of the Paleozoic era. The best-adapted marine reptile was *Ichthyosaurus* of the late Mesozoic era, which looked a great deal like our present-day porpoises and sharks. *Pteranodon,* a flying reptile with a wingspread of more than 25 feet (7.6 meters), flew across the Cretaceous skies.

Other developments of the greatest importance occurred during the Mesozoic Era. Birds had evolved from one line of flying reptiles. The first true bird was *Archaeopteryx,* whose fossil remains are found in rocks of the Jurassic period (Fig. 19-12). Mammals had evolved from mammal-like reptile stock during the early Mesozoic era, and by the Cretaceous period five types of mammals had evolved. Of the five, only one survived into the Cenozoic era, in which mammals became the dominant life form.

Extinction of the dinosaurs. Dinosaurs were the dominant land animals for more than 130 million years. They became extinct during the Cretaceous period of the Mesozoic era, which ended about 70 million years ago. There has been much speculation as to why dinosaurs disappeared so suddenly. It has been suggested that the widespread climatic changes that occurred between the Mesozoic and Cenozoic eras were responsible for changing the environments of the creatures. It is also possible that the increasing numbers of birds and mammals were responsible for the wholesale destruction of dinosaur eggs and young. Perhaps future study will provide additional information.

Cenozoic era

The Cenozoic era encompasses the last 70 million years of geologic time, including the present. During this era, the earth and its life have evolved to the familiar forms we know today. How much longer the Cenozoic era will last is unknown, but it will probably close with a mountain-building period, as have the other eras. It could be argued, in fact, that a revolution has already occurred, for the Ceno-

A

B

Fig. 19-11. A, The flesh-eating dinosaur *Tyrannosarus* (left) and the horned variety, *Triceratops*. **B,** Skeleton of *Tyrannosarus*. (Courtesy American Museum of Natural History, New York, N.Y.)

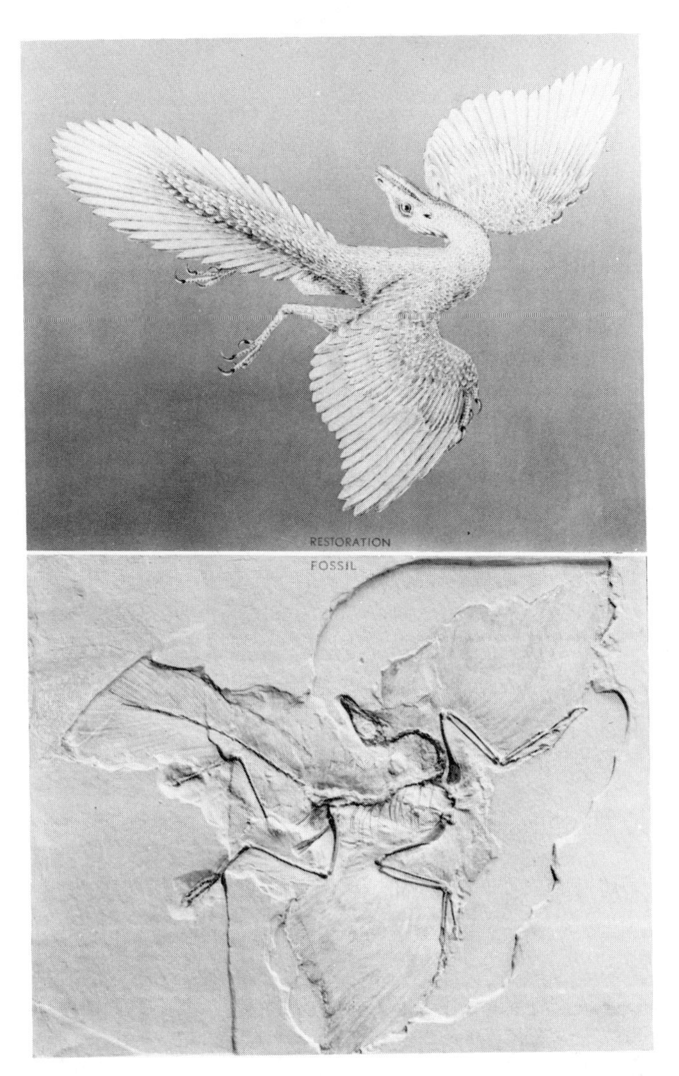

Fig. 19-12. Restoration and fossil of *Archaeopteryx,* the first true bird. (Courtesy American Museum of Natural History, New York, N.Y.)

zoic era has seen the uplift of major mountain ranges. Perhaps in the near future a major reevaluation of the Cenozoic era will be undertaken. It would not be surprising to find the Tertiary and Quaternary periods of the Cenozoic era elevated to the rank of eras. Because the Cenozoic era is still continuing, it is difficult to speak of its early, middle, and late periods. Therefore our discussion of the Cenozoic era will be based on the two periods into which it is now divided.

Tertiary period

The Tertiary period, consisting of the Paleocene, Eocene, Oligocene, Miocene, and Pliocene epochs, was a period of erosion and gentle epeirogenic uplift in eastern North America. The Atlantic and Gulf Coastal areas were often under water and subsiding under the weight of sediments eroded from the land. Florida, whose limestone base was laid during the Cretaceous era, began to take on a distinct character as more and more limestones were deposited from the south while clastic sediments were deposited from the north. Developments similar to the early history of the Florida Peninsula can be observed today in the Bahamas. The uplift of the Rocky Mountains, which began during the Cretaceous era, continued until the Eocene epoch. Erosion and subsequent epeirogenic uplifts have dominated the remainder of the Cenozoic era in this region. It was during the Miocene epoch that the Colorado River began to carve out the Grand Canyon in the southwestern United States, and the Columbia lava plateau was formed in the northwest. During the Pliocene epoch the coast ranges and the Cascade Mountains, including such famous volcanic peaks as Mt. Rainier and Mt. Hood, were formed. Elsewhere in the world during the Pliocene epoch, the Alps and the Himalayas were uplifted.

The Tertiary period is known as the Age of Mammals. While the reptiles were at their evolutionary peak, the mammals were an insignificant group of small, insect-eating animals. With the disappearance of the dinosaurs and the decline of the reptiles, many types of mammals appeared on the geologic scene, filling environments, or ecologic niches, once dominated by the ancient ancestors of the snakes, alligators, crocodiles, lizards, and turtles. The success of the mammals in surviving the widespread climatic changes of the late Mesozoic era and their rapid evolutionary radiation in the Cenozoic era can be traced to certain physical characteristics that, in combination, are unique to mammals:

1. The development of warm blood, maintaining body temperature through internal rather than external heat
2. A larger, more complex brain
3. Specialized teeth that led to more varied diets
4. Hair for body temperature control
5. A four-chambered heart for the separation of body and lung circulation, leading to more efficient respiration
6. Limited body growth
7. Internal gestation of the young
8. Parental care of the young after birth

Today, there are three groups of mammals: the *monotremes* (spiny anteater and duckbill platypus), the *marsupials* (opossum and kangaroo), and the *placentals;* 95% of all mammals are in this latter group.

Dominating the Tertiary period were hoofed, grazing mammals known as *ungulates*. The appearance of the ungulates apparently coincided with the evolution of grasses, the most important land plant development of the Tertiary period. Common in the Paleocene epoch was *Phenadodus,* a primitive, hoofed mammal about the size of a sheep. It had five hoofed toes, short, thick legs, a long tail, and a long, narrow head. *Unitatherium,* hoofed mammals about the size of a present-day mature rhinoceros, *Eohippus,* the earliest known ancestor of the modern horse, and a group of primitive, flesh-eating mammals called *creodonts* abounded during the Eocene epoch. The Oligocene epoch saw the development of *Indricotherium,* an early ancestor of the modern rhinoceros that grew to a height of 17 feet (5.2 meters), and *Miohippus,* a more advanced ancestor of the horse. In the Miocene epoch the horse took still another evolutionary step forward with the appearance of *Hypohippus.* Also appearing in Miocene times were *Trilophodon,* an early mastodon-like creature that was an ancestor of the elephant, *Pliopithecus,* an early primate, and *Proconsul,* an ape that may be a link to the evolution of man. During the Pliocene epoch *Teleoceras,* a rhinoceros, *Synthetoceras,* an ancestor of the deer, giraffe, and sheep, and *Amebelodon,* a mastodont, appeared.

Quaternary period

The Quaternary period, subdivided into the Pleistocene and Recent epochs, encompasses less than 2 million years. The Quaternary period has seen the development of land forms and life forms that are familiar to us today, for relatively few major changes have occurred in such a short interval of time. In North America the only noteworthy orogenic movements since the Tertiary period affected the extreme western edge of the continent. The present-day Sierra Nevada Mountains and the San Andreas fault line in California were formed during the Pleistocene epoch. In the West the volcanic activity that was so prevalent during the Tertiary period continued at a somewhat reduced level into the Pleistocene epoch. On most of the continent, erosion was the dominant geologic process, as the Grand Canyon approached its present majesty.

The Pleistocene epoch is known as the Great Ice Age. Although glacial conditions had existed during earlier times (as in the Precambrian era and the Permian period of the Paleozoic era), never before had ice been of such worldwide influence as in the glacial age of the Pleistocene. During this epoch glacial ice covered as much as 25% of the land surface and extended an unknown distance out to sea as *ice shelves* or *sea ice*. Glacial erosion shaped the land in Canada and the northern United States and also affected extensive areas in higher elevations of western North America. Elsewhere in the world similar conditions prevailed, as continental and valley glaciers grew to impressive proportions. Four glacial stages of the Pleistocene have been identified,

Table 20. Characteristics of major geologic time divisions

Common name	Geologic era or period	Dominant life	Selected events in North America
Age of Man	Recent	Man	Present erosion, Grand Canyon, San Andreas Fault, and Sierra Nevadas formed
Great Ice Age	Pleistocene		Four glacial stages
Age of Mammals	Tertiary	Mammals	Erosion and uplift in the east, frequent submergence of Atlantic and Gulf Coastal Plains, mountain building and volcanic activity in the west
Age of Reptiles	Mesozoic	Dinosaurs	Uplift of Rockies, much of North America submerged, erosion in the east
Age of Amphibians	Permian	Amphibians	Climates cool, some glaciation, continued uplift and erosion
Great Coal Age	Pennsylvanian		
Age of Fish	Devonian Mississippian	Fish	Simple plants continue to flourish, continued uplift and erosion in Appalachian Geosyncline
Age of Invertebrates	Silurian Ordovician Cambrian	Invertebrates (trilobites, graptolites, eurypterids)	First uplift of Appalachian Geosyncline, first simple plants on on land, Appalachian and Cordilleran Geosynclines forming
	Precambrian	Simple marine plants	Formation of earth, oceans formed, first appearance of life, mountains uplifted and eroded

and the Recent epoch, in which we are now living, is probably the latter part of the fourth interglacial. While the climate during most of the one-half billion years since the beginning of the Paleozoic had been warm to moderate, the climate of the Quaternary period has been one of extremes, with the worldwide temperature averaging about 20° F (11° C) below the levels of the Paleozoic and Mesozoic eras. The major characteristics of these periods are summarized in Table 20.

Man: a naked ape. The Quaternary period is known as the Age of Man, for this period saw the rise of the human species. Man belongs to the animal kingdom, phylum chordata (for he has a backbone) the class mammalia, the order primate, genus *Homo,* and species *sapiens*. He is clearly a product of organic evolution and is closely related to anthropoid apes, today represented by the orangutan, gibbon, gorilla, and chimpanzee.

The evolutionary history of man and his ancestors is not at all clear today. The reason for this is the rarity of fossil remains. When fossils are found, they are fragmentary—part of a skull, a few teeth, a piece of bone, and occasionally some stone tools or charred wood. The last two bits of evidence are very important in interpreting man's cultural development.

Even though we do not know the exact evolutionary history of man at this time, what evidence we do have indicates that his development could have proceeded along a path such as the following. Approximately 25 to 35 million years ago a line of ancestral anthropoid apes left their tree habitats to compete with well-adapted ground dwellers. The earliest fossil record of this primitive ape form, *Proconsul,* dates from the Miocene epoch. Proconsul was found in East Africa and may represent a common ancestor of both man and the modern apes.

Encountering a new environment, our early ancestors either had to become better killers than the well-established meat-eating mammals or better grazers than the plant eaters. He survived by adopting the best of both diets. Man, then, is a vegetarian who learned to also eat meat, or an omnivore. After an evolution of about 20 million years, ancestral man reached the level of *Australopithecus,* whose fossil remains were first found in early Pleistocene deposits in South Africa. Clearly apelike in most characteristics, *Australopithecus* would have to be considered a manlike ape, for the structure of the limbs and teeth and the nearly vertical posture are strongly suggestive of the human body form.

Our ground-dwelling ancestors were hardly a match for the stronger, faster, better-adapted animals with whom they found themselves competing for food and territory. The great equalizer was our ancestors' ability to first use and later make tools and weapons. The first of our ancestors to do this apparently was *Pithecanthropus,* the first known representative of man's genus, *Homo* (Fig. 19-13). *Pithecanthropus,* also known as Java man, must have been a tool user, and there are some indications that he was also a toolmaker.

Early man took a giant step forward with the discovery that he could control and use fire. He probably knew about fire from such natural sources as grass fires started by lightning. He may have accidentally learned to make fire himself as he struck two rocks together while making a tool or weapon. Fire meant warmth, light, safety from predators, and cooked food. Our first known fire-making ancestor was *Sinanthropus,* or Peking man. Hearths have been found in caves together with his fossil remains. Like *Pithecanthropus,* Peking man was a representative of *Homo erectus,* although a slightly advanced form.

Modern man, *Homo sapiens,* differs physically from

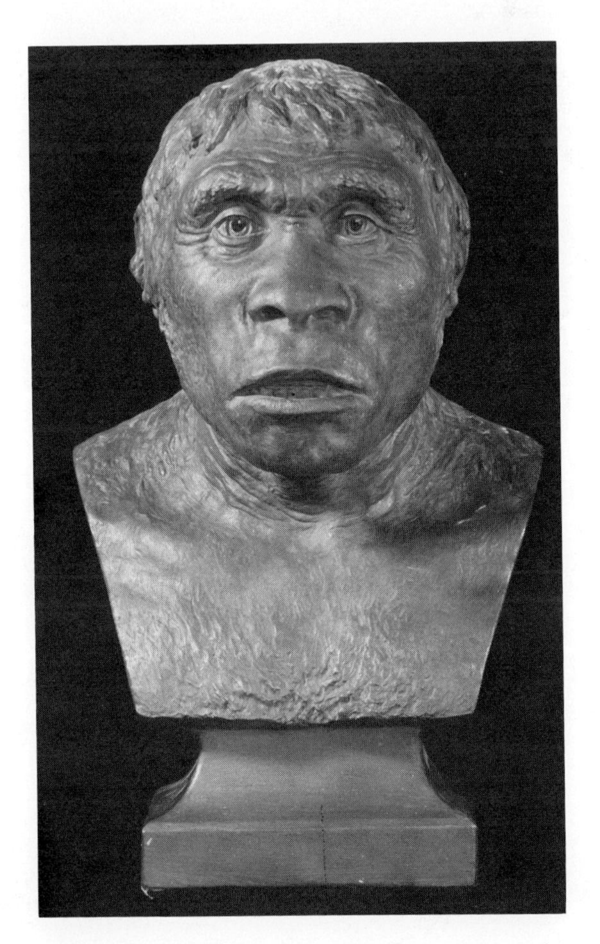

Fig. 19-13. *Pithecanthropus*. This first representative of our genus, *Homo,* had learned to use tools to equalize the odds in his confrontation with a hostile environment. (Courtesy American Museum of Natural History, New York, N.Y.)

Fig. 19-14. Cro-Magnon man. This last step in human evolution could pass for any other man on the street if he were dressed in modern clothes, an indication of how little we have changed in the intervening 52,000 years. (Courtesy American Museum of Natural History, New York, N.Y.)

early man in that he has a larger brain, an expanded pelvis to support his body in an upright position, and a rigid walking foot with stiff toes, an arch, and a heel. The earliest representative of the lineage leading directly to modern man was *Homo neanderthalensis,* or Neanderthal man. He was the first hominid to attach a point to a stick to make a spear, to bury his dead, and to develop a crude religious system. Neanderthal man, who first appeared about 150,000 B.C., was definitely not human, but the stage was set for modern man.

The final evolution to *Homo sapiens* has indeed been complex. Many noted anthropologists believe that modern man was the product of inbreeding. In light of the way man is inbred today, this belief certainly seems within the realm of possibility. It also has been proposed that Neanderthal man was produced as a result of inbreeding between early types of *Homo sapiens* and that he disappeared because he was inbred out of existence—absorbed by a more dominant type. The fossil record does show that Neanderthal man was replaced about 50,000 B.C. by an anatomically modern homonid group called Cro-Magnon man (Fig. 19-14).

The Cro-Magnons, related to the Neanderthals, made sophisticated tools and weapons and were skilled artists. Cro-Magnon man was definitely the first true representative of *Homo sapiens.* Why the Neanderthals died out is not clear at this time. It is interesting to note that the Cro-Magnons, a group of humans so biologically modern that with proper clothes and training they could pass as members of our society today, appeared rather quickly on the scene that had been dominated by the brutish, primitive Neanderthals.

It has been suggested that the Age of Man will be succeeded by the Age of Insects, biologically among the most successful animals ever to inhabit the earth, for their evolution has extended from the Devonian period to the Recent. Never before has a more "advanced" life form been replaced by a simpler one. Man, however, is a special animal. More than just a product of evolution, he is also a product of himself. Of all the creatures ever to live on this planet, he alone has had the power to alter history—not only his own but the history of life. He can make lower life forms extinct, as he has done with some 100 species, through overkill, ecologic encroachment, and pollution. He can make himself extinct slowly by ruining his environment or rapidly by pushing a button. Although the recent history of man would seem to indicate that he is a cruel, thoughtless creature bent on self-destruction, he also possesses the almost magical power to reverse a trend he has set. If *Homo sapiens* is to survive, he must return to a greater harmony with nature.

part three OCEANOGRAPHY

I pace the sounding sea-beach and behold
How the voluminous billows roll and run,
Upheaving and subsiding, while the sun
Shines through their sheeted emerald far unrolled.

And the ninth wave, slow gathering fold by fold
All its loose-flowing garments into one,
Plunges upon the shore, and floods the dun
Pale reach of sands, and changes them to gold.

Milton
Henry Wadsworth Longfellow

The Rance River project, with its dam across the river's mouth, uses the power of ocean tides to generate electricity. (Courtesy French Engineering Bureau, Washington, D.C.)

20 *unlocking the secrets of the sea*

Oceanography

They that go down to the sea in ships,
That do business in great waters,
These see the work of the Lord,
And His wonders in the deep.

Psalms 107:23-24

The earliest attempts to study the sea were motivated less by abstract curiosity than by practical needs—food at first and later navigation and trade. At some prehistoric moment someone would have discovered the sea, standing like a child on that long-forgotten shore in awe of the enormous expanse of water. That individual no doubt walked to the water's edge to touch or taste it. He would have observed its motions and would have found in time that it contained plants and animals that could be used as an easy source of food.

Early man then became a fisherman, first from the shore and then, as his courage and curiosity grew, from rafts and boats. He ventured farther and farther from shore, expanding his horizons and coming in contact with other men. He studied bottom topography to prevent his ships from grounding and wind effects on the sea surface to see how they might inhibit or assist him. Oceanography, the science of the sea, was born therefore of man's needs for navigational skills, his growing interest in geography, and his desire for trade.

HISTORY OF OCEANOGRAPHY

Oceanography as a science did not really begin until systematic observations were made and records were kept. The cradle of oceanography would have to be the Mediterranean countries, for prior to the fifteenth century the Egyptians, Greeks, Minoans, and Phoenicians dominated the waters of the known world. (The Vikings in the far north were the only exception.) After the fifteenth century, it was the emerging maritime nations of western Europe who dominated oceanographic thought.

Early Mediterranean seafarers

Many early Mediterranean cultures led the way in developing the body of thought we call oceanography. An ancient Egyptian carving tells of a voyage undertaken by Queen Hatshepsut as early as 1500 B.C. Exploratory in nature, this voyage probably took its participants as far south as the Somali Republic.

Both the Phoenicians (Fig. 20-1) and the Greeks explored the Mediterranean Sea and its shores, and by 800 B.C. both groups were venturing regularly into the unknown seas beyond the Pillars of Hercules, now called the Strait of Gibraltar. By 600 B.C. daring Phoenician sailors were rounding Africa, entering the Southern Hemisphere by sea for the first time. They probably explored areas as far west as the Sargasso Sea, in the middle of the Atlantic Ocean.

In the fourth century B.C. Pytheas, a contemporary of Alexander the Great, sailed past Gibraltar and discovered Great Britain. He also gave the name Thule to certain lands north of it, probably either the Scandinavian countries or Iceland.

Three hundred years later Hippalus discovered the pattern of the monsoon winds of the Indian Ocean. Knowing that these winds blow from the southwest in summer and northeast in winter made shorter journeys across this important ocean possible—it was necessary only to choose the proper season for a crossing. Nonetheless the Indian Ocean was the last to be completely explored, and study of the monsoon winds and their related ocean currents is not complete even today.

Even though nothing that could be called "oceanographic research" by modern standards occurred on these

Fig. 20-1. Despite their tiny ships, the Phoenicians sailed past the Strait of Gibraltar out into the open Atlantic, earning their reputation as a nation of expert men of the sea.

early voyages, they must be considered significant because of the wealth of oceanographic knowledge they inadvertently added to man's store.

The first known attempt to make our knowledge of the size and shape of the oceans more precise was Poseidonius, a Greek geographer, around the time of the birth of Christ. Poseidonius used a weighted rope to measure the depth of the Mediterranean Sea near Sardinia and obtained a reading of 1000 fathoms.*

During the Dark Ages that followed the fall of the Roman Empire, the study of the sea, like the study of nearly everything else, came to a halt. It was during this hiatus, however, that the daring men of the North, the Vikings, took the lead. Journeying over some of the stormiest seas on earth, they discovered and explored Greenland, Iceland, and North America, where they sailed at least as far south as Martha's Vineyard, off the coast of Massachusetts.

The Golden Age of exploration

As the Renaissance approached, oceanographic study gained momentum as European nations sought a shorter trade route to India. To help their countries achieve this goal, a long list of well-known explorers expanded the limits of the known world.

Bartolomew Diaz carried the Portuguese flag around the Cape of Good Hope (the southern tip of Africa) in 1488. Columbus opened America to southern Europeans in 1492. Vasco da Gama completed the trip Diaz started when he sailed around Africa to reach India in 1497. Magellan settled the question of the earth's shape once and for all when he circumnavigated the globe in 1520.

*The fathom, equal to 6 feet (2 meters), is roughly the distance between the fingertips of the hands when the arms are fully outstretched, and it is a convenient measure for gathering up rope dropped overboard.

Having found a way to reach India either from the east or from the west, the next problem tackled was the question of what to do about the inconveniently placed continent (North America) that blocked the western passage. Frobisher, Davis, Hudson, and Baffin spent their lives and fortunes searching for a Northwest passage big enough for ocean-going vessels. The next century saw Cook and Bering searching for the passage finally found by Sir Robert McClure in 1850. As with earlier oceanographers, it was not scientific curiosity that motivated these men but rather the drive to beat the Portuguese to India. Nevertheless, their findings added to the growing body of knowledge that would someday be called oceanography. These men's journals are treasure troves of information on surface phenomena, salinity, temperature, and the behavior of tides and currents—information they needed for voyages over extended distances.

Early oceanographers

With the voyages of Captain James Cook, an Englishman who was a scientist as well as a great seaman, came the application of scientific methods to the study of sea—the birth of oceanography as we know it. Even though his first journey (between 1768 and 1771) was primarily aimed at observing the 1769 transit of Venus and confirming the existence of a continent near the South Pole, Cook's voyages gave him the opportunity to make depth soundings and frequent observations of winds, currents, and water temperatures. Perhaps more important, the publicity given to his findings served as the spark needed to direct talented younger men to the study of the ocean.

Cook was not alone, however, for the nineteenth century saw more progress in this field. In 1818 Sir John Ross, also an Englishman, made the first accurate deep-sea sounding at a depth of 1050 fathoms in Baffin Bay west of Greenland.

Sir John also introduced the technique of *sampling* when he pulled a sample of mud from the ocean floor at that depth and brought it home to study. He also sounded the South Atlantic and the Cape of Good Hope, recording depths of 2435 and 2677 fathoms, respectively. These findings may seem unimportant when compared with the data obtained by modern electronic gear such as the precision depth recorder (PDR), but 2600 fathoms is nearly 3 miles (5 km) of hemp rope, and it would require a full 12- to 16-hour workday to complete one measurement.

What about the men who studied the life forms within the ocean? The founder of this aspect of oceanography was Carl Von Linne, a Swedish botanist. He classified plants and animals into orderly groups, providing the impetus that sent biologists scurrying all over the world, looking on land and sea for more plants and animals to classify! It was from Von Linne's groups, moreover, that later scientists developed the subscience of *marine ecology,* the study of the interrelationships of such groups with each other and with the physical and chemical forces around them.

Baron Alexander Von Humboldt, a German scientist, is the next gentleman to earn a place in our history. Observing sea life off the west coast of South America, he was intrigued by the abundance of forms he found in the cold ocean current that now bears his name. He discovered that guano, or bird droppings, made excellent fertilizer and reported this fact in Europe. (In the process he created an export for Peru!) He observed tropical storms and mapped the volcanoes of the New World, discovering in the process their linear pattern.

The voyage of the *Beagle* from 1831 to 1836 created a furor that has not completely subsided even today, for it was aboard this vessel that Charles Darwin served his apprenticeship as a naturalist. It was the variety of life forms he studied on this voyage that set his thinking on the course that led to *On the Origin of Species* and *The Descent of Man*. The evolutionary theories presented in these two books forced a reevaluation of theories in many scientific fields as well as causing the better-known commotion among the general public. In addition to Darwin's more startling discoveries, he is also credited with pioneering knowledge on the origins of coral reefs.

Most fields of study have a "father," that is, a person who has made outstanding contributions to the fledgling area. Oceanography had two such men, Edward Forbes, an Englishman, and Matthew Maury, an American. Forbes was an instructor of marine biology at the University of Edinburgh. Among his contributions were his improvements of the naturalist's dredge and his extensive collections of the flora and fauna near the Isle of Man in the Irish Sea. He noted that the abundance of life forms decreased with increasing water depth, and he classified eight vertical zones of the sea based on the abundance of life forms found in them. His studies, unfortunately, led him to conclude incorrectly that life did not exist below 300 fathoms. (He was apparently unaware of the work of his countryman Ross, whose studies, published 20 years earlier, showed the presence of worms and starfish 1050 fathoms below the surface of Baffin Bay.) Forbes gave a modern orientation to oceanography; his attempts to relate both plants and animals to their environments made him one of the earliest marine ecologists.

Matthew Maury's field was physical oceanography. As a lieutenant in the United States Navy, Maury was injured in action and reassigned to the Depot of Charts and Instruments, rising through the hierarchy to eventually head this division. From the assortment of naval and merchant logbooks he supervised in this capacity, he acquired an interest in the daily weather reports, tides, currents, and wind measurements. He was the first to note the correlation between the ocean currents and atmospheric circulation. Since the publication of his *Physical Geography of the Sea* in 1855, this work of Maury's has saved commercial firms untold millions of dollars simply by predicting the current patterns that help them to run their ships more efficiently.

Notable voyages

Since man's capacity to study the ocean unaided is necessarily limited, the history of oceanography is also the study of ships.

The *Challenger*

Probably the most important voyage in the history of oceanographic exploration was the *Challenger* expedition of 1872 to 1876. *Challenger* was a 226-foot, 2300-ton steam corvette that had been converted into an oceanographic research vessel (Fig. 20-2). Chief scientist for the cruise was Sir Wyville Thompson, and his orders were to "investigate the conditions of the deep sea throughout the great oceanic basins." He discovered his first "condition" at the first official "station," just west of Teneriffe; virtually every animal that appeared in his dredges was previously unknown!

During the 4-year adventure the scientists of the *Challenger* covered 68,890 miles (114,816 km) (more than 140 million square miles of surface) and logged 362 separate stations, accumulating data on animal life, temperature, the chemistry of sea water, water depth, winds, currents, and tides. Thus it was the *Challenger* that began the series of record-keeping voyages that continues even today. Despite

Fig. 20-2. The 226-foot, 2300-ton *Challenger* collected so much data on her 4-year cruise that the deepest part of the ocean, the Challenger Deep, is named in her honor.

technologic advances in both ships and instruments since the 1870s, this converted corvette still holds records for the longest continuous scientific undertaking of its kind and also for overall significance of any one voyage. Some of the achievements behind this record are the following.

1. *Challenger* scientists sounded the Marianas Trench, one of the deepest parts of the ocean. The 26,850-foot (8189-meter) depth they logged approaches the 36,000-foot (10,980-meter) depth of the deepest part of the seafloor, now named the Challenger Deep in the ship's honor.
2. *Challenger's* systematic records on temperature, salinity, currents, and other data formed the basis of modern physical oceanography. C. R. Dittmar's determination of the constituents of seawater was based on 77 water samples taken on *Challenger's* cruise. Dittmar's conclusion, that seawater does not vary regionally in relative composition, is now one of the basic premises of oceanography.
3. Perhaps the biologic findings were the most interesting of this cruise. *Challenger* scientists discovered that the abundance of life forms in the sea is controlled not so much by depth as by distance from shore. They found 4717 species and 715 new genera. They defined the world of plankton in terms of species and distribution, and correctly theorized that the concentrations of whales in the Antarctic regions were due to similar concentrations of plankton there.
4. They found manganese nodules on the ocean floor, sparking modern interest in mining the seafloor.
5. Their meteorologic observations substantiated earlier research into the relationship between the ocean and the atmosphere and opened new interest in phenomena of the sea-air interface.
6. Even in today's electronic world, the methods of trawling and dredging that the *Challenger* crew developed hold their own.

The *Albatross, Vitiaz,* and *Hirondelle*

During the 1880s Alexander Agassiz, an American, helped to organize several voyages of the *Albatross,* the first vessel ever designed specifically for oceanographic research. Over a period of more than 25 years the scientists on the *Albatross* accumulated a mass of oceanographic data that formed a base for future study. Investigating the abundance of the manganese nodules that had been discovered by the *Challenger,* they found both numbers and distribution encouraging.

Between 1886 and 1889 the Russians entered the field of ocean exploration with the cruise of the *Vitiaz,* under the leadership of Stepan O. Makarov. Scientists on the *Vitiaz* studied density and temperature in the North Pacific, but their countrymen failed to appreciate the importance of their data, and the Soviet Union dropped out of the oceanographic community, to reenter only in modern times.

In 1885 Prince Albert of Monaco fitted out several yachts as research vessels—first the *Hirondelle I* and *II,* then the *Princess Alice I* and *II*. Concentrating understandably on the Mediterranean and the North Atlantic, the prince's crews accumulated an extensive collection of deep-sea animals. They developed several new types of equipment, including trawls, nets, and traps, some of which used lights to attract the animals. He also founded and endowed the Musée Océanographique whose most famous alumnus is the Frenchman Jacques-Yves Cousteau, the inventor of scuba gear who is better known for his oceanographic research and films.

Twentieth-century expeditions

The twentieth century has seen a dramatic increase in the number of oceanographic cruises. From 1925 to 1927 the Germans on board the *Meteor* crossed the South Atlantic, systematically studying the seafloor with echo sounders. Their findings astonished oceanographers; the seafloor was not smooth and flat as previously believed, but varied and extremely rugged.

In 1925 the English again took to the sea in the *Discovery I* to study Antarctic whales. In 1926 this ship was joined by the *William Scoresby,* and in 1930, rather than scrap the project, the English replaced *Discovery I* with *Discovery II*. The three ships sent back a great deal of information not only on whales but also on the water they inhabit.

From 1920 to 1922 and 1928 to 1930 the *Dana,* an American ship supported by a Danish brewing company, carried out two research expeditions to collect information needed by the fishing industry. Coincidentally, these expeditions managed to piece together the story of the freshwater eel. Adult eels swim thousands of miles from the rivers that are their homes to the Sargasso Sea in the middle of the Atlantic Ocean. The adults spawn there and die, but their larvae drift on ocean currents back toward the freshwater streams where they mature, taking nearly 3 years to complete the cycle.

After a brief hiatus during the Second World War, oceanography again gathered momentum as the Swedes fitted out their own *Albatross* and sailed her around the world in 1947 and 1948. This expedition emphasized marine geology, and Dr. Borje Kullenberg and other ingenious crew members devised an instrument that allowed them to collect some remarkable core samples from the ocean floor. Dredge hauls from depths as great as 25,900 feet (7899 meters) were taken in the Atlantic.

In 1950 and 1951 the Danes entered the field of oceanography with a round-the-world cruise on the *Galathea.* The Danish specialty was deep sea dredging, and fish and anemones were dredged from depths of 23,200 and 32,800 feet (7076 and 10,004 meters), proving the existence of life even at those incredible depths.

In May of 1950, a new *Challenger* set out on a 2-year voyage to make precise measurements of depths in the Atlantic, Pacific, and Indian Oceans and the Mediterranean Sea. Her scientists found the deepest part of the ocean, the Pacific's Challenger Deep, which they named in honor of the first *Challenger*. Depth soundings that had required 2½ hours on the original *Challenger* because of the time necessary to raise and lower the rope could be made in seconds on her descendant, simply by bouncing sound waves off the sea bottom (Fig. 20-3). In addition, it was this *Challenger's* data that led to the discovery of the rift valley in the Mid-Atlantic Ridge.

A strange oceanographic expedition took place in 1960, one that was vertical rather than horizontal. Lieutenant Donald Walsh and Jacques Piccard descended into the Challenger Deep in the bathyscaph *Trieste* (Fig. 20-4). On the bottom they saw a fish resembling a sole casually slither away, apparently upset by their unexpected company.

Today, due to the great amount of accumulated data and the expense of outfitting an expedition, many nations participate in joint expeditions. Through the cooperation of 22 nations, the International Geophysical Year (1957 and 1958) and the International Indian Ocean Expedition of 1959 to 1965 were highly successful. Even though only small amounts of information were accumulated by any one ship, when all participating ships combined the results of their work the final outcome was quite satisfactory. Joint

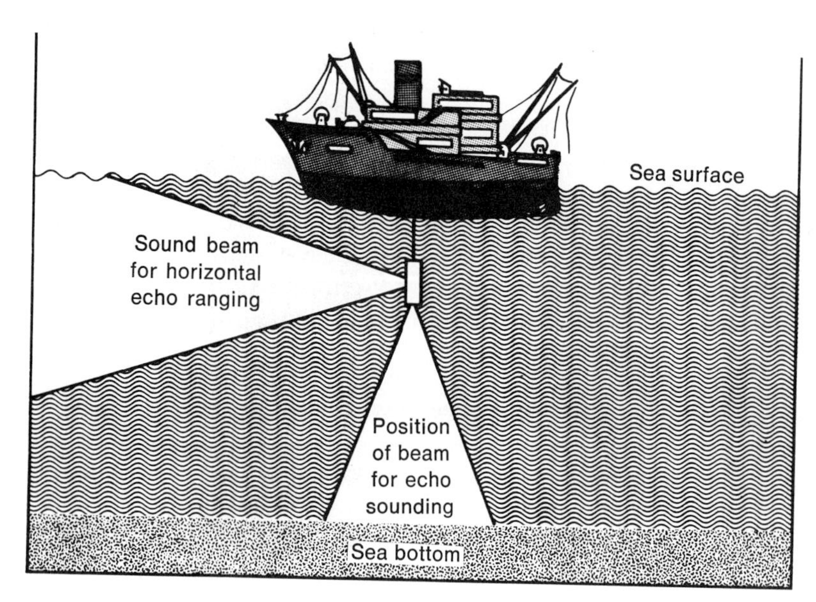

Fig. 20-3. Echo sounding has replaced the back-breaking rope handling of earlier depth measurements. Sound waves are bounced off the ocean floor and their echos can be timed to calculate distance. The echo of a sound impulse travels five times faster through water than through air.

Fig. 20-4. This photograph taken from the bathyscaph *Trieste* shows that certain life forms can exist even in the awesome environment of the deep ocean floor. These little creatures find life 35,800 feet below the ocean's surface quite normal. (Courtesy United States Navy, Washington, D.C.)

expeditions such as these have led to such discoveries as the Gulf Stream Countercurrent, which flows southward 6500 feet (1982 meters) below the northward-flowing Gulf Stream.

In 1965 and 1966 American, Soviet, French, and Brazilian ships cooperated in the Equilant Cruises to study the tropical Atlantic. Through their efforts, potentially productive fishing grounds were found off the west coast of Africa.

The summer of 1974 again found the international cooperation of 12 nations paying off in the GARP* project, whose purpose was to study the sea-air interaction in a broad region of the tropical Atlantic. Masses of meteorologic and oceanographic data were accumulated, but at this writing the final results and implications of this work are incomplete.

*GARP stands for Global Atmospheric Research Program.

OCEANOGRAPHY TODAY

The word ''oceanography'' means many things to many people. If you were interested in the biologic world, you might naturally think of oceanography in terms of dazzling coral reefs, mighty whales, or exquisite flying fish. If your interests were chemical or geologic, on the other hand, you might think of oceanography in terms of salinity, dissolved oxygen, sedimentation, or underwater mining. If you were fascinated by physical phenomena, waves and currents might play in your thoughts. The point is that all of these—whales, flying fish, salinity, sediments, waves, and tides—are pertinent to the study of oceanography. Unlike the pure or basic sciences such as chemistry, biology, geology, or physics, ''oceanography'' is a term used when an expert in one of the pure sciences applies that knowledge to the study of the sea. Not only must marine scientists be expert in one of these four basic sciences, they must also possess a knowledge of the others and how they relate and interact with their own discipline. Henry Bryant Bigelow, one of the pioneers of oceanography, described the science in this way:

> Oceanography has been aptly defined as the study of the world below the surface of the sea; it should include the contact zone between the sea and atmosphere. According to present-day acceptance, it has to do with all the characteristics of the bottom and margins of the sea, of seawater and inhabitants of the latter.

Divisions of oceanography

Since an oceanographer must be an expert in at least one of the basic sciences in order to apply it to the ocean, it is convenient to classify oceanographic branches according to the basic sciences on which they depend. This results in four branches—physical oceanography, chemical oceanography, biologic oceanography, and marine geology and geophysics.

Physical oceanography

When carcinogenic asbestos was discovered in Lake Superior, the governments of three surrounding states filed pollution suits against a local iron mining company that regularly dumps its waste products into the lake. The company claimed that its wastes were harmless and that the asbestos was the result of natural seepage from surrounding rocks. By studying current patterns, tides, etc., it will be a physical oceanographer who decides who is right.

Physical oceanographers study such phenomena as wave motion, currents, tides, heat in the ocean, light penetration, and density variations. Today's research efforts find physical oceanographers utilizing very sophisticated monitoring devices at various depths in the sea to help determine the rates of dispersion and means of travel of various pollutants. With devices designed to measure currents, physical oceanographers have shown that the ocean, once thought to be capable of diluting pollutants indefinitely, may now be accumulating appalling concentrations of a wide variety of harmful or unpleasant substances. The old adage, "The only solution to pollution is dilution," apparently has never been applicable. In addition to studying the open ocean, physical oceanographers have focused much attention on our shorelines, particularly on estuaries (former stream valleys that are now submerged). These are among the most biologically productive shoreline regions and are therefore the areas in which most pollutants have their greatest effect.

Chemical oceanography

To maintain our present level of civilization, we must utilize many of the world's resources. Chemicals such as iodine, bromine, potassium, and magnesium are just a few examples of those that are obtained directly from sea water. The water itself, without which life would not exist, is all around us, but is tainted with salts. The chemical oceanographer has helped develop purification techniques in an attempt to supply the ever-increasing demands for both water and the chemicals dissolved in it.

Chemical oceanographers are interested in areas such as salinity, dissolved gases, and reactions that occur both in the sea and on the seafloor. Utilizing techniques borrowed from physical chemistry, chemical oceanographers study such things as water "mass," water "type," the carbonate cycle, and the geologic history of seawater. Marine chemists have devoted much of their time to applied research, particularly to corrosion analysis. (Various types of pollutants, especially industrial ones, can enhance or negate the corrosive effects of sea water.) Marine chemists have also aided biologists by developing techniques in nutrient analysis that enable them to make more accurate surveys with respect to marine life.

Biologic oceanography

Life stirred first in the sea or at the rim of the sea, which cradled and nurtured it throughout eons of time. When geologic oceanographers remove samples from the ocean basins, it is biologic oceanographers who study them, searching for fossil forms of early oceanic life to piece together segments of early biologic history.

Marine biology, probably the most glamorous and exciting of all areas in oceanography, has received widespread publicity. Many students starting careers in oceanography are first attracted by marine biology. It is unfortunate that only the more exciting and dynamic aspects of this world are generally viewed. Surely no one mentions such unromantic and routine tasks as taking plankton samples from Long Island Sound in the middle of January!

The biologic oceanographer studies such phenomena as the distribution, occurrence, and speciation of plants and animals in the oceans. One might call "mariculture," or "farming the sea," an area of applied marine biology. Although this field is often romanticized, it is important to realize that this area is just as rigorous a discipline as the others. For example, consider marine biologists who want to study a body of water with respect to *primary productivity.** They must measure carbon dioxide, light, nutrients, turbidity, and temperature, and in the process they may apply techniques from physical, chemical, or geological oceanography.

Biologic oceanographers today are particularly involved in the development of techniques to assess the damage pollution does to plant and animal populations. To do this, they first study plants and animals called "indicator species," which need relatively constant environments to flourish and reproduce. Then they must try to relate changes in the populations of such species to specific changes in the environment. Just as canaries were used to indicate the presence of noxious gases in mine shafts many years ago, many marine animals have been discovered that point out given pollutants in an ecosystem. From their studies of such species in estuaries, many marine biologists fear that this most important asset from the sea is being exploited, overfished, and polluted right out of existence.

Geologic oceanography

From time to time, parts of our earth experience tremendous shifts of the crust, which we feel as earthquakes. As a direct consequence, tsunamis are often formed. Since so many of our population centers lie in the coastal areas most vulnerable to these horrifying phenomena, one tsunami may kill thousands of people. The geophysical oceanographer studies earth movements, and perhaps some day these destructive occurrences may be predicted far enough in advance to help give warning and save lives.

The geologic oceanographer, marine geologist, or marine geophysicist is concerned with the ocean floor and mineral and oil deposits located there as well as with the important concept of continental drift. Several decades ago many scientists felt that the deep ocean floor was quite barren and uninhabitable, and that the technology needed to delve into the deepest parts of the abyss could not be devel-

*Primary productivity is the total living matter accumulated through all the assimilated energy of the sun.

oped until well after the year 2000. It is now known, however, that there are rich deposits of cobalt, manganese, and copper, not to mention offshore deposits of petroleum. Shortages of land sources of these commodities have made it profitable for nations to develop the technology for their retrieval. Many marine geologists have contributed their expertise to discovering these rich deposits. Other interests of these scientists include sediment deposition, destruction of beaches due to wave motion, and the general topography of the ocean floor.

Through the efforts of marine geologists, the route that domestic pollutants, namely garbage and sludge, take across the continental shelves is being researched. Many marine geologists feel that the idea of dumping such materials onto the continental shelves with the subsequent "funneling" down to the seafloor is foolish. Much of this research occurs off the coasts of our large seaboard cities, which are responsible for dumping millions of tons of garbage onto the continental shelves each year.

• • •

From the preceding discussion, one might wonder why there are such divisions of oceanography at all. In the past few decades the field of oceanography has grown tremendously, and it would be impossible for any one man to learn all there is to know about the sea. We have indeed learned a great deal about the ocean, not from any one person, but from groups of men and women working as teams aboard ships or in laboratories on some remote or nearby ocean. Even though such names as Darwin, Forbes, Maury, Agassiz, and Cousteau often symbolize the glamour of oceanography, one must constantly remember that these men possessed the skill, stamina, and desire to learn about the sea. They had to amass men, money, materials, and equipment to make their discoveries. They were first of all leaders with a common purpose in life—a love of the sea.

Although its variety makes oceanography a fascinating area for study, it is also a necessary one if we are to survive. Not only must we study this ocean world, however, we must also learn to harness and utilize its resources much more intelligently than we have in the past. Therein lies the importance of this young science to our future.

21 *the elixir of earth*

Ocean water

O fair green-girdled mother of mine,
Sea, that art clothed with the sun and the rain,
Thy sweet hard kisses are strong like wine,
Thy large embraces are keen like pain.

The Triumph of Time
Algernon Charles Swinburne

In the veins of each of us flows a tiny ocean—the blood—perhaps a remnant of a time when our distant ancestors lived in the sea. The importance of ocean water—past, present, and future—cannot be overestimated. It not only nurtured the beginnings of life, but through its vital presence life is maintained. This vast accumulation of water moderates the earth's climate by absorbing and circulating radiation from the sun. In addition to providing a major source of the free oxygen that land animals breathe, it absorbs tremendous quantities of pollutants from our modern technologic world. Finally, the sea contains dissolved substances that make it an intriguing storehouse of food and minerals. The sea is the last frontier, man's hope for prolonging life, the true elixir of earth.

SALINITY

Ocean water is salty, as everyone knows, but most people fail to recognize the enormous variety and amount of matter we loosely call "salt." This marvelous solution contains not only the compound we call table salt, but many other inorganic compounds referred to chemically as "salts." The proportion of this dissolved matter in sea water is referred to as salinity.*

Source of salts

Since the beginning of time the seas have received abundant supplies of soluble minerals through runoff from the land. This is not the sea's only source of salts, however, for rocks and sediments on the seafloor also dissolve, adding their own distinctive contributions to the composition of the sea. Another source of materials is volcanic action, which adds several unique salts to the sea's supply. Seawater today contains (in varying proportions) almost all of the 92 naturally occurring elements that have been identified.

Identification of salts

How have we managed to identify all of these elements in the sea? Most of them have been isolated through direct chemical analysis, but sometimes indirect methods are more useful. Even elements that can be found only as tiny traces may be concentrated somewhere, and the scientist needs only to know how to look. The *tunicate,* or sea squirt, is an excellent example. This animal actively extracts the element vanadium from seawater and uses it in its oxygen-carrying fluid, much as we use iron in the hemoglobin of our blood. Unlike the sea squirt, we have not learned to extract usable quantities of vanadium, but this ability of the tunicate is evidence that this mineral occurs in seawater and that it can be extracted.

Variations in salinity

Although the worldwide average of salinity in the sea is 35‰, highly significant variations occur in certain regions. The most important of these is the surface water, which is constantly subject to the effects of such processes as *evaporation, freezing, melting, precipitation, runoff,* and *mixing.* Of these six, however, only evaporation, precipitation, and

*Salinity is expressed in parts of dissolved matter per thousand parts of seawater. The symbol for salinity is ‰. Therefore a salinity of 10‰ would mean 10 parts of dissolved salts in 1000 parts of seawater.

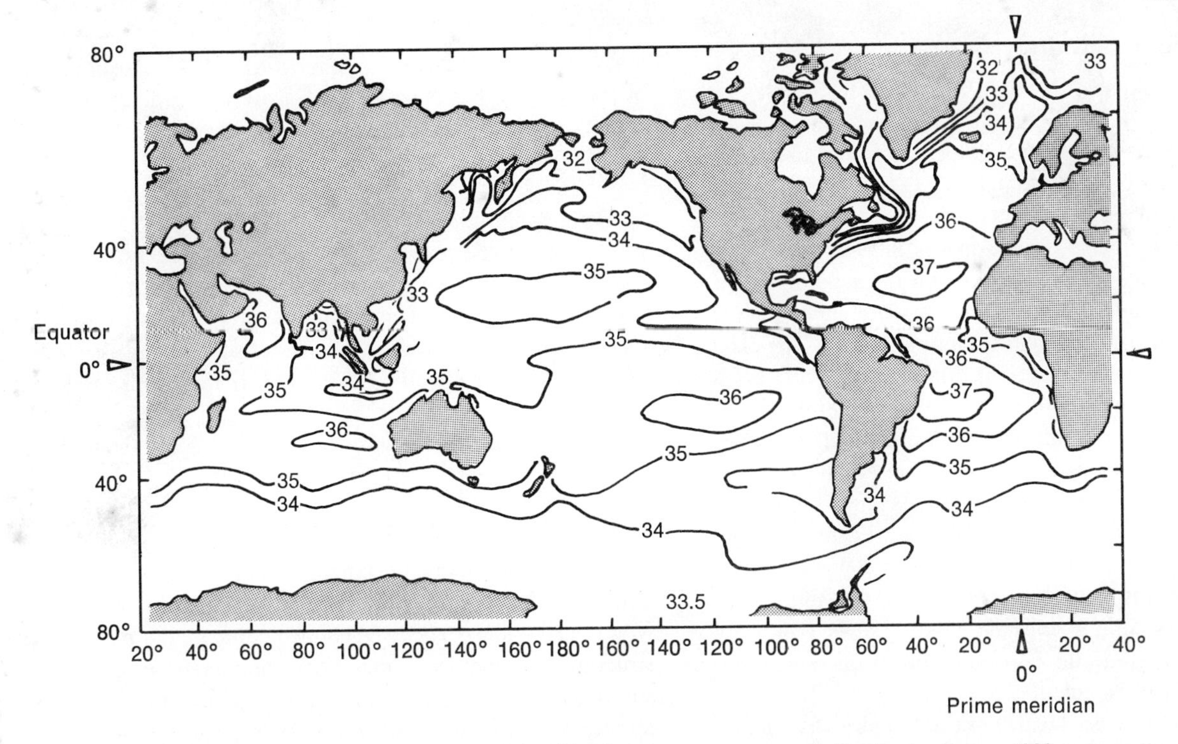

Fig. 21-1. Worldwide distribution of surface salinity. Numbers represent parts of dissolved salts per 1000 parts water.

mixing produce significant effects. The increase in salinity that comes with the winter freezing of seawater, for example, balances out when the ice melts again in the summer. Runoff, of course, affects only coastal areas.

Variations in surface salinity, then, are primarily the result of changes in precipitation, evaporation, and mixing rates. Since these three factors vary with global climatic changes, surface salinity is also related to latitude (Fig. 21-1). Generally speaking, ocean-borne salts are found in the smallest concentrations near the equator, where average temperatures are highest and precipitation rates are greatest. Maximum salinity rates are found around 20° N and 20° S latitude, where evaporation is greatest. Salinity decreases closer to the poles because temperatures and precipitation and evaporation rates are lowest. It is these variations in surface salinity that produce the changes in surface density, which in turn are partly responsible for oceanic water movements. These movements eventually mix the surface water with that below, eliminating salinity variations over time (Fig. 21-2).

At greater depths, salinity varies much less than at the surface. As density variations cause continual mixing, salinity below a depth of 3000 feet (1000 meters) approaches the general average, 35‰, as shown in Fig. 21-2. At these depths salinity is more closely affected by temperature than by any of the other factors we have discussed.

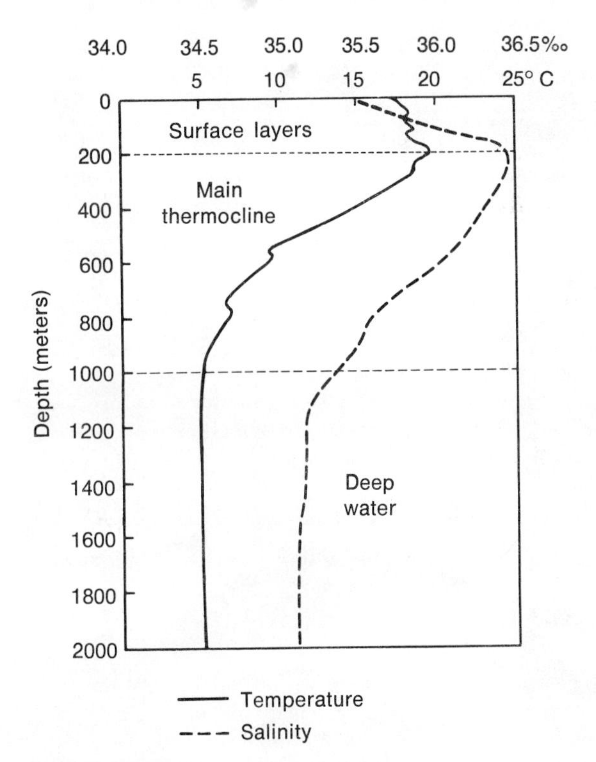

Fig. 21-2. Surface anomalies in salinity and temperature are eventually erased by water currents; below a depth of 300 meters, both salinity and temperature stabilize.

Determination of salinity

At first it would seem an easy task to determine the salinity of seawater. Simply place 1 kg (or any other known quantity) of ocean water over a flame, wait until the water boils away, and weigh the remaining salts (1 kg of water should yield about 35 gm of salts). This method, however, is not sufficiently precise, since some of the salt is also lost in the process. Oceanographers, requiring a precision of ±0.001 gm/kg, have developed much more sophisticated techniques and instruments for measuring salinity. The classic method is called *chemical titration*. In this process a chemical is added drop by drop to a given volume of seawater until an awaited result, usually a precipitate or color, appears. Different chemicals and different amounts of any chemical produce different results, so the precipitate, the amount of it, or the color that appears are all clues as to the chemicals present in the original water. With William Dittmar's 1884 confirmation that "the major constituents of seawater exist in almost constant ratios," it is now only necessary to determine the concentration of one substance and any others can be calculated.

Today, faster devices measure salinity indirectly, negating tedious, time-consuming chemical analyses. One such instrument, called the *salinometer,* utilizes the measurement of the electrical conductivity of seawater to determine its salinity (Fig. 21-3).

TEMPERATURE OF SEAWATER

Perhaps the greatest gift the sea gives us is its moderating effect on the earth's climate. Seacoasts generally have warmer winters and cooler summers than noncoastal areas in the same latitudes. This phenomenon is explained by the *specific heat* of water, its ability to absorb or release heat. With a very high specific heat, water both absorbs and releases heat slowly, so it absorbs more heat in the summer than surrounding land masses and then releases it so slowly that its warmth lasts well into the winter.

General thermal characteristics of water

Water is a unique substance, in that it has much higher boiling and freezing points than its chemical structure would lead one to expect. When we compare the molecular structure of water with that of a chemically similar substance such as hydrogen sulfide (H_2S), we find that pure water should freeze at −238° F (−150° C) and boil at −148° F (−100° C). At first glance, then, the chemical composi-

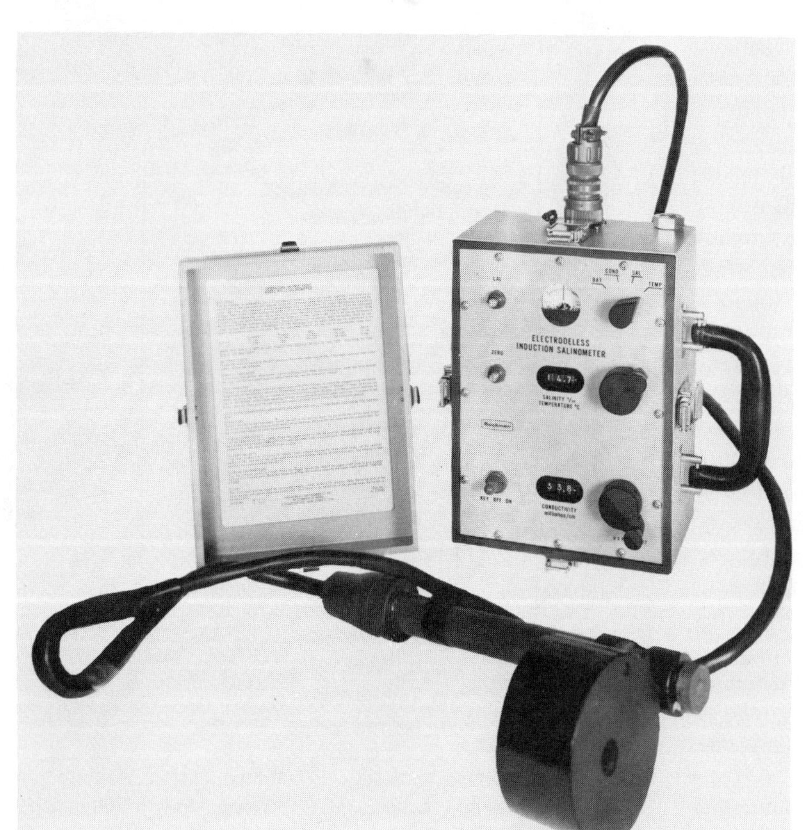

Fig. 21-3. The salinometer, a relatively new instrument, determines the salinity of ocean water by measuring its electrical conductivity. (Courtesy Beckman Instruments, Inc., Cedar Grove, N.J.)

tion does not explain why the freezing and boiling points of water are actually 32° F (0° C) and 212° F (100° C), respectively. Recently scientists have considered the possibility that water may act as a polymer, that is, a chain of water molecules linked together. To change from one state to another, then, the water molecules must first break the chain and then acquire enough energy to accomplish the state change. Since this is a two-step process, it takes more energy (a higher temperature) than a similar process involving H_2S, in which the molecules are not in chains. It is the need for this higher energy that keeps water from changing state at the temperatures predicted on the basis of isolated molecules.

Temperature distribution in the sea

The temperature of seawater is highly variable both horizontally and vertically. Surface water temperatures vary almost entirely with air temperatures, making latitude the dominant factor in surface water temperatures. With few exceptions, therefore, the coldest surface waters lie in the polar regions, while the warmest are near the equator. However, remember that water warms and cools more slowly than air, so although the warmest water lies in the same area as the warmest air, the air and water are *not* the same temperature. The average range in worldwide surface water temperatures, then, is from 86° F (30° C) to 28° F (−2° C), much narrower than the range for the atmosphere. The highest oceanic surface temperature ever recorded was 96.8° F (36° C) in the Persian Gulf. The minimum low temperature is 28° F (−2° C), which is the freezing point of salt water.

Vertically the temperature range of seawater varies even less. To explain vertical temperature distribution in the sea, oceanographers have developed a three-layered model based on the occurrence of *thermoclines,* regions in which the temperature drops abruptly with increasing depth (Fig. 21-2). The first 660 feet (220 meters) comprise the *surface layer,* which is strongly influenced by meteorologic conditions and thus exhibits a variable temperature profile that often fluctuates daily. Between 660 and 3300 feet (200 to 1000 meters), on the other hand, is a temperature region known as the *main thermocline,* where the temperature drops slowly with increasing depth; although the temperature of this layer may show seasonal variations, it does not vary from day to day. Finally, below 3300 feet (1000 meters) is the region of deep water that has a very narrow temperature range between 32° F and 39° F (0° C and +4° C). This range in temperature, uniform over the entire globe, makes the deep layer the most stable environment on earth and one of great importance to the distribution and speciation of plant and animal populations.

Measuring temperatures in the sea

Tying a thermometer to a string and dangling it in the ocean would appear to be the simplest way to measure the temperature of seawater, and this is in fact the way in which surface temperature readings are taken. For depth readings, however, more sophistication is needed, for a thermometer would register temperatures all the way down and then all the way back up, making it impossible for the scientist to know just what temperature had been measured at the desired depth. To solve this dilemma, oceanographers employ a device called a *reversing thermometer* (Fig. 21-4), which is usually used with a water-sampling device called a *Nansen bottle*. The reversing thermometer works much like a regular mercury thermometer, except that just above the bulb there is a constriction and then a loop. As the Nansen bottle is overturned to collect the water sample, the mercury column in the tube breaks and flows into the loop. It is here that the thermometer is read. Further changes in temperature cannot be registered because the constriction does not

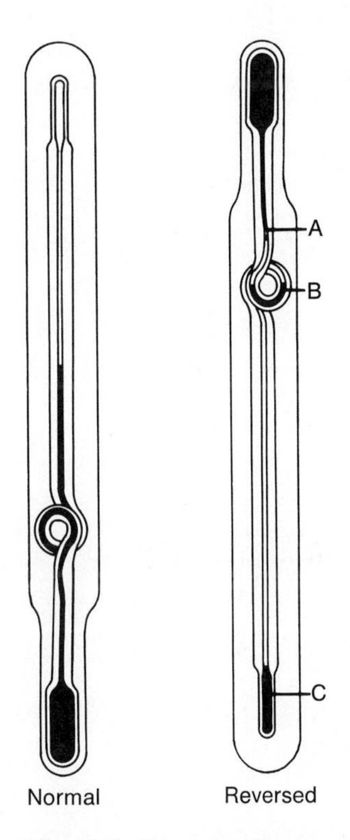

Fig. 21-4. Reversing thermometers are used to measure temperatures at depth. Above the bulb is a constriction *(A),* then a loop *(B)*. When it is used the mercury column breaks and mercury flows into the bulb *(C),* where this type of thermometer is read. This prevents the thermometer from registering subsequent readings as it is pulled to the surface.

allow the mercury to go back into the tube. The temperature registered on the device when it is raised to the surface is therefore the temperature at the desired depth, not that at any of the layers it passed through on the way up.

PRESSURE IN THE SEA

Of all the variables in the sea, the most obvious is pressure. Diving times must be carefully calculated to allow for it. Because of the immense pressure at greater depths, the sea is the environment most hostile to man, more hostile in many ways than space. This becomes apparent if we consider the meaning of pressure in the sea. In the atmosphere, pressure is the weight of the column of air starting at the top of an object (the head of a person, for example) and extending all the way to the top of the atmosphere. Water pressure similarly is the weight of the column of water starting above an object and extending to the water surface. But when calculating pressure in the sea, we have to consider both atmospheric pressure and water pressure, because that column of air is pressing on the surface of the sea, which is the top of the water column. At a depth of 100 feet, the water pressure is about 43 pounds per square inch (psi), but when we add the air pressure (14 psi) our total comes to 57 psi, and that is only 100 feet down!

In oceanography as in meteorology, pressure is measured in a unit called a *bar*.* Measurements have shown that for every 33 feet (10 meters) of depth, pressure in the sea increases 1 bar. This approximation is close enough to use for all routine oceanographic calculations. To know depth, then, is to know pressure. Since the sea is so deep, clearly we have enough feet for a great many "bar" increases. Pressures at a depth of 36,000 feet (10,920 meters) in the Marianas Trench were measured at more than 1000 bars. Even at intermediate depths of 33,000 feet (10,000 meters), pressure exerts a force 1000 times greater than it does at the surface.

Unlike temperature, pressure varies very little horizontally in the sea. Those variations that do occur, moreover, are due almost entirely to density variations, which will be discussed shortly.

Hydrostatic pressure: man's limit in the sea

Despite the great pressures there, many life forms exist on or near the ocean floor. Although these forms are not as abundant or as varied as those in surface waters, many species of both plants and animals have adapted to this environment and consider it as normal as our surface pressures are to us. Since adaptation to unusual pressure involves an organism's ability to match outside pressure with its own internal pressure, this adaptation is slow if not impossible in most creatures. This is why a diver is so aware of pressure changes in the sea.

To carry the comparison between the environment on the sea bottom and that in space further, we can consider the problems of building vehicles for travel in the two areas. Clearly a vehicle for traveling at depths in the sea must be able to withstand tremendous pressures on its hull; a space vehicle, by contrast, will be subjected to no pressure at all once it leaves the earth's gravitational force. An "aquanaut" could not take the romantic walks in the sea as his counterpart enjoys walks in space because the pressure would crush him if he ventured out of his protective shell. His ability to directly perform useful work is therefore limited to whatever can be done in the top 264 feet (80 meters) of the oceans. Even at these depths, he fatigues easily and has difficulty in maintaining adequate body temperatures. Thus underwater exploration presents every bit of the challenge of space exploration, and promises the same reward—a whole new world to explore.

*One bar is very nearly equal to 14.5 psi, approximately the same as atmospheric pressure.

DENSITY

Density is defined as mass per unit volume. To visualize density, imagine a class of 30 identical students, each representing a water molecule. Suppose that this class is forced to hold sessions in a room half the size of its regular classroom. To fit into the room, the students must sit closer together than they do in their own room. When they return to their regular room, they can spread out again. Density, then, is a measure of the distance between molecules and is controlled by the volume or cubic space a given number of identical molecules is allowed to occupy.

A term often confused with density is *specific gravity*. Specific gravity, however, is not a measurement like density but rather a comparison of measured density with a constant standard, pure water, at its minimum volume temperature of 40° F (4° C). The specific gravity of water in this state has a value of 1 because water has a density of 1 gm/cm^3. A substance with a specific gravity of 4, then, is four times denser than water, and, therefore, has a density of 4 gm/cm^3.

Density variations in seawater

Unlike pure water, whose density is influenced only by its temperature and pressure, the density of salt water is also influenced by its salinity. Therefore *the density of seawater increases with both pressure and salinity but decreases when temperature increases* (Fig. 21-5). Since all three of these factors are operating at any given time, it is

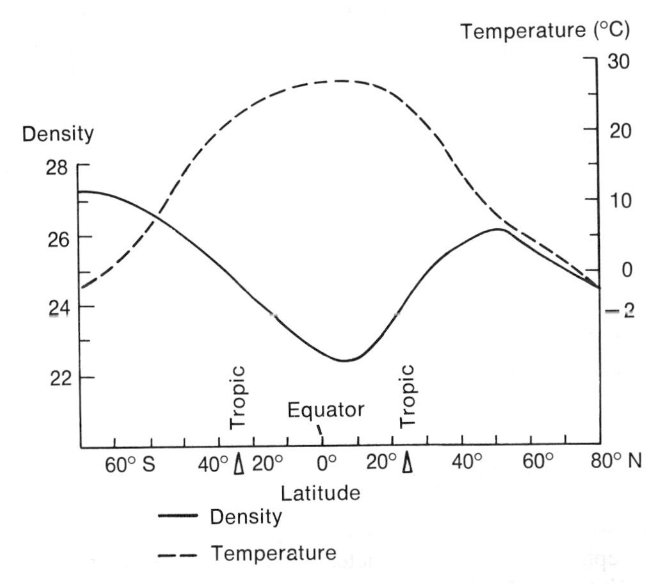

Fig. 21-5. Because water density is so dependent on temperature, it varies with latitude, as shown here. Note the drastic drop in density approaching the equator, as that is where temperatures are highest.

important to remember that the influence of temperature and salinity on density is much more significant than that of pressure, particularly in surface waters. This means that ocean water will be particularly susceptible to the climatic and geographic variations that affect temperature and salinity.

Density of seawater in the open ocean ranges between 1.02400 and 1.0300 gm/cm³. (In 1 cm³ you would find about 1.02400 to 1.0300 gm of water and other matter.) The density is lowest near the equator where the temperature is highest, and peaks at around 60° N and 60° S latitude. At these locations the increase in salinity predominates over the decrease in temperature. Farther north and south, density decreases slightly because again the balance between salinity and temperature tips slightly in favor of salinity, and density therefore drops. (Since the temperatures are low, density influenced only by temperature would be high.)

Seawater also separates into layers according to density. The denser layers, logically enough, are on the bottom because the denser layers, being heavier, naturally sink; second, the bottom layers are compressed into a smaller volume by the weight of the layers on top of them.

DISSOLVED GASES

Since the contact zone between ocean and atmosphere constitutes such an enormous area, one would expect there to be some exchange between air and water molecules. Fortunately this is indeed the case, and both plant and animal ocean life need the dissolved gases that enter the sea in this way to survive. Not a single marine animal has the ability to break down the water molecule to use its constituent oxygen, nor is a single known plant able to manufacture its own carbon dioxide. Both therefore depend on these dissolved gases for life.

Nitrogen

Nitrogen, comprising 78% of the lower atmosphere by volume, supplies only 64% of the volume of dissolved gases in the sea. This anomaly is due to its low solubility; nitrogen dissolves only about half as readily as oxygen, for example. Due to its chemical inactivity, the concentration of nitrogen in the sea remains largely constant and has little effect on the creatures living there. The one exception is a bottom-dwelling bacteria that manufactures nitrates and ammonia salts from the nitrogen in the sea.

Oxygen

Unlike nitrogen, oxygen is more abundant in seawater than it is in the atmosphere. While only 21% of the lower atmosphere is oxygen, oxygen accounts for 33% of the dissolved gas content in the ocean. This difference is explained first by the high solubility rate of oxygen, and second, by the fact that it enters the water from two sources. Oxygen comes not only from the atmosphere, as do other gases, but also from marine plants, the phytoplankton, who give it off as a part of their life processes. These plants alone are responsible for up to 80% of the free oxygen we need. Following close on the heels of this relatively recent discovery come current suspicions that the presence of DDT in ocean water may be interfering with the ability of these phytoplankton to function. Care must be taken to protect these microscopic benefactors.

Because both of the oxygen-producing processes are associated primarily with the surface, we would expect oxygen concentration to decrease with depth. Assuming that other factors such as water mixing and plankton population remain constant, the oxygen content of seawater does indeed decrease in this way. This decrease, however, is not uniform. The oxygen concentration decreases to a depth of 300 to 2400 feet (100 to 800 meters). This level is called the *oxygen minimum layer* (OML). Below this layer the oxygen concentration actually increases slightly and then decreases dramatically. The OML is believed to coincide with the main thermocline and probably overlies a region of active mixing, which brings in oxygenated water from other sources. After the brief increase in oxygen levels, the concentration drops off because animal respiration and bacterial decay deplete available supplies rapidly. The longer water remains below the OML, the less oxygen it retains, making oxygen at depth a good tracer for water motion.

Carbon dioxide

Carbon dioxide is second only to oxygen in importance in the sea. Comprising a tiny 0.03% of gases in the lower atmosphere, carbon dioxide makes up 1.6% of the dissolved gases found in seawater. Lacking oxygen's large surface source, most carbon dioxide is supplied by a chemical system in the sea itself. Bicarbonates of sodium, potassium, and calcium are brought to the sea by streams, and here they undergo chemical reactions that change them into carbonic acid and free carbonate ions. The carbonic acid then dissociates, releasing carbon dioxide in gaseous form. The reaction is reversible, so excess carbon dioxide can be stored as carbonates in rocks and shells. This permanent storehouse can be called on to maintain equilibrium in the oceanic system.

Carbon dioxide concentrations vary with depth in an almost perfect mirror image of oxygen concentrations. Any increase in oxygen is usually related to the photosynthesis of phytoplankton, which reduces carbon dioxide levels in favor of oxygen. Conversely, a decrease in oxygen levels can usually be traced to marine animal respiration, which tips the balance in favor of carbon dioxide at the expense of oxygen.

LIFE IN THE SEA

The chemical and physical properties of seawater are key factors in the development, evolution, and fate of an impressive array of life forms. Thus marine life may easily be considered an integral part of ocean water. As such, it is subject to the same natural laws that affect other aspects of the sea.

Marine environments

The marine environment consists of four related subenvironments. The *benthic* encompasses the entire ocean floor, while the *pelagic* is composed of the water mass itself (Fig. 21-6). Each may be further subdivided; the dividing line is arbitrarily set at a depth of 660 feet (200 meters). The benthic is subdivided into the *littoral* and the *deep-sea* systems. The littoral system is the shore environment, and the deep-sea system covers that part of the sea floor lying under 660 feet or more of water. The pelagic environment is likewise subdivided. The *neritic* province of the pelagic environment includes the shoreline waters overlying the littoral system, and the 660 feet or more of water that bury the deep-sea system are called the *oceanic* waters.

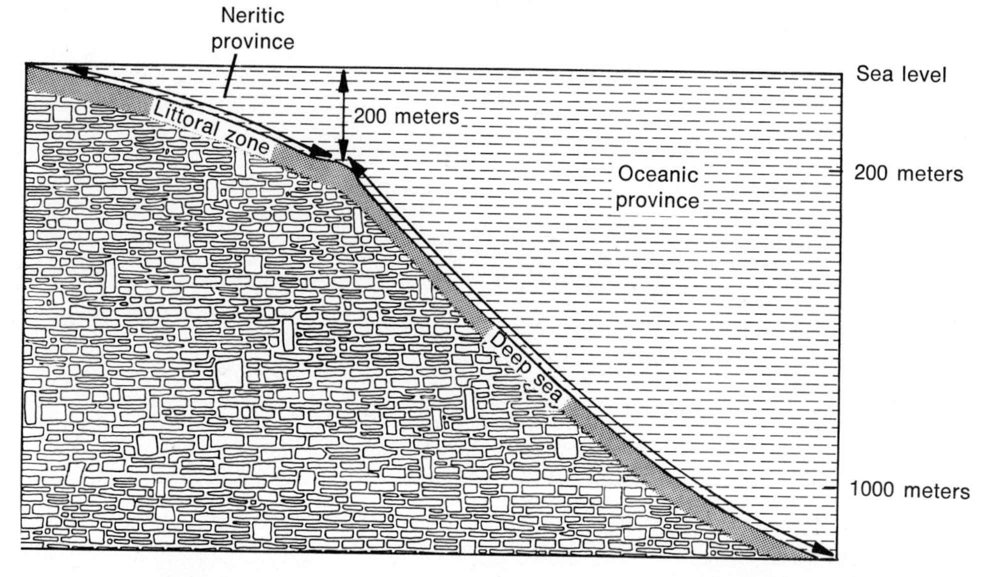

Fig. 21-6. To allow for more precise study, the ocean and the land beneath it are divided into two zones each. The land that adjoins the shore is called the littoral zone, and the water over it is called the neritic province. The land covered by 200 meters or more of water is called the deep-sea zone, and those 200 meters of water are the oceanic province.

Marine plants

Although fossils of marine algae indicate that plants have lived in the sea for nearly 3.2 billion years, such species are amazingly simple in structure and limited in variety. Clearly land plants have evolved and diversified more than their marine cousins. What marine plants lack in numbers of species, however, they more than compensate for in numbers of individuals. It is their immense population rather than their variety that gives these plants their major role in the earth's ecosystem.

Relationship between marine plants and man

Initially the relationship between marine plants and people seems rather limited, for plants provide direct benefit to man in only two ways. However, these two direct benefits, oxygen and food, are our most basic needs.

Both animal life in the sea and animal (including human) life on land are dependent on the oxygen produced through a plant process known as *photosynthesis*. While we will not explore the complexities of this process here, basically it involves a plant taking in carbon dioxide and water from its environment and using energy from the sun to convert them into a sugar for the plant's own nourishment and into oxygen. Because the plant has no use for the oxygen, it returns this vital gas to its surroundings as waste, which marine animals take in and circulate through their own systems. The animals expel carbon dioxide as waste, and plants use this gas to continue the cycle. As discussed earlier, much of the oxygen that marine animals do not use eventually finds its way to the sea-air interface and thereby into the atmosphere for use by land animals.

Sea plants provide food both directly and indirectly. Microscopic forms, known as phytoplankton, are too small to be efficiently harvested by us, but if we sit back and let fish or other marine animals do the harvesting, we can still benefit from their nutritional qualities by eating the larger animals.

Larger plants such as algae serve as food for marine animals, and when they are harvested they are devoured enthusiastically by land animals as well. In the Orient, many types of seaweed are regularly used as food items and provide a supply of iron and protein in these chronically meat-short societies.

Description of marine plants

Most marine plants are primitive forms with little or no specialization of parts; this means that such "parts" as roots, stems, and leaves are not well developed. Included in this group are marine algae (usually called seaweed) and fungi, including bacteria. The marine algae are usually beautifully colored, and can be classified according to this property as blue-green, green, brown, red, and yellow-green algae (Fig. 21-7).

There are certain more complex plants that have true roots, stems, and flowers; since they live in the seawater they are pollinated with the aid of water currents in very much the same way that their land cousins are helped by wind currents. If this type of plant sounds a good deal more like the plants you are familiar with, it should; (Fig. 21-8)

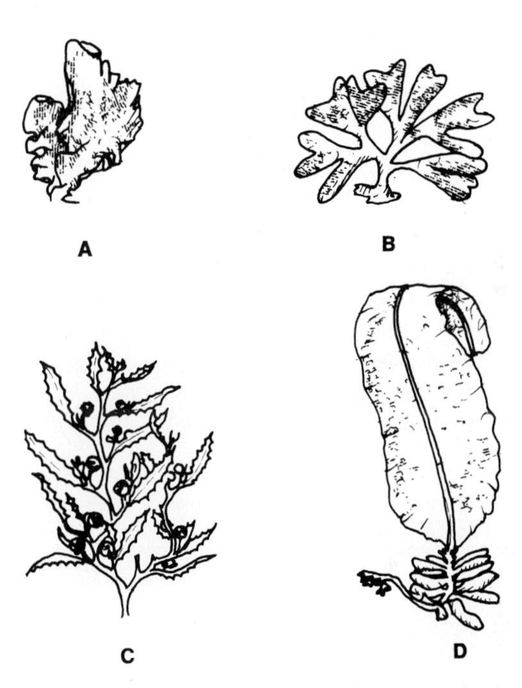

Fig. 21-7. Seaweed. Note that despite their diversity, these plants are relatively simple and appear rather "leafy" without much in the way of roots or stems. **A,** Ulva; **B,** fucus; **C,** red sargassum, from which the Sargasso Sea takes its name; **D,** alaria.

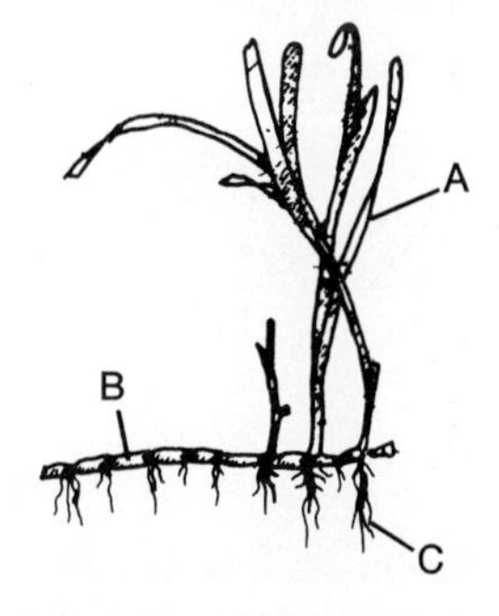

Fig. 21-8. Eel grass shows the plant development we are familiar with in land plants. Note that unlike the lower algae, this plant has *A,* leaves; *B,* rhizome; and *C,* roots.

scientists believe that after the land plants evolved from marine plants, certain of the more complex plants reversed the process and reestablished themselves in marine environments. Supporting this theory is the location of many such plants off the coasts of Europe, North America, Asia Minor, and Eastern Asia, where these *angiosperms,* or flowering plants, flourish close to the land they left.

Marine animals

Unlike plants, animals have attained wide variety as well as abundance in the sea as on land. Evidence indicates, moreover, that even those species that now live on land originated in the sea. Let us look at the variety of animals that still do inhabit this complex world.

Microscopic animals

The tiniest animals in the sea are one-celled animals of either microscopic or minute size. They may swim freely about, creep along the ocean floor, or attach themselves to the bottom. The free-swimming varieties are generally classified according to their means of free swimming. Some forms, called *flagellates,* move by whipping themselves through the water with a tail-like appendage called the *flagellum.* Some row themselves through the water with *cilia,* or little hairs, that cover parts of their bodies. Still others virtually ooze along like the monsters in science fiction movies by creating an internal flowing motion in their bodies. In all, there are more than 6000 variations in these simple animals, including one group that overlaps into plant forms. One type, Radiolaria, has skeletons that form some of the delicate and beautiful patterns associated with this strange undersea world (Fig. 21-9).

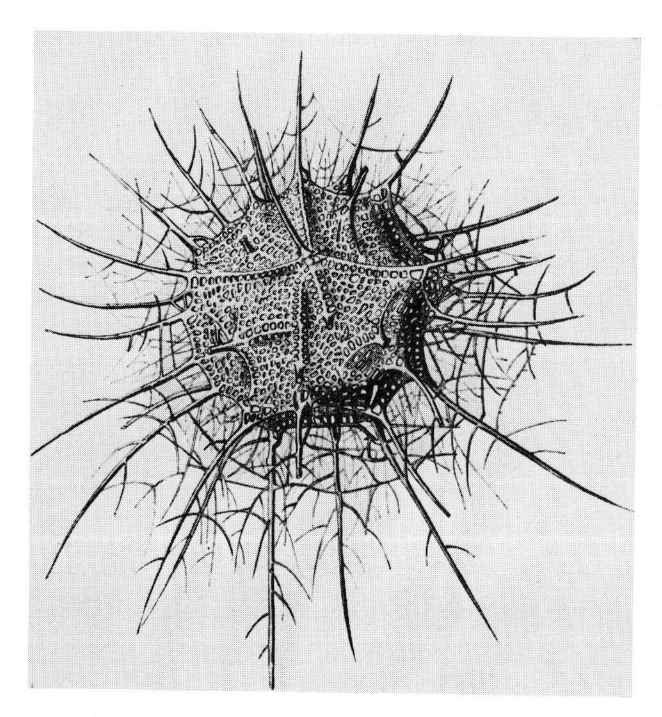

Fig. 21-9. Microscopic lacy radiolaria are among the most beautiful animals found in the sea. (Courtesy the *Valdivia* expedition.)

Colonial animals

Sponges. One of the more familiar colonial animal groups is the sponges, multicelled creatures living together in colonies. It is the strength of the conglomeration of skeletons that supports each animal (Fig. 21-10). Unable to swim after prey, sponges obtain their food by flushing surrounding water through tiny pores in their body walls, straining out microorganisms and other appealing debris as the water passes through. Sponges can be found at all depths in the sea, but the more abundant forms are found only in the deep-sea abyss.

Corals. Another fascinating animal that has found "safety in numbers" is the coral. These are small marine animals that live on the ocean bottom at virtually all depths in the sea. Corals are most abundant in warm, semitropical and tropical waters, but certain cool-water species are found as far north as Cape Cod. Coral have hollow, saclike bodies with a single opening, called the *mouth,* at the top (Fig. 21-11). Surrounding the mouth are a number of armlike

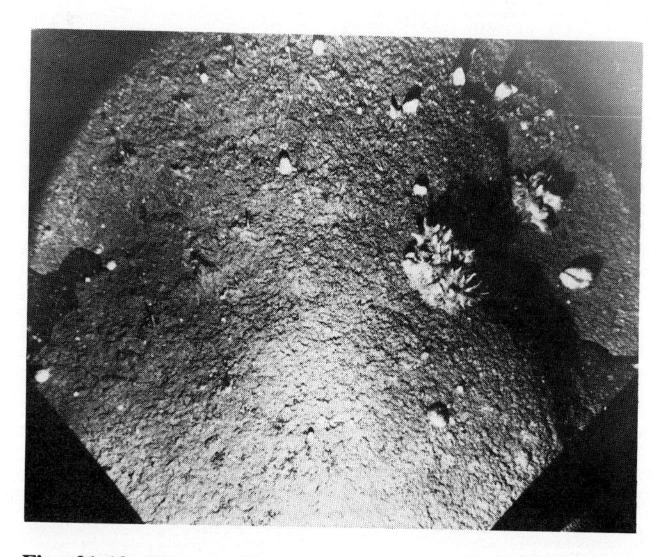

Fig. 21-10. Sponges like these live in colonies, each gaining support and protection from the skeletons of its neighbors. (Courtesy Lamont-Doherty Geological Observatory of Columbia University, Palisades, N.Y.)

tentacles that contain stinging cells. The creature uses these to collect the plankton it feeds on.

The coral that most interest us, however, are those that build reefs. These differ from the others in that they are able to secrete a calcium-based skeletal "cup" into which they can withdraw for protection. As older coral die, these skeletons remain, providing a base for new individuals. At an average rate of only about 1 inch (2.5 cm)/yr, these shells eventually grow into the huge barriers we call coral reefs.

FACTORS INFLUENCING REEF GROWTH. Reef coral require that the water temperatures remain in a fairly narrow range, with an average of 68° F (20° C). Coral grow most profusely at temperatures of about 75° F (24° C), but this growth ceases if the water approaches 87° F (30° C). This narrow temperature range leaves coral vulnerable to damage from both natural and artificial causes. It is now feared, for example, that the only remaining live coral reef in the continental United States is being adversely affected by the vast quantities of warm water released into the ocean by a nuclear electric generating plant at Turkey Point, Florida. On the other hand, reefs normally show their greatest development in the western parts of tropical oceans because on the eastern sides a natural upwelling of cold bottom water lowers the temperature below the required range.

Fig. 21-11. Cutaway view of a coral shows its internal structure. *1,* Tentacles; *2,* stinging cells; *3,* mouth; *4,* stomach; *5,* tubes for water vascular system. Since these animals generally live in colonies, the coral you are likely to see consists of groups of animals like this one.

Another factor influencing the locations and growth of reef coral is sunlight. Reef coral are dependent on huge quantities of plankton, which grow prolifically only in brightly illuminated water less than 200 feet (60 meters) deep. Anything such as sediment or pollutants that affects the clarity of the water affects the reef's growth. Scientists studying the reef off Turkey Point, Florida, are now quite concerned about the effects of the sewage Floridians regularly dump into the sea.

Salinity in the normal range (35‰) and a rich supply of free oxygen in the water are also prerequisites to successful coral reef development, so that pollutants that affect salinity or oxygen concentration are also problems.

TYPES OF CORAL REEFS. The different locations in which all of the conditions necessary for reef growth are found together determine the type of reef that is produced. There are three such types—fringing reefs, barrier reefs, and atolls.

Fringing reefs usually grow parallel to the rocky coastlines they border. In fact, they frequently grow directly attached to such a shore. Excellent examples of fringing reefs include the reef just off the southeast coast of Florida, a part of which has been preserved as the John Pennekamp Coral Reef State Park, and the reef off Runaway Bay in Jamaica.

A closer look at a fringing reef can provide us with a better knowledge of the reef environment (Fig. 21-12). Moving inland toward the coast, the *forereef* first appears, rising steeply from deep water at an angle rarely less than 45° and often as great as 70°, and culminating in great masses of branching, treelike corals. These corals break the force of the incoming waves; they are a living breakwater protecting their neighbors behind. Cutting perpendicularly through the forereef are numerous sand channels that drain sediments from the reef to keep it from drowning in its own debris.

Just shoreward of the forereef lies the *reef-flat,* where corals grow as high as the low-tide mark. Here, in very shallow water, the reef flat harbors a rich variety of living coral species in addition to the dead reef rock. A myriad of colorful tropical fish, sponges, sea fans, sea urchins, and calcerous red algae provide natural "interior decoration." Here and there in the reef flat one encounters sand islands called *cays,* composed of coral sediments. These cays are very numerous in the Bahamas.

Further shoreward lies the slightly deeper *back reef,* or *lagoon,* which usually terminates at a steep limestone cliff. This is a zone of sedimentation, which usually involves sand-sized, calcerous skeletons. In the lagoon, isolated coral groups that are called *patch reefs* (because of their small size) appear.

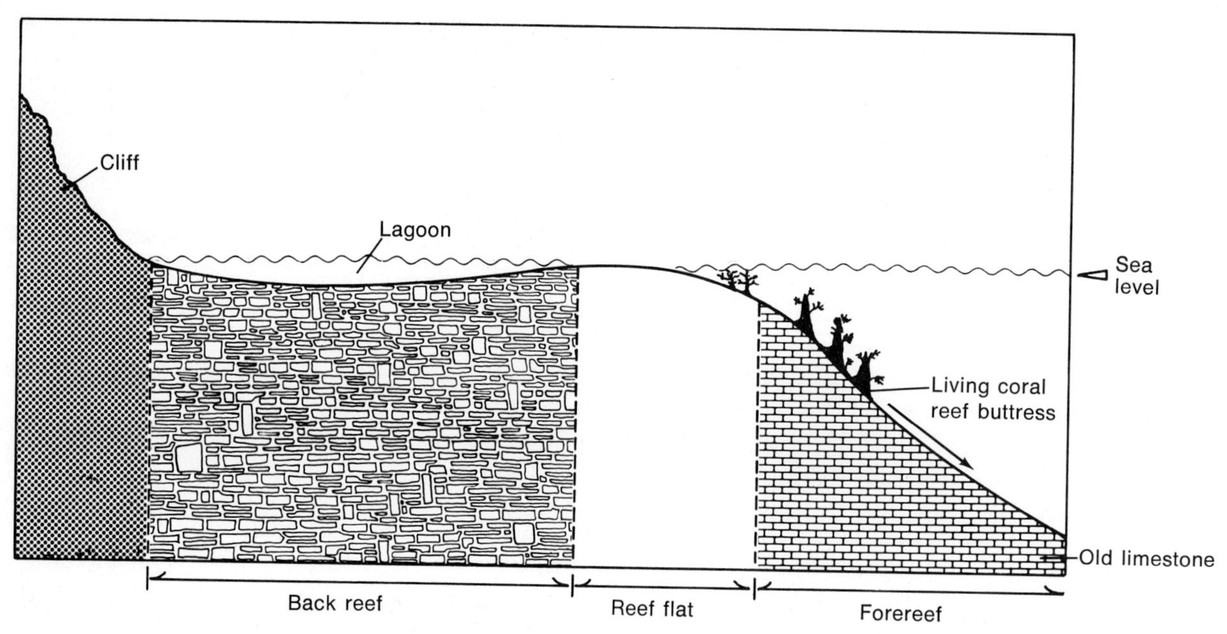

Fig. 21-12. Cross section of a fringing reef shows the forereef protecting the reef flat. Arrow shows the direction of the sand channels that drain off sediments and other debris. The back reef supports a lagoon.

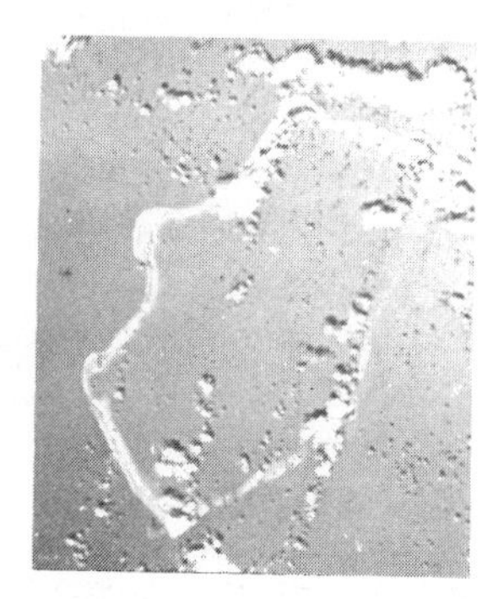

Fig. 21-13. Satellite photograph of an atoll, a circular reef that encloses its lagoon. (Courtesy National Aeronautics and Space Administration, Washington, D.C.)

Like fringing reefs, *barrier reefs* grow parallel to the shore, but they are separated from the land by lagoons too deep to allow intermediate coral growth. Barrier reefs form a breakwater that protects areas on the continents as well as isolated volcanic islands. The largest reef of this type is the Great Barrier Reef, extending for more than 1000 miles (1610 km) along the northeast coast of Australia. Closer to home, another barrier reef protects the Florida keys, which may themselves be the remains of such a reef.

The word ''atoll'' entered the vocabulary of most Americans during World War II, as American and Japanese soldiers fought their way across the chain of Pacific islands formed by the protruding tips of these coral complexes. Kwajalein, the largest, with an area of 840 square miles, Truk, Tarawa, and Ypa became as familiar to American households as Asian jungles seem today. Since the end of the war, these islands continue to appear in the news as the sites for extensive thermonuclear testing.

An *atoll* is a circular reef enclosing a central lagoon (Fig. 21-13). Most are isolated structures found in very deep water and have a series of small, low, coral islands spaced unevenly along them. These islands, varying in width from a few feet to 1 mile or more, are composed principally of unconsolidated debris that is the result of the erosion of the reef itself. Low dunes may reach inland from the beaches. Large areas of these islands are covered with the coconut palms, pines, and bay cedar thickets we mentally associate with island paradises. The central lagoon usually has a flat floor, but here and there knolls of patch coral may break the surface. With many outlets connecting it with the sea, the lagoon is an area of rapid sedimentation.

Coral atolls rest on volcanic foundations that keep the coral close enough to the surface for it to continue to grow. Charles Darwin, the famous naturalist, proposed this volcanic theory in 1842, but it was not confirmed until 1952, when drilling in the Eniwetok Atoll in the Marshall Islands did indeed unearth volcanic rocks.

Worms

Large numbers of worms live in the sea. Some are *parasitic,* meaning they live off other animals, while others are capable of supporting themselves. Since so many of them burrow into the ground, photographs of them are quite rare (Fig. 21-14). Flatworms are interesting, primarily because they are an evolutionist's dream: they have complicated body characteristics that are not present in the animals immediately below them but that are characteristic of all those above them. This would seem to indicate their place in a progression from simple to more complex animals, with the added features falling in all along the way. The features specifically that "fell in" in flatworms are the following.

1. Flatworms have *bilateral symmetry.* Simpler animals are "blob shaped." There is no particular relationship between the right and left sides of their bodies. Worms and more complex animals have definite shapes, and their left and right sides are almost mirror images.
2. Flatworms have a third body layer. Lower animals have only two layers, the *ectoderm* (or outer layer) and the *endoderm* (or inner layer). Flatworms and their descendants have a middle layer called the *mesoderm,* which makes them better able to adapt to environmental changes and therefore increases their chance of survival.

Shelled animals

Shelled animals are probably those sea animals with which you are most familiar. The starfish, sea urchin, sand dollar, serpent star, brittle star, sea cucumber, and sea lily are among the shells that appear in any beachcomber's collection. All 5000 species of shelled animals are marine and none are parasitic. Their lovely shapes are due to calcareous skeletons made up either of shaped plates or scattered spicules (points). Not so obvious is these animals' unique circulatory system, known as the water vascular system, in which seawater reaches all parts of their bodies through a series of tubes. One such animal, the sea cucumber (Fig. 21-15), seems to have developed the ultimate defense against enemies; he literally expels his internal organs, growing replacement parts when the danger has passed.

Mollusks

Another group that is probably familiar to you is the mollusks. Their value is nutritional rather than aesthetic, however, as you find so many mollusks on menus. The more than 80,000 known species include clams, mussels, snails, slugs, scallops, oysters, chitons, octopi, and squid. Their range in body size is enormous—from near micro-

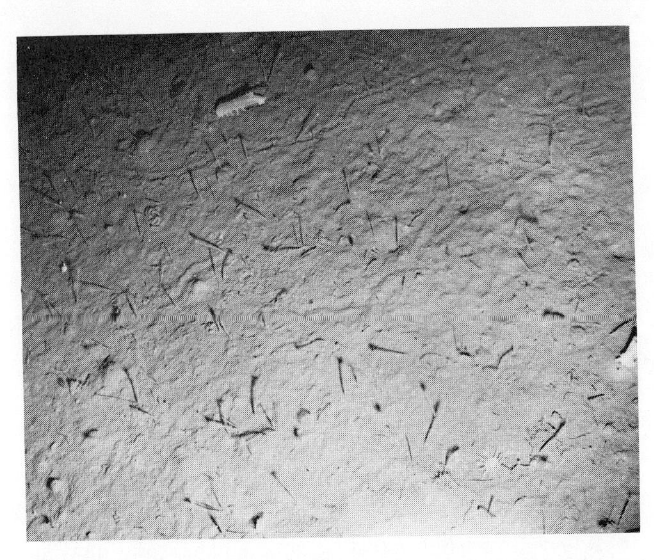

Fig. 21-14. A type of worm found on the seafloor by the crew of the research ship *Eltanin.* (Courtesy National Science Foundation and the Smithsonian Institution Oceanographic Sorting Center, Washington, D.C.)

Fig. 21-15. The sea cucumber and brittle stars show the pointed skeletons characteristic of this type of animal. (Courtesy Lamont-Doherty Geological Observatory of Columbia University, Palisades, N.Y.)

scopic snails to the giant squid, whose tentacles may reach 35 feet in length. Marine mollusks are known to inhabit all depths in the sea; some live on or in the sand, mud, or rocks on the bottom and some float on top of the water. Like sea cucumbers, mollusks have considerable powers of regeneration. Octopi, for example, can grow whole new tentacles.

Marine vertebrates

The *vertebrates,* or animals with backbones, are the most complex animals in the sea. We spoke before of allowing animals to harvest plankton and then harvesting the animals; it is generally these vertebrates that we harvest. This group includes all the varieties of fish, marine mammals such as whales and dolphins, and also snakes, eels, and even certain birds who live near the sea, though not in it.

It is with vertebrates that we generally have our most direct contact, from the fish deep-sea fishermen pursue so avidly to the dolphins the United States Navy is studying with the hope of uncovering the secrets of their communication. Unfortunately, it is also in our dealings with sea vertebrates that we have been most thoughtless, and both whales and several types of seals are now on the endangered species list.

MAN AND THE SEA

Perhaps it is the hostile environment so near at hand, perhaps it is the variety of living forms that have adapted to this cold world, or perhaps it is just simple curiosity about one of the few frontiers left to explore that draws us to the sea and the study of it. Perhaps we realize that in many subtle ways not only our way of life but also our continued existence is linked to that of those strange creatures who stare back through the windows of our submarines. Whatever it is that attracts us, it is indisputable that we have always been drawn to the sea, this great storehouse that needs our protection as much as we need protection from it.

22 *the hidden earth*

The land beneath the sea

Nor on the white tops of the glistening wave,
Nor down, in the mansions of the hidden deep,
Though beautiful in green and crystal, its great
caves of mystery . . .

The Lost Pleiad
William Gilmore Sims

Standing at the ocean's edge or on the deck of a ship, we tend to forget that below this almost flat though ever-undulating surface lies a world virtually unknown to us. This submarine world, however, is more familiar than we might expect, because most geologic processes affecting the land we inhabit also affect the land below the sea. We know that continents experience earthquakes and volcanic eruptions, undergo erosion, are characterized by mountains, valleys, plains, and plateaus, and most important, maintain life. The land beneath the ocean waves also exhibits all of these qualities—and more. That the features below sea level differ in appearance from their counterparts on land is due to the very nature of the submarine environment relative to the familiar subaerial environment.

SUBMARINE TOPOGRAPHIC REGIONS

As was pointed out in Chapter 9, it is hardly coincidental that the oceans are located where they are. The basic composition and structure of the rocks making up the ocean basins is vastly different from those comprising the continents. This difference in composition explains the difference in density, which in turn explains why the ocean basins overall lie at lower elevations than continents. Water naturally flowed to the lowest possible levels and filled the basins, at times even overflowing their boundaries to drown large areas of the continents.

Evidence from the geologic record indicates that oceans have existed to some extent for more than 3 billion years. Although no man was present to see those primordial seas, the oldest fossils, those of marine algae, are evidence of the early existence of the sea. By the same token, evidence indicates that neither the ocean basins nor the continents have always appeared as we know them today but have changed both in size and in location.

The ocean basins may be divided into several distinct parts, some of which are based on definite structural and compositional differences. Other divisions may be arbitrarily designated on the basis of nothing more than water depth. Adding to the confusion are the names given by geographers to various sections of the oceans (Mediterranean Sea, Gulf of Mexico, Pacific Ocean). All parts of the ocean basins are interconnected, and for the most part water passes freely among them. The areas of the ocean we shall discuss here include the continental shelf, continental slope and rise, the deep ocean basins, and the midocean ridges and trenches (Fig. 22-1). Several of these have associated features that will also be examined.

Continental shelf

American Maritime Law defines the continental shelf as the seaward extension of the coast to a depth of 100 fathoms (600 feet). Geologically, the shelf is indeed more closely related to the continent than to the ocean basins because it is underlain by continental rocks and smothered in sediments eroded from the land and deposited in this shallow water by such agents of erosion as wind, glaciers, and above all, streams. Representing about 7% of the total ocean area, the continental shelf is believed to be a part of the continent now submerged in the sea.

The width of the shelf varies from almost nothing to

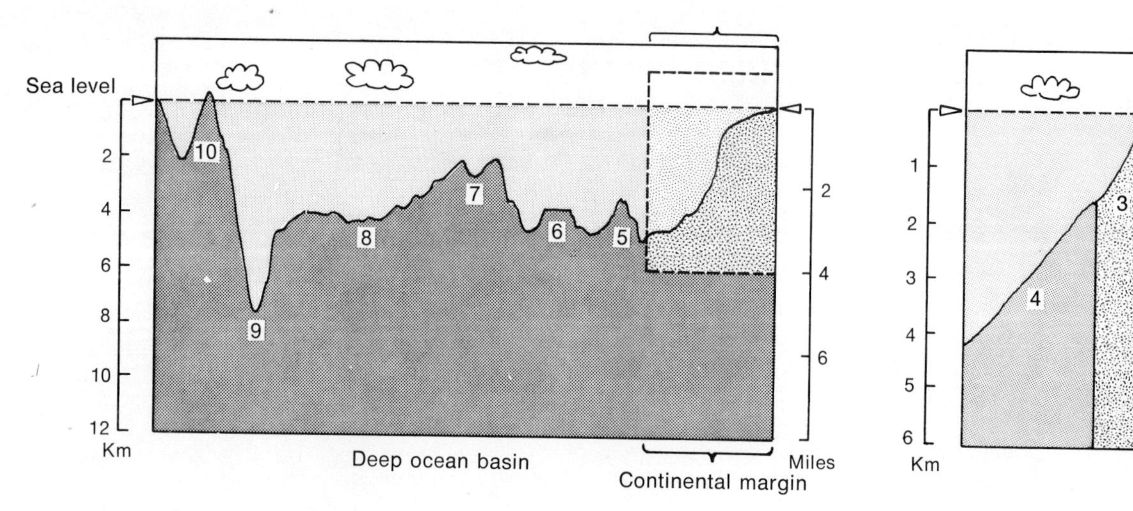

Fig. 22-1. Profile of various geologic features of seafloor. Inset at right is enlargement of portion of the area of continental margin enclosed by a dashed line. *1,* The coastal plain is not actually sea bottom but is geologically related and therefore included. *2,* The continental shelf, a relatively shallow area before the bottom drops off sharply at the continental slope, *3*. At the bottom of the continental slope lies the continental rise, *4,* which connects the shallower areas with the deep-sea bottom. Typical features of the deep-sea bottom include a seamount, *5,* which is a little mountain under water. Such a small mountain that was once uplifted and subjected to erosion is called a guyot, *6*. *7,* A midocean ridge (a big mountain under water) with a rift valley (note small dip in center). *8,* An abyssal (undersea) plain. If the sea bottom can have mountains and plains, it should come as no surprise to find valleys *(9)* there as well. Such valleys are called trenches. *10,* A volcanic mountain rising out of the seafloor; aside from the fact that it rises out of water, its structure is just like those of volcanoes on land.

hundreds of miles, but a good global average would be about 40 miles (67 km). Even as it slopes toward the ocean depths, its average drop in elevation is only about 12 feet (3.7 meters)/mile. (The elevation on land often changes by 100 to 500 feet or 30.5 to 152 meters in a mile.) Its small basins grade into canyons and ridges, but the relief rarely exceeds 60 feet (18.3 meters).

Water is about 1000 times more opaque than air. This means that sunlight can penetrate only to a certain depth in the ocean before it is completely absorbed. Although the depth of penetration is a function of the water's clarity and the sun's angle, among other factors, it is safe to say that the continental shelf is the only part of the ocean where sunlight can penetrate to the bottom. Although light intense enough to support life only penetrates seawater to a depth of about 300 feet (91.5 meters), some light does manage to filter down as low as 1000 feet (305 meters). This makes the shelf, then, the area of the sea which has the greatest variety and abundance of both plant and animal life.

Continental slope and rise

Comprising about 12% of the ocean area, the continental slope drops away from the edge of the shelf to depths averaging 9000 feet (2745 meters). Its average slope, approaching 100 feet (30.5 meters)/mile, more nearly approximates the elevation changes found on land. However, at some points (for example, off the southeastern coast of Florida), it drops off as steeply as 2500 feet (762 meters)/mile.

The continental slope not only acts as an oceanographic transition zone between the shallow water of the continental shelf and the water of the deep ocean basin, but its composition indicates that it is also a geologic transition zone. While the continents, including the continental shelves, are composed mainly of lighter granitic rocks, the deep ocean basins are heavier basaltic rocks. The continental slope is underlain by rocks of intermediate composition and density, such as the igneous rock andesite.

Submarine canyons

One of the more interesting features on the continental slope is its submarine canyons. These long, deep, narrow gorges resemble submerged stream valleys, and are in fact generally found offshore from some great stream such as the Congo or Ganges Rivers. The Ganges River Canyon, for example, is 4 miles (6.7 km) wide, 240 feet (73 meters) deep, and 1000 miles (1610 km) long. In the western Atlantic, the 200-mile (322-km) Hudson River Canyon is the one most closely studied by oceanographers. The Monterrey Canyon off the Carmel, California coast, however, is the most spectacular, with its Grand Canyon–like profile. This canyon winds through 50 miles (80.5 km) of ocean and drops in places more than 1 mile (1.6 km) from its rim.

Submarine canyons are usually steep walled in their upper portions, and they usually show the V-shaped profile characteristic of stream erosion. Their bottoms slope outward toward the sea bottom after winding sinuously through their long channels. Some join with tributary canyons to form drainage patterns much like those of land streams. These great canyons, so common along continental coasts, may also be found associated with islands in the sea.

Although the exact origin of submarine canyons is still being debated, it is generally believed that they are erosional features cut into the continental slope by *turbidity currents*. These bottom currents occur when sediments on the shelf are dislodged either by underwater disturbances such as earthquakes, or when they exceed their angles of repose so that the sediments move down the slope under their own weight. A current is thus created by the increased density of the sediment-laden water. Once in motion, a turbidity current dislodges more sediments and moves along at speeds of up to 50 miles (80.5 meters)/hr. The sweeping motion of these currents washes the slope, keeping it relatively free of loose sediments. The sediment moved in this way eventually settles out of the water in horizontal layers on the deep ocean basins, forming *abyssal plains*.

Continental rise

In many areas of the ocean the continental slope extends down into the very deep waters of the basins. In other areas, however, there is a smoother, less steep zone connecting the two regions. This zone, called the continental rise, is believed to be of depositional origin, probably deposited by the currents just discussed.

Deep ocean basins

Of the total area of the ocean, 76% is composed of large, oval-shaped depressions called *deep ocean basins* in which the water depth exceeds 9000 feet (2745 meters). With the exception of the deep-sea trenches, which we will discuss shortly, these are the deepest parts of the ocean.

Each ocean may have several basins, separated by ridges or rises. Within these basins, a wide variety of features may be found.

Abyssal plains

The abyssal plains, flat, smooth surfaces on the ocean floor, are probably the flattest regions on earth. They are featureless not because of a lack of underlying relief, but rather because heavy sedimentation has filled in the low spots, blanketing the entire region to depths as great as 7500 feet (2500 meters). The Argentine Plain off the coast of Argentina, for example, has less than 10 feet (3.05 meters) of relief over an 800-mile (1288-km) distance. Information obtained through the use of precision depth recorders (PDRs), indicates that only rarely does the plain rise or fall more than 3 feet (1 meter) from the general level.

Core samples of sediments taken from abyssal plains confirm that these features are indeed depositional rather than erosional. The cores are composed of loose sediments exhibiting graded bedding (vertical sorting, in which the coarser layers are at the bottom of the core and the finer ones on the top). Furthermore, sediments contain the shells of shallow-water marine animals that normally inhabit the waters of the continental shelf. It is therefore believed that these plains were formed by the deposition of sediments on the deep ocean floor by turbidity currents.

Littering the floor of the abyssal plains are great deposits of phosphate minerals and manganese nodules (Fig. 22-2). The manganese nodules, which also contain nickel, copper, cobalt, and molybdenum in smaller percentages, are present in concentrations of up to 5 to 7 lb/ft^2. Only a few years ago, it was not thought that such "wet mining" could be competitive with the costs of mining similar minerals on land, but these submarine deposits are now considered a national resource. Forming around a nucleus such

Fig. 22-2. Manganese nodules, which litter the floor of the abyssal plains, tantalize scientists. (Courtesy Lamont-Doherty Geological Observatory of Columbia University, Palisades, N.Y.)

as a volcanic rock, a shark's tooth, the ear bone of a whale, or even an old artillery shell, manganese nodules have occasionally grown as large as 2000 pounds (908 kg). It was announced in 1974 that a consortium of American mining companies had completed the first plans for the development of a joint program to mine these nodules. This is a significant development, for the United States presently imports 98% of its cobalt, 95% of its manganese, 74% of its nickel, and 11% of its copper. Mining these nodules could eventually make this country self-sufficient in these important minerals and thus less vulnerable to political pressure of the sort that has been practiced in the petroleum industry.

Isolated mountain peaks

The most spectacular scenery on land is provided by mountains—great ranges running almost continuously for hundreds of miles or isolated volcanic peaks occurring singly or in larger numbers arranged along linear tracts. The sea also has its mountains, both ranges and isolated peaks. Submarine mountain ranges will be discussed later in this chapter; we will discuss the peaks now because they tend to develop in the deep ocean basins.

Mountains in the sea are of volcanic origin—outpourings of basaltic lava from fissures in the ocean bottom. Accumulating over millions of years, these great piles of volcanic rock may eventually reach the ocean surface, where they break through to form islands. On land the greatest mountain is Mt. Everest, whose summit is more than 29,000 feet (8845 meters) above sea level. The sea's greatest mountain, however, rises a full 1000 feet more; it is 30,000 feet (9150 meters) from the sea bottom, but only the last 14,000 feet is above the water. This top 14,000 feet (4270 meters) forms the island of Hawaii. The principal islands in the Hawaiian chain, in fact, are all the projecting tops of submarine volcanoes, growing from a fissure in the floor of the Pacific Ocean.

Seamounts and guyots

A count of volcanic islands projecting above sea level, however impressive, would pale when compared with the number beneath the waves. Submarine volcanic peaks are classified as seamounts or guyots depending on their shape (Fig. 22-1).

Seamounts are isolated underwater mountains rising at least 3000 feet (915 meters) above the surrounding topography. They have characteristic pointed tops, the shape that is expected of volcanic peaks. They may be found singly or in groups, clustered or linearly arranged. Seamounts are found in all ocean basins, and 1400 have been cataloged in the Pacific Ocean alone. The Kelvin Seamount Group in the Atlantic off the coast of New England and the Emperor Seamount Chain in the North Pacific are both good examples.

Guyots, on the other hand, are submerged volcanic peaks with flat tops. Much rarer than seamounts, guyots were probably once seamounts that were exposed to erosion by a rise in the sea bottom, a drop in the sea level, or both. The rising sea bottom theory gains credence from the fact that (1) guyots are virtually restricted to the Pacific Ocean, which is the most seismically active of the ocean basins, and (2) all guyots are now 2000 to 3000 feet (610 to 915 meters) beneath the surface. Examples include the Fieberling Guyot off the coast of Southern California and the Horizon Guyot south of Hawaii.

Deep ocean canyons

Earlier in this chapter we discussed the origins and location of submarine canyons, primarily those cutting across the continental slope. We now turn to the great canyons frequently found in the deep ocean basins. Deep-sea canyons, also known as *troughs* and *trenches,* are usually located on the deep ocean side of volcanic island *archipelagoes* (chains of closely spaced islands more or less paralleling the coastline). These canyons, the deepest parts of the oceans, are believed to be *subduction zones* in which one crustal plate is sinking beneath another (see Chapter 18). Therefore, unlike submarine canyons, the trenches were produced by diastrophism. Their great depth is due to their geologically recent origin; they have yet to be filled with sediments.

The deepest part of an ocean canyon yet discovered is the Challenger Deep in the Marianas Trench in the western Pacific. The trench itself is more than 1500 miles (2415 km)

Table 21. Major oceanic trenches

Name	Maximum depth (meters)	Length (km)	Width (km)	Ocean
Marianas Trench	11,030	2550	70	Northwest Pacific
Tonga Trench	10,880	1400	55	Southwest Pacific
Kuril-Kamchatka Trench	10,542	2200	120	Northwest Pacific
Phillippine Trench	10,497	1400	60	Northwest Pacific
Kermadec Trench	10,047	1500	60	Southwest Pacific
Puerto Rico Trench	9220	800	—	Atlantic
South Sandwich Trench	8260	—	—	Antarctic
Atacama Trench	8060	—	—	Eastern Pacific
Sunda Trench	7455	—	—	Indian

long and about 45 miles (72.5 km) wide—easily of geosynclinal proportions. Its bottom is 36,198 feet (11,946 meters) below sea level. Other major trenches are listed in Table 21.

We have briefly entered the world of the deep ocean canyons. On January 23, 1960, the bathyscaph *Trieste* descended into the Marianas Trench to a depth of 35,800 feet (10,919 meters). This descent took 4 hours, 48 minutes. Safe in their steel bathyscaph, Jacques Piccard and Lieutenant Donald Walsh of the United States Navy were able to look through glass inches thick at a world of perpetual darkness and under pressures as great as 7 tons/in². Only a few free-swimming animals shared this world with them. Forever trapped by their adaptation to this hostile environment, these predatory creatures live off each other or fight for the few bits of food that filter down from the world above.

Perhaps an understanding of the trenches will bring the relationships between the geosynclinal theory, mountain building, and continental drift to light. Could the trenches, for example, be geosynclines without sediments? When filled with sediments, could forces uplift and weld these sediments to the continents as mountain ranges? The fact that relatively young geosynclinal mountains such as the Andes, Rockies, Alps, and Himalayas are found today in areas with active crustal plates could lend credence to this hypothesis.

Midocean ridges

The most interesting and probably the most important parts of the land beneath the sea are the midocean ridges. These ridges have provided the key that has virtually rewritten the history of the ocean basins and indeed much of geologic history.

The ridges form a discontinuous volcanic mountain range some 40,000 miles (64,440 km) long and hundreds of miles wide. Tracing the ridge location and that of its associated rift valleys and fault systems provides interesting information. Follow the path on the map (Fig. 22-3). Starting in the Atlantic *(1)* where it is appropriately known as the Mid-Atlantic Ridge (the system's most closely studied part), it veers eastward about midway between Africa and the Antarctic continent to enter the Indian Ocean *(2)*. There it splits into two parts, one branch heading into the Red Sea *(3)* and the other continuing on into the Pacific Ocean, following a course about midway between Antarctica and Australia *(4)*. The Red Sea branch, meanwhile, has associated rifts that extend inland to form the great Rift Valley of southeast Africa and the Holy Lands in the Middle East. In the eastern Pacific, the ridges head north as the East Pacific Rise *(5)*. They turn landward again at the Gulf of California, slicing through the state of California in a series of faults, the most notable being the San Andreas fault. They then head out to sea south of Eureka, California *(6)*. In the North Pacific, the track becomes badly contorted in complex rifts but finally emerges again in the Arctic Ocean *(8)* after apparently passing through the Verkhoyansk Mountains of Siberia *(7)*. From the Arctic, the system reenters the Atlantic Ocean and is temporarily interrupted by the great mass of lava that forms the island of Iceland *(9)*. This completes a most amazing journey around the world beneath the sea.

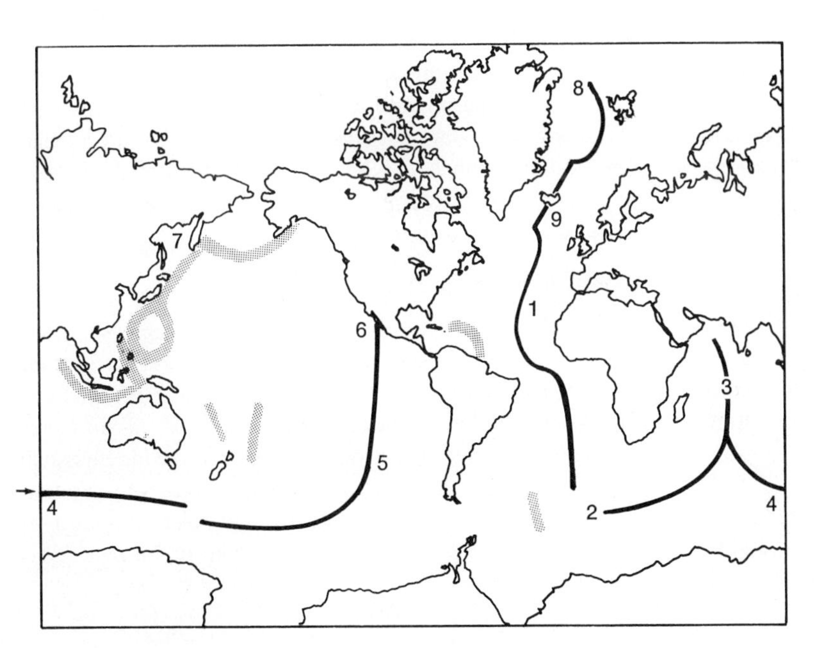

Fig. 22-3. Midocean ridges (heavy lines) and deep ocean canyons (shaded areas). To trace the path of the ridges, use this map and the discussion on this page.

Scientists now believe that these midocean ridges mark the places in which crustal plates are colliding or pulling apart, driven by upwelling magma from deep inside the earth. As evidence, consider the fact that these ridges today are the sites of frequent outbreaks of volcanic activity. Recent examples include the igneous activity on Tristan da Cunha on the South Atlantic Ridge in October of 1961, Surtsey near Iceland on the North Atlantic Ridge in November of 1963, and Heimaey, again near Iceland, in January of 1973. The Heimaey eruption was the fourteenth volcanic eruption in Iceland in this century alone.

In addition to volcanic activity the ridge zones show direct evidence of earth movement. Most shallow-focus earthquakes (less than 60 miles) are either associated with the axes of these ridges or lie in their attendant rift zones. The lack of continuity in the ridges, then, is due to earth movements—different plates moving at different rates and in different directions. The huge faults that mark the breaks are similar to stretch marks and are probably a very good clue as to their history. It was evidence of the movement of these huge chains, enlarging or shrinking the basins they border, that provided the earliest support for the theories of seafloor spreading, plate tectonics, and continental drift (see Chapter 18).

The typical profile of a submarine ridge differs from the profiles of continental ridges in that a submarine ridge resembles an M rather than an inverted V. The dip in the middle is called a *graben,* and its formation is related to these plate movements. As separating plates pull the ridge apart, any blocks left in the middle simply drop into the hole vacated by the blocks moving away on either side. Graben floors may be as much as 1 mile below the peaks on either side.

OCEANIC SEDIMENTS

The ocean serves as the great catch basin of almost everything on earth—human wastes, industrial chemicals and other pollutants in great variety, meteorite dust, volcanic ash and dust, the remains of marine organisms, precipitates from seawater, and debris from weathered rocks. Regardless of origin, most of these materials eventually settle on the ocean bottom. The fate and effects of synthetic materials in the sea are largely unknown at this time. Marine sediments of natural origin fall into one of five main categories, which are discussed in the following paragraphs.

Terrigenous sediments

Those sediments that had their origin on land are known as terrigenous sediments, and they include the forms studied earlier, namely, gravel, sand, silt, and clay. Formed primarily through weathering processes, then reworked and transported to the sea by the various erosional agents, these sediments are generally restricted to the continental shelves, where they lie in graded beds. Unconsolidated sediments as thick as 12,000 feet (3660 meters) have been measured near large land masses.

Sediments from the land on the parts of the abyssal plains nearest land usually consist of more resistant minerals such as quartz, feldspar, and mica. Accumulation rates of such sediments are so variable that no meaningful figures can be given. Terrigenous sediments in deep water far offshore, on the other hand, are rather scarce, but when found they usually take the form of red clays whose color comes from iron oxides. This material also contains about 20% aluminum and small amounts of copper. Red clay deposits are believed to accumulate at a maximum rate of about ¼ inch (50 mm) each 1000 years, although they usually accumulate much more slowly.

Volcanic sediments

Volcanic activity has recurred throughout our discussion of undersea phenomena. In many places the debris from these volcanic eruptions now extends above sea level, forming volcanic islands. The finer pyroclastic debris, particularly dust, may be carried great distances by the wind or by ocean currents, finally settling out as volcanic sediments on the bottom. Like the terrigenous sediments discussed above, volcanic sediments fall out in wedge-shaped beds—thicker deposits near the source, thinner deposits with increasing distance.

The analysis of a core of volcanic sediments from the sea enables the oceanographer to reconstruct the sequence of volcanic events in an area. Such information as the location of the source, the approximate time of the eruption, its severity, and the prevailing wind direction at the time can be ascertained. Data such as this from the Mediterranean Sea has led scientists to wonder whether the fabled lost continent of Atlantis was not a continent at all but a volcanic island that erupted itself off the map.

Extraterrestrial sediments

Each day, as the earth orbits through space, it collides with innumerable small particles. As a result, a constant rain of meteorite dust falls on the surface of the sea, settling gradually to the bottom. This dust accumulates at such a slow rate that it mixes with other sediments such as the red clay rather than forming layers of its own. Such extraterrestrial sediments usually take the form of small black spheres of magnetic iron or brown glass spherules.

Organic sediments

The sea teems with life, which reaches its greatest concentration in the top few hundred feet, where sunlight penetrates. Most of the animal life in the sea consists of invertebrates, creatures with no backbones, and most of these are very small animals called *zooplankton*. Of necessity marine invertebrates develop some external covering, or shell, both for support and for protection. Unlike the more complex animals, they have the ability to draw from the surrounding water the dissolved minerals they need. (Certain simple marine plants share this ability.) The most important of these minerals are calcium carbonate and silica, particularly calcium carbonate, because of its greater solubility. When these organisms die, their remains drift to the sea bottom, returning the chemicals to their source.

Since the shells and plates of these organisms are so small, sediments composed of their remains are called *oozes*. (The term also covers a variety of other deep-sea sediments, all of which are fine grained.) Just as terrigenous sediments dominate the shallow water, it is these organic oozes that dominate the deep. Calcium carbonate is one of the first compounds removed from seawater by organisms, and it is one of the first to return. Silica, on the other hand, resists dissolution. As a result, the composition of organic sediments is directly proportional to water pressure (increasing with depth) and water temperature—both factors that influence solubility. In other words, organisms with siliceous shells produce organic sediments at greater water depths than those with calcium-based shells, simply because it is only at these greater depths (pressures) that silica dissolves at all. Siliceous oozes cover about 9% of the ocean bottom. Calcareous oozes, on the other hand, are believed to cover about 29% of the ocean floor in those areas in which the water depth averages 1200 feet (366 meters), chiefly clustering around the equator.

It has probably occurred to you that if you can wade a few feet into the water at the beach and dig shells out of the sand, not all of the organic matter in the sea finds its way to the deep seafloor. What about organic matter that for one reason or another ends up closer to the shore? Such matter includes mollusk shells, fish bones, and the remains of coral and worms. Since most of these creatures either are bottom dwellers or feed on bottom dwellers, they die where they lived—on the bottom. Moreover, most bottom-dwelling life forms cluster in the shallow waters of the continental shelf, since that is the only region in which light penetrates all the way to the bottom. It is on the bottom of the continental shelf, therefore, that organic deposits other than oozes are most common.

Location is not the only difference between these de-

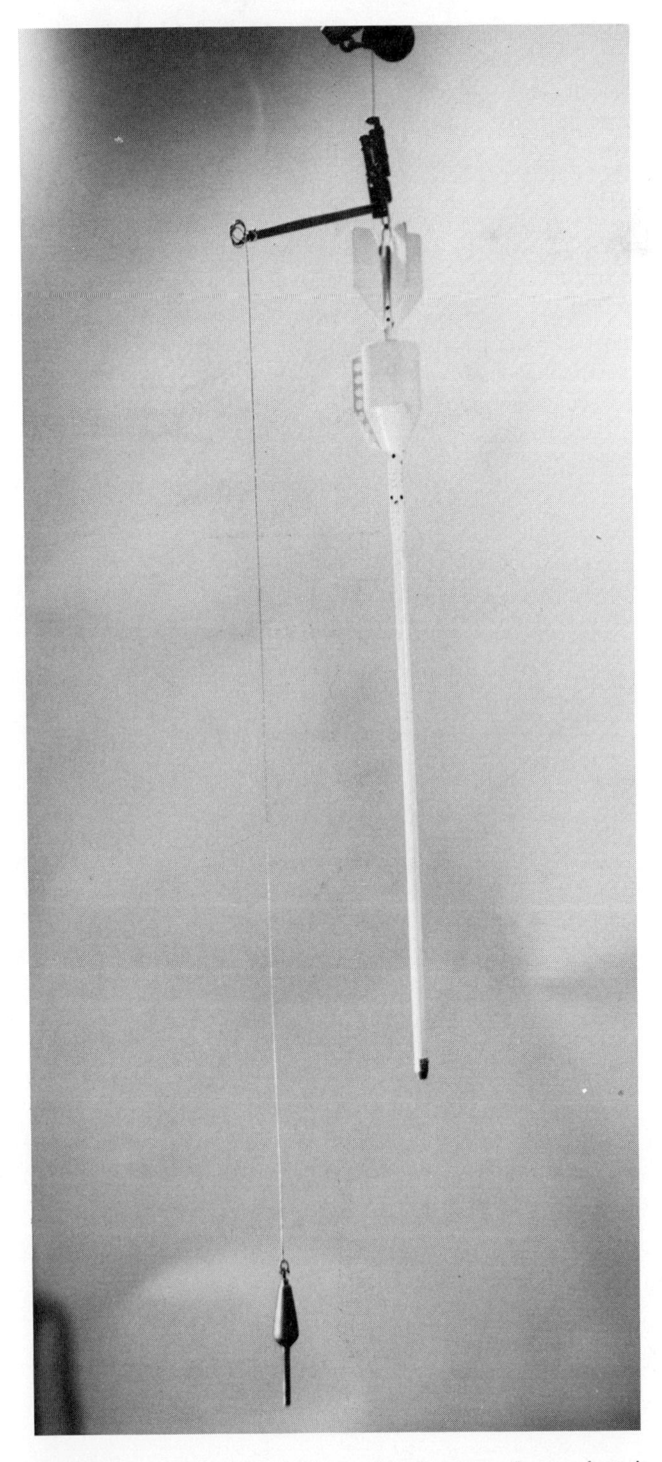

Fig. 22-4. A piston corer is lowered to the ocean floor, where it penetrates the sediments. When it is full, it is pulled up to a ship, where the core (or sample) can be analyzed. (Courtesy Lamont-Doherty Geological Observatory of Columbia University, Palisades, N.Y.)

posits and oozes, however. Since shallow-water remains are not as abundant as those that find their way to the seafloor, they do not form separate beds like oozes but rather mix with the terrigenous sediments already on the shelf.

Inorganic precipitates

Inorganic precipitates form sediments whenever a chemical dissolved in seawater reaches it saturation point or when that water cannot hold any more of that particular chemical. Under these conditions, the available chemical that the water cannot hold drops to the bottom and accumulates. This excess is called a *precipitate*. Inorganic precipitates with the one exception of calcium carbonate (which may be either organic or inorganic, depending on its source) are quite rare.

Rarity, however, is a relative term when discussing something as large as the ocean, and these precipitates may still help to satisfy our growing need for mineral resources. Littering as much as 50% of the deep ocean floor are nodules of manganese, phosphate, and barite, all precipitated out of the seawater when the supply of the chemical exceeded the amount that the water could hold. The volume of manganese nodules alone has been conservatively estimated at 1 trillion tons.

SEDIMENT SAMPLING DEVICES

Sediment samples from the ocean floor are obtained through a variety of devices and techniques. Oceanographers use a simple *grab sampler* to recover material from the surface in only one location. To obtain material from the surface over a wide area, on the other hand, a *dredge* is used. To examine a cross section of sediment layers, a gravity-operated device called a *piston-corer* is used (Fig. 22-4). Consisting of a long, thin, heavily weighted pipe attached to the research ship by a wire, the piston corer is dropped to the bottom, where it penetrates the sediments. Once back on board ship the coring tube, which is filled with sediments, is opened, and the sediment core is removed and examined.

23 *the eternal struggle*

Waves, ocean currents, and coastlines

Ever drifting, drifting, drifting on the
shifting currents of the restless main;
Till in sheltered coves and reaches of
sandy beaches all have found repose again.

Seaweed
Henry Wadsworth Longfellow

Miami Beach, the southern tip of Florida's "Gold Coast," contains some of the most valuable real estate in the world. Owners of this property soon discovered, however, that the tourists on whom this area's economy depends were not impressed with natural beaches—they preferred their beaches precleaned and sifted. Therefore in order to protect the value of their property, hotel and apartment owners artificially sanded their beaches. This process did not fit in well with the ocean's natural tide and current processes. Normally the ocean keeps sand moving down the coast, picking it up on one beach and depositing it farther down. In other words, everyone always has a beach, but it isn't always composed of the same sand. Fine, but back to those artificial sanders. Suppose you are a beach owner, and you do not want the ocean trading your expensive artificial sand for plain old natural sand picked up further up the coast. You build what is called a *groin,* a concrete or wood structure jutting out into the sea perpendicular to the shore. This slows down the flow of water, and as you remember from our earlier discussion of stream mechanics, any time the water slows down, it starts to drop its sediment load. The ocean, then, will drop all your sand on your property, and all you have to do is pick it up and put it back where it came from.

The next problem is your neighbor. The ocean is still absconding with his sand, and it is supposed to be replacing it with yours. But you won't let the ocean take yours off your property, so he ends up with no sand. He catches on quickly and puts up a groin of his own to hang on to what he has left. Next thing you know, the whole beach is lined with groins (Fig. 23-1). However, the ocean still has its sand-carrying capacity; what is it going to do, now that no one will let it carry off beaches? The answer is that the ocean's erosive forces are intensified, and it digs deeper and deeper into the beaches and the offshore sand. The result is that Miami Beach is now stuck with a million-dollar bill to try to restore the beaches to their original condition.

WAVES: ARCHITECTS OF THE SEA

As this example clearly illustrates, the sea is capable of performing great quantities of geologic work, usually very slowly but sometimes very rapidly and violently. To us spending most of our lives on the land, the most important part of the ocean is the narrow zone where the bottom of the sea becomes the surface of the land. This area, the *coastline,* is a region of constant change.

The nature of waves

There is a close similarity between a sound wave, a seismic wave, a light wave, and a water wave. Each represents energy moving from one place to another. In its general oceanographic sense, a wave is simply a surface disturbance of the water. In reality, however, ocean waves are not simple but complex and are of varying types in varying locations. Some ocean waves do not even occur at the surface.

Surface waves in the sea are generated in three basic ways; all place stress on the surface of the water and thereby create a disturbance. The first method of wave formation is *frictional drag* as the result of moving air (wind), although

Fig. 23-1. This section of the Miami Beach front shows what happens when everyone on the coast must build groins to protect his beach from the effects of his neighbor's groin! (Courtesy Miami Beach Tourist Development Authority, Miami Beach, Fla.)

the exact nature of this energy transfer is not fully understood at the present time. Second, and less common, is an energy transfer from the ocean basin itself to the water. Examples of this are waves produced by underwater earthquakes, volcanic eruptions, and landslides. These seismic ocean waves (tsunamis), the most feared and potentially destructive of all waves, fortunately occur infrequently. Finally, there are the waves produced by the gravitational action of one celestial body on another, such as the moon and the sun on the earth. Actually, these are the true *tidal* waves.

Wave components

Regardless of the form a wave takes, all waves basically share the following similar components, although the terms used to describe them may differ somewhat. All waves possess a *wavelength,* which is the horizontal distance from the crest of one wave to that of the next. Second, all waves have an *amplitude,* or height, which is defined as the vertical distance from the crest to the trough (Fig. 5-1). Waves also have a *frequency,* which is the number of wavelengths passing a given place in a given period of time. Still another similarity shared by waves is their *period,* the time required for one complete wave to pass a given point. In addition, waves also have a *velocity,* which deals with their speed and direction. Unlike other types of waves, however, water waves are affected by *fetch,* or the horizontal distance over which the wind is blowing.

Wavelength

Ocean waves with the greatest wavelengths are the tides, or tidal waves, whose length equals half the circumference of the earth, or 12,500 miles (20,125 km). Their crests produce high tide, while their troughs form low tide. *Tsunamis* have much shorter wavelengths than the tides, averaging about 100 miles (161 km). Wind-generated waves in the open sea have wavelengths that average 600 feet (153 meters), a length that decreases as the waves approach the shore.

Wavc hcight

Waves range in height anywhere from 100 feet (30.5 meters) to less than 1 inch (2.5 cm). The highest wave ever reported in the open sea was recorded on an American tanker, the U.S.S. *Ramapo,* on February 7, 1933, to have reached a height of 112 feet (34 meters). While there have probably been tsunamis higher than those on record, their fury undoubtedly prevented any fortunate survivors from remaining in the area long enough to measure them. Tsunamis in the open sea, on the other hand, may be so tiny that ships can sail right over them without noting them.

The second-highest wave on record occurred on April 1, 1946, as the result of an earthquake in the Aleutian Trench just off Unimak Island in the Aleutians. Within minutes of the initial quake, a lighthouse 57 feet (17.4 meters) above sea level was destroyed by a tsunami estimated to have a height of 100 feet (30.5 meters). In contrast, the height of a seismic sea wave in deep water is only 2 to 3 feet (about 1 meter).

Wind-generated waves as high as 50 feet (15 meters) usually develop only during severe cyclonic storms, while waves of 25 feet (7 meters) are more frequent. Normal, nonstorm waves rarely exceed 12 feet in height. As a general rule, however, the height of a wave in the open sea does not exceed half of the wind speed in miles per hour, or one twentieth its wavelength. The limit to this length-height ratio, then, is one to seven, the point at which the wave breaks and whitecaps form.

Periods

The period of a wave is the time required for it to pass a given point. Tides have the longest possible periods, half a day. Due to their great wave speed, tsunamis generally have a period of 15 minutes to 1 hour. In contrast, wind-driven waves approaching the coast usually have shorter periods of from 1 to 3 minutes.

Velocity

Theoretically, the fastest wave is the tide. The earth, rotating through the tidal bulge, moves on its axis at a rate of 1000 miles (1610 km)/hr at the equator, and at progressively slower rates toward the poles. The ocean, located on the surface of the spinning planet, also rotates, even though frictional drag retards it somewhat. As a result, the movement of the tide at any one location is hardly detectable, except over a period of time. Notable exceptions, however, are *tidal bores* such as those found in the Bay of Fundy in Nova Scotia, the Ghien-tang River of northern China, and the Amazon River of South America. The bore of the Amazon river, for example, is 24 feet (7 meters) high and travels

Table 22. Beaufort's wind scale and sea conditions

Beaufort number	Height of sea (feet)	Descriptive term	Wind speed (miles/hr)	Sea's appearance
0	0	Calm	0	Mirrorlike
1	Less than 1	Light air	1-3	Formation of ripples
2	1-2	Slight breeze	4-7	Small wavelets, do not break
3	2-3	Gentle breeze	8-11	Large wavelets, begin to break
4	3-4	Moderate breeze	12-16	Small wave
5	4-8	Fresh breeze	17-22	Moderate waves, chance of spray
6	8-10	Strong breeze	23-27	Large waves, some spray
7	10-12	Moderate gale	28-34	Sea piles up, spray forms light streaks parallel to wind direction
8	12-15	Fresh gale	35-41	Moderately high waves, well-marked streaks
9	15-20	Strong gale	42-48	High waves, dense streaks, spray may affect visibility
10	20-30	Storm	49-56	Very high waves, sea takes on white appearance, tumbling of sea is heavy, visibility affected
11	30-40	Violent storm	57-67	Exceptionally high waves, sea completely covered with long white streaks of foam, wave crests blown into froth
12	Over 40	Hurricane	Above 67	Waves mountainous, air filled with foam and spray, sea completely white, visibility very poor

at a speed of twelve knots* (14 miles/hr), 300 miles (483 km) up the river from its mouth. Although the speed of tsunamis often exceeds 434 knots (300 miles/hr), most wind-generated waves normally travel at less than 21 knots (24 miles/hr).

There are three primary factors that determine the length, height, and velocity of wind-generated waves. The first and most obvious factor is the wind velocity. In fact, estimates of wind speed can actually be made by observing the appearance of the waves at sea. The Beaufort wind scale (Table 22) is usually used in this calculation. A second factor is *fetch*. Although the 112-foot (34-meter) wave mentioned earlier was generated by top winds of only 68 knots (78 miles/hr), its height was nearly twice that expected at that wind speed. This record wave can be explained by the wind, which was blowing over a fetch of several thousand miles. The final factor to be considered here, the *length of time* for which the wind blows, requires no further explanation.

Types of surface waves

Studying simple wave motion in a wave tank is entirely different from studying the motion of actual ocean waves, which come in all sizes and often converge from all directions. Through our study of the ocean, we have classified waves into two basic types, *progressive* and *standing*.

Progressive waves

Progressive (or running) waves are those whose waveform moves across the water's surface. They are further classified as sea, swell, or surf waves.

Sea waves include nearly all wind-generated waves or those under the direct influence of the wind. Although "sea" and "ocean" are often used interchangeably to describe a large body of salt water, technically the term "sea" is used to describe wave categories. Unlike swells and surf, *seas* are complex and typically show a pointed crest and a more rounded trough. The Beaufort scale actually describes this category of wave.

Swells are waves not under the direct influence of the wind. They are formed by water motion caused by winds elsewhere. Swells are much simpler than seas and tend to have more rounded peaks and troughs. Seas are usually superimposed on swells. Swells have recently come under investigation in an attempt to determine the distance over which they travel, their speed, and the amount of energy dissipated en route. Waves generated by Antarctic storms, for example, have been tracked by wave-recording stations across half the globe. It was discovered that 10 days after their birth near the frozen continent, the swells died on the distant shores of Alaska. This trip also included movement across the equator, which surprisingly caused no detectable loss of energy.

The last category of waves is *surf,* the best known of the three wave types and the only one found in shallow water. As a sea or swell approaches water less than half its wavelength in depth, it begins to "feel bottom," slows down because of friction, steepens, and then breaks, producing surf, or *breakers*. It is in this form, then, that the energy originally imparted by the wind to the water is finally expended.

Water motion in a progressive wave. Progressive waves give the impression that water is moving from one place to another—from Antarctica to Alaska, for example, yet in reality this is true only of surf. Elsewhere it is merely the waveform moving forward through the water, not the water itself that moves. Research has shown that water particles in a wave exhibit a circular, orbital motion by means of which they return to their original location (Fig. 23-2). As a wave advances in an open ocean, the individual water particles move upward and forward under the crest and then downward and backward under the trough, thus completing a circular orbit that decreases in diameter with depth. At the surface the diameter of the orbit is equal to the wavelength of the wave. However, with increasing depth in the wave the diameter of the orbit decreases until finally, at one-half wavelength, the motion of the particle is only 4% what it was at the surface. Below one-half wavelength, particle motion is practically negligible. The waveform then extends downward into the water to a depth of one-half wavelength; this is why we said earlier that a wave "feels bottom" at this depth.

Standing waves

Instead of moving along the surface of the water, standing waves simply slosh back and forth. You may be familiar with this phenomenon for it occurs in a container of water that is being carried. The tide is probably the best known of all standing waves. Elsewhere, standing waves may form where progressive waves strike a vertical surface such as a sea wall and are thrown back. However, if this reflected wave meets an incoming wave, a standing wave may be produced for a few seconds.

Wave refraction and alignment of breakers

Like other wave types, water waves can be reflected, diffracted, or refracted. Wave reflection, for example, is il-

*A knot is 1 nautical mile/hr. The nautical (or sea) mile is 6082.66 feet, as compared to the statute (or land) mile, which is 5280 feet.

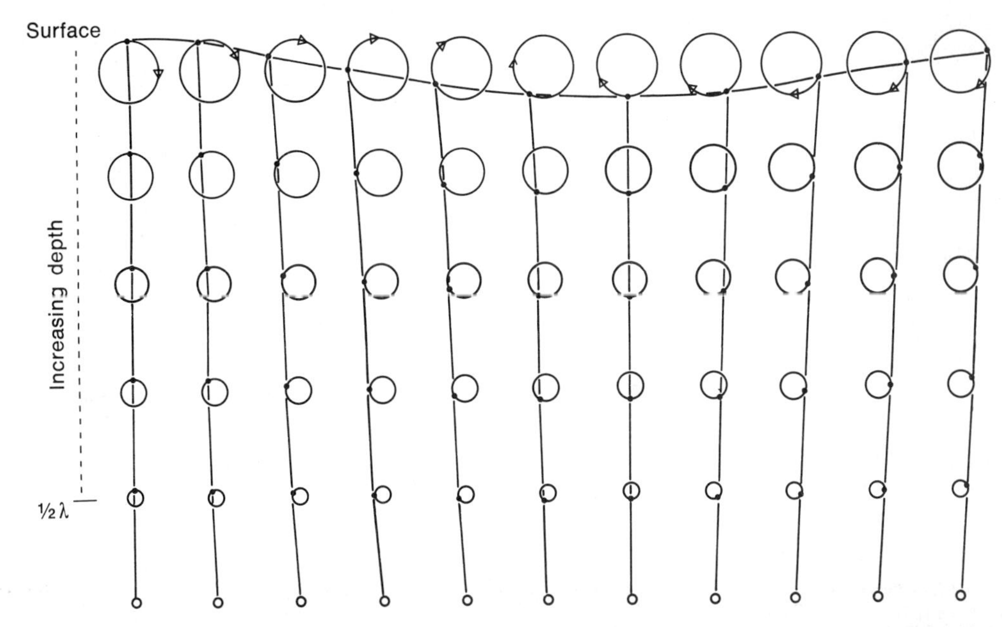

Fig. 23-2. Each water molecule in a wave makes its own small circle, returning to its original position. It is the pattern of adjoining circles that forms the characteristic wave bulge. Note that the molecules' paths decrease in size with water depth.

lustrated in the formation of standing waves at a seawall. Water waves may also be diffracted; a sideward propagation of waves around a barrier such as a breakwater tends to build the disrupted waves back into a single wave. In addition, waves formed in the sea may be refracted, or bent. Of these three effects, however, refraction is the most significant.

Although waves rarely approach parallel to the shoreline, they all break in that direction, an irony explained by wave refraction. A combination of the wind blowing from different directions and the irregular shape of the coastline virtually ensures that a wave will seldom approach parallel to the shoreline. As a result, the leading part of an advancing wave reaches shallow water before the trailing part, "feels bottom," and is slowed down. The remainder of the wave in deeper water continues on until it enters shallow water, where it too is slowed down. This continual "catching of the head by the tail" causes waves to break almost parallel to the shoreline.

Wave energy and damage

Since the sea is never still, its energy is constantly being spent, or dissipated, at the shoreline. Because the *kinetic* energy (energy of motion) of a wave is enormous, its power can be terrifying. A 4-foot (1.2-meter) wave with a period of 10 seconds, for example, will expend more than 35,000 horsepower of energy/mile of shoreline where it strikes. While this may seem impressive, in reality the potential power of a wave may best be realized when it is viewed in terms of its resultant devastation. For instance, at Cherbourg, France, a wave was reported to have hurled a 2-ton boulder over the 30-foot (9.2-meter) breakwater during a storm, while pieces of the structure weighing up to 65 tons were moved many feet. In another instance, a piece of breakwater along the coast of Scotland that weighed 1350 tons was broken off and moved by a storm wave. Its replacement, weighing 2600 tons, was carried away 5 years later. The beacon of Tillamook Rock lighthouse, situated on a small island off the coast of Oregon, has been broken by wave-hurled rocks several times, even though the beam is located 140 feet (42.7 meters) above sea level. Finally, a 135-pound (61-kg) rock was reportedly thrown through the roof of the lightkeeper's house 100 feet (30.5 meters) above the waves at the same location.

Internal waves

Before leaving the subject of waves, it is important to note again that all waves are not water waves at the sea-air interface. Those waves that form beneath the surface of the ocean are called internal waves. The only condition necessary for their formation is the presence of water layers of different densities, the lighter water resting on top of the heavier water. Differences in both salinity and temperature can produce the *interface,* or boundary, along which internal waves are transmitted in much the same manner that surface waves are formed along the sea-air interface. Just as the surface of the ocean is in constant motion, so are the internal layers of transition between water of differing densities, making the eternal unrest of the sea complete.

OCEAN CURRENTS: RIVERS OF THE SEA

Earlier we pointed out that the passage of waves is associated with virtually no horizontal movement of water. There are, however, instances in which ocean water does flow horizontally, producing an ocean current. Therefore ocean currents may be thought of as "rivers in the sea," great masses of water flowing through relatively "stable" water. The banks and beds of these oceanic rivers are interfaces, formed by differences in density that in turn are caused by differences in salinity and temperature. Because they are prone to minute-by-minute changes in characteristics, many ocean currents are highly variable in their width, depth, speed, and direction. From time to time, a section of current will break away from the main mass, forming an *eddy*.

Formation of surface currents

The top of a surface current is at the air-sea interface, while the bottom and sides are at a submerged density boundary. Often surface ocean currents are thought to be produced by winds; although the wind does play a vital role, there are also other forces, both external and internal, that come into play. First these forces set the sea in motion and then keep it moving and guide it on its way.

External forces

As the name implies, external forces originate outside of the mass of water itself. They include wind drag, the Coriolis effect, and air pressure. There are also the forces of friction with the sea bottom, gravity, and tidal forces.

Wind drag. The first force, wind drag, would have to be listed as the primary external force in producing the initial movement of a current. The friction set up between the moving air and the water's surface is capable not only of forming waves but also of setting up this horizontal flow.

Coriolis force. Another important external factor in the formation of surface currents is the Coriolis force, which develops as a result of the earth's rotation. It is not a force, as its name implies, but an effect that can never initiate ocean currents, only steer them. Under the influence of the Coriolis effect, moving particles such as water or air are deflected to the right in the Northern Hemisphere and to the left in the Southern Hemisphere.

The earth spins eastward on its axis at a rate of 1000 miles (1610 km)/hr at the equator, the rate of spin decreasing with increasing latitude until there is no spin at all at the poles. For example, as water or air flows due southward in the Northern Hemisphere, an observer standing near the equator would be changing his position relative to this straight-line flow. Since the observer has advanced eastward, however, relative to the flow, the water or the air would appear to curve to the right of the direction of motion. Particles moving northward from the equator also would appear to curve to the right. The Coriolis effect also carries objects to the left in the Southern Hemisphere because reasoning dictates that a person standing near the South Pole will appear to spin in a direction opposite someone standing near the North Pole.

Other external forces. Even though the effect of other external forces such as air pressure, friction with the sea bottom, gravity, and tidal forces are relatively unimportant to ocean currents, we can still look at them briefly. Differences in *air pressure* between two different locations in the sea can cause a slight difference in sea level, thus setting up a flow from the region of low pressure (higher sea level) toward the region of higher pressure (lower sea level). *Friction* has little effect on surface currents in deep water but takes on greater importance in shallow water and channels. Even there, however, it only acts as a brake to retard the flow of the water. The oceans, lying on the surface of the earth, are under the direct gravitational control of one celestial body, the earth. The main influence of the earth's gravity is only to compress the water layers. Like friction, the differential gravitational force on the ocean exerted by the moon and the sun have their most noticeable effects in shallow water.

Internal forces

Internal forces, which originate within the water mass itself, include both vertical and horizontal pressure and internal friction.

Vertical pressure. Ocean water, like all matter under gravitational pull, has weight, which exerts vertical pressure. The deeper one goes in the ocean, the greater the pressure, because greater amounts of water extend overhead. This weight pressure has been calculated at more than 7 tons/in^2 at the bottom of the Marianas Trench, which is 35,800 feet (6000 fathoms) deep. Since the weight of the overlying water causes compression of the water below, a vertical equilibrium usually exists that prevents vertical movement, thus minimizing its influence on ocean currents.

Horizontal pressure. Although an equilibrium of vertical pressure prevails in the ocean, the same situation does not exist horizontally. Two bands of water at the same depth are not likely to have equal pressures because of such factors as the subtle variations in salinity and temperature that determine the water density. The change in pressure along a horizontal line at any given depth in the sea is called the pressure *gradient.* When such differences in pressure exist at the same level, water tends to flow from the region of higher pressure to that of lower pressure. Once this type of flow has begun, however, the Coriolis force takes over and the flow moves between the two pressure areas. Together with wind drag, pressure gradients would have to be considered a major force in initiating ocean currents.

Internal friction. Internal friction between water particles and water layers, like the Coriolis force, affects currents in motion. As adjoining particles or layers move at different speeds, each has a distinct effect on the other. The faster ones are braked by the slower ones, while the slower ones are simultaneously pulled along by the faster ones. What is actually occurring is a transfer of velocity from one layer of water to another.

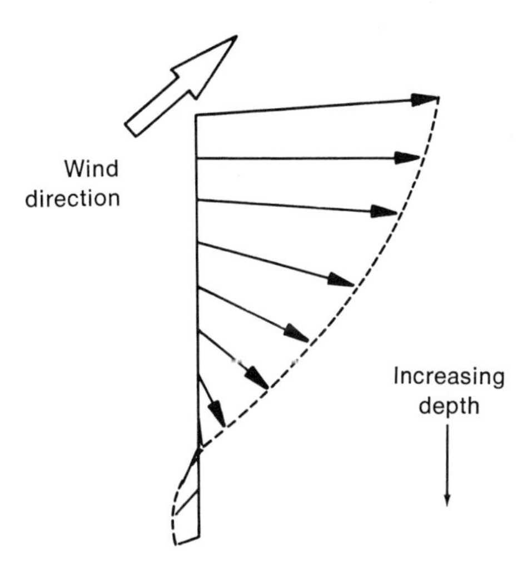

Fig. 23-3. V. Walfred Ekman theorized that once a surface current was set in motion, it would drag along the next water layer under it. This layer would bring along the layer under it, and so on, all the way to the bottom. He also predicted that the Coriolis effect would influence each layer more than the layer above it, until finally layers at a depth of 300 to 500 feet would swing so far that they doubled back on themselves.

Types of surface currents

Earlier in this chapter it was pointed out that there are two primary forces capable of initiating the movement of surface waters; these are wind drag and pressure gradient. Thus the currents produced by these two forces will logically correspond to them. If a current is produced primarily by wind drag, it is called a *drift* current. If produced primarily by differences in pressure gradient, the current is called a *gradient* current.

Drift currents

The complexity of drift currents in the ocean makes it almost impossible to present one adequate model to explain them. The one that comes closest, however, was developed in 1905 by a Scandinavian physicist, V. Walfred Ekman, who found that fully developed surface currents were deflected by the Coriolis force 45° to the right of the wind direction in the Northern Hemisphere and 45° to the left in the Southern Hemisphere.

As a result of his discovery, Ekman predicted that currents initiated by wind drag in the surface waters would set up current flows in the adjacent lower layer through the forces of internal friction and that these in turn would induce further flow in the next succeeding layer. Ekman further hypothesized erroneously that if the water were homogeneous (having the same characteristics throughout), the velocity of flow in successively deeper layers would deviate more and more toward the direction of the Coriolis force. Finally, he calculated that at a depth of frictional resistence, generally between 300 and 500 feet (92 and 153 meters), the current would flow in the direction opposite the wind and at a velocity equal to only 4% of the velocity of the surface layers. This relationship is called the Ekman spiral (Fig. 23-3). While a complete spiral has not been observed in the sea, currents nevertheless do tend to deviate to the right of the wind direction in the Northern Hemisphere (left in the Southern Hemisphere), and countercurrents do indeed exist.

Gradient currents

Unlike drift currents, which are created and controlled primarily by external forces such as wind, gradient currents are formed and controlled by internal forces, chiefly by differences in internal pressure. In contrast to drift currents, which because of the nature of internal frictional resistance are usually restricted to the top 500 feet (152 meters) of water, gradient currents can extend to at least twice this depth.

Unlike the gradient flow of surface streams, which move from higher to lower *elevations,* ocean water and air (Chapter 29) flow from areas of *high* to areas of *low pressure* but soon come under the influence of the Coriolis force (Fig. 23-4). In the Northern Hemisphere the high pressure lies to the right of the flow and the low pressure lies to

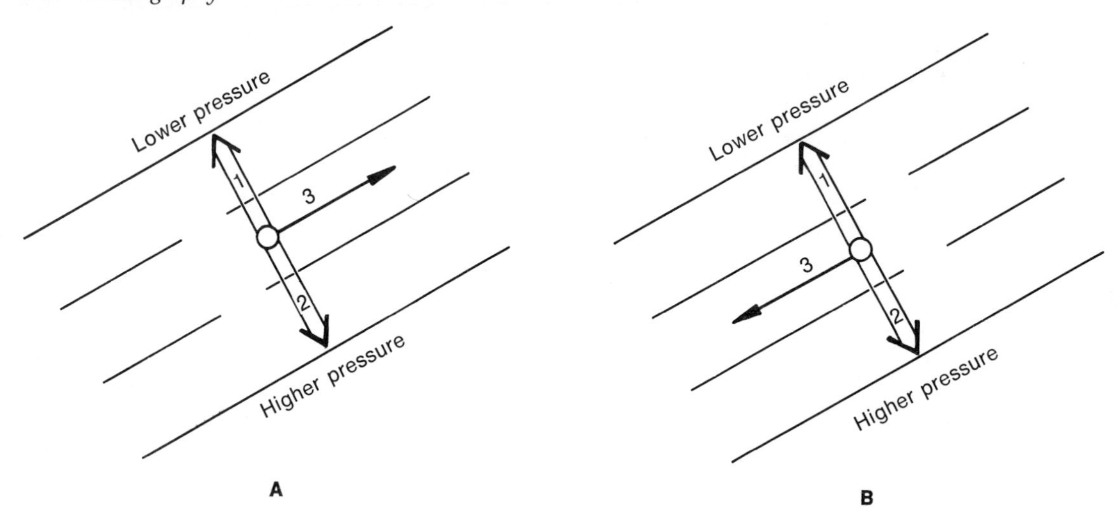

Fig. 23-4. The direction of current flow results from a balance of the conflicting forces at work. **A,** Northern Hemisphere. **B,** Southern Hemisphere. Arrow *1* (in **A** and **B**) shows the direction a current would follow if the only force acting on it were the gradient force. If the only force were the Coriolis force, the current would flow in direction indicated by arrow *2*. Since both forces play a part, the actual direction of current flow is shown by arrow *3*. The difference between direction of current flow in **A** and **B** stems from the fact that the Coriolis force pulls to the right in the Northern Hemisphere and to the left in the Southern Hemisphere.

the left. Conversely, in the Southern Hemisphere the situation is reversed.

Worldwide surface circulation

The ocean's surface waters are the site of relatively well-defined, permanent circulation patterns of both drift and gradient currents. *Surface currents* consist of complex, separate filaments whose motion is irregular. They have eddies that break off and dissipate, while the currents themselves are constantly changing their speed, direction, and location.

In both the northern and southern parts of each ocean there is a great circulation of water in the form of surface currents called *gyres*. In the Northern Hemisphere the flow is clockwise, while in the Southern Hemisphere it is counterclockwise. This general circulation pattern, which is fairly well known, is initiated by the trade winds that blow the surface waters westward (see Chapter 31). Deflected to the right by the Coriolis force (to the left in the Southern Hemisphere) and by land masses, the currents swerve into higher latitudes and come under the influence of the westerly winds, which blow the waters eastward. The water ultimately moves back down to the equator, where it goes through its gyration pattern once more.

North Atlantic gyre

The gyration of water in the Atlantic Ocean north of the Equator is probably the most closely studied and thoroughly understood of any in the world. This gyre is called the *Gulf Stream*. Tracing the flow in this gyre can be an interesting adventure if you keep in mind that of necessity we are oversimplifying reality (Fig. 23-5).

The Gulf Stream gyre begins as the North Equatorial Current, a drift current, and flows toward the west in the belt of the Northeast Trade Winds. Added to this volume of water is part of the flow from another drift current, the South Equatorial, driven across the equator by the Southeast Trade Winds. The currents are turned northward, into the Gulf of Mexico, by both the Coriolis force and the shape of the South and Central American coasts. Here excessive pressure builds and vents itself through the Florida Straits as the Florida Current, a gradient current.

The discharge of water through the Florida Straits is estimated to be 26 times that of the combined flow of all streams on land. Additional reinforcement is given to the Florida Current by the Antilles Current, found offshore from Vero Beach, Florida. This tributary current, also known as the Bahama Current, is a branch of the North Equatorial Current and is therefore a drift current. Remaining fairly close to shore until it reaches Cape Hatteras, North Carolina, the Florida Current then gradually veers away from shore toward the northeast, where it becomes the Gulf Stream, a gradient current. Continued movement in a northeasterly direction carries the current near the Great Banks of Newfoundland, where its warm waters meet the cold waters of the Labrador Current. As the North Atlantic Drift Cur-

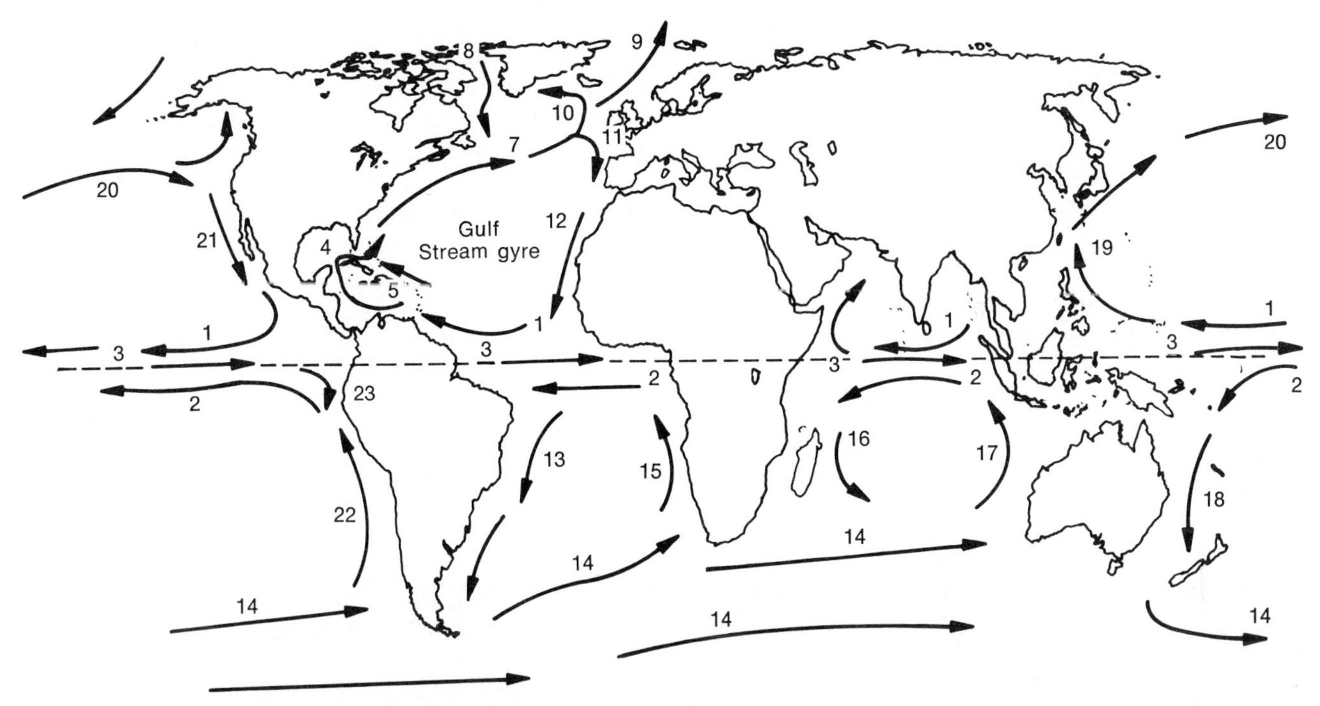

Fig. 23-5. Major ocean currents. *1,* North Equatorial Current. *2,* South Equatorial Current. *3,* Equatorial Countercurrent. *4,* Florida Current. *5,* Antilles Current. *6,* Gulf Stream. *7,* North Atlantic Drift. *8,* Labrador Current. *9,* Ireland Current. *10,* Irminger Current. *11,* Portugal Current. *12,* Canary Current. *13,* Brazil Current. *14,* West Wind Drift. *15,* Benguela Current. *16,* Agulhas Current. *17,* West Australia Current. *18,* East Australia Current. *19,* Kuroshio Current. *20,* North Pacific Drift. *21,* California Current. *22,* Peru Current. *23,* El Niño.

rent, it then continues on its course across the Northern Atlantic Ocean to Norway.

When the North Atlantic Drift approaches Europe, one branch travels north of Ireland as the Ireland Current, while the other branch, the Irminger Current, turns toward Iceland. Most of the stream's mass, however, continues on toward Norway, where it turns south as the Portugal Current. Then, off the coast of Africa, it becomes the Canary Current, which flows toward the equator, finally completing the gyre as the North Equatorial Current.

In the central part of the North Atlantic lies the Sargasso Sea, an area of relative calm whose name is derived from the vast amounts of *Sargassum,* a seaweed that thrives there. Like the central parts of other oceans around which currents flow, the Sargasso Sea has been found to consist of low-density water, which may be thought of as "hills" in the sea, due to their slight elevation in sea level with respect to the surrounding denser water. These "hills" form a gentle slope down which water flows (under gravity) toward the "valleys" of the currents. Such gravity movement as this is called a *geostrophic flow.* (Air is subject to the same phenomenon.)

Water in geostrophic flow moves "downhill" and is deflected by the Coriolis force, until a balance of forces exists between this force and gravity. Water movement in geostrophic flow is therefore nearly parallel to imaginary contour lines on the slopes of the "hills."

Other gyres

Although we cannot take a detailed look at all of the gyres in all parts of the ocean, it is important to note that basically most are very similar to the Gulf Stream system just described. Every gyre has a poleward-flowing warm current that is analogous to the Gulf Stream. In each case these warm currents occur in the wastern parts of the basins and in each case they are gradient currents. In addition to the Gulf Stream, these currents include the Kuroshio or Japan Current, the Brazil Current, the East Australia Current and the Agulhas Current.

Subsurface circulation and currents

Surface currents, induced by wind drag or pressure gradient, pierce the surrounding, stable water like an arrow. (In this case the term "stable" is relative to the time factor because if enough time passes, there is no such thing as "stable" water in the sea.) Generally speaking, there are actually two different, interrelated subsurface circulations that eventually exchange water. Since both are produced by the conditions of heat and salinity within the water, they may both be called *thermohaline* (thermo = heat and haline = salt) circulations.

Shallow circulation

The first thermohaline circulation involves the top few hundred feet of water in warm tropical and subtropical regions extending from 45° latitude to no more than 55° latitude. This is the so-called *stable* water through which the surface currents pass. In the top layers of the sea in equatorial regions, warm, less dense water flows slowly toward the subtropical areas and returns, cooler and more dense, at deeper levels. This movement, however, is by no means restricted to the tropical-subtropical exchange.

Mediterranean-Atlantic interchange. The exchange of water between the Atlantic Ocean and the Mediterranean Sea is an excellent example of shallow water thermohaline circulation. While there are slight temperature differences between the two bodies of water, they are negligible compared to the differences in salinity. Due to excessive evaporation, the salinity of the Mediterranean may be as high as 38‰ compared to the Atlantic's 35‰.

The Mediterranean and Atlantic basins are separated by a sill at the straits of Gibraltar, which lies only 170 fathoms (1020 feet) beneath the surface. Even though both ocean bottoms extend below 170 fathoms, the Atlantic is the deeper of the two. The highly saline waters of the Mediterranean sink and flow over the Gibraltar sill and on into the Atlantic, where they subside to a depth of about 550 fathoms (3300 feet) before reaching equilibrium with the surrounding water. Meanwhile, less dense water from the Atlantic flows over the outflowing water, through the straits and into the Mediterranean Sea.

Deep circulation

The second thermohaline circulation, much more extensive and important than the first, involves the slow movement of all remaining cold water, regardless of whether it is at the surface in the polar regions, or in the abyss everywhere. The deep-water circulation is so slow, in fact, that from 500 to 2000 years elapse before surface water actually returns to the surface. This figure has been determined by the amount of dissolved oxygen measured in water samples.

The temperature of the air at the poles is so cold that surface waters often reach a temperature low enough (28° F or 0.2° C) to allow sea ice to form. This constant removal of fresh water from the sea in the form of ice makes the polar waters, already dense due to the low temperatures, even denser because of their increased salinity. After sinking to the bottom, this dense water eventually starts to flow toward the equator.

The south polar region is the coldest part of the earth, the continent of Antarctica. Its cold, dense waters sink to the bottom and then very slowly circle the Antarctic continent, mixing gradually with adjacent cold-water masses. These cold waters eventually flow northward, filling deep ocean basins all over the world. Antarctic bottom water has been recognized in all parts of the oceans by its temperature of 31.2° F (−0.4° C) and salinity of 34.66‰.

Deep ocean circulation appears to be primarily oriented north to south, and like the surface circulations it is strongest on the western sides of the basins. Deep waters of the abyss continue to move because they are being pushed by the continual movement of new, cold, dense water from the polar regions. The movement of deep water is also greatly influenced by bottom topography such as the mid-ocean ridges.

The waters of the abyss must return to the surface eventually, and may do so at least in part with the aid of the wind.

Vertical movement of water induced by wind

The winds are capable not only of producing horizontal water flows in the form of currents but also of forming vertical movements of water that are known as *sinking* and *upwelling*. These vertical movements occur where prevailing winds blow parallel to the coastline, strongly illustrating the Coriolis effect and the Ekman spiral.

The upwelling (Fig. 23-6) can be explained as follows. Initially the wind causes the surface waters to move. Then the Coriolis force causes the waters to flow toward the right of the wind direction, in this case, offshore. The void that is left after the water is removed is filled by the slow, upward movement of deeper, colder water.

This type of wind-induced upwelling has great significance for us because it brings rich supplies of dissolved nutrients into the surface water. These nutrients support the lush growth of phytoplankton, and the abundance of these small plants in turn guarantees the vital well-developed food chain and the resultant fish population. The richest fishing areas in the world are found in such regions of upwelling as the offshore waters of Nova Scotia, California, Peru, and Northeast Africa. In fact, in order to reap the abundant harvests indigenous to these upwelled waters, many nations send their fishing fleets halfway around the world. In addition to the comfortable cooler climates that

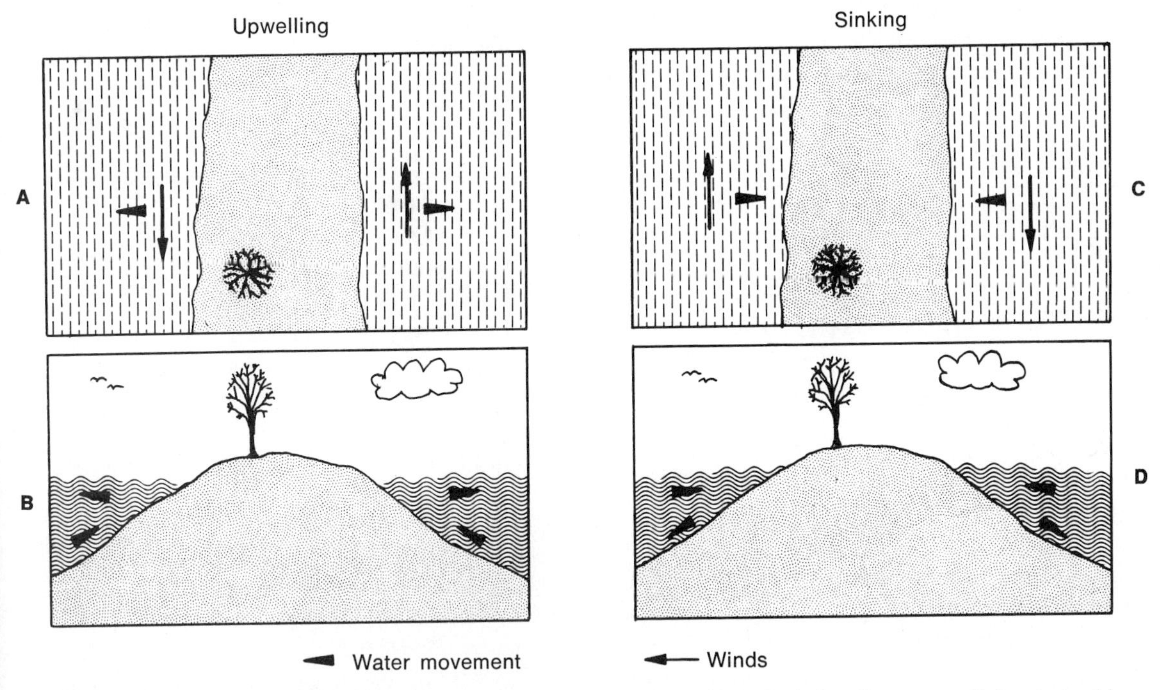

Fig. 23-6. When the wind blows parallel to the shore, as in **A,** it and the Coriolis effect may collaborate to skim off the warmer top layers of water. **B,** To fill in the void left by the departing waters, the deeper, cooler water flows in. This is known as "upwelling." **C,** An opposite reaction that takes place when the Coriolis effect does not help the wind in moving excess water offshore. Here the water skimmed off the top simply accumulates until there is so much around that the whole mass sinks under its own weight. In this case, the underlying cold water is forced back and away from shore, as shown in **D.**

usually prevail in these regions, there is also a much greater incidence of fog.

Wind-induced sinking of water occurs wherever the net flow of water is toward the coast. In such areas an excess of surface water piles up and sinks along the coast. Although less noticeable than upwelling, sinking can radically affect the distribution and abundance of marine life in a given local area.

El Niño

The ocean current called *El Niño* (the child), which develops from time to time along the coasts of Ecuador and Peru, is an excellent example of the effects that the replacement of upwelling with sinking can have on a local region. The dominant current is usually the cold, northward-flowing Peru Current. Upwelling along the coast of Peru makes this one of the richest fishing grounds in the world. Nearly every year around Christmas along the coast of Ecuador, a branch of the Equatorial Countercurrent passes over the equator. Although this southward-flowing, warm current does not usually reach the Peruvian fishing grounds, if it does it can prove to be disastrous. When upwelling of cold, nutrient-rich water is replaced by the subsiding of warm, nutrient-poor water, fish die by the millions, rot, and turn the water black with hydrogen sulfide. This substance further coats the hulls of ships in a phenomenon described by local residents as the "Callao painter," after the coastal town of Callao.

Local currents

Most of the currents and types of circulation we have discussed so far have been of far-reaching and in some cases worldwide importance. However, there are still more local currents, and while they are limited in extent they have great significance to their local areas.

Longshore currents

Earlier in this chapter we discussed *wave refraction* as the bending of waves in a way that causes them to strike parallel to the coastline. Since one part of a wave breaks before another part, an excess of water tends to pile up along the shore, producing a very slight difference in water level. Therefore in shallow water along the coast the water flows in the direction of the moving waves. This movement of water is called the longshore current and is of the utmost importance in the transportation of rock debris along the coastline.

Rip currents

Even more localized than longshore currents are rip currents, narrow, fast-moving ribbons of water that extend from the beach through the surf zone into deeper water (Fig. 23-7). These extremely strong currents, not to be confused with the undertow (which is simply the continual return of the surf water under the incoming waves), form either where two longshore currents converge or where the current is deflected seaward by a breakwater or rocky headland that juts out into the water.

From the beach, rip currents may be recognized by (1) the general absence of surf; (2) their darker color, which usually indicates slightly deeper water because the rip usually cuts a channel in the sandy bottom; and (3) a line of agitated or confused water running straight out from the shoreline.

Swimmers beware! Rip currents have taken a tragic toll of both experienced and inexperienced swimmers. The natural tendency when being swept out to sea by one of these currents is to try to swim back toward the shore against the current's flow. Here, however, the shortest distance is not a straight line, since rips are rarely more than 100 feet wide. If you are caught in a rip current, then, you should (1) swim parallel to the shore until you leave the current or (2) float with the current until you pass the surf zone where the rip weakens. In either case you can then make it safely to shore.

Current measurements

Surface ocean currents can be measured in a number of ways, many of which are quite ingenious because of the usual lack of a fixed point of reference at sea.

The speed of a surface current may be measured by comparing it to the speed of a ship. If a vessel were steaming north at 5 knots (6 miles/hr) against a southward-flowing current and making no headway, this would then indicate that the current had a speed of 5 knots.

The speed of a surface current may also be measured by a ship at anchor through the use of a current meter. This is a device with a propeller that is rotated by moving water. The revolutions are recorded by a counting mechanism. Current meters may be either of the simple Ekman type or of the sophisticated type like the GEK (Geomagnetic Electrokinetograph), which reacts to electrical forces generated by geomagnetic fields in moving seawater.

Still another more dramatic and historically popular technique for measuring the speed and extent of surface ocean currents is the use of simple, weighted drift bottles. These are usually soft-drink bottles filled with sand for ballast. They contain a note and a self-addressed, stamped postcard requesting the finder to record the date and location on which the bottle is found. (The return rate of this method generally runs at approximately 10%!)

The longest undisputed drift on record was by a bottle released on June 20, 1962, at Perth, Australia, and re-

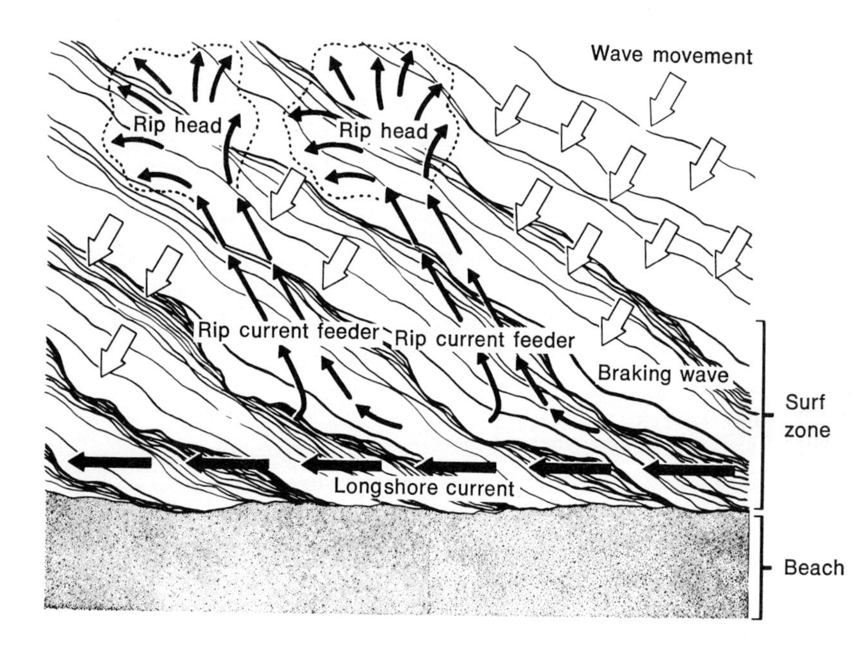

Fig. 23-7. Powerful rip currents like these form where two longshore currents converge or where a current is thrown seaward by a breakwater or headland. (From Zottoli, R.: Introduction to marine environments, St. Louis, 1973, The C. V. Mosby Co.)

covered near Miami, Florida in 1967. This 16,000-mile (25,760-km) drift probably took the bottle around the Cape of Good Hope, northward along the African Coast, across the Atlantic to the Coast of Brazil, north again along the South and Central American coasts and into the Gulf of Mexico, and finally through the Straits of Florida to Miami.

The speed of surface currents as well as those in deeper water may also be calculated by using a Swallow Float, or Pinger This free-floating device gives out a constant radio signal that can be tracked from onboard a ship and its speed computed. It was through the use of the Swallow Float that the Gulf Stream Countercurrent was discovered.

SHORELINES: THE BATTLEGROUND

The shore is the narrow zone in which the bottom of the sea becomes the surface of the land. Since our shorelines are transient features, future generations will undoubtedly be doing their sunning, surfing, boating, and swimming along a shore that will either be far inland or seaward from its present location. The changing nature of the coast is the work of energy released by breaking waves and of the movement of sand by currents.

Geologic work of waves and currents

Oceans are enormous, yet most of their geologic work is done in shallow waters near land. Here waves reach water with a depth of half a wavelength or less, break, and expend their energy. It is here also that wave refraction develops the important longshore current. The sea, like the other geologic erosional agents discussed earlier, performs three geologic functions: erosion, transportation, and deposition.

Erosion

Most of the erosive action of the sea is done by wave action. Like streams on land, the sea performs erosion by impact, abrasion, and solution. Each cubic foot of sea water weighs 64 pounds. When one considers the thousands of cubic feet of water striking the coast continually at any given point, in addition to the kinetic energy supplied by motion, it is easy to see how the force of breaking waves can do tremendous quantities of geologic work on anything they strike. This is erosion by impact, which destroys rocks, breakwaters, foundations, and beaches. This effect is magnified when the waves increase in height, as during a storm. If a way could be found to harness it, wave energy could undoubtedly provide an inexpensive, pollution-free source of vitally needed power.

Added to the force of impact are the erosional processes of abrasion. All waves contain sediments, mostly the size of sand, that have been delivered earlier by both streams and the process of impact. The constant back-and-forth agitation of the water containing these sediments batters away at the surfaces that are struck, and then the particles themselves wear away through abrasion. Gravel-sized sediments, broken loose by impact, clearly show the effects of abrasion in their smooth, rounded appearance.

Solution, the dissolving of rocks by water, is considered a minor erosional force on most coasts. If the rocks along the coast are limestone or dolostone, however, the force becomes more effective because these rocks are most susceptible to solution. At best, then, the sea's contribution to its own mineral content would have to be considered negligible, since most of this work was done previously by streams and groundwater.

Erosion by wave action produces many beautiful and interesting landforms that are discussed in a later section of this chapter.

Transportation

The movement of sediment along a coastline is accomplished by the two active forces found there, wave action and the longshore current. Most of the material being transported is sand.

As each wave breaks, sand along the bottom is moved shoreward, with the orbital motion described in our earlier discussion of waves. First the sand is picked up by waves. Then it is moved forward; finally it is set down, where it awaits the next breaking wave. In the surf zone, therefore, there is a general onshore-offshore movement of sand by the breaking waves; the motion, however, tends to be seasonal and related to storms. Low waves of the typical nonstorm type, for example, gradually cause sand to migrate shoreward, thereby increasing the amount available for shore processes. Higher waves, usually caused by local storms, break rapidly, one after another and with more force, creating a strong turbulence that keeps the sand in suspension. Thus little sand can settle to the bottom before another wave breaks. The powerful countercurrent required to return the greater volume of water to the sea causes the sand to be moved offshore for some distance. The beaches take on a somewhat starved appearance, a look that is only temporary, however, because soon the gentler waves will restore the beach, again producing a state of dynamic equilibrium.

The other major mover of sediment along the shore is the longshore current, which is induced in the surf zone by wave refraction. This current, together with the continual upwash of breaking waves, sets up a downcoast transport of sand. At the beach face, where the breaking waves carry water up and back, the transport follows a zigzag path in the direction of the longshore current. In the surf zone, however, the movement of sand is almost lateral. This along-shore movement of sand is the major problem in shoreline conservation.

Deposition

The sediment carried in the to-and-fro motion of the waves and the lateral movement of the longshore current is deposited wherever this motion or movement is slowed down or stopped. Such a reduction of speed occurs either when an obstruction is encountered or when shallow water is entered. At these locations the sediment, nearly always sand, settles to the bottom where it forms layers. At every seawall and jetty, in every cut and bay, on every coast, sand is being dropped, filling in and smoothing out the shoreline. Landforms produced by wave and current deposition will be discussed later in this chapter.

Classification of coastlines

It is very difficult to classify coasts. Much of the western coast of North America, for example, was shaped by diastrophism and volcanic activity, while the coasts of Scandinavia, Greenland, and Alaska were modified by glacial action. Other coastal areas were built by organisms such as coral. In addition, there are also emerged coasts,

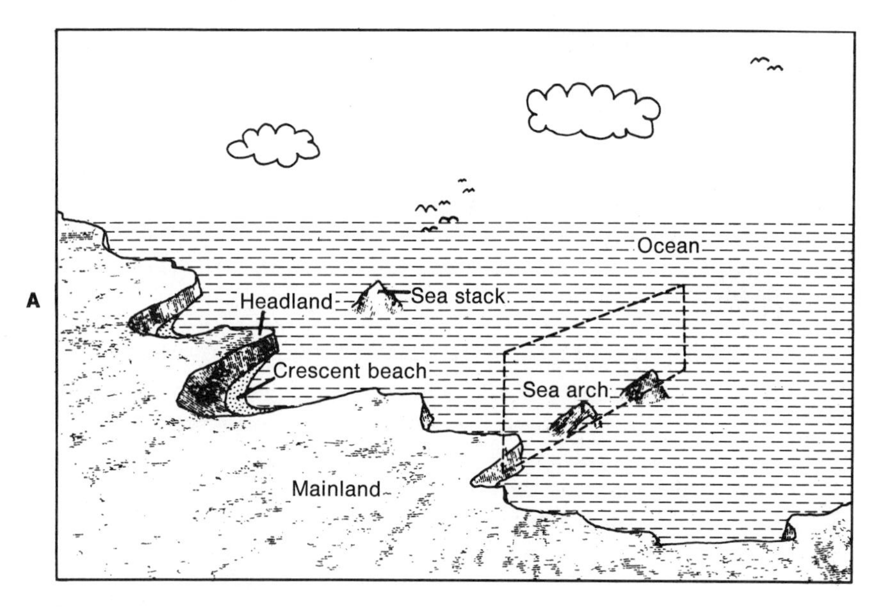

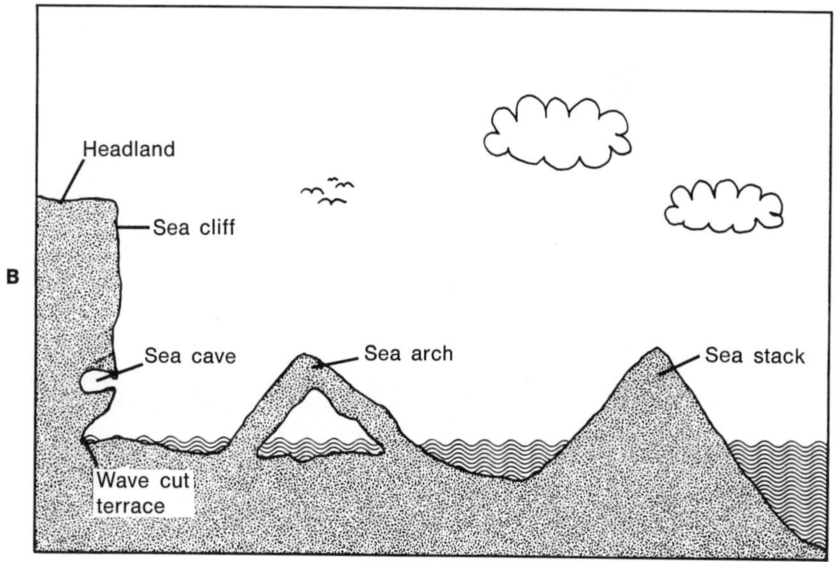

Fig. 23-8. A, Generalized features of a rocky coast. **B,** Generalized cross section of a rocky coast. Sea caves are often submerged at high tide. Stacks and arches are remnants of former headlands.

submerged coasts, rocky coasts, and sandy coasts. The possibilities seem almost limitless, but even so, common threads do exist and these will be used here to weave an oversimplified but workable picture of shorelines.

Rocky coasts

Rocky coasts by their very nature tend to be high-energy coastlines where erosion is the predominant form of geologic action. Because rocky coasts have deep water offshore, the waves do not reach the one-half wavelength depth until they are virtually on the shore. Since the waves possess large quantities of kinetic energy, they strike the shore with great force, carving the most beautiful coastlines in the world, primarily through impact and abrasion. Waves, like other geologic erosional agents, attack differentially, that is, by destroying the weaker rocks first. Rocky coasts are not necessarily composed of only one type of rock, and even in small areas there are variations in rock types and therefore in erosional resistance. As a result, rocky coasts are irregular, with many ins and outs (Fig. 23-8).

A *headland* is the place at which a wave first comes into contact with a rocky coast. Jutting out into the sea, headlands undergo concentrated erosion and therefore over periods of many years they recede toward the shore. This recession leaves a number of curious features behind; over time these features themselves are completely eroded away. One of the first erosional remnants of a headland is a *sea arch,* which is a natural bridge carved by the waves. Another type of headland remnant is the *sea stack,* which is nothing more than an isolated rock mass that projects out of the water. Many seals make their homes on sea stacks along the California coast.

Still another erosional feature found on rocky coasts are *sea cliffs*. Rising almost vertically out of the water, sea cliffs are maintained by resistant rock as waves cut the softer rock from their bases, causing them to recede. (Sea caves are frequently carved by erosion into the face of sea cliffs. Although high and dry at low tide, many sea caves become sources of danger to the unwary explorer because they are underwater at high tide.) At the base of sea cliffs, an erosional notch called a *wave-cut terrace* is eroded by the concentrated wave action. Such terraces, found well inland, are usually indicative of a change in sea level, either through a rising of the land or a decrease in the level of the sea, or both.

Although erosion is the dominant feature on a rocky coast, transportation and deposition also occur. With erosion concentrated on the headlands, the longshore currents move the resulting sediments into small indentations in the shore, usually called *bays*. Here, in relatively shallow water, sand is deposited in crescent-shaped beaches. Virtually every rocky coast has at least one crescent beach.

Sandy coasts

A sandy coast is a low-energy shoreline with shallow water where low waves usually break at some distance offshore. Deposition dominates this type of coastline, smoothing and straightening the shore into a more-or-less regular pattern (Fig. 23-9). Sandy coasts are not spectacular to view, but in their own way they possess many features that are unique and interesting.

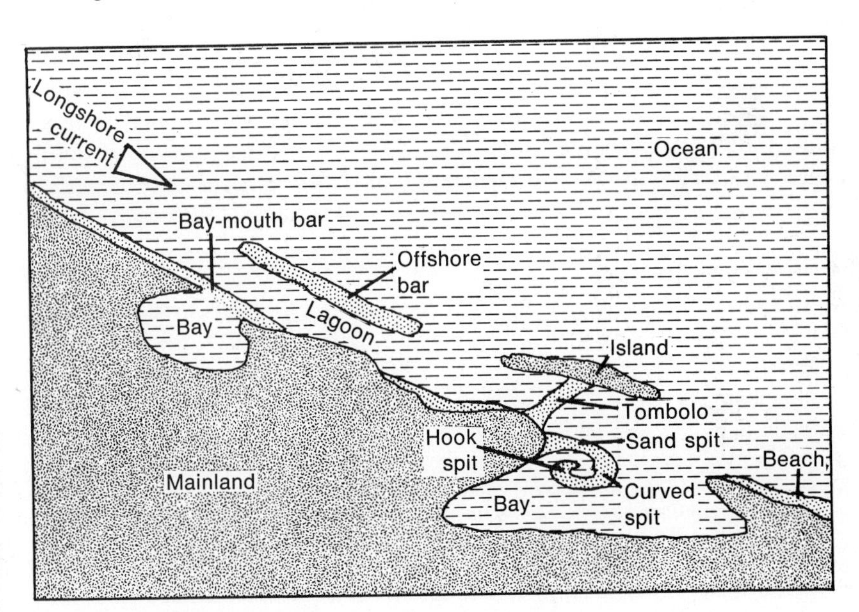

Fig. 23-9. Generalized features on a sandy coast. Coastline is more regular than a rocky coast. Sand tends to move from left to right because of the trend of the longshore current.

Sandbars. Sandbars are common features along sandy coasts. One type is the *offshore* bar, which develops wherever a submerged obstruction, either natural or artificial, lies parallel to the coast, causing waves to break prematurely. The obstruction might be an irregularity in the ocean bottom, a coral reef, or a pile of discarded automobiles used as a breakwater or an artificial fishing reef.

Between the offshore bar and shoreline lies an area of quiet, protected salt water called a *lagoon,* a natural site of deposition. As the lagoon naturally fills in with sediments over time, the land expands seaward. The east coast of Florida shows some good examples of offshore bars; Miami Beach, for example, is an offshore bar built on an old coral reef. The part of Biscayne Bay situated between Miami Beach and Miami is a lagoon (Fig. 23-1). In fact, many lagoons found along the Florida coast are mistakenly called rivers, for example, the Indian River.

Another type of sandbar characteristic of sandy coasts is the *bay-mouth* variety. Like the erosional sequence involving headlands on rocky coasts, bay-mouth bars are often the end result of a depositional sequence in a bay. Longshore transport carries sand toward the bay and drops it in the shallow water, where it gradually builds a *sand spit.* If the bay is rather large and the currents are strong, the sand spit will curve into the bay and form a *curved spit.* If the currents are very strong, the end of the curved spit will develop a hook, forming a *hook spit.* In this way, sand slowly narrows the entrance to the bay until it closes entirely to produce a *bay-mouth bar.* Once blocked, the bay is eventually filled with silt and becomes part of the mainland.

Still another type of sandbar develops in shallow waters between an island and the mainland or between islands themselves. These features, called *tombolos,* are natural causeways of sediments. At low tide it is often possible to walk through mud from island to island across these natural causeways. Through the development of tombolos, islands are welded to islands and eventually become part of the mainland.

The beach: a stream of sand. The most dominant of all features along a sandy coast is the *beach.* There is a great deal more to a beach than meets the eye. The *backshore zone,* for example, includes sand dunes, or in the case of a rocky coast, the sea cliff. Here also is the familiar *berm,* the gently sloping sand terrace commonly used for sunbathing; this area is underwater only during exceptionally high tides or storms.

The *foreshore zone* lies between normal high and low tides. This is the zone of breakers and the location of the longshore current. There may also be a *beach face* (an area that slopes steeply toward the water), a *runnel* (a hollow that carries away water draining off the beach face), and a *ridge* of sand, just offshore from the runnel, which is frequently exposed at low tide.

The *offshore zone* usually consists of a trough, or depression, and a submarine bar. Both the trough and bar are permanently submerged. The sand making up this bar is the source of sand needed for the foreshore and backshore zones.

On the coast

Over billions of years, waves and currents have been pounding away at coastlines, building, modifying, and destroying, yet always remaining in a constant state of dynamic equilibrium. Since we are unable to change the natural patterns existing between land and sea, we can only interfere in them. It is essential, then, for us to try to understand the complexities of marine processes. If and when we learn to accept their ultimate outcome, we may be able to hold onto our investments in coastal real estate. With or without us, however, the sea will continue its eternal processes.

ICE IN THE SEA

An icebreaker once encountered a block of ice 100 miles (161 km) wide and 100 miles (161 km) long in the South Shetland Islands near the frozen continent of Antarctica. Weighing an estimated 6 billion tons, this magnificent iceberg illustrates the magnitude that ice in the sea is capable of reaching.

Under normal atmospheric pressure, surface ocean water will begin to freeze at 28° F (−2° C). At this temperature the salt content is left behind and the resultant ice is composed of frozen fresh water occasionally contaminated by pockets of salt water. This is called *sea ice.* Here is another oceanographic irony; icebergs are ice in the sea, yet they are not sea ice.

Sea ice is always lighter than seawater, so, regardless of its manner of formation, it is found at the ocean surface. Since the formation of ice acts as an insulator against further freezing, sea ice is of limited thickness. Arctic ice, for example, averages 9 to 10 feet (2.7 to 3 meters), while Antarctic ice is about 15 feet (4.6 meters) thick.

Arctic sea ice

The north polar region, with the exception of some islands, is predominantly water, namely, the Arctic Ocean. Here sea ice occurs in three distinct areas. First and most important is the *Arctic ice pack,* a great central core of the Arctic Ocean that permanently contains sea ice. In summer there are broad lanes of open water, yet ice is always present. Even in winter, the ice is not frozen solid from coast to coast but often contains cracks and even some open water.

The second form of sea ice is *drift ice,* huge pieces of sea ice that drift in a slow, zigzag, clockwise motion around the borders of the Arctic pack. These ice islands, ranging from 5 to 18 miles (8 to 24 km) wide and from 10 to 20 miles (16 to 32 km) long, have served for weeks as the homes of scientific expeditions.

The last type of Arctic sea ice is called *fast ice,* since it tends to remain attached (fast) to the Arctic coast for as long as 9 months of the year. Great slabs of fast ice, breaking off and drifting with the Arctic-circling currents, are probably sources of the ice islands.

Antarctic sea ice

Unlike the north polar region, the south polar region is dominated by a continental land mass. As a result, there is much less restriction in the movement of sea ice. *Pack ice* of the Southern Hemisphere actually consists of floating ice fields that are more-or-less flat slabs of sea ice. They tend to circulate in a general counterclockwise direction around the continent of Antarctica. During the winter, icebreakers often have to cross as much as 600 miles (966 km) of pack ice before landing on the Antarctic continent. Thus the amount of pack ice, while highly variable with the season, is always great.

The other type of Antarctic sea ice, very similar to the fast ice of the north polar regions, is called *shelf ice* because it spreads out over the continental shelf from several areas of Antarctica, primarily the Ross Sea and the Weddel Sea. (''Little America'' is located in the vicinity of the Ross Ice Shelf.) The shelf ice, a transition zone between sea ice and the glacial ice of the interior of the continent, is the source of the great Antarctic icebergs.

Icebergs: floating mountains

Icebergs are great masses of glacial ice floating in the sea. They are rarely contaminated with salt, so when they melt they release almost pure water. Icebergs form by *calving* (breaking off) from glaciers or other icebergs. Of all the glaciers on earth, 98% are concentrated in two regions: the island of Greenland and the continent of Antarctica. In these areas glaciers reach the sea in two basically different ways, which explains why icebergs from the two regions differ so radically in appearance. In Greenland, glaciers from the great interior ice sheet reach sea level as rather small valley glaciers through the coastal mountains on the western side of the island. Calving of these glaciers produces the classic irregular, jagged iceberg. In the Antarctic, however, so much glacial ice has accumulated that it reaches to sea level and extends out into the ocean where it joins with shelf ice. Calving here produces icebergs of the *tabular* type, which can be enormous.

Icebergs do not always remain at their points of origin but may move across great distances, using the energy of ocean currents. Greenland icebergs, for example, follow two principal paths out of Baffin Bay. One is westward across the bay and then southward in the Baffin Current, which flows into the greater Labrador Current. The other path is northward, in the West Greenland Current, then southward in the Baffin Current. No more than 20% of the icebergs calved from Greenland survive to float as far south as Newfoundland, where they become a threat to shipping. The number of icebergs in the North Atlantic varies greatly; 1000 were noted in 1912, the year of the Titanic disaster, and none were seen in 1966. The average yearly number of icebergs tracked by the International Ice Patrol is between 200 and 400.

The great tabular icebergs of the south break off from the ice shelves and float westward into the seas near the frigid continent. Sailors often report seeing as many as 400 icebergs within a distance of 20 miles (32 km). Poised as they are at the ''head'' of some northward-flowing currents, Antarctic icebergs have been seen as far north as 26°, 30′ S, 25°, 40′ W, almost at the Tropic of Capricorn between South Africa and South America.

24 *the deep challenge*

The future and the sea

Darkness settles on roof and walls,
But the sea, the sea in the darkness calls,
The little waves, with their soft, white hands,
Erase the footprints in the sand,
And the tide rises, the tide falls.

The Tide Rises, The Tide Falls
Henry Wadsworth Longfellow

On June 18, 1973, a 23-foot, 9-ton minisubmarine, the Johnson *Sea Link,* became entangled in the wreckage of a scuttled destroyer 15 miles southeast of Key West, Florida (Fig. 24-1). After 31 hours, the submarine was finally freed, but the incident cost the lives of two of its four crewmen. One of the marine scientists who lost his life was E. Clayton Link, the son of Edwin A. Link, an inventor and, ironically, a pioneer in oceanographic research. As this tragic event once again demonstrated, the sea is an environment to which we adapt even less easily than space. Many men and several women have gone into space, some staying for more than a month before returning safely to earth. During that time they adapted well to weightlessness and found living and working in space a relatively easy and exhilarating experience. The marine environment, on the other hand, presents phenomena we cannot adapt to and constantly provides new and frightening surprises.

Space has often been called "the last frontier," and this is no doubt accurate in terms of distance and time. There is another unexplored frontier closer to home, however. This is the great storehouse that surrounds our shores. We can hardly continue to rely on 29% of our earth's resources, disregarding that other 71% hidden beneath the sea. To our existence, then, the sea is our frontier, and our challenge.

FOOD IN THE SEA

If the world's population continues to increase at its present rate, there will be 6 billion people to feed by the year 2000. Even today, with a population half that size, many people suffer from malnutrition and die of starvation. Unexpected meteorologic or other phenomena such as the prolonged droughts that have struck the Sahale Desert in Africa push additional millions of people over the thin line that separates their normal diet from starvation.

At present only about 1% of our food supply comes from the sea, but more than 25,000 species of fish are waiting to enrich the underdeveloped world's grain-staple diet with swimming protein. Fish, however, are not the only source of protein in the sea. Crustaceans (lobsters, crabs, shrimp, and prawns), mollusks (clams, oysters, scallops, mussels, cockles, and squid), and many other forms of sea life are available food sources.

In addition to animals, the sea supports plants able to contribute their share to global nutrition. *Kelp,* a brown alga, is often referred to as "super spinach" because of its nutritional value, and *sol,* a red alga, may substitute for many more familiar vegetables. Seaweed production is on the increase, but still less than 1 million metric tons are produced each year—a quantity about equal to the annual prune production in the United States.

Marine life not only provides an excellent direct source of food but also indirectly supplies an abundance of products for human consumption. Trash fish such as the red hake, for example, may be ground to a powder—head, scales, bones, and all—dried, and processed to remove the fishy odor. The resultant neutral-tasting white meal, called *fish protein concentrate* (FPC) can be added to less nutri-

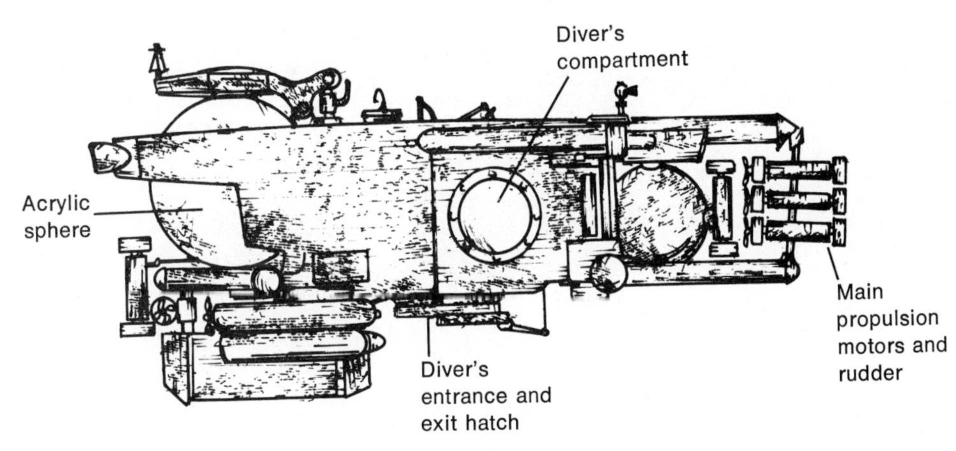

Fig. 24-1. The *Sea Link,* a minisubmarine owned by the Smithsonian Institution, carried two divers to their deaths when she became entangled in undersea wreckage. Two other divers in the acrylic sphere survived.

tious staples like wheat or rice to provide increased levels of animal protein. FPC also adds nutrients to feed for pigs, cattle, and chickens.

Extracts of seaweeds are already used in a number of food products. *Agar* and *carrageenin,* both extracted from red algae, are used as a gel in ice cream, pies, icings, and meringues. Brown algae provide *algin,* which is used extensively in the food industry in ice cream, soups, mayonnaise, sauces, and sausage casings. It is also widely used in the dairy, rubber, textile, and pharmaceutical industries. Seaweed itself may be used as a fertilizer to help increase the production of other foodstuffs.

The utilization of food from the sea has economic, political, and religious ramifications. Nations with seacoasts, for example, have a natural delicatessen at their front doors. Landlocked nations, on the other hand, have no access to the sea, so they will have to cooperate economically with those nations that do. Religious beliefs often prohibit the consumption of certain types of food, and accommodation will be required if maximum use is to be made of marine food supplies. International cooperation is both necessary and expedient, for all nations have a stake in the world food problem; a world of full stomachs is the first prerequisite to political peace.

MINERALS FROM THE SEA

Industrial nations need minerals to maintain and improve living conditions for their people. Historically, mineral resources have been mined on the land, but recently we have become aware of the mineral wealth dissolved in seawater, lying on the ocean bottom, or buried in the rocks and sediments beneath the ocean floor.

Salts

Vast quantities of the salts in the sea can be easily removed through simple evaporation. One such salt, sodium chloride, or table salt, is widely used as a food seasoning and in such lesser-known fields as meat and fish packing, canning, refrigeration, metallurgy, water treatment, and the production of dyes and other chemicals, soap, leather products, and textiles. Other important although less familiar salts include magnesium chloride, potassium chloride, magnesium sulfate, and the bromides. About 75% of the bromine used in the United States comes from the sea. This chemical is used extensively in the production of gasoline.

Metals

Seawater also contains many metals of economic value, including magnesium, gold, silver, and platinum. These may be either dissolved in the water itself or concentrated in solid deposits in the sediments on the bottom.

Whether these minerals are removed from the sea and how this would be done are matters of economics. For example, 100% of the magnesium used in the United States comes from seawater because it can be extracted more cheaply than it can be mined on land (Fig. 24-2). Gold is present both in dissolved form and as placer deposits on the bottom. Until recently, land deposits were developed because it is cheaper to mine gold on land than to either dredge the bottom for pieces or to distill the gold out of water. For example, 1 cubic mile of seawater contains about 50 pounds of gold, but it also contains a tremendous amount of water. To obtain a mere ounce of gold, it would be necessary to process 5 million tons of water. Today, however, some gold

Fig. 24-2. The addition of oyster-produced lime to seawater forms magnesium hydroxide, which is collected in these huge tanks until it can be processed into magnesium. The size of the tanks gives you an idea of the mineral wealth waiting in the sea. (Courtesy Dow Chemical U.S.A., Midland, Mich.)

is being dredged up from the ocean bottom offshore from known gold-producing regions. This method will begin to look far more promising when the land deposits are exhausted and there is no other choice.

Other minerals

Manganese, nickel, iron, copper, and cobalt are found on the sea floor in lumps called *nodules*. Diamonds, sand, gravel, pearls, platinum, phosphate, tin, thorium, and titanium are also regularly obtained from the sea. Petroleum, on which industrial nations depend so greatly, is being found in the sediments and rocks beneath the sea in ever-increasing quantities. As land sources of these and other minerals become unavailable for political reasons or are exhausted, the sea is a storehouse of many resources we need. We need only apply our ingenuity to obtain them, and our ingenuity will be spurred by need.

WATER FROM THE SEA

On February 23, 1961, in a special message to Congress, President John F. Kennedy emphasized the national and international importance of the sea as a source of fresh water when he said:

> No water resource program is of greater long-range importance for relief not only of our shortages but for arid nations the world over than our efforts to find an effective and economical way to convert water from the world's greatest, cheapest, natural resource —the oceans—into water fit for consumption in the home and by industry. Such a breakthrough would end bitter struggles between neighbors, states, and nations and bring new hope for millions who live out their lives in dire shortage of usable water and all its physical and economic blessings, though living on the edge of a great body of water throughout their parched lifetimes.

Desalinization, the process of making saltwater potable, is not yet an efficient method compared to the cost of

Fig. 24-3. The desalinization plant at Key West, Florida, provides enough water for 35,000 people and the industries that employ them. (Courtesy Key West Chamber of Commerce, Key West, Fla.)

purifying fresh water. The techniques are known, but the cost of the energy to convert salt to fresh water makes the price of the purified water too high to be useful on a large scale. Today, the cost of producing 1000 gallons of fresh water from seawater is about $1.00. The cost of purifying an equal amount of fresh water is less than half that amount. Nevertheless, if the need is critical enough, the whole question of cost becomes secondary.

There are two basic ways of converting saltwater to fresh: to remove the salt from the water through electrodialysis, or to remove the water from the salt through distillation or freezing. A form of the latter method, called *flash distillation,* seems to be very promising at this point. This technique involves pumping hot saltwater into low-pressure chambers. There it instantly turns into steam, which is then condensed. Since the salts dropped out when the water changed to steam, the resultant water is now fresh. The real problem is the development of an economical system for heating the water. At San Diego, California, a flash distillation plant uses nuclear power to produce 1 million gallons of water daily. A similar plant in Key West, Florida, provides the entire water supply for this city of 35,000 people (Fig. 24-3).

In regions with plentiful precipitation, the wise use and recycling of available fresh water is no doubt the answer to the ever-increasing thirst of the world; in the United States alone 1.5 million gallons disappear every hour. In arid regions without this choice, however, desalinization holds great promise, not only to provide drinking water and water for industry, but also to provide the irrigation such countries need to maintain independent food supplies.

Irrigation in arid countries is creating another desalinating headache. The water for such irrigation comes either from streams or from underground water supplies, and both contain salt, though in much smaller concentrations than ocean water. Nevertheless, this salt is now accumulating in the land being irrigated, a process that is very expensive, or perhaps impossible, to reverse. If irrigation is to increase productivity in barren lands, rather than turning them into "barrener" ones, we may soon be desalinating fresh water as well as seawater.

POWER FROM THE SEA

The world today is crying for power. Long accustomed to economical fossil fuels such as petroleum, coal, and natural gas, we become a little unnerved at the possibility that these fuels are becoming unavailable. Alternate sources of power are badly needed, and it is imperative that some be found.

It is at least theoretically possible to produce large quantities of power from the sea by taming three aspects of the ocean's power. All of these are now under serious consideration.

Surf power

At present there is no feasible method for harnessing the surf, although the amount of energy expended each second, even under normal wave conditions, is enough to inspire scientists to keep looking for one. In certain areas such as the north shore of Hawaii, incoming waves regularly climb 20 feet (6.1 meters) or higher. This is a perfect site for experimenting with wave-turned dynamos.

Current power

Another technique that looks much more promising at this point involves the great ocean currents. If some kind of great dynamo could be suspended in a current such as the Gulf Stream, tremendous amounts of power could be generated as the water moved through. With so many strong ocean currents lying just offshore from population centers, the correlation between the need and the opportunity is too close for scientists to pass up. Clearly many problems need to be solved: methods for suspending the dynamos and channeling the power thus created are only two. Yet the Gulf Stream, which sweeps by the heavily populated East Coast of the United States, the California Current, which passes the sprawling megalopolis of Southern California, and the Kuroshio Current, which tantalizes the energy-pressed country of Japan, may one day be the sources of these areas' energy supplies.

Tidal power

The last technique, and the only one currently operational, involves harnessing power from the tides. Tides affect all coastal regions, with most locations experiencing two high and two low tides every 24 hours. In most regions the *tidal range,* or difference between high and low tide, amounts to less than 10 feet (3 meters). Occasionally, however, the tidal range becomes enormous; tides in excess of 30 feet (9 meters) can be found around the world, and in the Bay of Fundy in Nova Scotia, the high and low tides may differ by as much as 50 feet (15 meters).

Several tidal power stations are already complete and in operation today. Like desalted seawater, however, electric power from the sea must not only be available, it must be as inexpensive as other sources. This balance has been achieved at Kislaya Bay near Murmansk in the Soviet Union and on the Rance River in Brittany, France. A description of the Rance River project illustrates how tidal currents are used.

Tides near the mouth of the Rance River reach a maximum of 40 feet. To convert this vast amount of energy into electric power, Electricity of France, the state-owned power company, placed 24 turbine units across the river to catch the tidal flow (Fig. 24-4). Each unit was equipped with devices to increase its efficiency and with reversing capability to draw power from both the ebb and the flow of the tidal cycle. Annual output from this plant has reached 500 million kilowatt-hours.

Fig. 24-4. The Rance River project, one of a very few successful attempts to harness oceanic power, uses the energy of tidal flow to generate electricity. To increase its capacity, reversing units were installed to wring power from the outgoing as well as the incoming tides. (Courtesy French Engineering Bureau, Washington, D.C.; photograph by Baranger.)

THE UNIVERSAL DUMP

The world needs a dump into which all of its wastes can be thrown; the ocean has played this role since the beginning of time. In the light of burgeoning world populations and new knowledge about the effects of these waste products on the sea, however, we must now ask how much and what kind of material can the sea be expected to absorb safely.

Among the contaminants allowed to enter the sea virtually unchanged are heat from power plants, chemical wastes from factories, waste and sewage from cities and towns, petroleum wastes from many sources including oil spills, insecticides and fertilizer from land runoff, and low-level radioactive wastes from reactors, hospitals, and laboratories. This, of course, does not include such occurrences as the United States Army's plan several years ago to dump surplus poisonous gases into the ocean. Due either to their sheer volume or to their chemical stability, many such wastes are not being degraded, diluted, or dispersed. The fact that wastes do not just go away can no longer be ignored, for the sea is far too critical to our survival to risk the consequences of such neglect.

POLITICS OF THE SEA

Simply because of its size, the sea is involved in world politics. Add to this the resources just discussed, and the sea takes on even greater significance in the affairs of nations. The sea is international—owned by no one, yet owned by everyone. Some fortunate nations have ocean access, most do not. Does land-locked Switzerland, for example, have the same right to the products of the sea as Great Britain, which is wholly surrounded by it? Should a state like Wyoming, without a coastline, share in the marine products of Florida or of California's long coastline?

Beyond geography, there is the ever-present question of economics. Trade agreements are sometimes used as a form of sharing, but what about the nations who have nothing of value to trade? Should "have-not" nations receive special considerations from the "haves" regarding the benefits of the sea? What if an international organization such as the United Nations should head up an international commune? Could nations contribute products they have taken from the sea, and all nations take out what they need? Could the sea meet these "needs" indefinitely?

Any time one deals with nations, one inevitably runs into the problem of sovereignty. On land, most disputes can be conveniently settled with a border agreement and a surveyor. In the sea, however, things become more complicated. Who owns what? Historically most nations have claimed the water (and the land under it) adjoining their shores and outward to a distance of 3 miles. This not only protects their fishing and mining rights but gives them a more or less comfortable border between them and unfriendly ships. Recently, however, many nations have unilaterally extended their domain into what once was open ocean. Certain countries claim 4, 5, 9, 10, 12, or even 200 miles as their territorial holdings in order to preserve fishing or mineral rights they consider their own. Clearly this is another matter that will have to be settled by international cooperation.

Other political issues directly or indirectly relating to the sea include keeping the seas open to all shipping, arms controls that prohibit emplacement of weapons on the sea floor, the establishment of fishing rights and controls, the development of aquaculture, or "farming of the seas," and the prevention and control of spills of oil and other wastes. To a great extent our future is linked to political decisions about the sea.

We are capable of destroying the life-support capabilities of the sea and may have already begun this deadly process. If the oceans should "die," this would be the beginning of an end to life as we know it. The probable outcome of tragic, step-by-step suicide has been described by Jacques-Yves Cousteau:

> With life departed, the ocean would become, in effect, an enormous cesspool. Billions of decaying bodies, large and small, would create such an insupportable stench that man would be forced to leave all the coastal regions. But far worse would follow . . .
>
> The ocean acts as the earth's buffer. It maintains a fine balance between the many salts and gases which make life possible. But dead seas would have no buffering effect. The carbon dioxide content of the atmosphere would start on a steady and remorseless climb and when it reached a certain level, a "greenhouse effect" would be created. The heat that normally radiates outward from earth to space would be blocked by the CO_2, and sea level temperatures would dramatically increase.
>
> One catastrophic effect of this heat would be melting of the icecaps at both the North and the South Poles. As a result, the oceans would rise by 100 feet or more, enough to flood almost all the world's major cities. These rising waters would drive one third of the earth's billions inland, creating famine, fighting, chaos, and disease on a scale almost impossible to imagine.
>
> Meanwhile, the surface of the ocean would have scummed over with a thick film of decayed matter, and would no longer be able to give water freely to the skies through evaporation. Rain would become a rarity, creating global drought and even more famine.
>
> But the final act is yet to come. The wretched remnant of the human race would now be packed cheek by jowl on the remaining highlands, bewildered, starving, struggling to survive from hour to hour. Then would be visited upon them the final plague, anoxia

(lack of oxygen). This would be caused by the extinction of plankton algae and the reduction of land vegetation, the two sources that supply the oxygen you are now breathing.

And so man would finally die, slowly gasping out his life on some barren hill. He would have survived the oceans by perhaps thirty years. And his heirs would be bacteria and a few scavenger insects.*

*From Cousteau, J.-Y.: Letter of appeal for membership in The Cousteau Society, Danbury, Conn., 1974, The Cousteau Society.

We are notorious for our failure to act until tragedy is about to strike. Then, with a great surge of energy and money, we finally defeat the enemy—whoever or whatever it may be. Tragedy is once again about to strike. At stake is the survival of all human beings and all other life on earth. We must have an immediate ''inner space'' program to correct the abuses, solve the problems, and exploit the riches of the marine environment. The strategy will not belong to the Americans, Russians, Chinese, or to any other national group; the strategy will be one of all mankind.

part four

METEOROLOGY

I am the daughter of Earth and Water,
And the nursling of the Sky;
I pass through the pores of the ocean and shores;
I change, but I cannot die.
For after the rain when with never a stain
The pavilion of Heaven is bare,
And the winds and sunbeams with their convex gleams
Build up the blue dome of air,
I silently laugh at my own cenotaph,
And out of the caverns of rain,
Like a child from the womb, like a ghost from the tomb,
I arise and unbuild it again.

The Cloud
Percy Bysshe Shelley

Satellite photos such as this allow scientists to understand and more accurately predict the world's weather. Latitude and longitude lines are superimposed. (Courtesy United States Department of Commerce, National Oceanic and Atmospheric Administration, Rockville, Md.)

25 story through the ages

History and applications of meteorology

But there went a mist from the earth
and watered the whole face of the ground.

Genesis 2:6

Meteorology is the study of the invisible gaseous shell that surrounds the earth and all its attendant characteristics, including wind, temperature, pressure, and moisture. Short-term changes in these characteristics are commonly grouped into a category called *weather*. On the other hand, atmospheric changes over time periods greater than 1 year lie within the related area of *climatology*.

In order to understand meteorology it is necessary to also have a good understanding of many other subjects. Known for many years as the physics of the atmosphere, meteorology is very closely interrelated with mathematics, oceanography, geology, and the newer environmental sciences. Because they are so intertwined, no earth science can easily be isolated for study. This fact is illustrated by the term "meteorology," which was derived from an early four-volume treatise that was written by the Greek scholar Aristotle entitled *Meteorologica*. This great work, to be discussed later, not only encompasses the areas previously mentioned, but also physical geography and astronomy.

HISTORY OF METEOROLOGY

Early peoples undoubtedly realized that many types of weather phenomena affected their daily lives. While low temperatures, rain, or snow caused discomfort, warm temperatures and sunny skies benefited them greatly. Their inability to logically explain the weather that affected them so greatly forced them to seek the answers in religion and superstition. Being in the good graces of the gods meant a reward of fair weather, but storms were seen as a warning or punishment from the wrathful dieties that controlled all physical events. In ancient civilizations, prayers were offered to many gods in hopes of good weather. Gods such as Boreas, Greek god of the north wind; Thor, Norse god of thunder; and Ra, Egyptian god of the sun, exercised their influence on the daily lives of these ancient people. The Babylonians attributed their weather problems to the stars instead of a series of gods. Many cultures today still believe that weather is controlled by gods or by the stars.

The early Greeks

Greek scholars between 600 B.C. and the birth of Christ showed a remarkable interest in their environment. Empedocles (500 B.C.), for example, separated the universe into four distinct elements—*earth, fire, water,* and *air*—and believed that all matter was made up of these elements in varying combinations. Aristotle (384-322 B.C.) later added a fifth substance called quintessence, which was a perfect heavenly material not related to the earlier four (Fig. 25-1).

The Greeks' love of abstract reasoning and theoretical deduction aided their remarkable advances in the sciences. However, they relied chiefly on observation and largely neglected experimentation. Thus their theories were neither proved nor rejected until the beginnings of modern scientific thought and study in the sixteenth century.

Two Greek philosophers stand out as important early atmospheric scientists. Aristotle, the earliest known theorist of meteorology, completed the *Meteorologica,* which remained the undisputed authority on the physics of the atmosphere for more than 2000 years. Since it was virtually heretical to challenge Aristotle's work, post-Greek meteorology progressed very slowly until after the Dark Ages.

Hippocrates (460-377 B.C.), the "Father of Medicine," wrote extensively on medical well-being. One such work on medical meteorology, *Airs, Waters, and Places,* was published around 400 B.C. Hippocrates observed that one must first study the weather prior to contemplating its medical ef-

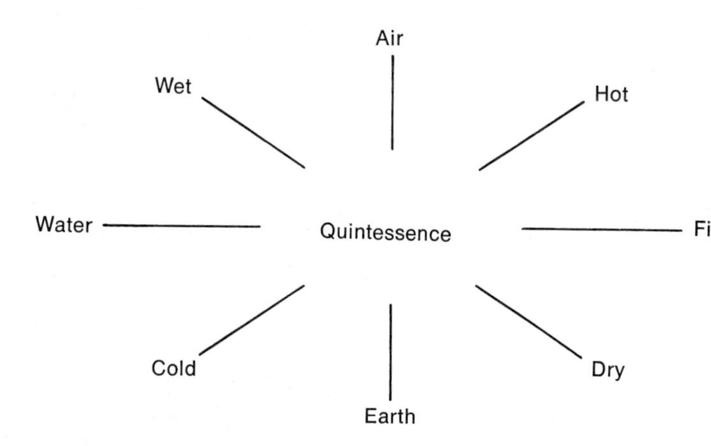

Fig. 25-1. The early Greeks believed that all matter in the universe was made of four elements: earth, air, fire, and water. Aristotle later added a fifth element, which he called quintessence, a perfect material from which the others developed.

fects on people. One of his outstanding contributions was the observation that abnormal amounts of sunshine on the skin resulted in lesions. Many centuries later this statement was proved valid with the discovery of a connection between skin cancer and solar radiation.

During the remainder of the Golden Age of Greece, few notable achievements were made in the atmospheric sciences. However, one invention, the *wind vane,* proved to be an important tool for the ever-growing Greek merchant fleet. Today the wind vane is still in use, basically unchanged through the ages.

Age of nonaccomplishment

The Golden Age of Greek science came to an end during the second century B.C., when Greece was overwhelmed by the conquering Romans. The Romans did very little to encourage the study of meteorology. One notable exception was a handbook for Roman scientists entitled *Natural Questions*. Written by Seneca (ca 54 B.C.–39 A.D.), this scientific work was eventually passed on to the western world and was found to be quite accurate.

Just as the Greek influence had come to an abrupt end, so did Roman domination. Hordes of barbarians descended on the disintegrating empire and thus began the Dark Ages. Although there were numerous technical advances such as iron plows and waterwheels, study of the sciences, especially the field of meteorology, was totally neglected.

After the Dark Ages, the Church dominated what was to be called the Middle Ages (1100-1500 A.D.). During this period it was difficult for scholars to carry on scientific studies that were not first approved by the Church. Since the Church felt that nature was the work of God, not man, scientists were discouraged from close study of the natural world and were shamed by having to retract their findings or face being sentenced to die as heretics. This attitude prevailed until the Renaissance and the accompanying scientific revolution (1500-1900).

Birth of a science

Until the 1500s, most scientific work was based on Aristotle's *Meteorologica*. However, the fourteenth century saw a change in this approach. Instead of relying on observations based only on the senses, scientists began to invent a variety of instruments to measure weather variables. The invention of instruments is the first step in a discipline's becoming a true science; until this time meteorology did not qualify.

The first instrument developed during this period was the *hygroscope*. Now called the *hygrometer,* the hygroscope was the brainchild of Leone Battista Alberti (1414-1472), an Italian painter and philosopher, who discovered that humidity could be measured by weighing an absorbing substance. The difference between the dry and wet weights of the material equaled the amount of moisture in the air. Alberti used a sponge set out in the free atmosphere and a crude scale for measuring the weight of water (Fig. 25-2).

Immediately after the invention of the hygroscope, two important instruments were developed in rapid succession. These marked the turning point for meteorology as a science. Galileo Galilei (1564-1642) invented a device he called a *thermoscope,* a forerunner of today's thermometer (Fig. 25-3). The instrument used air trapped in an inverted tube, which was dipped beneath the surface of a quantity of water. As the temperature of the outside air changed, the temperature of the inside air would react accordingly, thus changing the water level.

Many notable scientists of the seventeenth century, including Galileo, believed that there was a void in space called a vacuum. In 1640 Gasparo Berti performed the first in a series of experiments that led to proof of the existence

Fig. 25-2. Alberti's hygroscope worked on the principle that the difference between the dry and wet weights of a substance must be the weight of the water. By leaving a sponge in the open air, he could measure the amount of water it absorbed and calculate how much moisture had to be in the air.

of a vacuum and to the eventual invention of the barometer (Fig. 25-4). Berti's experiment involved attaching a long lead tube with a glass flask at the top to a building. Inside the flask was a screw plug that could be taken out to fill the tube with water. There was a stopper at the bottom of the tube, which was immersed in a large cistern of water. After the tube was filled with water from the top and the screw plug was replaced, the stopcock at the bottom was opened. The water traveled downward into the cistern, leaving a soon-to-be discovered vacuum. After securing the bottom stopcock, the plug was loosened with a resulting loud whoosh, which indicated that air was filling a void or vacuum.

In 1644 Evangelista Torricelli (1608-1647), a noted physicist and mathematician, used Berti's information to invent the barometer, an instrument used to measure air pressure. As a student of Galileo, Torricelli was greatly interested in why ordinary suction pumps could not lift water higher than 34 feet above the level of free water. Torricelli explained the phenomenon as being the result of the weight of the air that exerted pressure on a liquid surface. This pressure pushed the water upward, but it exerted a force that was capable of supporting a column of water only 34 feet (10.4 meters) high.

Torricelli constructed a simple instrument, later called a Torricellian barometer, in order to prove his assumption. The instrument consisted of a glass tube 1 meter long that

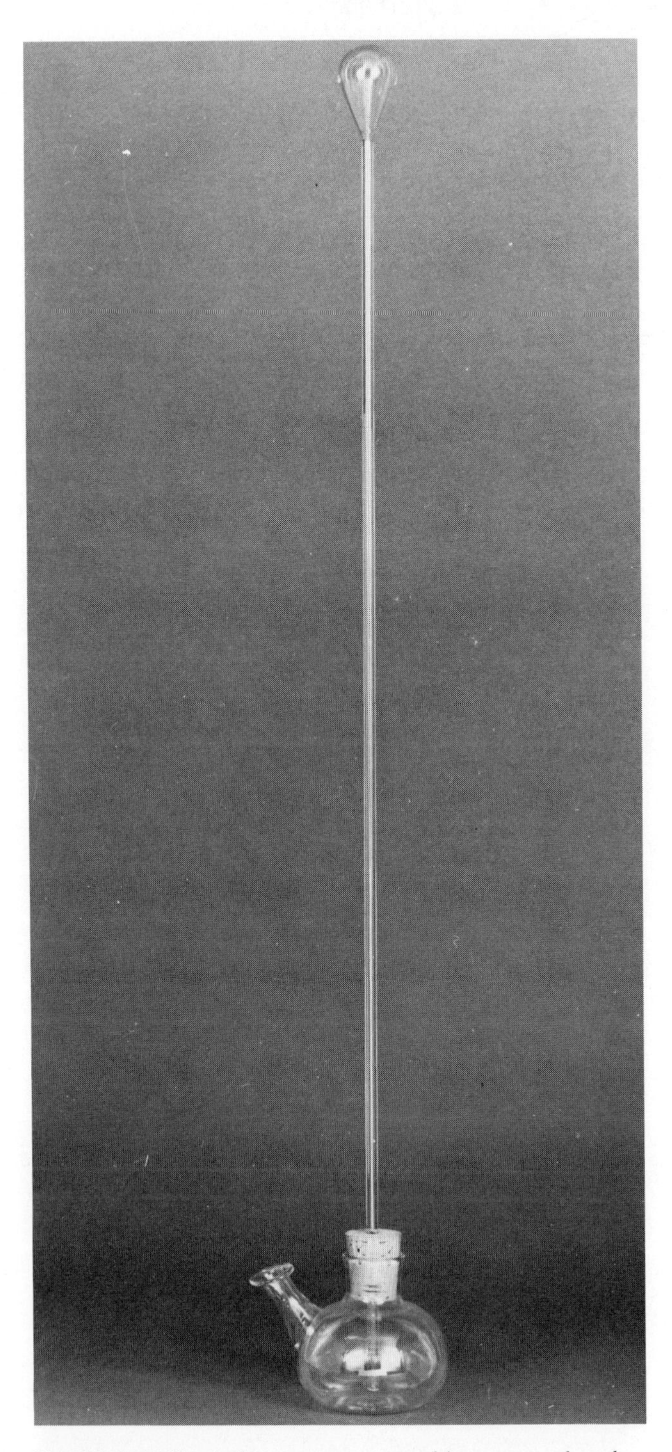

Fig. 25-3. Galileo's thermoscope operated in a manner based on the characteristics of water. When the air temperature outside the tube rises, the water is heated, so it expands. Creeping up the tube, the water registers a higher temperature. The reverse happens when the air temperature drops. (Courtesy Smithsonian Institution, Washington, D.C.)

Fig. 25-4. While he solved the problems involved in constructing this barometer, Torricelli theorized that it was not the weight of a water column that forced it to fall but the weight of the air pressing on the column. (Courtesy Smithsonian Institution, Washington, D.C.)

was sealed at one end. The tube was filled with mercury and immersed, open end down, in a vessel that was also filled with mercury. It was immediately observed that the mercury fell into the reservoir until the level of the column reached 29.92 inches (76 mm) above the reservoir level. The free space above the mercury again proved the existence of a vacuum, since there was no air in the tube initially. The ability of the air to hold up a column of mercury proved that the weight of the atmosphere exerts a downward force on all objects.

The eighteenth and nineteenth centuries saw many rapid developments in meteorology and related fields. The hygrometer, used to measure water content, was greatly improved by H. B. deSaussure in 1783, and the telegraph, important for rapid weather communication, was invented in 1830. Many theories were also advanced to explain atmospheric phenomena, while the classification of cloud types resulted in the publication of the first *International Cloud Atlas,* which is still the basis for cloud classification today. Scientists such as Bernoulli and Lavoisier attempted to explain how water could become invisible vapor and reappear again as a liquid raindrop. Furthermore, observational networks were developed to acquire surface weather data, and kite-flying stations provided useful data on the upper air. It was with this last development that meteorology finally became the complete science we know today.

Other major discoveries in related areas contributed to the ever-expanding base of knowledge. For example, the four major gases of the atmosphere were identified, and the physical laws governing these gases were proposed by Charles, Boyle, and Dalton. Benjamin Franklin discovered electricity in the atmosphere, and others developed theories on rising air and rain and hurricane formation.

Coming of age

With the advent of the twentieth century there were major advances in both theoretical and applied meteorology. In 1918 the Norwegian meteorologist Bjerknes postulated his famous theories on cyclones and the polar front. Both of these theories are still the basic tools of our daily weather forecasts. For these achievements, Bjerknes became known as the father of modern meteorology.

Undoubtedly the airplane has proved to be the twentieth-century invention that was most important meteorologically, for it became necessary to obtain meteorologic data that were as complete and accurate as possible. Airplanes also made it possible for scientists to be up where weather conditions were developing.

Other outstanding achievements that constantly influence our daily lives were the invention of the radiosonde, weather radar, weather satelites, and high-speed computers, to name just a few.

Radiosonde. The radiosonde, an instrument developed by a Russian meteorologist in the late 1920s, is essentially a radio transmitter sent aloft attached to a helium-filled balloon. The transmitter sends weather data from the lower atmosphere back to earth (Fig. 25-5) and gives the meteorologist an inexpensive means of obtaining weather observations from the upper air.

Weather radar. The application of radar to weather observations provides meteorologists with the ability to detect and track severe storms. By bouncing radio signals off water-droplet targets, echos or blips could be seen on a receiver. At present, the use of sophisticated equipment makes

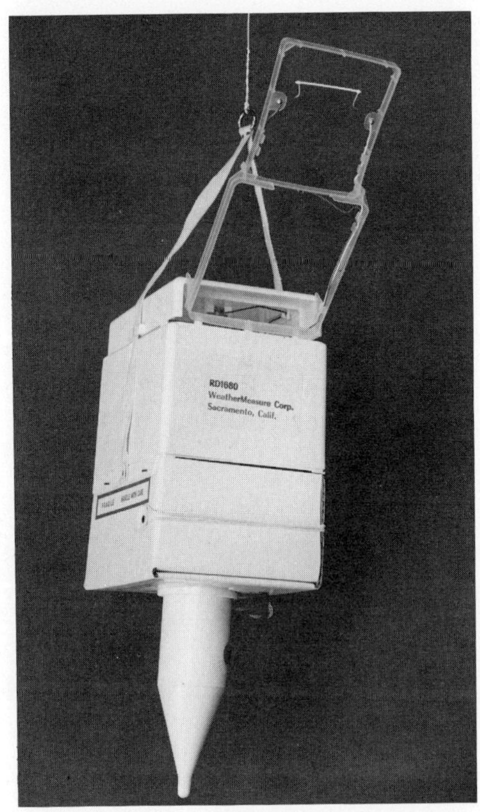

Fig. 25-5. The radiosonde that soars aloft on a helium-filled balloon sends data about the upper atmosphere back to earth. (Courtesy WeatherMeasure Corp., Sacramento, Calif.)

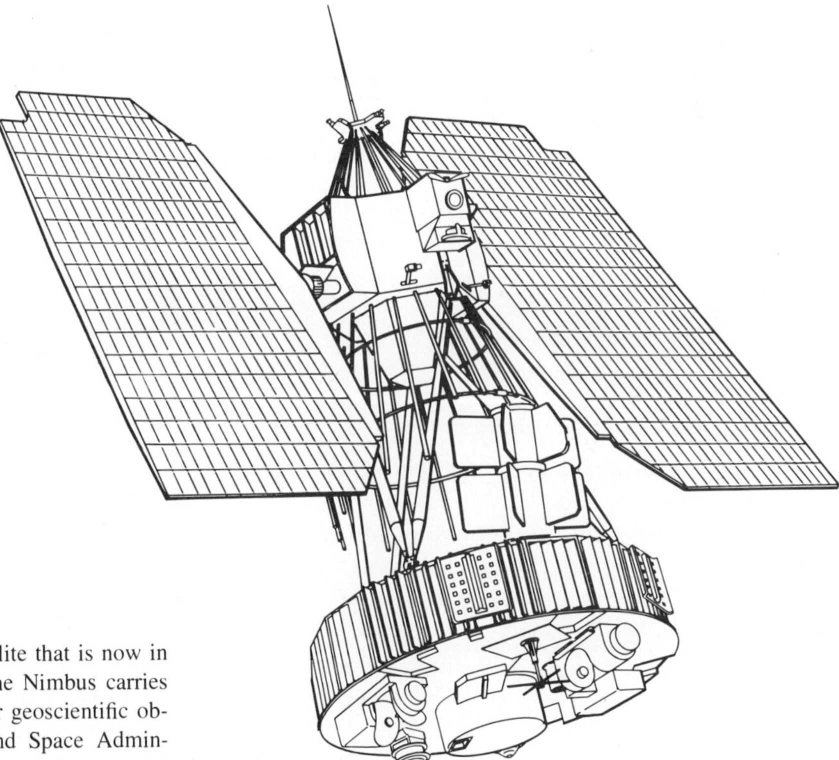

Fig. 25-6. Sketch of the Nimbus weather satellite that is now in a nearly circular polar orbit over the earth. The Nimbus carries advance sensors for weather research and other geoscientific observations. (Courtesy National Aeronautics and Space Administration, Greenbelt, Md.)

it possible to detect any type of storm from light rain to hurricanes and tornadoes many miles away. The storm's intensity, location, and path can now be determined with great accuracy, and warnings can be sent to the areas most likely to be affected. Since its inception, weather radar has saved many lives and millions of dollars in damage.

Weather satellites. On April 1, 1960, years of research culminated in the successful launching of the first meteorologic satellite, *Tiros I*. This small but capable spacecraft transmitted back to earth the first pictures of the earth's cloud cover from outer space. This marked the beginning of a new frontier in meteorology. The initial success was quickly followed by a series of nine new Tiros satellites, each of which was equipped with improved instrumentation and photographic equipment. On the heels of the successful Tiros series came the launching of the much improved Nimbus, ESSA,* and applied technology satellite (ATS) series (Fig. 25-6). Each of these has made significant contributions to our ever-increasing knowledge of the atmosphere by providing information on clouds, radiation data, and the course of major weather disturbances.

High-speed computers. More than 50 years ago, basic equations were developed to describe the behavior of the atmosphere. Not until recently, however, could these equations be applied to weather forecasting. True progress came with the advent and improvement of high-speed computers. Meteorologists now feed in and store data on weather observations worldwide. These data are used to automatically analyze daily weather maps as well as forecast, from known models of the atmosphere, short- and long-term weather. These predictions are then distributed to the general public and to specialized users. In computers, scientists now have a tool that can handle their most complicated theoretical problems and utilize greater quantities of raw data than ever before dreamed possible.

METEOROLOGIC APPLICATIONS

Weather influences our daily lives in many ways. It usually determines the clothes we wear, the foods we eat, and the activities in which we participate. Common sense tells us to check the latest weather report before planning our daily schedules.

Because of mass-media forecasts, many people have the mistaken impression that meteorology involves only weather forecasting, but in reality it is a complex subject involving many areas of study. Specialized areas of applied meteorology involve work in agriculture, forestry, health, air pollution, aviation, architecture, and industry. Meteorology is also spreading into more diversified fields such as land, sea, and air interactions, weather modification, and global weather predictions. The future of meteorology lies within these newer specialities, and other and more beneficial areas of meteorology will surely be discovered.

Agricultural meteorology

In order to live, we must eat and drink, and the world's population is dependent on agriculture. The world we live in is faced with widespread hunger, and the available technology is presently being used to increase agricultural production. Producing enough food for our mushrooming populations, however, is an ever-increasing problem. It is in this area that the agricultural meteorologist best uses his knowledge, so that crop yields can keep up with the birth rate.

An important ingredient in our survival on earth is the weather. Not only does it directly affect our dietary and nutritional needs, but clear skies, vast amounts of sunshine, and ample precipitation are most important in bringing crops to maturity. Unfortunately weather does not always meet our needs, and it can in certain instances produce a major catastrophe such as a hailstorm or tornado that destroys an entire area's crop production. Even the more subtle weather elements such as frost, weather-dependent diseases, or drought can have damaging effects equal to or greater than those of a violent storm.

In the mid-1960s, for example, a drought hit the northeastern part of the United States. Crops were totally destroyed in many farm areas and consumers were forced to pay higher than normal prices. As we continue to squander good, usable, clean water, natural drought periods will lengthen. This will result in greater hardships to both the agricultural community and the world's population as a whole.

Realizing the enormity of these problems, agricultural meteorologists are constantly investigating new methods. They seek to improve harvest efficiency, increase rainfall, correlate disease and pest outbreaks with weather conditions, and issue special types of agricultural forecasts such as the Fruit-Frost Service (Fig. 25-7). It is estimated that food production will have to be doubled by the year 2000 just to maintain the present low level of nutrition that exists in the world. It appears that our greatest challenge is to adequately feed the world's population, and the agricultural branch of meteorology is trying to aid in this task.

Biometeorology

Biometeorology is the branch of meteorology that deals with the effect of weather on the health of living organisms.

*ESSA stands for Environmental Science Services Administration.

Fig. 25-7. These wind machines can be used to mix higher warm air with cooler lower air to protect sensitive crops from cold weather. Fruit-Frost Service bulletins allow farmers to ready their equipment when bad weather is expected. (Courtesy University of Florida, Institute of Food and Agricultural Sciences, Lake Alfred, Fla.)

Although this is a relatively new interdisciplinary science, the effects of weather on comfort, health, and physical stature were known as long ago as 400 B.C.

Recently anthropologists have grouped people into so-called racial regimes. By studying these groups in relation to their geographical locations, they have hypothesized that genetic differences and behavioral patterns are strongly linked to weather and climate. It would at first seem strange that people living in high-temperature areas tend to be tall and thin, while those living in cold climates are relatively short and stocky. A logical explanation for these physical differences is adaptability of the body to weather. A tall, thin person has a greater skin area over which evaporation of body fluid can occur. This results in the cooling of the body, while the small surface area of shorter, stockier individuals allows body heat to be conserved and more easily regulated.

Comfort results when a person has adjusted to the environment. If that environment changes, an immediate physical readjustment process begins in order to bring the body into equilibrium with the new environment. Physical factors such as heat, water vapor, wind, and sunshine are a few of the important weather parameters that affect the area of the atmosphere in which we live. The sun's rays cause photochemical reactions in the skin that result in sunburn, skin aging, and even cancer. Heat and water vapor combine to produce perspiration, while discomfort is linked to the amount of water vapor in the air at a set temperature. The *discomfort index* can be utilized to forecast how much work will be accomplished within a certain range of temperature and humidity. When discomfort occurs, moods change drastically and performance is often greatly reduced.

The effects of weather and climate on diseases ranging from the common cold to mental illness are quite extensive. The more common effects of temperature and humidity are an increase in the occurrence of lung diseases such as tuberculosis, asthma, and bronchitis. Solar radiation affects the eyes and skin, while infectious diseases such as flu have been linked to temperature and wind. Mental disorders such as schizophrenia are said to be influenced by thermal stress and general unrest.

Although research linking many areas of human and animal behavior with weather is being carried out by scientists throughout the world, the surface of this intriguing subject has not even been scratched. It may be possible some day to create an atmospheric environment that will eradicate or reduce drastically many bothersome and dread diseases.

Aviation meteorology

Aviation meteorology is closely related to *synoptic* meteorology, which is the art of weather forecasting through the use of a series of weather maps. As mentioned previously, aviation has been clearly the single most important factor in the growth of meteorology. The need for fast and accurate weather forecasts increased as airplanes flew further and higher to meet the demands of the world's mobile population.

Aviation forecasting is done mainly by the weather bureaus of different nations. In the United States, this task is accomplished by the National Oceanographic and Atmospheric Administration (NOAA). NOAA is responsible for accumulating, updating, and disseminating observations and weather forecasts. Hundreds of observations, including data on moisture, wind, cloud cover, visibility, and pressure, are taken every hour and updated whenever drastic changes take place. This information is teletyped across the nation. Coupled with surface observations are unmanned upper-air radiosonde reports. These pieces of raw data are used to draw a three-dimensional picture of the atmosphere by which pilots can be briefed as to the weather they may expect to encounter en route. With this information pilots can then preplan flight altitude, fuel consumption, and many other important considerations. Extremes of weather such as turbulence, thunderstorms, and icing conditions can thus be avoided, reducing inconvenience as well as the risk of disaster.

Air pollution meteorology

Recently the subject of air pollution has received much attention from people in all walks of life. Realizing that pollution is a threat to our health, safety, and prosperity, action has been demanded and the topic of our national environment is thus emerging as an important policy issue.

Since meteorologists deal largely with the portion of the atmosphere closest to the earth, it is only natural for them to enter this field. Air pollution meteorologists measure weather parameters such as wind and temperature and their effects on pollutants. But their primary objective is the accurate forecasting of weather conditions that could contribute to severe air pollution problems (Fig. 25-8).

Wind plays the most critical role in the dispersion of pollution in the atmosphere. For centuries natural pollutants have entered the air during such catastrophic events as earthquakes and volcanic eruptions. Small forest fires and the evaporation of seawater have added their particulate matter to the air we breathe. However, the amount of material entering the atmosphere was relatively small and the movement of air dispersed the pollutants fairly easily. The problem today is that the increased amounts of particulate matter and noxious gases we are introducing into the atmosphere can no longer be handled by the moving air.

Recently, joint studies involving meteorologists and other scientists have focused on the problems of air pollution injury to plants and the harmful effects of pollution on people as well as other animals. Special attention has been given to the damage done to the human respiratory system and the economic disasters that can result from increasing the quantity of foreign bodies in the atmosphere. However, little attention has been paid to the destruction of many ecologic riches and to the loss of great natural beauty in much of the world.

Meteorologists have an important role to play in keeping

Fig. 25-8. Air pollution meteorologists are consulted when sites for industrial complexes are selected. In the case of this plant either this step was bypassed or the meteorologist's advice was ignored, for the gaseous and particulate emissions from this factory stream down the hillside, creating numerous air pollution episodes in the valley below. (Courtesy United States Department of the Interior, Environmental Protection Agency, Washington, D.C.)

the earth we live on safe and beautiful. It also seems that the tasks of sampling the atmosphere, recognizing pollution potential, and forecasting air pollution episodes are of vital importance to all mankind.

Industrial meteorology

The United States Weather Bureau provides numerous forecasts for the general public, but these forecasts are seldom specific or fast enough to be useful to industries. Therefore, industrial meteorology was developed to fill the void left by the weather bureau.

From the outset, the planning of a new industry depends on factors such as availability of raw materials, labor, building, topography, and most important, meteorologic considerations. Weather factors influence air pollution problems, employee comfort, and wintertime accessibility to the industrial complex. Many times weather factors are ignored when making a site selection and numerous problems result.

Therefore many private meteorologic firms have come into existence in the last 20 years. Areas such as retail sales, manufacturing, advertising, construction, and sporting events all use specialized weather forecasts. These forecasts aid industry in decisions that involve many thousands of dollars. Tourism is one industry that depends almost entirely on weather conditions. Sunny, warm areas such as Florida capitalize on their good fortune in the wintertime, while northern ski areas hope for snowfall and cold wave predictions.

Strange as it may seem, our buying habits are partially weather dependent, and therefore retail stores take advantage of special weather forecasts. Raincoats and umbrellas seem to disappear from stock during a prolonged rainy period, while bathing suits remain on the shelves. A recent study demonstrated that people were more apt to buy one type of pastry than another, depending on the average daily temperature. This type of buying reaction could also be applied to other marketable items such as coffee versus cold drinks or restaurant short orders versus full-course dinners.

Municipalities also use the services of industrial meteorologists. Forecasts of a heavy snowfall are very important to the highway and city personnel who have the responsibility of keeping roads open. Knowing the starting time, the area to be affected, the amount, and the duration of a snowfall are essential for planning adequate snow clearance, which can save thousands of dollars. In the northern part of the United States very few state and city governments are without this type of service.

Other areas such as the oil and natural gas industry, picnic suppliers, concrete pourers, handlers of perishables, and the transportation and fishing industries all have their own peculiar needs for the weather information that industrial meteorologists supply.

Weather modification

The possibility of eliminating adverse weather conditions and their effects has been of great interest to scientists for many years. If weather could be controlled, huge losses from violent storms, fog, and drought would be reduced or even completely eradicated.

Almost from the very beginning of our history on earth we have been unknowingly modifying the weather. Fires add gases such as carbon dioxide and carbon monoxide and particles to the atmosphere, and clearing some land and cultivating other areas increases different atmospheric components such as oxygen. Not until recently, however, has this type of alteration gotten completely out of hand. Today, inadvertent weather modifications are caused by cement jungle cities and expanding industrial complexes that spew out pollutants, creating havoc with our health and economy. Out of this situation arises the question of whether the atmosphere can be manipulated by us in order to modify our environment for our own benefit.

Many civilizations have tried in one way or another to alter the weather. American Indians were said to have doused their holy men with water and buried their children up to their necks in order to placate the gods and make it rain. In the late 1800s, rainmakers roamed the southwestern part of the United States plying their trade. Since the result of their work was unpredictable, this occupation was very hazardous to say the least, for rainmakers were often arrested or shot if rain did not fall.

During the twentieth century, giant steps were taken in the field of weather modification. These included the construction of windbreaks to control drifting snow, frost protection, artificial production of rain by cloud seeding, hail suppression, short-term dissipation of fog, and hurricane research.

Cloud seeding to stimulate precipitation is one of the most vital areas of weather modification. Frequent drought conditions make it imperative that we either produce water, conserve it, or perish. Potable (drinkable) water can be obtained by the use of desalinization plants located close to large bodies of salt water or by the cloud-seeding process. The first method results in fresh water at a cost far greater than that of transporting water from natural or artificial reservoirs. Therefore it would presently appear that cloud seeding to increase rainwater reserves is the most economical method for supplying water.

In the mid-1940s two scientists, Dr. Vincent Schaeffer and Dr. Bernard Vonnegut, discovered that supercooled

Fig. 25-9. This hole, which was opened in a supercooled cloud layer by seeding with Dry Ice pellets, proves the feasibility of dissipating clouds by mechanical means. (Courtesy United States Air Force, Cambridge Research Laboratories, Bedford, Mass.)

Fig. 25-10. Hundreds of forest fires are triggered each year by lightning. Photograph is of an area in which eight separate fires began during this particular storm. (Courtesy United States Department of Agriculture, Forest Service, Washington, D.C.)

clouds would produce large ice crystals if they were seeded with Dry Ice or silver iodide. These crystals melted as they fell and ended up as raindrops at the surface. This left the cloud with less moisture and sometimes even caused a hole to appear in it (Fig. 25-9). Thus cloud seeding became a reality and is currently being used in areas with low water tables.

Hail is an international problem and its control is of worldwide interest. The United States alone sustains losses of over $300 million annually in agricultural and property damage due to hail. The Soviet Union has been carrying out an extensive program to check the formation of hailstones. The Russian technique is to locate the supercooled liquid zone in a cloud and fire artillery shells into the center of the cold zone. The bursting shell disperses silver iodide into the area and turns the droplets into ice crystals, which cannot then form hail. This method has also been used successfully in other countries such as Kenya, Canada, and England.

The following abbreviated list of other important aspects of weather modification includes a brief description of each type of project.

1. The rate of water evaporation can be reduced through the use of a harmless chemical layer that floats on the surface of a body of water. This layer acts as a cover, reducing water loss due to evaporation by as much as 50%. Reducing wind speed over water bodies by using shelter belts has also proved to be a reliable method of reducing evaporation.
2. Snowfall can be increased by seeding clouds over land areas whose temperature is below freezing. This can increase the amount of snow available for recreation purposes. If the snow melts slowly, this will add significantly to the yearly water supply in reservoirs in such areas.
3. Project Skyfire, conducted by the United States Forest Service to prevent lightning-caused forest fires, is another specialized cloud-seeding project. Scientists seed thunderstorm clouds with large quantities of silver iodide crystals in the hope that cloud-to-cloud lightning will be produced instead of the destructive cloud-to-ground discharges (Fig. 25-10).
4. Project Stormfury, carried out by the combined forces of the United States Navy and NOAA, has yielded positive results in our effort to reduce the destructive power of hurricanes. In 1971 a moderate reduction of wind speed within a hurricane was achieved after seeding the clouds with silver iodide crystals, but little change in precipitation patterns was observed. This is important because wind speeds at hurricane force are quite destructive, but rainfall is badly needed for supplies of fresh water.
5. The use of artificial devices to stir or warm the air can prevent frost. Smudge pots have been used for years but their dense, vile-smelling smoke created air pollution problems. Current methods include smokeless orchard heaters, sprinkling, propeller-driven wind machines, and the flooding of low-lying areas.

26 *the invisible ocean*

The atmosphere

Over earth and ocean, with gentle motion . . .

The Cloud
Percy Bysshe Shelley

The word "atmosphere" is derived from two Greek words: *atmos,* meaning vapor, and *sphaira,* meaning sphere. The atmosphere is a mixture of invisible gases held close to the earth by gravity. It extends from the earth's crust to an indefinite height. It is impossible to state the exact upper limit of the atmosphere, but 18,600 miles (30,000 km) is sometimes used as an average. It is at this point that the earth's gravitational attraction can no longer keep a moving gas molecule in orbit. Therefore the gas molecules above this level can easily reach their escape velocity and be lost to outer space.

THE EMERGENCE OF AN ATMOSPHERE

Our atmosphere is believed to have evolved to its present state after billions of years of change. Although there are several theories about its birth and history, the most widely accepted explanation is the "four-stage" theory, which is based on atmospheric composition and temperature.

Stage I: formation of the earth. Scientists through the ages have asked the same question: "How did the earth form?" One theory is the *coalescence,* or *nebular,* theory, which suggests that primordial gases and solid particles came together, or *coalesced,* to form a hot, molten earth. An alternative hypothesis suggests that gases and debris from some cataclysmic event in outer space became the molten material that in turn formed the earth. There have been dozens of other ideas about the earth's origins, but they all end up with the same result: the earth began as molten rock called magma, and temperatures on the earth's surface must have been in excess of 14,500° F (8060° C).

Geochemical reasoning indicates that the gases present at the formation of the earth were the volatile vapors of chlorine, bromine, iodine and ammonia. Because of the earth's high temperature, these gas molecules became hot enough to reach their escape velocities, so that the original atmosphere of the earth was lost to outer space after only a few million years.

Stage II: cooling. As the earth cooled, the molten magma formed a solid crust that allowed its trapped gases to escape. Thus a new atmosphere of carbon dioxide, nitrogen, and water vapor was formed; the most predominant gas was the water vapor. At this point, temperatures on the surface were estimated at between 300° to 500° F (148° to 260° C), a range at which free oxygen could not exist.

Stage III: continued cooling. The most outstanding feature of the third stage of the earth's formation was the cooling of the earth's surface temperature below the boiling point of water, 212° F (100° C). This important substance could now exist as a liquid as well as a gas. This led first to the formation of clouds and eventually to precipitation. When the rains started, they probably continued for more than 49,000 years, inundating the earth's entire surface. These vast expanses of water permitted the origin and evolution of one-celled marine plants and animals. With plant life came photosynthesis and the liberation of oxygen, a gas new to the atmosphere. The atmosphere in stage III therefore consisted of large amounts of water vapor and carbon dioxide and smaller quantities of nitrogen and oxygen.

Stage IV: present atmosphere. The temperature of the earth continued to fall toward its present average of 60° F (15° C). During this period of cooling, complex life forms appeared, and plants continued to liberate oxygen in great quantities. A second process added to the increasing oxygen supply. Water vapor (H_2O gas) at high altitudes dissociated into the hydrogen atom (H) and the oxygen molecule

(O_2). In the upper atmosphere the lighter hydrogen atoms escaped into space while the heavier oxygen molecules remained within the atmosphere. Thus oxygen became more abundant than carbon dioxide. Argon, formed by the radioactive decay of an isotope of potassium, also appeared in the atmosphere.

COMPOSITION OF THE ATMOSPHERE

The atmosphere consists of basically three main types of substances.

1. Dry gases (permanent and variable)
2. Water vapor (gaseous water)
3. Solid particles (condensation nuclei)

Each of these three components plays a special role in the weather process close to the earth's surface.

In its dry state, the atmosphere is composed of some 17 gases. This mixture is quite homogeneous up to a height of 50 miles (80 km) because of the thorough mixing of the air by heat from the earth's surface and by wind. Table 23 lists the major components of the atmosphere in order of abundance. The four most abundant atmospheric components are nitrogen, oxygen, argon, and carbon dioxide, all dry gases. Nitrogen and oxygen alone make up 99.03% (by volume), of the dry, uncontaminated air; when argon and carbon dioxide are added, the four make up 99.99% of the total volume of the atmosphere. It is no wonder that the other 13 gases are called *trace gases.*

Table 23. Average composition of the atmosphere below 50 miles

Component	Chemical symbol	Volume (%)
Nitrogen	N_2	78.08
Oxygen	O_2	20.95
Argon	A	0.93
Carbon dioxide	CO_2	0.03
Neon	Ne	0.0018
Helium	He	0.0005
Methane	CH_4	2×10^{-4}
Krypton	Kr	1.14×10^{-5}
Hydrogen	H_2	5×10^{-5}
Nitrous oxide	N_2O	5×10^{-5}
Xenon	Xe	8.7×10^{-6}
Ozone*	O_3	$0\text{-}7.0 \times 10^{-6}$
Sulfur dioxide*	SO_2	$0\text{-}1 \times 10^{-4}$
Nitrogen dioxide	NO_2	$0\text{-}2 \times 10^{-6}$
Ammonia*	NH_3	0-trace
Carbon monoxide*	CO	0-trace
Iodine*	I_2	Trace

*Indicates elements whose quantity is variable.

On the average, the concentration of most of these 17 gases remains fairly constant over time and area. Carbon dioxide and a few of the trace gases such as ozone and sulfur dioxide are quite variable horizontally because of emissions from industrial plants and mass transportation in major population complexes.

The upper atmosphere has a composition vastly different from that of the lower part. At heights greater than 50 miles (80 km), large-scale vertical mixing does not take place. Therefore gases are distributed according to their densities, with the heavier gases on the bottom. Molecular nitrogen and oxygen are still the most abundant gases at these levels, but here the sun's ultraviolet radiation also produces charged particles, called *ions,* by stripping electrons from these nitrogen and oxygen molecules. The maximum concentration of these charged particles is at 250 miles (400 km), where they form a layer called the *ionosphere,* which is discussed in greater detail at the end of this chapter.

Nitrogen

Nitrogen, the most abundant gas in the atmosphere, was first identified in 1772 by a Scottish botanist named Daniel Rutherford. It is a tasteless, odorless, colorless gas that is for all practical purposes chemically inactive. Neither people nor other animals can use atmospheric nitrogen directly even though it produces protein, which is vital to both growth and life. Since we cannot manufacture protein from nitrogen, we must depend on plant life to produce usable nitrogen through the *nitrogen cycle* (Fig. 26-1).

The cycle begins with an unlimited supply of atmospheric nitrogen. This gas is transformed into simple nitrite and nitrate compounds by bacteria that live among the roots of certain cover crops such as clover, soybeans, and alfalfa (Fig. 26-2). A small amount of nitrogen is also transformed into these simple nitrogen compounds during thunderstorms, when lightning produces the energy needed to initiate the process. Such nitrogen compounds are carried down to the earth by precipitation, and they replenish the earth's nitrogen supply. The cycle continues as plants use the nitrogen and convert it to protein, which is then ingested by animals and man. Eventually the nitrogen is returned to the earth either through the elimination of waste products or by decomposition after death. Finally, the earth gives up nitrogen to the atmosphere through the work of the ever-present bacteria, and the cycle is completed. There is no gain or loss of nitrogen within the cycle, and therefore quantities of this vital gas have remained relatively constant for millions of years.

The inactivity of nitrogen changes under certain severe conditions. Under high temperature and pressure, nitrogen

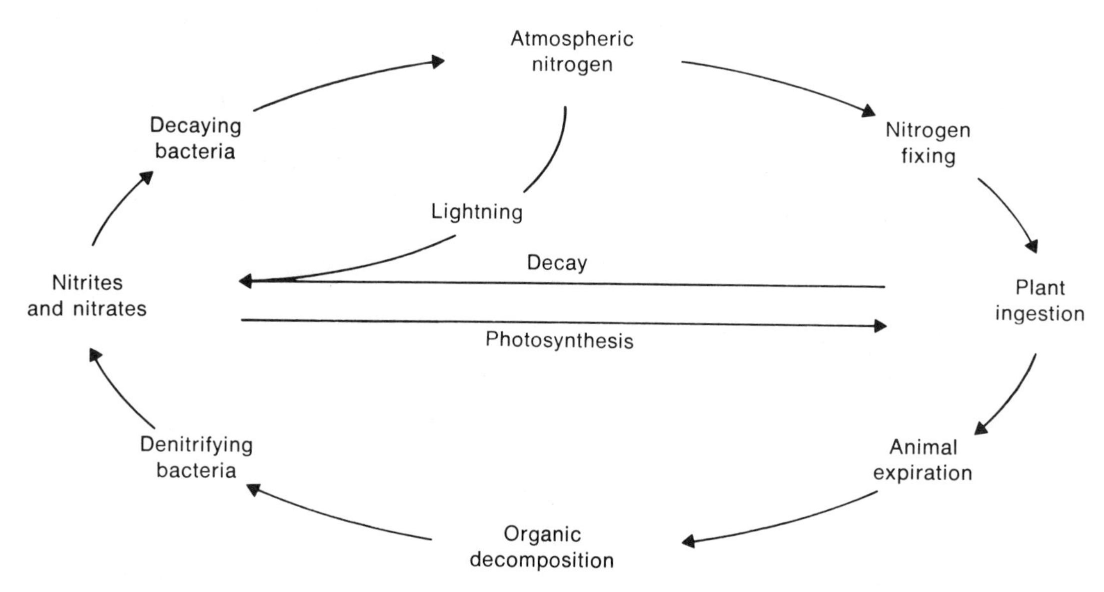

Fig. 26-1. Relationships between the different processes that convert nitrogen into nitrogen compounds and then reverse the process.

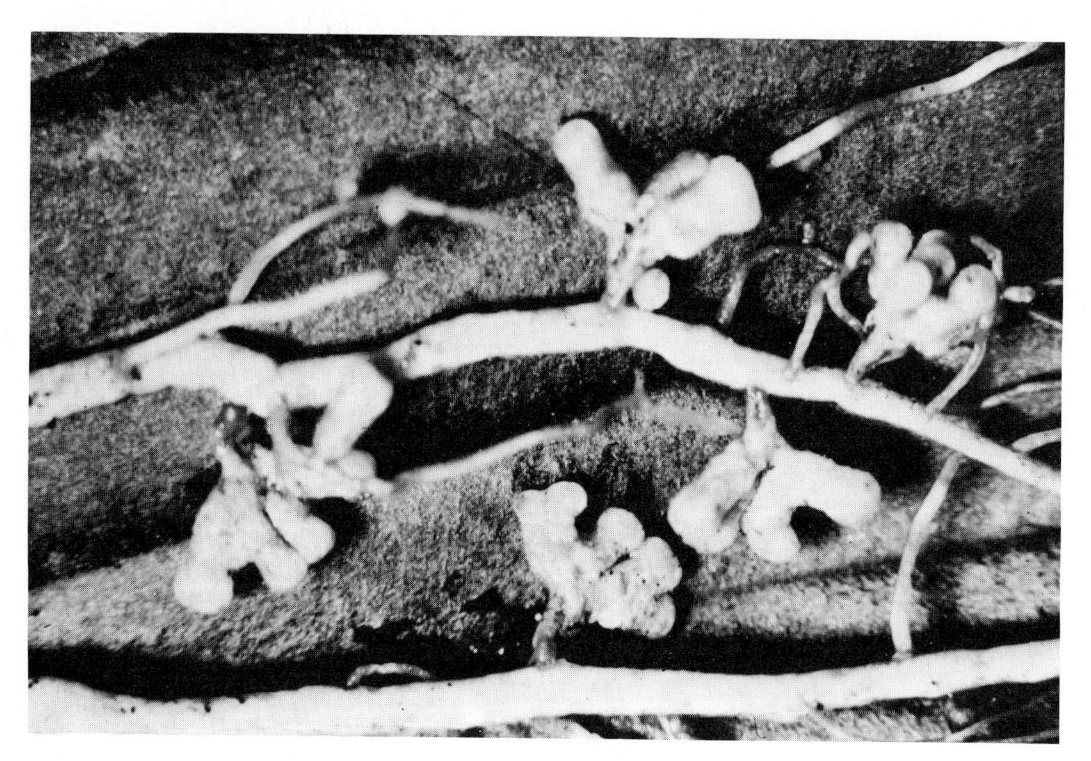

Fig. 26-2. The swellings on these clover roots are called nodules. In these nodules nitrogen-fixing bacteria grow among the plant cells. (Courtesy Carolina Biological Supply Co., Burlington, N.C.)

and oxygen combine to produce oxide products such as *nitric oxide* (NO) and *nitrogen dioxide* (NO_2). Both oxides are considered to be highly toxic. Although these harmful gases are produced as a by-product of all combustion processes, the largest single source in our modern society is without a doubt the automobile. The yellowish brown haze hanging over many populated areas during the morning hours is the result of nitrogen dioxide, formed from the emission of NO and its subsequent reaction with sunlight. To make matters worse, NO_2 reacts with water vapor and tiny suspended water droplets to produce *nitric acid* (HNO_3). This acid is highly corrosive, and even the small amount formed in the atmosphere can be injurious to plant life and human lung tissues. It should be stressed that oxides of nitrogen rate as one of the enormous offenders in our urbanized environment.

Oxygen

The second most abundant gas in the atmosphere (21% by volume) is oxygen, which was discovered by an English clergyman named Joseph Priestly. Oxygen is a gas that is essential to all air-breathing animals and accounts for 89% by weight of all the water found on the earth or in the atmosphere.

This colorless, odorless, tasteless gas is highly reactive and is capable of combining with almost all the other elements. This ability to combine easily results in the formation of many oxides. In addition to the previously mentioned *nitrogen oxides,* oxides of sulfur and carbon also produce major pollution problems for urban areas.

Sulfur is an abundant by-product of the incomplete combustion of fossil fuels; it is released into the atmosphere during combustion. There it combines with free oxygen (O_2) to form *sulfur dioxide* (SO_2). This colorless, acrid-smelling gas can in turn be easily transformed into *sulfuric acid* (H_2SO_4) by combining with the water vapor in the air. The end result is usually the yellowing of vegetation, corrosion of iron or steel, and upper respiratory diseases in humans. Breathing a mist of this highly toxic substance can severely injure lung tissues.

Argon

Argon is one of the six inert gases found in the atmosphere. It is relatively abundant when compared with the remaining gases in the air (Table 23). Since argon does not combine with other elements, its usefulness is quite limited. The gas is mainly used in prolonging the life of tungsten filaments in light bulbs and in welding.

Carbon dioxide

The last of the four most abundant constituents of the atmosphere is carbon dioxide. It is a colorless, odorless gas that has a sour taste when mixed with water. The most important function of CO_2 is its absorption by green plants, which combine it with water in the presence of sunlight to produce sugar and free oxygen in the process called photo-

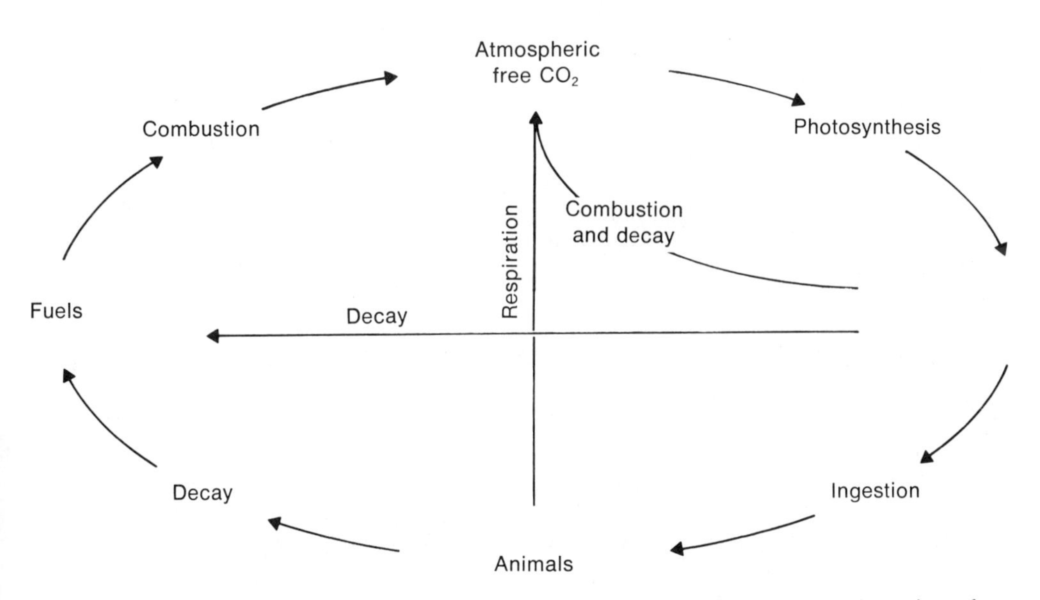

Fig. 26-3. Route of carbon dioxide from free carbon dioxide in the atmosphere through the various plants and animals that use it in their life processes and back into the air.

synthesis. The sugar is used by the plant producer, while the oxygen is released into the atmosphere (Fig. 26-3).

One of the most familiar properties of carbon dioxide is its ability to change directly into a solid without going through the liquid stage. This process, called *sublimation,* is reversible, and this makes solid CO_2 a valuable refrigerant, since there is no messy liquid left as an end product. *Dry Ice,* the trade name of solid CO_2, is a very important tool in scientific research, where very low temperatures are often needed.

The importance of carbon dioxide as a major pollutant of our already highly vulnerable atmosphere stems from the fact that carbon, like sulfur, is also released from the incomplete combustion of fossil fuels. Once again automobiles are the prime culprits in the formation of this gas; quite naturally, its relative abundance is dependent on the extent of urbanization in any given area. For this reason, extremely high concentrations of carbon dioxide are found in metropolitan areas, and these high concentrations have led to an increase in worldwide temperatures and an alteration of world climate.

Ozone

Ozone is found in the atmosphere in varying but very minute quantities (Table 23). Although this gas is relatively rare, scientists consider ozone critical to the existence of life on this planet. If this bluish-tinged gas did not exist at heights of 28 to 30 miles in the atmosphere, highly concentrated amounts of ultraviolet radiation would strike the surface of the earth, causing all plant and animal life to burn up in a matter of seconds. Fortunately, the ozone layer absorbs most of these potentially lethal wavelengths. The small amount of ultraviolet radiation that does eventually penetrate the earth's surface is still enough to produce a painful sunburn, and increased doses can result in skin cancer.

Ozone is produced at the upper levels of the atmosphere in a two-step process. Step 1 is the absorption of ultraviolet (UV) radiation by the oxygen molecule (O_2) and its subsequent splitting into 2 monatomic oxygen atoms (O).

1. $O_2 + \text{UV radiation} \rightarrow O + O$

The second step is the reaction of a monotomic oxygen (O) with a nearby oxygen molecule (O_2) in the presence of a third neutral body (M).

2. $O + O_2 + M \rightarrow O_3 + M$

The third neutral body is needed to satisfy the premise that total energy and momentum must remain constant throughout the process. Almost immediately after formation, ozone dissociates by a two-step process back into free oxygen. Again, energy in the form of ultraviolet radiation is needed.

3. $O_3 + \text{UV radiation} \rightarrow O + O_2$
4. $O + O \rightarrow O_2$

Throughout the formation and destruction of ozone, ultraviolet energy is absorbed, and a real biologic threat to the existence of life on earth is averted.

Close to the earth's surface, however, ozone does present a threat to man. This pungent, garlic-smelling gas is one of the most powerful air pollution irritants known. At low levels it can be created by lightning or automobile exhaust. The production of nitrogen dioxide (NO_2) and its subsequent reaction with sunlight yields ozone as an end product. This *photochemical* reaction helps to account for the dense smog problems in the greater Los Angeles, California, area. The possible effects of ozone on humans include eye irritations, headaches, coughing, and severe lung diseases. The effects on agriculture include reduction of crop yields, while industry suffers mainly from rubber deterioration.

Water

Water vapor, just as the name implies, is water in a gaseous form. It enters the atmosphere through the processes of *evaporation* from water surfaces and *evapotranspiration* by plant life (Fig. 26-4). Since the water available for conversion into water vapor is found on the earth's surface, it is quite understandable that 90% of all water vapor in the atmosphere is found in the lower several miles.

Water vapor is considered to be the most important variable in the atmosphere. The infusion of this gas (as much as 4% of the atmosphere by volume) into the air directly affects temperature, density, and humidity. Overall, water vapor is responsible for the different climates found over the earth.

Atmospheric impurities: solid particles

Spread throughout the entire lower atmosphere are solid particles called *aerosols*. Because they act as centers on which *condensation* of water vapor takes place, these wind-driven aerosols are also called *condensation nuclei.*

The main source of condensation nuclei is the earth's surface. Important microscopic salt crystals enter the atmosphere through the evaporation of ocean water, while dust, pollen, and volcanic ash are injected into the atmosphere by wind erosion, forest fires, and volcanoes.

After entering the atmosphere, the larger of the nuclei are usually washed out by snow or rain. The microscopic particles that are left are distributed around the world by

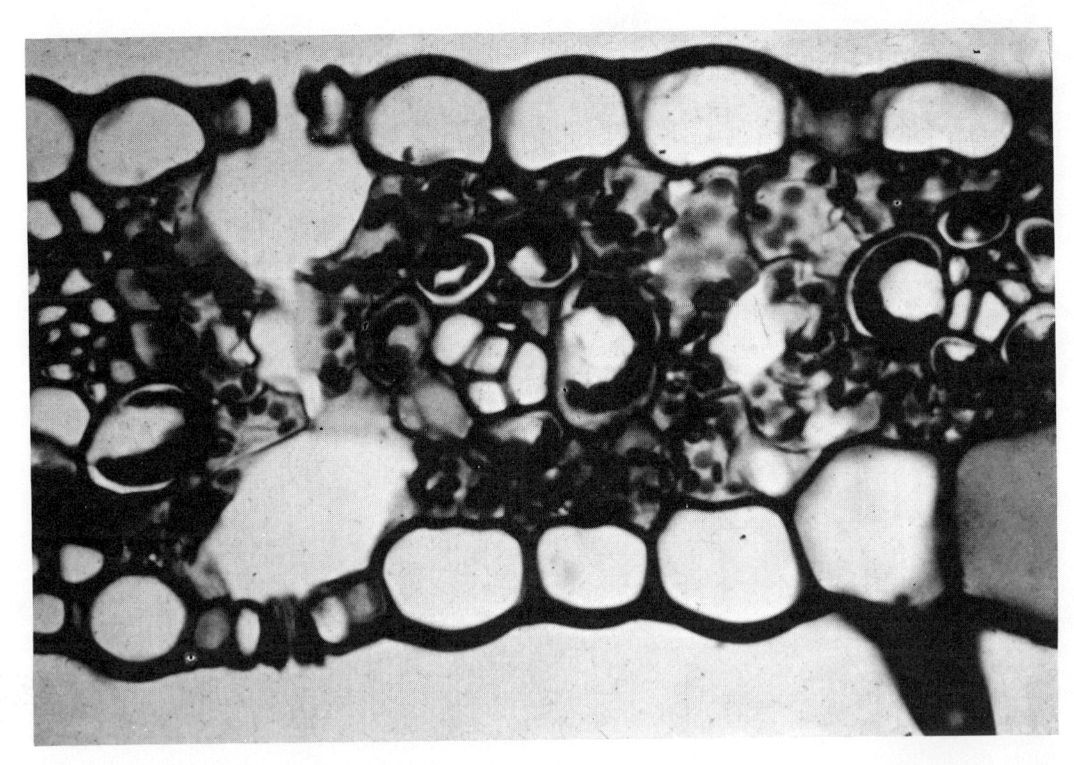

Fig. 26-4. Microscopic view of stomata openings in leaf. Water released from these tiny stomata through a process called transpiration then evaporates into the air. (Courtesy Carolina Biological Supply Co., Burlington, N.C.)

convection currents and wind after remaining suspended in the air for very long time periods. The violent volcanic eruption of Krakatoa (near Java) in 1883, for example, produced volcanic ash that remained suspended for nearly 2 years before settling back to earth. With this quantity of volcanic dust and ash in the air over such a long period, it is quite obvious that temperatures around the world could have been affected. As a matter of fact, the year following the eruption of Krakatoa was known as "the year without a summer." Temperatures fell drastically, rivers froze over in the late spring, crops were ruined, and the economies of many nations suffered. Although the following winter was extremely bitter, normal temperatures finally returned 2 years later.

Nuclei and air pollution

Condensation nuclei also play an important role in both visibility and air pollution. As the quantity of solid particles increases, visibility decreases and air pollution episodes are greatly increased. In urban areas great quantities of *particulates* are produced by incomplete combustion. This soot and smoke may become very active and attract large amounts of chemicals from the air. Many of these particles are then carried into the lungs, causing respiratory ailments and lung cancer. It is estimated that a city such as Los Angeles, California, puts 40 tons of particulates/day into the air from automobiles alone and that New York City releases more than 350 tons/day from combustion simply to produce heat for homes and industry.

DIVIDING THE ATMOSPHERE

Since the atmosphere stretches from the earth's surface outward into space, meteorologists find it quite difficult to study such a large body. For convenience they have divided the atmosphere into layers by using physical characteristics as the criteria for separation.

There are many ways in which to separate a large area so as to make it more convenient for study. Simply, the area might be broken in half or quarters. In the case of the atmosphere, pressure can be used, since we know that the pressure at the top of the atmosphere is 0. Dividing the atmosphere according to pressure, therefore, is quite realistic and easily done. Other methods such as composition or wind characteristics can also be used. Meteorologists,

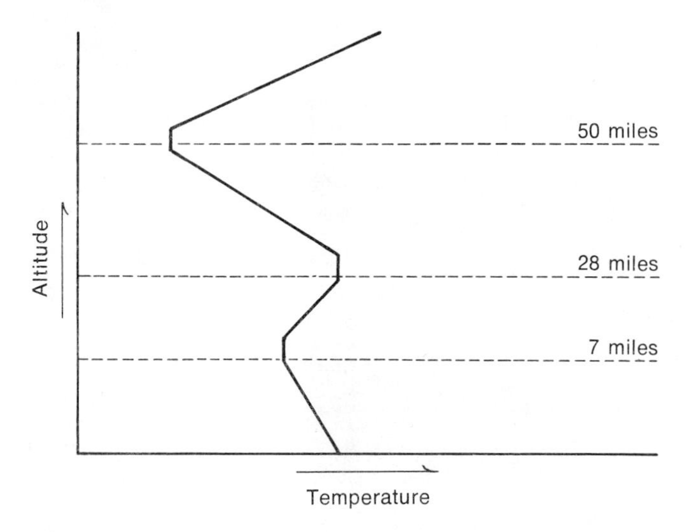

Fig. 26-5. This graph shows how temperatures decrease gradually with altitude except for certain small bands where it remains constant for a short distance. At an altitude of 28 miles, the decrease is no longer gradual, as temperatures drop rapidly.

however, prefer to use temperature as the criterion for subdividing the atmosphere.

Temperature measurements of the lower levels of the atmosphere have been made for many years by balloon and aircraft observations. Upper air temperatures are now measured by rockets, satellites, radio wave propagation analysis, and sound studies. Upper air observations were taken over many sections of the world as early as the 1940s, and when these observations were averaged over many areas and many years, a mean or standard temperature trace of the entire atmosphere was devised. Thus the United States *standard atmosphere* was created to represent the average features that existed within each level up to 120 miles.

It was thought for many centuries that as one left the earth and traveled upward, the temperature would decrease until absolute zero was reached. In 1902, however, a French meteorologist, Teisserenc de Bort, discovered that at the height of about 7 miles the temperature did not continue to decrease but remained constant or *isothermal* (from the Greek *isos,* meaning equal, and *thermē,* meaning heat) for a distance and then increased (Fig. 26-5). This amazed scientists and encouraged further investigation of greater heights in the atmosphere. Then it was discovered that at an altitude of about 28 miles the temperature stopped increasing, remained constant again, and then began to decrease. Continued investigations indicated that the temperature stopped falling at 50 miles and started to rise again. Thus meteorologists found perfect dividing lines for separating the atmosphere on the basis of temperature layers.

Troposphere

The troposphere, the layer of the atmosphere closest to earth, lies between the earth's surface and an average altitude of 7 miles (11 km). It reaches its greatest height of 11 miles (18 km) at the equator and its shallowest point of 5 miles (8 km) at the poles. These differences in height are caused by differences in temperature and the rotational spin of the earth. The name "troposphere" was suggested to indicate that this layer is very unsettled, or constantly overturning. *Tropos* is derived from the Greek meaning to turn, and this is the most important aspect of this lowest layer.

Since *convection,* or overturning, dominates this layer, moisture and solid particles such as salt and dust are constantly circulating. The results of this circulation are the following.

1. All of the dirt and dust in the atmosphere (except meteoric dust) is found in this layer.
2. Any moisture in great quantity is found in the troposphere.
3. All pollutants are found in this layer.
4. Almost all of the heat and moisture exchange of the atmosphere takes place here.
5. All cloud formation starts in the troposphere; with the exception of the tops of some thunderstorms clouds, clouds remain in this layer.
6. Since all dust and moisture as well as most of the clouds are found in the troposphere, we can say that all weather as we know it is found here.

Other characteristics of this layer are the following.

1. The wind speed increases with height because the frictional force produced by contact with the earth's surface decreases with distance from the earth.
2. Since the gases of the atmosphere are compressible, the troposphere contains three fourths of the atmosphere by weight (Fig. 26-6).
3. The earth acts as the major heat source for the atmosphere. Therefore at increased altitudes the distance from the source of heat increases, resulting in a decrease of temperature. This decrease is more or less uniform (−3.5° F/1000 ft), so at a height of 7 miles the temperature has dropped to −76° F (−60° C). This temperature drop of 3.5° F/1000 ft is called the *atmospheric lapse rate.*

Trouble in the troposphere

Since the troposphere is the layer that must accommodate thousands of tons of pollutants every day, it is the layer that should be treated with the greatest care. This very thin layer (7 miles thick as opposed to the hundreds of miles of

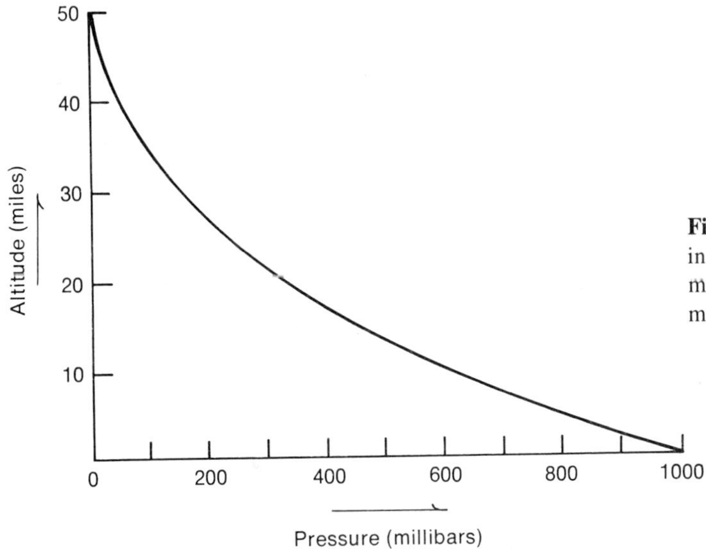

Fig. 26-6. Graph showing how higher pressures are concentrated in the lower layers of the atmosphere. Note how pressure at 50 miles up is not even measurable (reading of 0), while at only 30 miles pressure measures about 150 millibars.

the thickness of the entire atmosphere) is the only area in which life as we know it can be sustained. The processes of *dispersion, dilution,* and *natural washout* are nature's way of handling unwanted material, and for centuries such matter has been easily treated in this way. We have reached a level of pollution emission in our modern civilization, however, that is too great for this thin layer to cope with. Thus disastrous air pollution episodes are not only increasing in frequency but are also lasting for longer periods of time (Fig. 26-7).

Stratosphere

The second layer in the atmosphere, 7 to 28 miles (11 to 45 km) above the earth's surface, is called the stratosphere, and as the term (from the Latin word *stratum,* meaning layer) indicates, there is no overturning or convectional activity within this layer. This results in the following characteristics.

1. Moisture is not available, since evaporated water cannot reach this level.
2. Dust and other solid particles are rarely found in the stratosphere. However, extremely powerful volcanic eruptions such as Krakatoa have sent volcanic dust into the stratosphere, where it has remained for more than 2 years.
3. There are very few clouds, no precipitation, and no weather in the stratosphere.
4. Persistent horizontal wind circulation patterns and high wind speeds are typical.

Other characteristics of the stratosphere that are not associated with the lack of air circulation are the following.

1. Most of the ozone found in the atmosphere is located within the stratosphere. It is in this layer that much of the deadly ultraviolet energy from the sun is absorbed.
2. Since absorption of ultraviolet energy takes place here, it stands to reason that temperature readings increase dramatically, since the energy is being turned into heat. This now answers the riddle of why the temperature stops falling at the bottom of the stratosphere, remains isothermal for a short distance, and then increases; the upper part of the stratosphere is acting as a heat source.
3. A certain type of cloud called a *nacreous* or *mother of pearl* cloud is found in this layer. It has an irridescent look, appears to be stationary, and is seen only at sunrise or sunset in the northern latitudes. This peculiar phenomenon occurs only rarely and has not yet been explained.

More trouble in the stratosphere

Recently there has been much publicity regarding the United States supersonic transports (SST) as well as the English-French and Russian versions of these high-flying aircraft. It was pointed out that these aircraft fly in the stratosphere (as high as 23 km above the earth's surface), thus encountering no weather or turbulence and would therefore be much more economical to operate. This would result in shorter travel times and lower air fares.

However, one must look at the entire SST situation when considering their benefits. It is well known that dispersal of gaseous pollutants from the jet engines of these

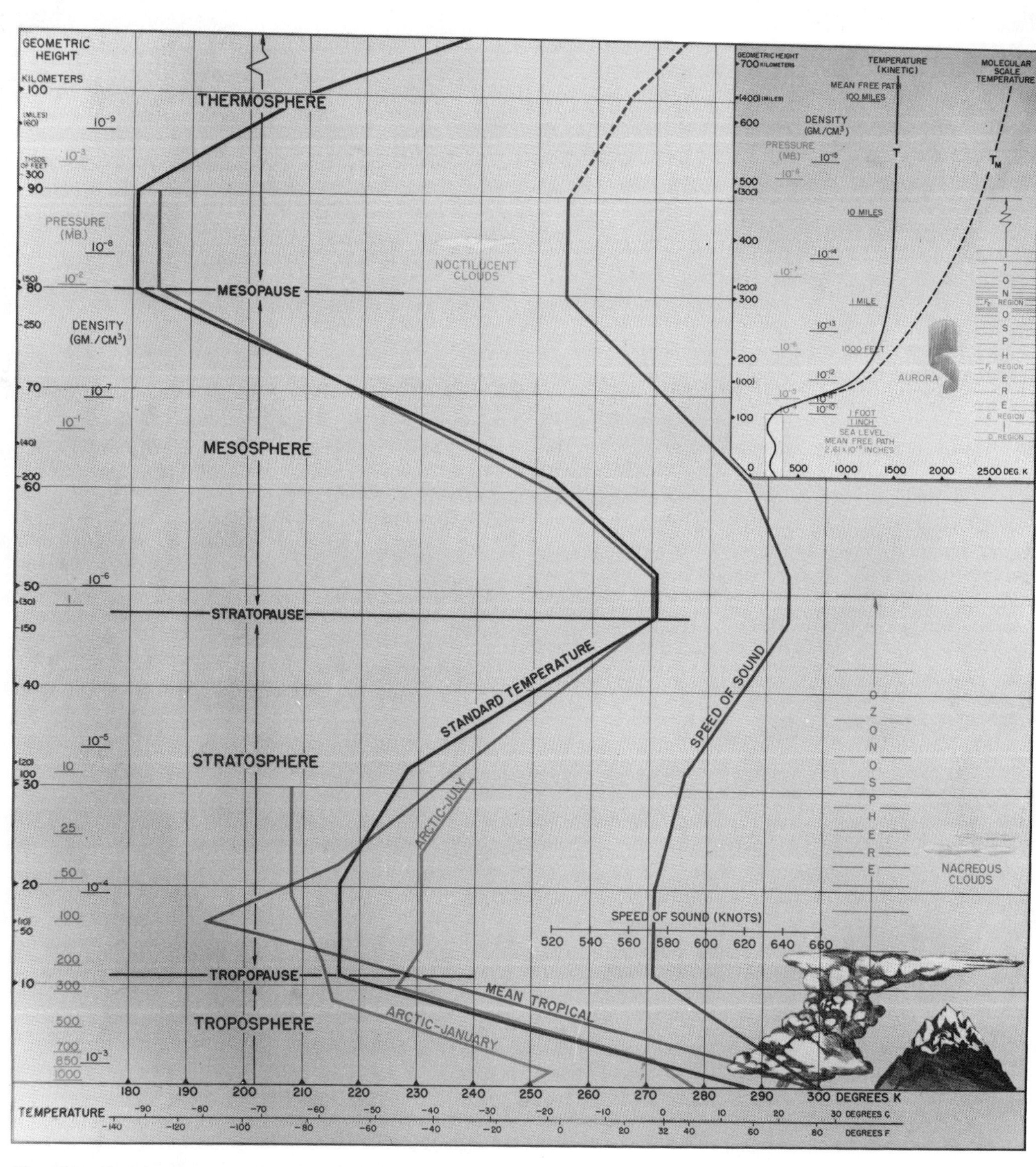

Fig. 26-7. Graphic summary of the atmospheric conditions characteristic of each layer of the atmosphere. (Courtesy United States Navy Weather Research Facility, Monterey, Calif.)

aircraft could conceivably cause trouble on a global scale. Critics of the SST argue that one of two outcomes could result. First, the sun's rays could be absorbed and reflected at higher altitudes and temperatures close to the surface of the earth would decrease. This could possibly trigger another ice age. Second, the heat energy reradiated by the earth might become trapped in the troposphere, thus increasing the temperature, melting the continental glaciers, and increasing the sea level. If this were to occur, many miles of our present coastlines would be lost to the sea and entire states and nations would be under water. Unfortunately it would not take drastic changes in temperature for either of these two alternatives to occur.

Warm layer

Located just above the stratosphere, 28 to 37 miles (45 to 60 km) above the earth, is a small isothermal layer called the warm layer. The warm layer lies just above the maximum region of the ozone concentration, and it is here that the highest temperatures of the upper atmosphere exist, up to approximately 30° F (−1° C).

Mesosphere

The layer located above the warm layer is called the mesosphere (from the Greek *mesos,* meaning middle). It lies 37 to 50 miles (60 to 92 km) above the earth. The mesosphere is the third of the four major layers of the atmosphere and is characterized by sharply declining temperatures that eventually reach −137° F (−83° C) at the top. The mesosphere is also sometimes named the *chemosphere* because of the chemical processes that take place quite frequently within the layer.

Thermosphere

The last layer of the atmosphere, which begins at 50 miles (92 km) above the earth, is called the thermosphere (from the Greek *thermē,* meaning heat). The thermosphere is characterized by high temperatures due to its extremely thin air. Temperatures increase as the layer approaches the great original heat source, the sun, with temperatures estimated to be as high as 2500° F.

Two rare and peculiar phenomena take place in the thermosphere. The first, the Aurora Borealis and Aurora Australis (the Northern and Southern Lights), are glowing, brilliant yellow, green, and red colors seen in the skies of the polar regions at night. These auroral displays result from showers of charged particles being thrown out from the sun into the upper atmosphere during solar magnetic storms (Fig. 26-8).

The second rarity is that of cloud formations called *noctilucent clouds* that are thought to be composed of cosmic dust particles. These clouds can be seen moving swiftly across the sky on long clear summer nights. Twilight is usually the best time to observe them.

Fig. 26-8. Eerie glow of the aurora borealis lights up the thermosphere. Both pictures were taken at the University of Alaska. (Courtesy Geophysical Institute, University of Alaska, Fairbanks, Alaska; photograph by V. P. Hessler.)

Zones of transition

Separating the layers of the atmosphere are zones of transition. These are regions no more than 1 mile thick that are called *pauses*. Thus the line marking the transition from troposphere to stratosphere is called the *tropopause*. Other transition areas are the *stratopause* and *mesopause*.

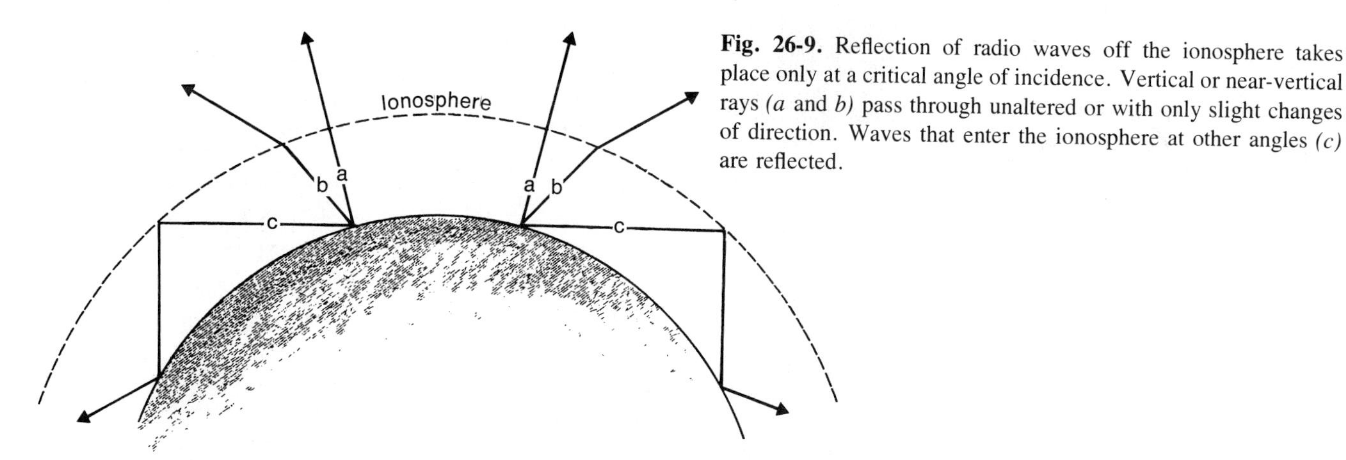

Fig. 26-9. Reflection of radio waves off the ionosphere takes place only at a critical angle of incidence. Vertical or near-vertical rays *(a* and *b)* pass through unaltered or with only slight changes of direction. Waves that enter the ionosphere at other angles *(c)* are reflected.

Ionosphere and ozonosphere

Superimposed on the atmosphere are two more important layers. The highest of the two is the ionosphere, which is located above the thermosphere. Here unfiltered X-ray and ultraviolet energy from the sun ionize the atoms of the gases in this layer. These charged atoms are called ions; hence the name "ionosphere." These ions interfere with short-wave radio transmission and repeatedly reflect these short waves and broadcast them back to earth. As a result it is possible to transmit and receive voice communications over great distances. Ham radio operators express this phenomenon as the radio wave skip. This skip is more pronounced at night than at any other time (Fig. 26-9).

The second important layer is the ozonosphere. This layer extends from the bottom of the stratosphere to the top of the warm layer. All the ozone of the upper atmosphere is found within the ozonosphere. It should be noted that if all the ozone within the ozonosphere were compressed and brought down to sea level, a layer only a fraction of an inch thick would result. This may seem to be only a trivial quantity, but without this all-important gas, excessive amounts of ultraviolet energy would penetrate to the earth's surface.

27 *a thermal blanket of air*

Heat in the atmosphere

Toil of the day displayed, sun-dust,
Aerial surf upon the shores of earth,
Ethereal estuary, frith of light,
Breakers of air, billows of heat,
Fine summer spray on inland seas . . .

Woof of the Sun, Ethereal Gauze
Henry David Thoreau

The sun is the primary source of energy for the earth and its atmosphere. Although we use this energy in many varied forms, it is not always consumed on impact. In many cases the energy is stored as *potential* energy for thousands of years. On release, it is used to do work or release heat. Energy that is being used is *kinetic* energy.

If we follow an individual ray of solar radiation in its path towards the earth, we observe that after it leaves the sun's surface, the ray travels first through space and then penetrates the earth's atmosphere. The energy arrives on earth in the same form as when it left the sun, but its magnitude has decreased greatly. As these rays strike various objects, the energy either performs work, as in a solar cell, or is stored for future use. Let us say that the ray falls on a plant, allowing photosynthesis to take place. The energy is now stored within the plant, which eventually will dessicate and decay. The plant in turn could eventually become part of a coal-producing sedimentary rock layer. When the coal is mined and burned, heat will be released into the environment while performing work that directly influences daily human activity and food production. Without incoming solar radiation there would be no weather on the earth's surface. With this thought in mind, it is reasonable to assume then that a thorough understanding of the sun and its energy transmission is a prerequisite to the study of weather.

THE SUN

Although we spent considerable time on the subject of the sun in Chapter 6, it is so important to the earth's weather processes that some of its physical features and processes should be reviewed at this time.

Although the sun is about 93 million miles from earth, the energy given off by this star is sufficient to sustain life here. The next closest star is some 250,000 times farther away than the sun—a distance that precludes the possibility of its energy influencing conditions on earth.

A continuous nuclear reaction takes place inside the center, or *core,* of the sun. This reaction results from the fusing of hydrogen nuclei into helium, a process that releases tremendous amounts of energy. This energy release produces temperatures in the sun's core of between 20 to 40 million degrees F (11 and 20 million degrees C). This energy is then transferred to the sun's photosphere (outside surface), where hot columns of rising gas are generated. As is the case with all heat sources, temperatures decrease drastically with distance from the origin. Temperatures of 10,340° F (5727° C) have been measured at this surface, and it is at this temperature that the sun's energy begins its long journey through space to our small planet.

ENERGY TRANSMISSION

One of the most interesting questions concerning energy is how it is transported from one place to another. This movement must depend somehow on the interaction between two surfaces or on natural flow through a vacuum such as outer space. There are four basic methods of energy transmission; these are conduction, convection, advection, and radiation.

Conduction

The slow transfer of energy from one medium to another by molecular means is called conduction. In order for conduction to take place, two substances of different temperatures must come in contact with one another. The higher the temperature, the faster the molecules making up a substance will move. The resulting collisions with adjacent molecules impart energy, and conduction through the area results. This process continues for as long as an imbalance of energy exists between the adjacent molecules. Eventually the temperature of the cooler object increases (Fig. 27-1) and the temperature of the warmer object decreases. The process stops when the temperatures of the two objects are equal.

Generally the best conductors are solids, although there is still quite a variance between the conductivity of metals and nonmetals. Even among the metals, distinct differences exist. Copper and silver are by far the best transmitters of energy, and their use in wiring and solid state circuitry attests to this fact. The other two states of matter, gas and liquid, do not conduct heat well. Liquids are much better conductors than gases because their molecules are closer together, and collisions between them therefore occur more frequently. In light of this, it should be fairly obvious that conduction plays practically no role in transporting atmospheric heat. Only the air layer closest to the earth's surface is affected by conduction and then only at night when the earth cools rapidly.

Convection

The transfer of energy vertically through the movement of a medium is called convection. Since movement of the medium is a requirement of this process, convection can take place only in fluids. Remember that earlier we discussed the ways in which air has some properties of fluids. Convection is one of these properties, and for the purpose of our discussion here will be considered a fluid. It is well known that hot fluids rise and cooler fluids sink because of the difference in their densities. While moving, the medium (gas or liquid) carries energy from one place to another.

In the troposphere, convection takes place rapidly as warm and cool air currents carry moisture and energy to various levels (Fig. 27-2).

Advection

A third method of transfer is called *advection,* which for all practical purposes is horizontal convection. This process is the most important method of heat conveyance in the atmosphere. In the Northern Hemisphere, cold arctic air advects southward to break up heat waves in the summer or to bring below-zero temperatures in the winter. Warm air, on the other hand, advects northward; the resulting interplay of these two air masses results in weather fronts and sometimes the spawning of violent tornadoes. However, advection and convection are also responsible for the distribution of heat and moisture throughout the troposphere.

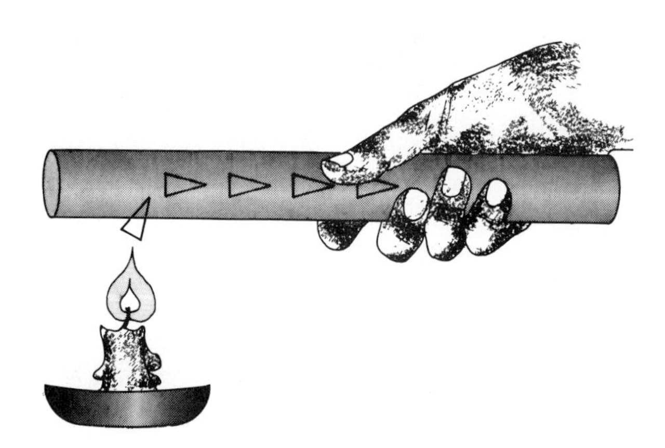

Fig. 27-1. The molecules in an object hand energy (heat) along from one to another until temperatures in all parts are equal. This process is called conduction.

Fig. 27-2. Since hot air is less dense than cool air, it rises to the top of the system. In this case it rises to the top of the chimney and escapes at the top.

Radiation

The fourth method of heat transfer is radiation. It is the process by which energy is transferred at the speed of light, with or without the help of a medium. Therefore it is the only method by which the sun's energy travels through the vacuum of space, and it is also the means by which the earth reradiates its heat energy into the atmosphere. If you were standing in front of the fireplace pictured in Fig. 29-2, the energy that warmed you would be the result of radiation.

Radiation theories

There are two schools of thought on the subject of the transmission of energy by radiation. Each theory attempts to explain how solar energy, especially light, travels from the sun to the earth's surface.

Max Planck (1858-1947), a German scientist, proposed the *quantum theory* in 1900. This theory introduced the concept that solar energy consists of noncontinuous bundles of energy called *quanta* that travel in straight lines from the sun to the earth.

The second theory, originally proposed by the Dutch scientist Christian Huygens (1629-1695) in the seventeenth century, suggests that solar energy moves at the speed of light in a series of troughs and ridges. These troughs and ridges would then make a wave pattern like that shown in Fig. 5-2.

An analogy to this *electromagnetic wave theory* is the propagation of surface waves on a large body of water (the only difference being that water waves need a medium while electromagnetic waves do not). If all the irregular motions of the water wave were removed, it would be pictured as a series of horizontally moving troughs and crests. The distance between two successive crests is known as the *wavelength*. The rate at which the wave crests pass a fixed point in a set time period is called the *frequency* (see Chapter 5 for a more detailed discussion).

In analyzing both theories, we find that less than one millionth of the total solar energy that reaches the earth's surface can be accounted for by the quantum theory, and most of the incident energy, especially light, is in the form of electromagnetic waves. Since the light energy portion of the *solar spectrum* dominates weather patterns, the wave theory is the most important to the meteorologist.

Electromagnetic spectrum. In order to understand electromagnetic energy, it is important to have a basic understanding of the physical laws that govern its initiation and propagation.

1. All bodies except those whose temperatures are at a theoretical point of absolute zero emit radiation to a surrounding space in the form of electromagnetic waves.
2. The amount of radiation and the wavelength at which this energy is transmitted depends greatly on the temperature of the body—the higher the temperature, the shorter the wavelength of energy.
3. All emitted energy travels in a straight path at the speed of light (186,000 miles/sec). When it is finally absorbed by another body, the energy is turned into *sensible heat,* warming the second body, which in turn reradiates this energy once again according to its temperature.

Electromagnetic energy can be broken down into two main categories: *solar* and *terrestrial* radiation. Solar radiation, or *insolation* (from *in*coming *sol*ar radi*ation*), includes primarily the short wavelengths of energy. The shortness of the wavelength is due to the high temperature of the sun. Terrestrial radiation is the energy given off by the earth's surface; by all standards, this is quite cool compared to the sun. Therefore terrestrial radiation is frequently called *long-wave* radiation.

Separation of solar and terrestrial radiation. The short-wave energy leaving the sun travels through space, penetrates the atmosphere, and falls on the earth's surface. However, not all of this energy reaches the earth, as will be explained shortly. The portion of solar energy that does reach the earth heats up its surface. The earth's surface in turn starts to reradiate its own energy back through the atmosphere to space. The main difference between the two types of energy transmission is that the incoming energy is *short wave* and the outgoing energy is *long wave*. The reason for this difference is that the bodies sending out the energy are at two greatly different temperatures, and within the energy spectrum the temperature of the substance is responsible for the type of energy that it produces.

Earthward path of solar radiation

Although energy arriving at the earth from the sun has traveled 93 million miles, it has diminished very little since leaving its source. Only a fractional depletion of energy has occurred on its journey, and this is so small that it can be ignored. Recently, measurements made by rockets and satellites at the fringes of outer space have indicated that a constant value of 2.00 calories of energy fall on each square centimeter every minute. This is called the *solar constant,* for it represents a constant amount of energy hitting the top of the atmosphere.

As the solar beam penetrates the atmosphere, five major processes take place, two of which deplete this radiation greatly before it reaches the earth's surface. As a matter of fact, less than half (48%) of the total solar energy reaching the top of the atmosphere ever reaches the surface.

Absorption of energy takes place mainly in the higher layers of the atmosphere. The gamma and X-rays (short wavelengths) are absorbed in the ionosphere, and shortly thereafter the ozonosphere captures most of the ultraviolet light. In the lower layers of the atmosphere *permanent gases, variable gases,* and *condensation nuclei* absorb small amounts of energy. The net result is a reduction of 19% in the total insolation.

Reradiation is the process by which matter retransmits the energy it has received. As was pointed out previously, all objects whose temperatures are not absolute zero emit radiation. If an absorber did not start emitting radiation on receipt, it would continue to heat up until it finally exploded due to the expansion of the heated area. It is wise to remember that *a good absorber is a good emitter*. The reradiated energy is usually given off in lower wavelengths due to the lower temperature of the second source. Since energy is reradiated in all directions, some is lost to space, a few units are absorbed by other particles, and the remainder travels earthward again. Depletion by reradiation is minimal.

Refraction is a third response that can take place when insolation hits the atmosphere. It is responsible for a direction change only and does not deplete the incoming solar radiation. On passing from one medium to another (for example, from air to a suspended water droplet) a light ray visibly bends. After this *refraction* the energy travels in a different straight-line path. *Mirages* are the result of terrestrial refraction, and rainbows are the product of solar refraction and reflection (Fig. 27-3).

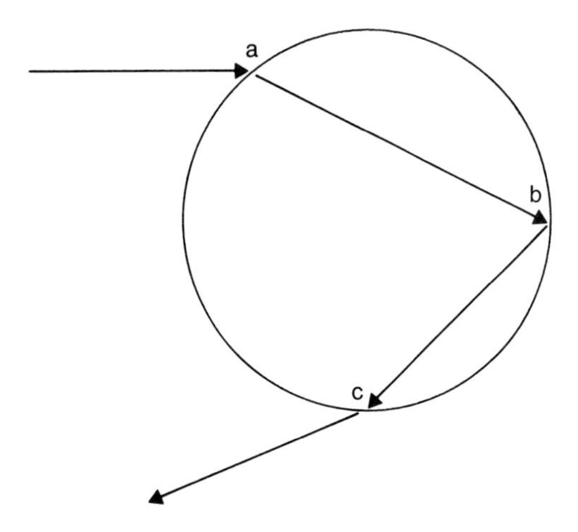

Fig. 27-3. Energy entering the earth's atmosphere is bent at point *a* by refraction. When the energy gets ready to leave *(b)*, it is again bent because it is passing once more from one medium to another.

Reflection is the phenomenon whereby a surface turns back part of the incident energy. The process is dependent on the composition of the surface and the angle at which the rays strike it. Out of the many particles in the atmosphere that are capable of reflection, visible water droplets, which form clouds, and floating condensation nuclei are the most important. They account for most of the 32% depletion of incoming short waves by reflection.

An additional 2% reduction takes place when radiation hits the earth. Freshly fallen snow, deserts, and ice are excellent reflectors, as is water at certain critical angles.

The latest research data indicates that the total reflective power of the earth-atmosphere system is measured at 34% (32% from clouds and 2% from earth). This is called the earth's *albedo,* which is its *degree of reflectivity.*

The scattering of visible light accounts for daylight as we know it and the blueness of the sky. Although there is no depletion of energy from this process, if there were no particles to scatter light (as in space), the sun would look like a bright white solar disk, and elsewhere the sky would look black. However, the particles in our atmosphere easily scat-

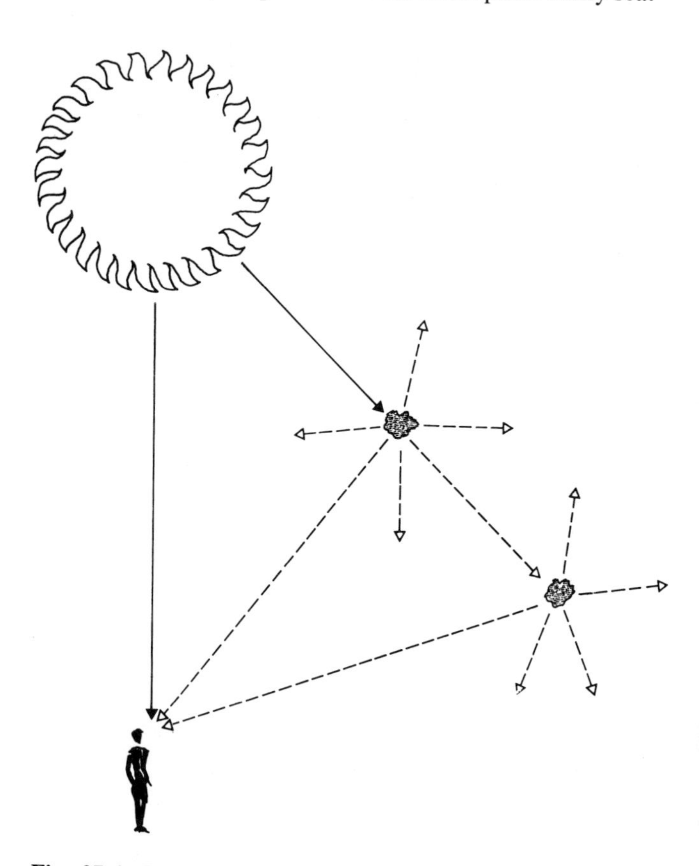

Fig. 27-4. Particles in space scatter solar light, making the sky appear blue instead of black.

ter the shorter wavelengths of visible light in all directions. This is especially so in the blue wavelength range. In the longer-wavelength red range, on the other hand, most of the energy is transmitted directly to earth. Therefore by looking away from the yellowish tinted sun, the eye receives the scattered blue light and the sky appears blue (Fig. 27-4).

Near sunrise and sunset the path that the sun's rays must take is longer than at any other time of the day. This allows more blue light to be scattered out and the sun appears as a red or orange disk in the sky (Fig. 27-5).

ATMOSPHERIC WARMING

Surprising as it may seem, the atmosphere is warmed directly by the earth and not by the sun. Of the four ways in which this warming takes place, the *greenhouse effect* is the most important. This atmospheric heating mechanism works on the same principle as a greenhouse. The glass of a greenhouse allows short-wave radiation to penetrate, warming the inside. However, as long-wave radiation is retransmitted, the glass stops it so that the air inside does not cool. Similarly, the atmosphere is *transparent* to solar radiation but *opaque* to long-wave energy. The gases of the atmosphere most responsible for this trapping of energy are water vapor, carbon dioxide, and oxygen. It is now possible to understand why the lowest temperatures occur on a cloudless night when there are no suspended water droplets to trap the heat, and why higher average minimum temperatures occur over urban areas where there is plenty of water vapor and carbon dioxide to act as a heat trap.

The second major heating process is the result of long-wave energy that is released when invisible water vapor changes to visible droplets. This change of state, called *condensation,* accounts for the stored energy in the water vapor being released as sensible heat. The entire process is called *the latent heat of condensation.*

The last two methods of atmospheric heating are by *conduction* and *convection.* Both of these minor methods transfer energy from the earth's surface to various levels in the troposphere. Conduction is accomplished by contact with the land and convection by the hot air rising from the surface.

HEAT VERSUS TEMPERATURE

One of the primary points to be made regarding heat and temperature is the difference between the two. Heat, or *thermal energy,* is the energy itself; temperature is the measurement of the heat of a given object. Temperature is measured by an instrument called a thermometer *(thermē* means heat and *meter* means to measure).

Heat always travels from areas or objects that are warmer to those that are cooler, but the transfer of heat does not necessarily result in a change of temperature. Objects transferring the heat always get colder but receipt of the heat does not always result in an increased temperature. An example of this heat flow without a temperature change is the heating of an ice-water mixture. As heat is added to the mixture the ice melts, but the temperature of the mixture remains constant at 32° F (0° C). Even though heat and tem-

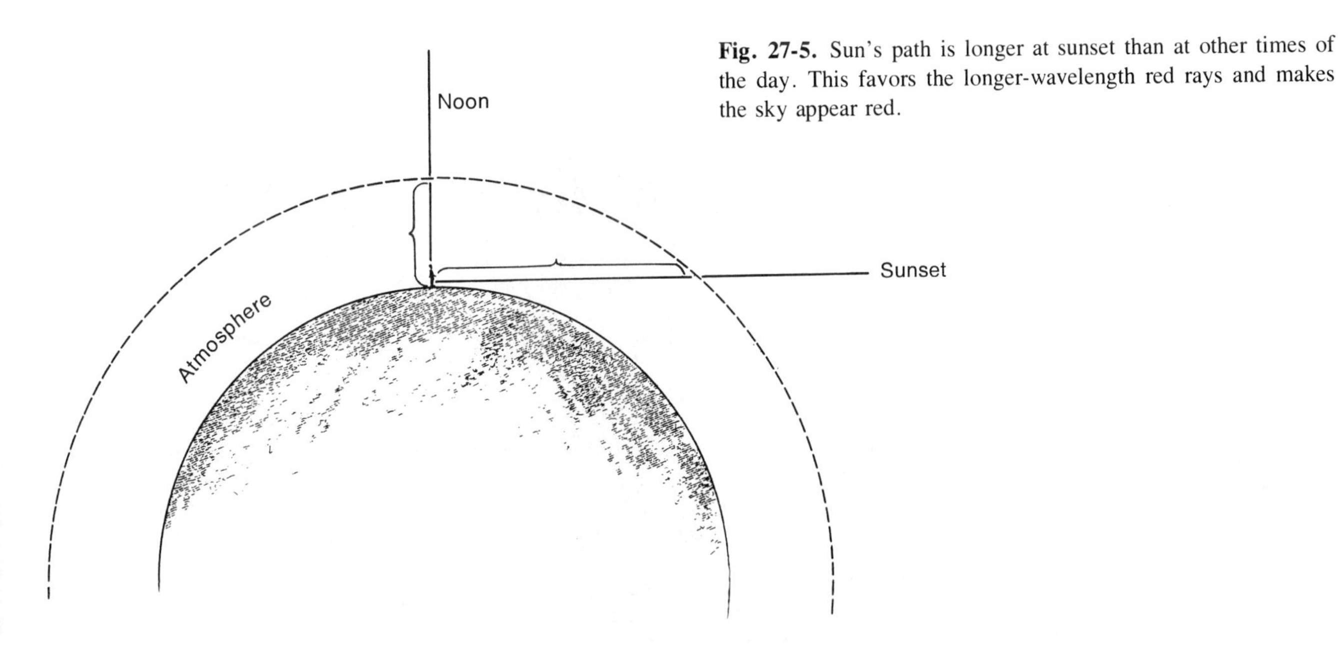

Fig. 27-5. Sun's path is longer at sunset than at other times of the day. This favors the longer-wavelength red rays and makes the sky appear red.

perature are closely related, it should be made clear that they are not the same.

MAJOR CONTROLS OF TEMPERATURE

Although the amount of energy reaching the earth's surface is fairly uniform, the resulting temperatures are not. It is therefore quite obvious that other factors must contribute to the irregular daily and annual temperature patterns found on the earth. Some of these factors are the following.

1. Angle of incidence of the sun's rays
2. Duration of daylight
3. The greenhouse effect
4. Physical characteristics of a surface area
5. Advection of warm or cold air
6. Convection and subsidence of air
7. Quantity of pollutants in the air

Of the factors listed above, the angle of incidence and duration of daylight are most important and therefore deserve a detailed consideration. Of the remaining factors, some have already been discussed in this chapter and others will be explored later.

Angle of incidence. The angle of incidence of the sun's energy is the most important temperature control known to man. When a ray of energy hits an object at various angles, the heating that takes place is determined by the area the ray circumscribes. A ray hitting at an angle of 90° would illuminate a circle, while a ray striking the surface at an oblique angle (>90°) would produce an ellipse. If both rays represented the same amount of energy, the resulting temperature of the ellipse would be lower than that of the circle because the same amount of heat would be spread over a greater area (Fig. 27-6).

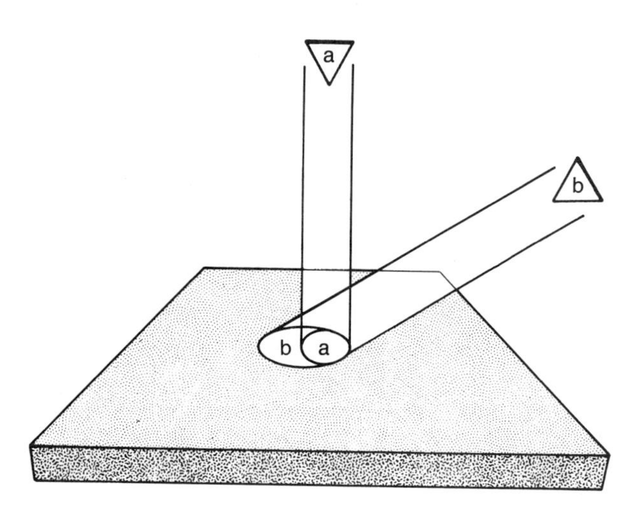

Fig. 27-6. The amount of heat an area receives depends on the angle of incidence of the sun's rays. Ray *a*, for example, falls on an area circumscribed by circle *a*. Ray *b*, however, falls on the area circumscribed by the larger ellipse *b*. This means that the same amount of heat is spread over more area by ray *b*, and no single spot retains as much heat as an equivalent spot in the smaller area *a*.

Duration of daylight. The amount of radiation available to an area is heavily dependent on the duration of daylight. The daylight period varies according to the season, for the length of the ecliptic or apparent path of the sun across the sky is not constant. The angle at which the sun's rays hit a surface in the Northern Hemisphere during the winter is oblique, resulting in a shorter path and a resultant shorter daylight period. In the summer the angle of incidence is more perpendicular, and this results in a longer path and a greater daylight period (Fig. 27-7).

TEMPERATURE VARIATIONS

Differences in temperature are usually broken down into two categories: *horizontal* and *vertical*.

Horizontal variation

Horizontal temperature variation relates to the heat distribution from place to place over the globe. The main controls of these temperature changes are:

1. Position or latitude
2. Season of the year
3. Irregularities of the earth's surface
4. Continentality versus oceanity
5. Air pollution

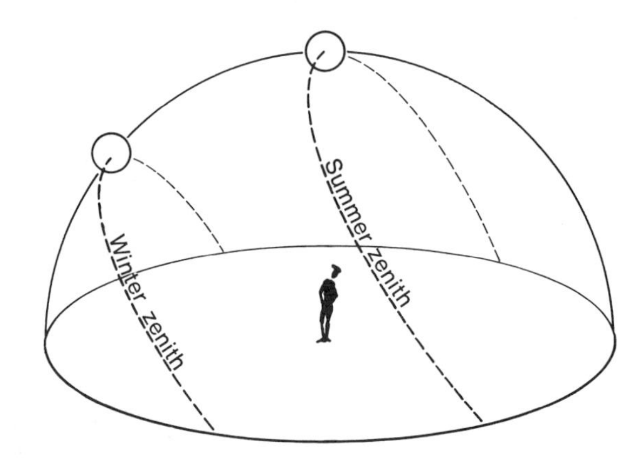

Fig. 27-7. In the summer the sun appears to cross more nearly overhead in a longer path. Since it travels at the same speed, it must stay above the horizon longer to travel the larger distance. Since a day is the time the sun stays above the horizon, the days are longer in the summer.

Latitude

As latitude increases, the sun's rays hit at ever-decreasing angles, the magnitude of which depends on the season of the year. This change in the angle of incidence is due chiefly to the curvature of the earth (Fig. 27-8). The latitude factor causes a slow decrease of temperature from the equator to the poles.

Seasons

With the seasons of the year, marked changes of temperature occur at various latitudes. A thorough explanation as to the cause of the seasons is given in earlier chapters. Since one of the major reasons for the seasons is the earth's position relative to the most direct rays of the sun, this is also a very important factor in horizontal temperature changes. Fig. 27-9, a one-ray representation of Fig. 27-8, shows the zenith position of the sun during summer and winter at the same latitude. The diagram shows that it is the angle at which the sun's rays strike a surface (the angle of incidence) that is the determining factor in temperature differences.

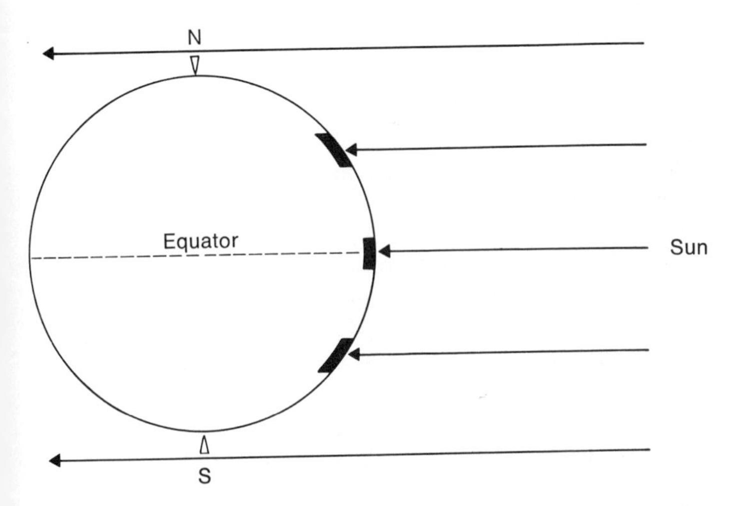

Fig. 27-8. Since the earth's surface is curved, the angle of incidence of the sun's rays increases with increasing latitude. Therefore each ray must cover more territory in the middle latitudes than at the equator. This is why it becomes cooler closer to the poles.

Irregularities of the earth's surface

The *nonhomogeneous composition* of the earth's surface leads to distinct temperature changes. Dark, wet soil heats and cools more quickly than light, dry soil because of the light soil's poor conductivity. The resulting temperature differences influence the air directly above the surfaces, so that air is warmer over the hotter soil. Additional differences result from the differential heating characteristics of land and water bodies and the physical features of the earth.

Continentality versus oceanity

Bodies of water heat up and cool down more slowly than large land areas. A city located in the interior of a continent, for example, would be much warmer in the summer than a city located along the ocean coast at the same latitude.

Nearly level land masses are constantly interrupted by large mountain chains whose alignment tends to control temperature fluctuations. A north-south alignment would not have as large an influence on temperature as a chain of mountains running east to west, for the latter arrangement would block large masses of cold air from colder latitudes or warmer air from the tropics. This influence can be illustrated by comparing the Himalayas in Asia, which are

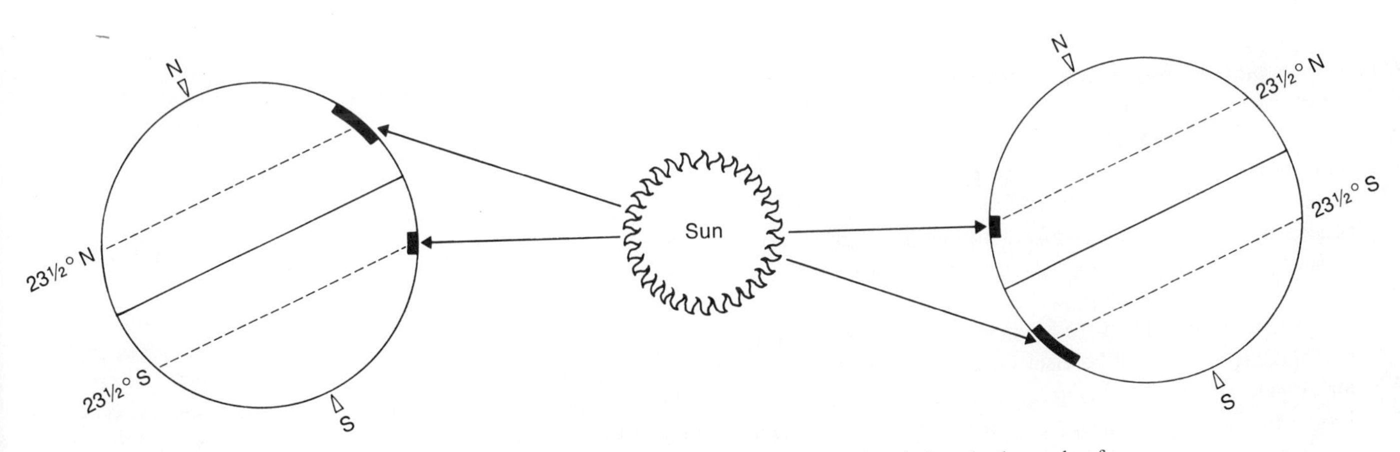

Fig. 27-9. This one-ray version of Fig. 27-8 shows the relationship between seasonal variations in the angle of incidence and temperature.

oriented east to west, and the Rocky Mountains in the United States, which are oriented north to south. In the United States cold masses of frigid Arctic air pour southward, sometimes penetrating as far south as the tip of Florida. However, cold air pooled north of the Himalayas is trapped and does not invade India to the south.

Air pollutants

The influence of pollutants on temperature is both large and complex. In order to keep this discussion simple, we will focus only on *carbon dioxide,* the pollutant that has the greatest effect on the air's temperature. Carbon dioxide is a natural constituent of the atmosphere, but it is also released into the air by the combustion of fossil fuels such as coal and petroleum products. Since 1880 the amount of carbon dioxide has increased by 15%, chiefly because of the mechanization and industrialization of our society. Recall that carbon dioxide is one of the gases chiefly responsible for the greenhouse effect, which tends to maintain the temperature in the atmosphere. If larger amounts of carbon dioxide enter the air, then it is quite obvious that a rise in worldwide temperature could result, bringing about the melting of the polar ice caps. However, an increase in temperature would also lead to an increased rate of evaporation; with more water vapor in the air, cloudiness would increase. This in turn would mean an increase in reflectivity of insolation, so that less of the sun's energy would reach the earth. The lower temperatures that would result could eventually produce another ice age. Thus we are left with the perplexing thought that increased pollution could cause either a glacial invasion or a worldwide rise in sea levels that could inundate millions of miles of dry land presently in use.

Vertical temperature variations

Vertical temperature differences result mainly from the fact that temperature decreases as altitude increases because greater altitude means greater distance from the earth, which is the atmosphere's direct heat source. Also, as atmospheric pressure decreases the air becomes less dense and there are fewer collisions among molecules, thus decreasing the temperature. In addition, water vapor, which retains heat, diminishes greatly with height. Vertical temperature variations are much greater than the horizontal differences just discussed.

In determining the vertical differences in temperature, two specific cases must be discussed: *environmental air,* or air already at specified levels, and *rising air,* or air that is forced to rise mechanically through the environmental air. Environmental air is often called the *surrounding air* because it surrounds the rising air.

Environmental air

The temperature decrease of environmental air varies greatly with time and location. Only measurements by airborne instruments can accurately record these short-term vertical fluctuations. The actual *plot* of the vertical temperature is called a *lapse rate.* Over a long period of time the lapse rates from many locations have been averaged to yield a *normal lapse rate,* which is 3.5° F/1000 ft (6.5° C/km). There are three types of lapse rates. The *positive lapse rate* is the decrease of temperature with height, the *isothermal lapse rate* represents no increase or decrease of temperature, and the *inversion* or *negative lapse rate* is an increase of temperature with height.

By far the most important influence on human comfort is the formation of a temperature inversion close to the ground. An inversion is formed wherever temperatures increase with altitude rather than decrease, which is the normal situation in the lower atmosphere. Inversions are usually formed either when warm air slides above cold air or when air near the ground cools faster than the air above. Whatever the cause, devastating air pollution episodes can result. Inversions retard the upward movement of air, and this traps all the noxious gases and smoke close to the ground. Not until the sun's radiation increases the temperature of the lower air layer will the inversion break up and the upward movement of air begin once again. The phrase coined for the breaking up of the inversion is "burning off," since it takes an increase in temperature to start the air moving.

Rising air

Air rises because of a variety of mechanical lifting processes (Fig. 27-10). These vertical air movements may result from air encountering a mountain (topographic lifting), the convergence of air, the collision of air of different densities, and heating of air (convection).

When air is forced to rise, it expands because of the decrease in air pressure. In order for this expansion to take place, energy must be used; since no energy is available from outside sources, the necessary energy comes from within; as a result, the rising air cools. A simple analogy to this process is the rapid expansion of air escaping from an automobile tire. The air feels cool and the metal part of the tire stem is quite cold. The process responsible for the cooling of the air due to its expansion is called the *adiabatic process,* and the rate at which this cooling takes place is called the *adiabatic lapse rate.* Since both dry and wet air exist in the atmosphere, there are two adiabatic lapse rates. The dry adiabatic lapse rate is larger than the wet because when the air is saturated with water vapor, the latent heat of con-

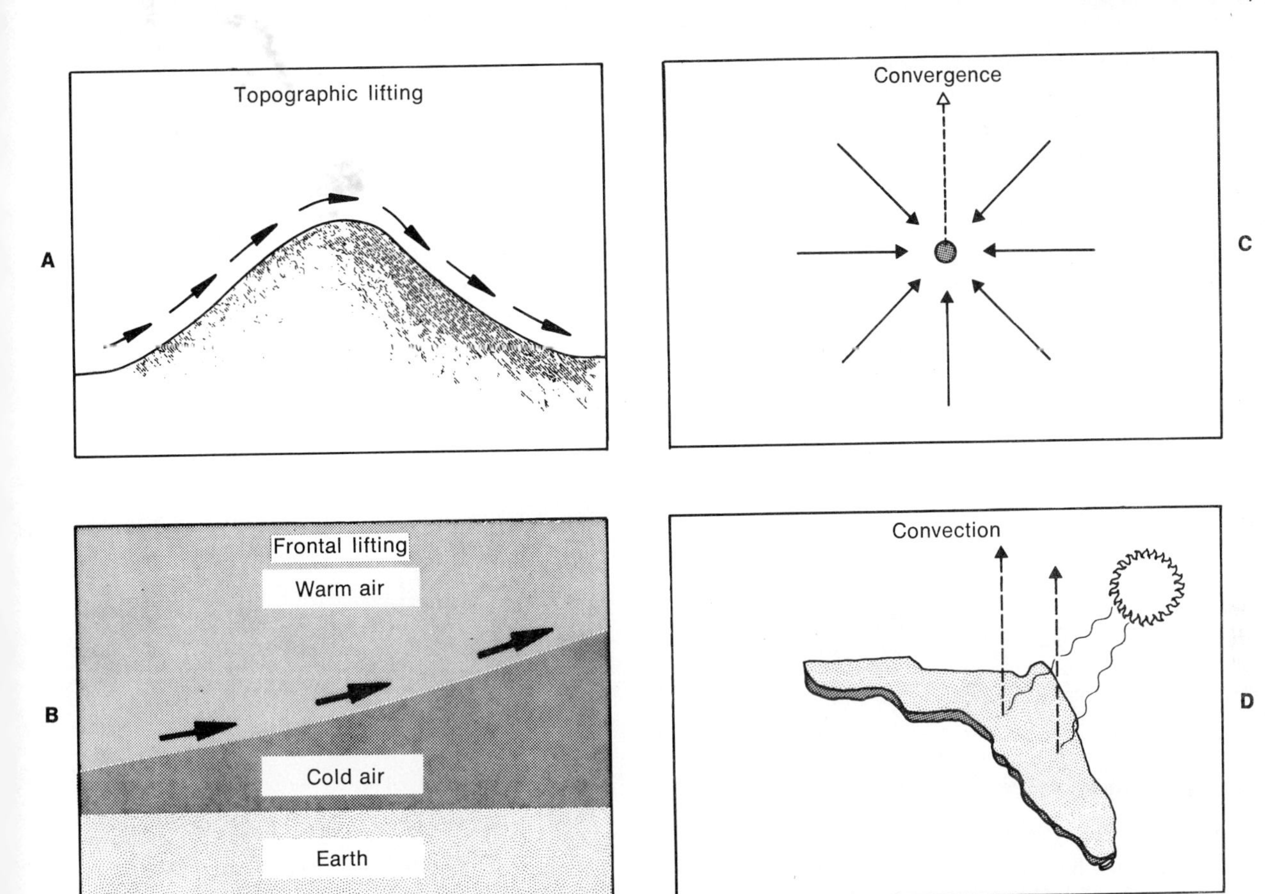

Fig. 27-10. Four methods of forcing air to rise. **A,** In topographic lifting air is forced up by a geographic barrier like a mountain range. **B,** In frontal lifting cool air moves under air forcing the lighter, less dense warm air out. **C,** In convergence, several air masses are all compelled to move into the same area. Since they cannot all occupy the same space, one or more of them must go elsewhere. Since they can't drop (the ground is there), they rise. **D,** Convection involves the natural rising of warm air.

densation is released into the atmosphere, warming the air and reducing the rate of cooling.

Dry adiabatic lapse rate	−5.5° F/1000 ft
Latent heat of condensation	+2.3° F/1000 ft
Wet adiabatic lapse rate	−3.2° F/1000 ft

TEMPERATURE SCALES

The degree of warmth or coolness of an object is a matter of arbitrary decision. The *relative* feeling for temperature depends first on the person doing the estimating and second on the warmth of the medium. Therefore *specific* scales were devised to numerically decide the *degree* of warmth. There are currently three temperature scales used throughout the world. They are the Fahrenheit, centigrade, and Kelvin scales.

Fahrenheit scale. The Fahrenheit scale, devised by Gabriel Fahrenheit (1686-1736) in the early 1700s, consists of 180 divisions. It utilizes the known fixed freezing and boiling points of water, which are 32° for freezing and 212° for boiling. This scale is used primarily in the United States.

Centigrade scale. The centigrade scale (also called the Celsius scale) was named to honor a Swedish astronomer, Anders Celsius (1701-1744), who developed this measuring system in the eighteenth century. The reference points for this scale are 0° for the freezing point of water and 100° for the boiling point. This temperature scale is used by the entire scientific community and generally in almost all the countries of the world except the United States. The United States, however, is slowly changing to the *metric system* of measurement, which includes the centigrade scale.

Kelvin scale. The Kelvin (or absolute) scale, developed by Lord Kelvin (1824-1907) of England, utilizes a starting point at which no molecular activity exists. Thus absolute zero, equivalent to −273° C, was chosen. This scale cannot have negative numbers because 0° K is the lowest number on the scale. The freezing point of water is 273° K (0° C), and the boiling point is 373° K (100° C).

TEMPERATURE CONVERSIONS

Centigrade to Kelvin

The easiest conversion between scales is from centigrade to Kelvin, or vice versa. The difference between the two is a constant 273°, with the Kelvin temperature always the highest (Table 24). Therefore:

$$K = C + 273°$$

and

$$C = K - 273°$$

EXAMPLE: What is the Kelvin equivalent of 20° C?

$$K = 20° + 273° = 293°$$

Centigrade to Fahrenheit

The conversion of centigrade to Fahrenheit may be done in three ways, all of which are correct. Use the method that suits you best (Table 24). The formula for converting centigrade to Fahrenheit is:

$$F = \frac{9}{5C} + 32° \text{ or } 9C = 5F - 160°$$

EXAMPLE: 10° C equals what temperature on the Fahrenheit scale?

$$F = \frac{9}{5}(10°) + 32°$$

$$F = 50°$$

or

$$9(10°) = 5F - 160°$$

$$90° + 160° = 5F$$

$$F = \frac{250°}{5} = 50°$$

To put this method into words:

1. Double the centigrade temperature
2. Subtract one tenth from the result
3. Add 32°

EXAMPLE: 20° C equals what temperature on the Fahrenheit scale?

Table 24. Conversions for Kelvin, centigrade, and Fahrenheit temperature scales

Kelvin	Centigrade	Fahrenheit
273	0	32.0
274	1	33.8
275	2	35.6
276	3	37.4
277	4	39.2
278	5	41.0
279	6	42.8
280	7	44.6
281	8	46.4
282	9	48.2
283	10	50.0
284	11	51.8
285	12	53.6
286	13	55.4
287	14	57.2
288	15	59.0
289	16	60.8
290	17	62.6
291	18	64.4
292	19	66.2
293	20	68.0
294	21	69.8
295	22	71.6
296	23	73.4
297	24	75.2
298	25	77.0
299	26	78.8
300	27	80.6
301	28	82.4
302	29	84.2
303	30	86.0
304	31	87.6

1. 20° + 20° = 40° (double)
2. 40° − 4° = 36° (minus one tenth)
3. 36° + 32° = 68° Fahrenheit (add 32°)

EXAMPLE: What is the Fahrenheit temperature for −20° C?*

1. −20° + (−20°) = −40°
2. −40° − (−4°) = −36°
3. −36° + 32° = −4°

Fahrenheit to centigrade

To convert from Fahrenheit to centigrade, either of the following formulas may be used (Table 24).

*Remember in doing these conversions that a minus and a minus equal a plus.

$$C = \frac{5}{9}(F - 32°) \text{ or } 9C = 5F - 160°$$

EXAMPLE: What is the centigrade temperature for 86° F?

$$C = \frac{5}{9}(86° - 32°)$$

$$C = 30°$$

or

$$9C = 5(86°) - 160°$$
$$9C = 270°$$
$$C = 30°$$

INSTRUMENTS FOR MEASURING TEMPERATURE

One of the most commonly used weather instruments is the thermometer, which works on the principle of an energy imbalance between two media. In response to this imbalance, energy will flow into or out of the thermometer, providing the resultant temperature measurement. Although there are many types of temperature measuring devices, only the most commonly used thermometers will be discussed.

Standard air thermometer. The temperature of freely moving air is usually measured by a liquid-in-glass thermometer. The liquid used is usually mercury, but alcohol, with a lower freezing point, is sometimes employed when readings lower than −39° C are expected. The thermometer is constructed in such a way that the liquid in the bulb end can expand or contract easily with an energy exchange. This movement is made into a long tube called the *bore*. One of the three temperature scales is etched onto the glass, and the thermometer is often given a metal backing for support. This instrument is capable of giving instantaneous readings (Fig. 27-11).

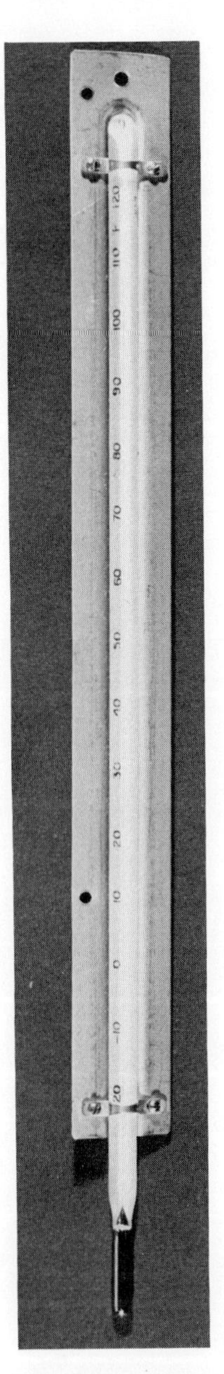

Fig. 27-11. Standard air thermometer measures air temperature by measuring the expansion or contraction of the liquid in the tube. (Courtesy WeatherMeasure Corp., Sacramento, Calif.)

Maximum thermometer. Often it is necessary to determine the highest temperature to occur in a given period of time. Therefore a *self-registering* liquid-in-glass thermometer was developed. The construction of this thermometer is similar to that of a clinical thermometer used to take a person's temperature. The actuating element is mercury, which resides in a round bulb and expands part way up the bore. Unlike ordinary air thermometers, self-registering thermometers have a *constriction* in the bore just above the bulb (Fig. 27-12). This constriction does not prohibit the mercury from advancing or retreating through the tube but impedes the flow somewhat. When the temperature increases, the force of expansion is great enough to force the mercury through the constriction and into the bore. However, when the temperature decreases the constriction prevents the contracting mercury from traveling downward. The weight of the mercury also prevents this retreat, because of the narrowness of the constriction. Thus the mercury remains at its highest level and the maximum temperature is recorded. The thermometer is reset by spinning or shaking it to force the mercury back through the narrow constriction.

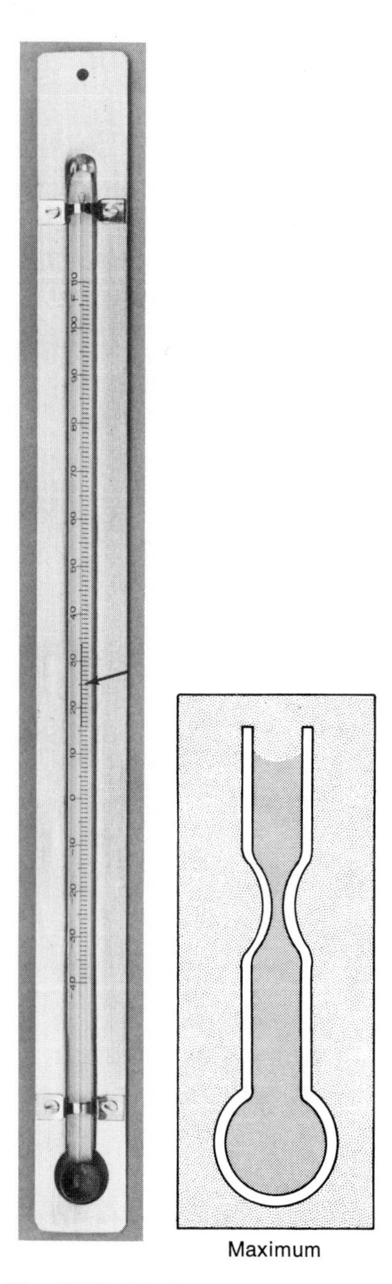

Fig. 27-12. Maximum thermometer involves a constriction in the bulb that allows the mercury to flow through and register the highest temperature in a given time span. Like a one-way door, however, the constriction does not allow the mercury to return when the temperature falls. The thermometer is thus stuck at the highest temperature until it is reset. (Courtesy WeatherMeasure Corp., Sacramento, Calif.)

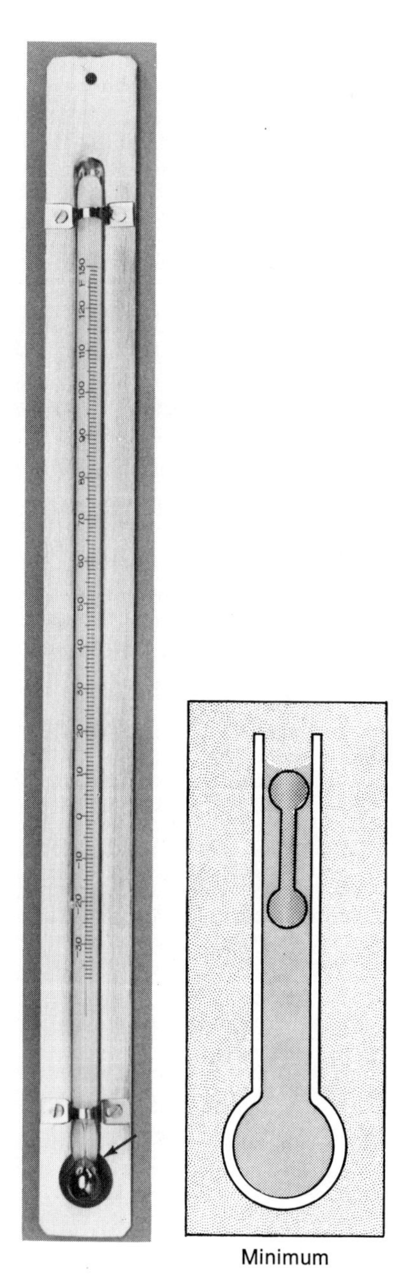

Fig. 27-13. A minimum thermometer has a small glass dumbbell in its center. As the temperature falls, the thermometer liquid (usually alcohol) retreats, carrying the dumbbell with it. As the temperature rises the alcohol flows into the tube, but since the liquid is unable to carry the dumbbell with it, leaves the dumbbell marking the lowest temperature. (Courtesy WeatherMeasure Corp., Sacramento, Calif.)

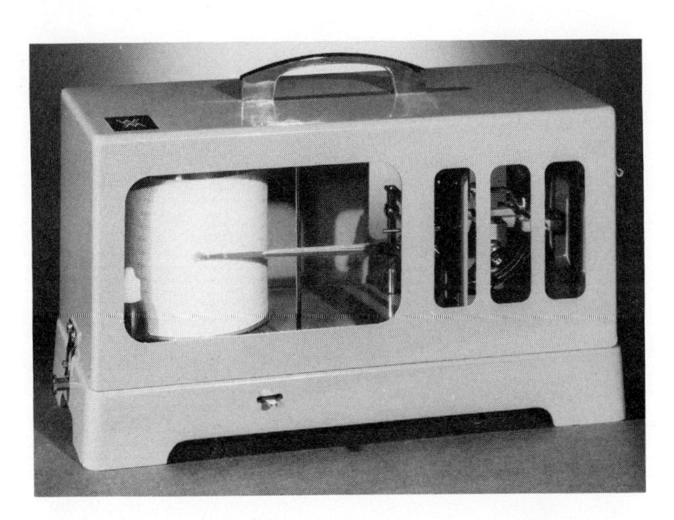

Fig. 27-14. Thermograph is used to provide a continuous record of temperature changes. The needle registers temperature-caused changes in a bimetallic strip on a rotating, paper-covered drum. (Courtesy WeatherMeasure Corp., Sacramento, Calif.)

Minimum thermometer. Just as the highest temperature of a given time period may be important, so may the lowest temperature. The minimum thermometer, constructed to record this data, resembles the maximum thermometer in physical appearance. However, the actuating element is usually colored alcohol or some other low-density liquid. Immersed in the alcohol is a small *glass dumbbell* that can travel through the low-density liquid with ease. When the horizontally placed thermometer is set, the glass *index* is resting against the edge, or meniscus, of the alcohol (Fig. 27-13). As the meniscus retreats with a drop in temperature, the index is dragged downward because the meniscus is too strong for the dumbbell to break through. When the temperature rises, the alcohol flows around the index, leaving it at its lowest point. Therefore the right side of the dumbbell indicates the lowest temperature reached, while the edge of the alcohol always records the current temperature.

Thermograph. Although the thermometers just discussed are excellent measuring devices, they do not record continuous temperature fluctuations. Therefore a *self-recording* temperature-measuring device called a *thermograph* was developed. The actuating element consists of either a sensitive bimetallic strip or a curved tube filled with liquid (a *Bourdon* tube). Both the bimetallic strip and the Bourdon tube react to temperature changes by bending slightly. Since this bending can be reproduced exactly for the same temperature changes, it can be converted by a series of linkages to temperature readings permanently recorded on a revolving chart. This instrument, however, is not as accurate as the standard air thermometer (Fig. 27-14).

28 *cycle of life*

Moisture in the atmosphere

I wield the flail of lashing hail,
And whiten the green plains under,
And then again I dissolve it in rain,
And laugh as I pass in thunder.

The Cloud
Percy Bysshe Shelley

For hundreds of years scientists sought without success the answer to a perplexing meteorologic problem: Where did water come from? The answer, however, baffled investigators, for they could not pinpoint its actual origin. For many years it has been known that water in the atmosphere came from the evaporation of surface water and that surface water in turn was the result of condensation from the moisture-bearing layers of air. This only led, however, to the familiar chicken-and-egg dilemma: Which came first?

It was not until recent times that a workable theory regarding the origin of water was presented. The scientific community largely agrees that water originated neither in the air nor on land. They now theorize that this most vital part of our life's existence began beneath the surface of the earth, in the crust and mantle.

Millions of years ago, when the earth was in its infancy, fiery outpourings of molten rock called *magma* were thrown onto the earth's surface by gigantic volcanic eruptions. Together with this magma came enormous quantities of steam that had been trapped in the hot molten material. The steam vaporized instantly because of the extremely high temperatures of the earth's environment. Eventually, the earth cooled sufficiently to allow raindrops to fall from the steamy, vapor-laden skies. This in turn allowed further cooling, and torrents of rain fell continuously for hundreds of years, completely covering the earth with water. Here water played its most important role, as the cradle of life itself.

PHYSICAL PROPERTIES OF WATER

At normal atmospheric pressure water is the only known substance that can occur in three separate forms—solid, liquid, and gas. Water is the most abundant substance found on or near the earth's surface.

Liquid water in its pure form is an odorless, colorless, and tasteless substance. However, natural water has a bluish-green color, and its taste varies according to the presence of mineral content and impurities. Liquid water is present in the atmosphere as *rain, drizzle, dew,* or *cloud droplets*. Water covers nearly 75% of the total surface of the earth.

The solid state of water, called *ice,* occurs very easily in the atmosphere as *snow, hail, sleet,* and *ice crystals* in clouds. On the surface of the earth ice is commonly observed as glaciers, frozen ponds or lakes, and frost.

Water vapor, the gaseous form of water, is the only state that is invisible. It mixes easily with the other gases in the atmosphere and is an important variable in weather. Immense quantities of heat energy can be stored by this gas. When released, it becomes the driving force behind such devastating storms as hurricanes, thunderstorms, and tornadoes. The circulation of this energy by the wind distributes heat throughout the world, tempering the cooler regions in the far northern and southern latitudes while moderating the torrid temperatures of the equatorial regions.

Changes of state

As previously mentioned, water in the atmosphere or on the surface can easily change from one state to another. This change is accomplished by the addition or subtraction of heat, which governs the movement of the water molecules. In some instances, especially the liquid-to-gas transformation, pressure also plays an important role. In this case a decrease in pressure allows the change to take place with less energy. This is why water boils at a lower temperature at higher altitudes.

Fusion. The change of state from a solid to a liquid is called fusion or, more commonly, *melting*. It takes place under normal atmospheric pressure at 32° F (0° C). When heat is absorbed by ice, the water molecules within the solid ice start to move more rapidly. This increased vibration causes the molecules to slip freely over one another, and the solid liquifies. The amount of heat needed to melt enough ice to yield 1 gram of water without raising its temperature is 80 *calories* (cal).*

Solidification. The process by which a liquid changes into a solid is called solidification, or freezing. It is the reverse of fusion and occurs at the same temperature. When the temperature of water decreases, contraction takes place as the water molecules slow down to a speed such that each molecule cannot pull away from the attraction of others around it. At 39.2° F (4° C), however, a slight expansion occurs and continues until the freezing point of 32° F (0° C) is reached; at this point a slight contraction occurs again. This expansion during freezing is peculiar to only a few substances and accounts for the fact that the density of ice is less than water. That is why ice floats and why, as water freezes at the surface of a lake (where the water interacts with the colder temperatures), the ice does not sink. We say, then, that water freezes from the top down. In order to prove to yourself that expansion does take place during freezing, look at a tray of ice cubes and note the small raised area on the upper surface of each cube.

The amount of energy released during solidification is exactly the same as that absorbed by fusion: 80 cal/gm. This release of heat on freezing enables farmers to use the sprinkling method for frost protection. As the fine water droplets freeze, heat is liberated into the air, lessening the drop in temperature and thus the danger of frost.

Evaporation. As energy is added to liquid water, the molecules speed up. Eventually, due to the collisions among adjacent molecules, speeds are attained that allow the molecule to escape into the air. (Thus only the slower-moving molecules are left in the liquid, and slower-moving molecules result in a lower temperature and cooling.) This cooling process is the reason for the cool feeling when water evaporates from the skin. Evaporation of water results in the absorption of 539 cal/gm of heat, which is now stored in the *water vapor*. The process will be explained in greater detail later in this chapter.

Condensation. The process that is the reverse of evaporation is called condensation. As energy is taken away from water vapor, its temperature decreases and the molecules slow down greatly. Now they have a greater attraction for one another and eventually, at a temperature that depends on pressure, the water molecules start to stick to one another until liquid water appears. Although there is still considerable motion and the molecules continue to slide over one another, they are also constantly colliding with and temporarily sticking to several neighboring molecules.

Sublimation. On a dry, windy day when the temperature is well below freezing, ice and snow can disappear without melting to form water. Blocks of dry ice grow smaller as carbon dioxide gas is given off without a liquid in evidence. At ordinary room temperature, napthalene (moth balls) disappears without leaving a liquid. Evidently, something must be acting on the solids so that they skip over the liquid stage while going from a solid to a gas. Such a transformation, or its exact reverse, is called sublimation.

Dry Ice, for example, is made by subjecting gaseous carbon dioxide to high pressures. The result is a solid that formed without going through a liquid stage. One of the most common sublimation processes that occurs in the atmosphere involves the formation of ice crystals at very high levels. These crystals eventually form snowflakes, which fall to the surface in colder climates.

Factors affecting changes of state

Temperature and pressure

By plotting points corresponding to a change in the state of water at different temperatures and pressure, a *phase diagram* for water can be drawn. This useful diagram shows visually how changes in temperature and pressure affect the changeover points of solids, liquids, and gases (Fig. 28-1).

At standard temperature and pressure (0° C, 760 mm Hg) the fusion and solidification of water takes place. The change that takes place is dependent on the direction in which the temperature of the substance is moving. It should be noted from the phase diagram, however, that a large pressure decrease to 4.6 mm Hg only results in 10% difference in temperature. Therefore pressure does not play an important role.

When the temperature is increased at a constant pressure of 760 mm Hg, water remains in a liquid stage until it reaches a temperature of 212° F (100° C). At this point the gas, or water vapor, appears. The diagram indicates that by decreasing the pressure the temperature required for water to turn into a gas is also greatly decreased.

There is one point on the diagram at which the two lines separating the three phases meet. This location is

*A calorie is the unit of measure for heat energy: 1 cal is the energy needed to raise the temperature of 1 gram of water 1° C.

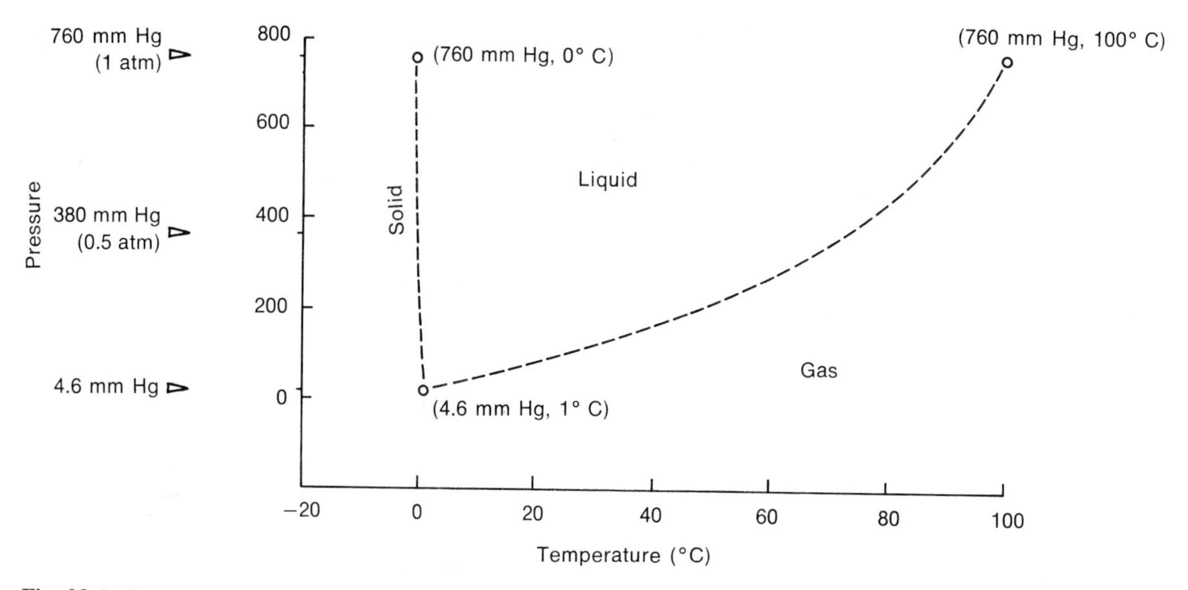

Fig. 28-1. Phase diagram showing the relationship between pressure and temperature in bringing about changes of state in water. Pressure has very little effect on the solid-liquid change, as even a drop from 760 to 4.6 mm Hg requires very little temperature change to compensate for it. In the liquid-gas change, on the other hand, every time pressure rises, a greater temperature is also required to effect the change.

called the *triple point* of water and clearly points out that at 4.6 mm Hg of pressure and 32.18° F (0.1° C) water can exist in equilibrium as a solid, liquid, and gas at the same time.

Temperature-moisture relationship

The amount of water vapor that a parcel of air can hold is dependent on the temperature of the air. *The higher the temperature, the more water the air can hold.* This is the case because an increase in temperature causes the molecules to move faster, so they cannot return to the liquid state. Eventually a point is reached at which the number of water molecules that are returning to the water surface is equal to the number that are escaping. At this equilibrium point the air is said to be at *capacity*. The capacity point is reached when air is holding as much water vapor as possible for a particular volume and temperature. If the temperature is once again raised, the molecules move faster, the capacity point increases, and more water can be evaporated into the volume of air.

This temperature-moisture relationship is not constant. The air's capacity increases at a much greater rate at higher temperatures (Table 25).

When capacity is reached, the air is said to be *saturated*. Based on the previous discussion, we can say that the saturation point of a parcel of air can be reached in one of two ways: either enough water can enter the air through the evaporation-sublimation process or the temperature of the air can be lowered, thus reducing the air's capacity to hold water.

Table 25. Water vapor values of air at capacity per cubic foot of air

Temperature (°F)	Water vapor (oz/ft³)
20	0.002
30	0.004
40	0.006
50	0.008
60	0.012
70	0.016
80	0.022
90	0.030
100	0.040

The *increase of water vapor* in the air results in the parcel holding as much water as it can possibly hold. Fig. 28-2, *A*, depicts a parcel of air in an unsaturated condition. Its temperature is 60° F and although it is capable of holding 5.7 grains of water vapor (1 grain = 0.002 oz), the amount it is actually holding is 2.7 grains. If the temperature of the air remains constant and more water is evaporated into the air, the parcel will eventually hold 5.7 grains of water

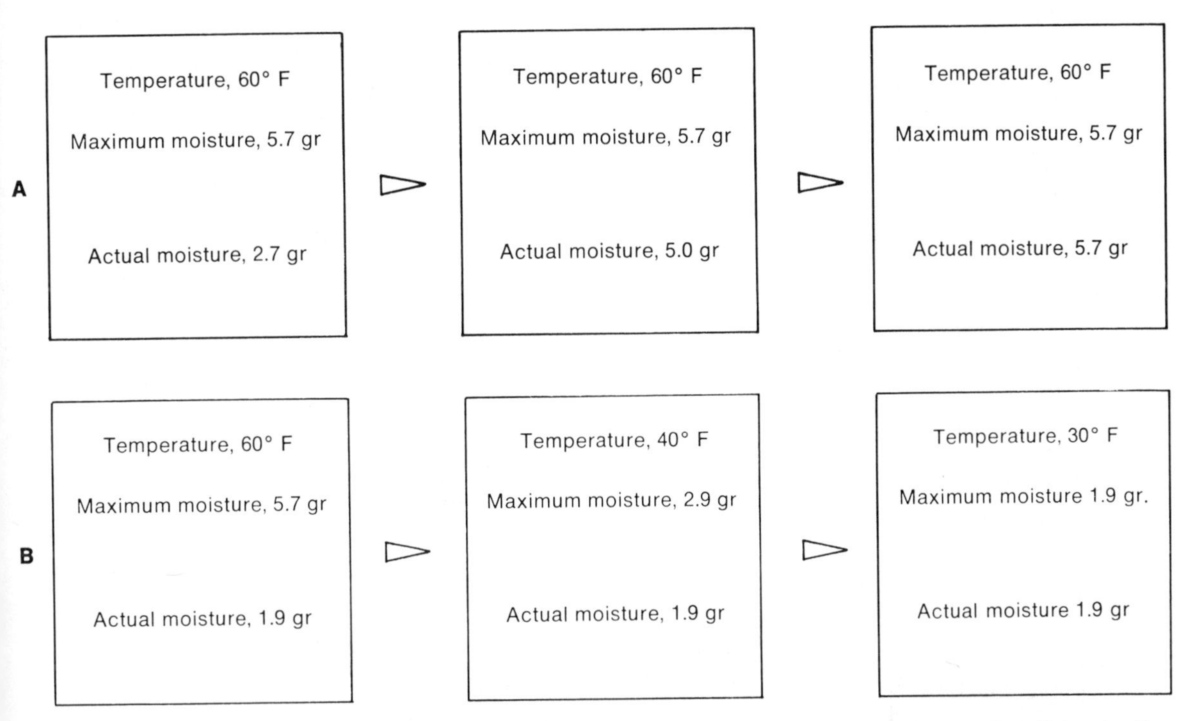

Fig. 28-2. Two ways of bringing about saturation. **A,** Amount of water vapor present is increased until the cloud's capacity is reached. **B,** Decreasing the temperature and thereby the amount of water the cloud can hold until it reaches the amount of water present in the cloud.

and its maximum will have been reached. At this point the air will be saturated.

The second method of reaching saturation is by lowering the air temperature. As the parcel becomes cooler its capacity to hold moisture decreases until the amount of moisture in the air equals the capacity point. Fig. 28-2, *B*, shows a parcel of air again in an unsaturated condition. The actual amount of moisture it is holding is less than its maximum. However, at 40° F (4° C) the capacity is low enough to be equal to the actual amount of water in the air.

Lowering the air temperature is the more important of the two methods in the atmosphere. This natural process of cooling is achieved primarily by rising air.

Terms used to discuss moisture

The amount of water vapor in the air at any one instant may be measured and then expressed in several ways.

Dew point temperature. Dew point temperature is that temperature to which the air must fall in order for it to become completely saturated. A parcel of air with a temperature of 60° F (15° C) has a dew point temperature of 30° F (−1° C) (Fig. 28-2, *B*). In other words, the present air temperature must fall to 30° F (−1° C) before the air will become saturated. If the air is further cooled, to 25° F (−4° C), for example, the vapor will condense into tiny water droplets as the maximum capacity of the air is exceeded. Under normal conditions, the dew point can never be higher than the temperature of the air.

Relative humidity. Relative humidity is one of the most commonly used concepts for the measurement of moisture in the atmosphere. Relative humidity is a concept that is easy to understand, for it is a percentage *ratio* of the amount of moisture in the air compared to the actual amount the air *could* hold at that same temperature and pressure. The relative humidity (RH) in the air in Fig. 28-2, *B*, at 60° F (15° C) is:

$$\text{RH} = \frac{1.9 \text{ grains}}{5.7 \text{ grains}} \times 100 = 33\tfrac{1}{3}\% \ (60° \text{F})$$

It must be kept in mind, however, that the ratio must be at the same temperature. If the temperature changes and the actual amount of moisture remains the same, the relative humidity will change:

$$\text{RH} = \frac{1.9 \text{ grains}}{2.9 \text{ grains}} \times 100 = 65\tfrac{1}{2}\% \ (40° \text{F})$$

Thus when the temperature of the air decreases, the relative humidity increases.

Relative humidities range from near 0% over desert areas to 100% in saturated air. Under ordinary conditions, relative humidity cannot be greater than 100%.

Absolute humidity. Absolute humidity is the weight of water vapor in a given volume of air. It is measured in grams per cubic meter or grains per cubic foot and has the same dimensions as the air's density (mass/unit volume). Absolute humidity is affected by changes in the amount of moisture in the air and by changes in pressure; the latter is the result of expansion or contraction of air. Because of this change in volume, the humidity will change even though no moisture has added or subtracted from the air. Thus absolute humidity is not the most desirable or accurate means of expressing or measuring atmospheric moisture, although it is more conservative than relative humidity.

Specific humidity. Specific humidity is the weight of water in the air compared to the unit weight of the entire parcel and is usually expressed in grams of water per kilogram of air. This is the most conservative of all the means of expressing measurements of moisture, for it is not affected by changes in temperature or pressure. Only an increase or decrease of moisture will change specific humidity.

Vapor pressure. Vapor pressure is the pressure exerted by the water vapor in the air. All gases have weight, and gaseous water, or water vapor, is no exception. It exerts a small force over some area and therefore contributes to the total atmospheric pressure. The vapor pressure depends on how much water is in the air at any one time. When the air is holding a maximum amount of water at any one temperature it is also exerting a maximum vapor pressure for that temperature. This is called the *saturation vapor pressure*. It is quite obvious that temperature affects the saturation vapor pressure because the higher the temperature the more water the air can hold and thus the higher the saturation vapor pressure.

Table 26. Temperature and saturation vapor pressure relationships

Temperature (°F)	Pressure (inches of mercury)
40	0.248
50	0.362
60	0.522
70	0.739
80	1.032
90	1.422
100	1.933

Since relative humidity is a ratio of actual to maximum conditions, it can also be determined by using a ratio of vapor pressure to saturation vapor pressure (Table 26). For example, to calculate the relative humidity if the air temperature of a parcel of air is 70° F and the dew point is 50° F, one could use the information in Table 26 as follows:

Vapor pressure at 50° F = 0.362 inch
Saturation vapor pressure at 70° F = 0.739 inch

$$\text{RH} = \frac{0.362 \text{ inch}}{0.739 \text{ inch}} \times 100 = 48.8\%$$

DEVICES FOR MEASURING HUMIDITY

Instantaneous measurements of the amount of moisture in the air close to the earth's surface can be made by an instrument known as a *sling psychrometer*. This instrument relates the difference in temperature, caused by moisture in the air, to the relative humidity and dew point.

The instrument consists of two identical mercury thermometers attached at two different levels on a strong metal backing (Fig. 28-3). The lower thermometer bulb is encased in a muslin wick that is saturated with water. The instrument is whirled in a circle, which allows water to evaporate from the wick. The temperature of the first thermometer (the air temperature) always remains the same and is called the *dry-bulb temperature*. As evaporation takes place, the temperature of the other thermometer decreases due to the absorption of *latent heat of evaporation*. The reading at its lowest point is termed the *wet-bulb temperature*. (The wet-bulb reading can never be higher than the dry-bulb reading.) Using a psychometric table such as that reproduced in Table 27, computed values for dew points and relative humidities can be derived.

For example, if the temperature of the air is 70° F (21° C) and after evaporation the wet bulb reads 62° F (17° C), what is the dew point temperature?

70° F (Dry-bulb reading)
−62° F (Wet-bulb reading)
8° F (Difference)

Therefore, using data from Table 27, the dew point would be 57° F.

An instrument used to make a continuous record of atmospheric moisture is called a *hygrograph*. The sensing element usually consists of several strands of blonde human hair, although other fibers sensitive to humidity changes can be used. As the moisture content of the air increases, the hair expands. This expansion is reflected by a series of

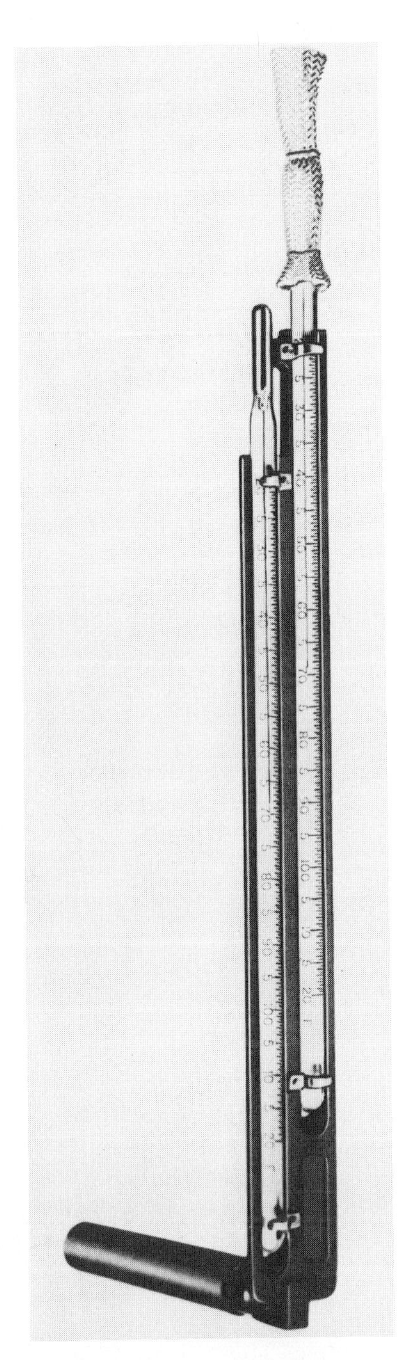

Fig. 28-3. Sling psychrometers like this one are used to determine the dew point of a given body of air. The lower thermometer bulb is wrapped in wet muslin. When the psychrometer is spun, the water in the muslin evaporates, allowing the thermometer to measure the amount of heat given off. The dry-bulb thermometer simply measures the temperature of the air. A comparison of the two readings can be compared to a psychrometric table to give the dew point of the particular body of air. (Courtesy WeatherMeasure Corp., Sacramento, Calif.)

Fig. 28-4. Hygrographs make a continuous record of changes in the moisture content in the air by registering changes in a human hair or other moisture-sensitive filament. (Courtesy WeatherMeasure Corp., Sacramento, Calif.)

springs and levers to a pen, which inscribes a trace of humidity reading on a rotating chart (Fig. 28-4).

EVAPORATION PROCESS

From a physical standpoint, evaporation is simply the transformation of a liquid into a gas. In the atmosphere, however, this mechanism takes place when the vapor pressure of the liquid exceeds the vapor pressure in the atmosphere. When the water molecules escape into the air, they displace air molecules; this results in an increase in partial pressure due to water vapor. Eventually, the vapor pressure of both water and air will be equal. This is a state of equilibrium in which as many water molecules enter the liquid as leave it. At this point the air is said to be *saturated* and exerts the force called the *saturation vapor pressure*.

During rain showers, evaporation takes place as rain falls, and the energy used in the process comes from the atmosphere itself. The raindrop thus becomes cooler and decreases the temperature of the surrounding air. This is why summer showers bring relief from oppressively high temperatures.

CONDENSATION PROCESS

As stated earlier, condensation is the change from a gaseous to a liquid state. *Sublimation* was defined as a change from a gas to a solid or solid to a gas without passage through the liquid stage. The factor that determines whether condensation or sublimation will occur is the temperature of the air at the time at which the change takes place.

Table 27. Saturation vapor pressure and temperature of dew point*

Air temperature (°F)	Saturation vapor pressure (inches of mercury)	Wet-bulb depression													
		1	2	3	4	6	8	10	12	14	16	18	20	25	30
0	0.038	−7	−20												
5	0.049	−1	−9	−24											
10	0.063	5	−2	−10	−27										
15	0.081	11	6	0	−9										
20	0.103	16	12	8	2	−21									
25	0.130	22	19	15	10	−3	15								
30	0.164	27	25	21	18	8	−7								
35	0.203	33	30	28	23	17	7	−11							
40	0.248	38	35	33	30	25	18	7	−14						
45	0.298	43	41	38	36	31	25	18	7	−14					
50	0.362	48	46	44	42	37	32	26	18	8	−13				
55	0.432	53	51	50	48	43	38	33	27	20	9	−12			
60	0.522	58	57	55	53	49	45	40	35	29	21	11	−8		
65	0.616	63	62	60	59	55	51	47	42	37	31	24	14		
70	0.739	69	67	65	64	61	57	53	49	44	39	33	26	−11	
75	0.866	74	72	71	69	66	63	59	55	51	47	42	36	15	
80	1.032	79	77	76	74	72	68	65	62	58	54	50	44	28	−7
85	1.201	84	82	81	80	77	74	71	68	64	61	57	52	39	19
90	1.422	89	87	86	85	82	79	76	73	70	67	63	59	48	32
95	1.645	94	93	91	90	87	85	82	79	76	73	70	66	56	43
100	1.933	99	98	96	95	93	90	87	85	82	79	76	72	63	52

*Data are based on a barometric pressure of 30.00 inches.

In order for condensation to occur, three ingredients must usually be present:

1. Condensation nuclei
2. Cooling of the air
3. Adequate quantities of water vapor

Condensation nuclei are the abundant microscopic particles that float in the air. They may be salt particles, smoke, pollutants, dust, or even volcanic ash. The particles have a chemical affinity for moisture and are therefore called *hygroscopic*. These abundant nuclei absorb water even before the relative humidity reach 100%. When this process occurs, the newly formed droplets are about the size of drizzle drops, and soon saturation is reached (100% relative humidity).

There are rare cases in which cooling and moisture are adequate but there are no condensation nuclei. In such a situation, the air can hold enough water so that the relative humidity is greater than 100%. This is called *supersaturation* and is only a short-lived condition.

Condensation products

There are four major forms of condensation. Keeping in mind that condensation results in only very small droplets, it should be obvious that this is not the process by which rain forms. The only direct end products of condensation are dew, frost, fog, and clouds. With the exception of frost, which is formed by sublimation, all of these types of condensation form in the same manner. It is only the relative position of the observer to the formation point that separates each category.

Dew and frost. The air in contact with the earth cools during the evening by means of conduction. If this thin layer cools sufficiently, the dew point of the air is reached (100% relative humidity) and small water droplets condense out and are deposited on available surfaces. This deposition is called dew. If this same condition exists when the air temperature is below freezing, sublimation takes place and tiny *ice crystals,* or frost, form instead of water droplets. Both processes happen most frequently on calm, clear, cool evenings.

Fog. A second type of condensation is fog, which is an accumulation of minute water droplets suspended in the air near the earth's surface.

Since there are two primary methods of reaching the saturation point, and since fog only occurs when the air is saturated, there must be only two main categories of fog;

that produced by cooling and that produced by evaporation.

COOLING FOGS. Cooling fogs are formed when the air has been cooled so that it reaches saturation (100% relative humidity). Cooling fogs are generated only over land, since a cool underlying surface is needed. There are four general methods by which this cooling can take place.

1. Air touches a colder surface (conduction)
2. Air itself cools (radiation)
3. Warm and cold saturated air collide (frontal mixing)
4. Air rises (adiabatic expansion)

Some of the more common cooling fogs include the radiation, inversion, upslope, and advection types.

Radiation fogs usually occur during the early morning hours after the earth has had a chance to lose the heat of the previous day by radiating it into the air during the night. (The air touching the surface is also cooled by conduction.) Radiation fog is most frequently found in low-lying areas close to the ground; hence its other name, *ground* fog. The meteorologic criteria needed for ground fog to occur are provided by a calm, clear, cool, moist night.

Inversion fog is a fog that forms when cold, moist air is trapped below an *inversion* layer (warm air above cooler air). Radiational cooling takes place every night, so that the layer at the surface eventually cools to its saturation point. Therefore this type of fog is essentially a nighttime phenomenon, for during the day the fog rises due to heating; it is then classified as a low cloud.

Upslope fog is formed by adiabatic expansion as air flows up over higher terrain. Because of the expansion of rising air, the temperature of the ascending parcel cools to its saturation point. Mountain fog is usually upslope fog.

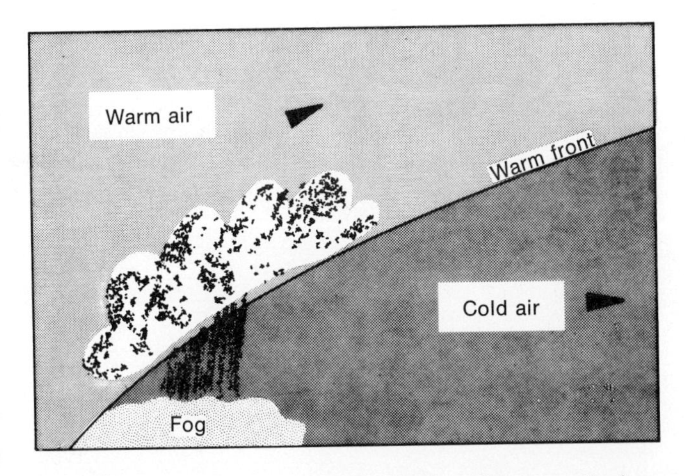

Fig. 28-5. Formation of a precipitation fog. The warm air rises over the cooler air mass, forming clouds at the zone of contact. If the clouds produce precipitation, it falls through the cooler air, producing fog.

Advection fog is formed when cool, moist air moves over a colder surface. Usually this type of fog is formed near the coasts of continents as cool, moist ocean air moves over colder land.

EVAPORATION FOGS. Evaporation fogs occur when enough water evaporates into the air to cause the air to reach the saturation point. Two of the more important evaporation types are precipitation and steam fogs.

Precipitation fog, a most important continental fog, is formed when warm air rises over humid cold air (as in a warm front). This in turn forms clouds. If the clouds produce rain, the raindrops then fall through the colder air and start to evaporate, allowing the cold air to reach saturation and form a fog (Fig. 28-5).

Steam fog results when cool, drier air overlies a warmer body of water. Water evaporates immediately so that the relative humidity reaches 100%. The water vapor does not condense, especially in the layers closer to the ground. Cooling takes place in the upper layers, and upward-moving moisture then becomes visible. It is because the movement of visible water droplets looks like steam from a boiling kettle that this is called steam fog. This type of fog is commonly found wherever cold, dry air moves over a warmer water surface, as in the Arctic regions and over ponds and other small bodies of water in the early morning hours.

Smog. The original meaning of smog, as defined by De Voeux in 1905, was a mixture of smoke and fog. Recently, however, the word has been used to indicate a much broader encroachment on the atmosphere. As a matter of fact, people commonly use the term "smog" to describe situations in which there is no natural fog. This is especially true over large urban areas where visible air pollutants are prevalent.

Conditions favorable to smog formation are the same as those for the formation of an inversion fog. The *aerosols* present in the atmosphere where the droplets are formed, however, originate from factories or automobile exhausts. These pollutants become the nuclei, or centers, of each water droplet, while other droplets convert harmful exhaust gases into deadly acids. The infamous smogs of Los Angeles, California, and the Meuse River Valley in Belgium are good examples of smogs formed by inversions.

Clouds. A cloud is a visible aggregate of minute particles of water or ice, or both, in the air. Again, a cloud is the product of the condensation process (or sublimation when temperatures are below freezing), just as are frost, dew, and fog. However, cloud formation takes place right above the earth's surface in the free atmosphere.

So many varied forms of clouds can be seen in the sky that at first it might seem impossible to devise an acceptable

Table 28. Criteria for cloud classification

Height of base (ft)	Classification	Cloud type
20,000	High clouds	Cirrus, cirrostratus, and cirrocumulus
6500-20,000	Middle clouds	Altostratus and altocumulus
Above surface-6500	Low clouds	Nimbostratus, stratocumulus, and stratus
In low cloud range	Vertical development clouds	Cumulus and cumulonimbus

classification system. By observing cloud characteristics over many years, however, meteorologists have been able to classify clouds according to their *appearance* and *height*. Using these criteria, ten types of clouds can be described (Table 28).

Based on their appearance, clouds can be separated into three major types.

Cirriform	Streaked clouds
Stratiform	Layered or sheet clouds
Cumuliform	Puffy or heaped clouds

Based on height of the cloud at the base, three further divisions appear. See Table 28 for details.

In order to more fully understand the names of the clouds, the following root words should be defined:

Stratus	Layered
Cumulus	Puffy or heaped
Nimbus	Dark and rainy
Alto-	Prefix to indicate middle cloud
Cirro-	Prefix to indicate high cloud

Using these root words, all 10 cloud types can thus be described.

CIRRUS CLOUDS. There are three types of cirrus clouds.

CIRRUS: A *high, streaky* cloud made up of delicate filamented ice crystals. They are often feathery or fibrous in appearance. These highest of all clouds usually form above 25,000 feet (7625 meters) (Fig. 28-6).

CIRROSTRATUS: A *high, thin-layered* cloud with a smooth appearance, forming a delicate whitish veil or sheet. These layered ice crystal clouds are so thin that the sun or moon

Fig. 28-6. Cirrus clouds, because of their great height, are composed of tiny ice crystals, which gives them their delicate appearance. (Photograph by Norman Mendelson.)

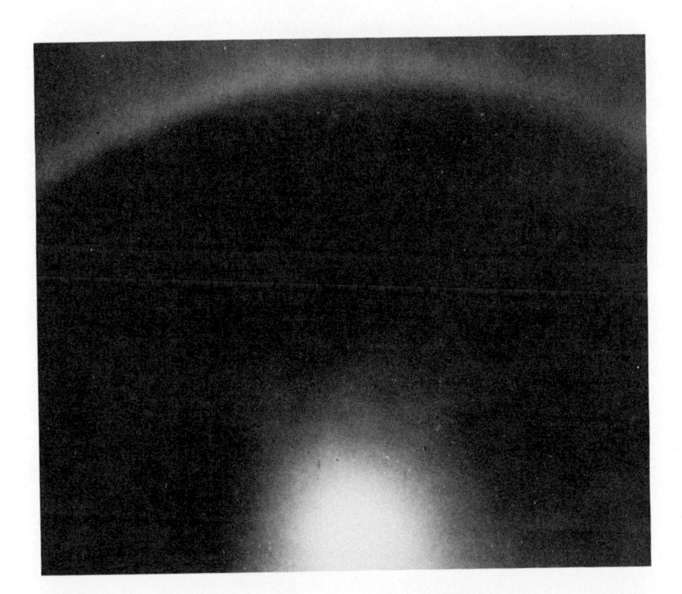

Fig. 28-7. These cirrostratus clouds can form a halo around the sun or moon, an occurrence often followed by rain. (Courtesy United States Department of Commerce, National Oceanic and Atmospheric Administration, Rockville, Md.)

is easily seen through them, forming a halo that to many amateur weathermen indicates rain within 24 to 48 hours (Fig. 28-7).

CIRROCUMULUS: A *high, thin,* whitish cloud composed of small globular puffs that look like sand ripples on a beach. These ice crystal clouds are usually banded in appearance and do not cover the entire sky.

STRATUS CLOUDS. There are three types of stratus clouds.

STRATUS: A *low,* rather uniform, grayish layered cloud that can sometimes be mistaken for fog. Usually only drizzle or snow grains fall from the stratus clouds (Fig. 28-8).

ALTOSTRATUS: A grayish, *middle-layered* cloud of uniform appearance. These clouds usually cover all or a large percentage of the sky, and the sun shows through only as a vague spot. Amateur meteorologists see this phenomenon as an indication of precipitation within 12 to 24 hours.

NIMBOSTRATUS: A *dark, shapeless* rain or snow cloud that is very gray in color. It is thick and ominous-looking and is associated with heavy, steady precipitation.

CUMULUS CLOUDS. There are four types of cumulus clouds.

Fig. 28-8. Stratocumulus clouds like these are often mistaken for fog because they lie so low over the ground in mountain areas. (Photograph by Norman Mendelson.)

Fig. 28-9. These cauliflower-shaped cumulus clouds float low over the ground and are usually associated with fair weather. (Photograph by Richard E. Jaffe.)

Fig. 28-10. This towering cumulonimbus cloud shows the anvil top characteristic of these thunderstorm clouds. (Courtesy United States Department of Commerce, National Oceanic and Atmospheric Administration, Rockville, Md.)

Fig. 28-11. Stratocumulus clouds like these often cover the whole sky in thick, threatening layers. (Photograph by Richard E. Jaffe)

CUMULUS: A *puffy, fair-weather* cloud with flat bases and cauliflowered tops. Many times they float across the sky only 500 to 1000 ft (152 to 305 meters) above the ground and are a brilliant white color (Fig. 28-9).

ALTOCUMULUS: A *middle cloud* composed of water droplets or ice crystals (depending on the temperature) that appears to be formed in *large, rounded puffs or rolls*. They are usually found in elongated, parallel bands and are sometimes called sheepback clouds because they look like a flock of sheep seen from above.

CUMULONIMBUS: A *dark, heaped rain* cloud with great vertical height. These towering masses usually form an anvil top characteristic of the well-known thunderstorm cloud. Usually, heavy showery precipitation and sometimes hail accompany the cumulonimbus cloud (Fig. 28-10).

STRATOCUMULUS: A low gray or whitish gray cloud formed by heavy *rolling masses* that are layered across the entire sky (Fig. 28-11).

PRECIPITATION TYPES

Although condensation may occur and tiny suspended water droplets or ice crystals may appear, precipitation does not necessarily have to be the result. The precipitation process takes place only if the particles grow sufficiently large for the pull of gravity to overcome the buoyant force of rising air currents.

Many varied types of precipitation are found throughout the world. Usually the form that the falling particle takes depends on the air temperature at the time of formation, the type of condensation nuclei in the environment, and the turbulence of the air.

Meteorologists use the term "hydrometeor" to describe in a general way all forms of precipitation or other products of the condensation process. Although there are 50 specific types of hydrometeors broken down into seven major categories, only the most commonly observed precipitation forms will be discussed here.

Liquid precipitation. Liquid precipitation is *drizzle* and *rain*. Drizzle consists of numerous small water droplets that appear to be floating in the air. Actually, they are falling very slowly and sometimes evaporate before they hit the ground. When this occurs, the drops should be called *mist,* although often the terms "mist" and "drizzle" are used interchangeably. Drizzle usually falls from low, stratus-type clouds.

Rain, which is the most common type of precipitation, is made up of larger water droplets. Because they are larger, they are not as numerous as those in drizzle. Rain normally falls from higher clouds such as nimbostratus and cumulonimbus; collisions of tiny water droplets are important in

Fig. 28-12. The weighing rain gauge measures the weight of the precipitation it collects and then converts it into inches. (Courtesy Science Associates, Inc. Princeton, N.J.)

raindrop growth and the higher the cloud, the better the chance of collisions.

One of the more common types of instruments used to record the amount of precipitation that has fallen is called a *weighing rain gauge* (Fig. 28-12). This instrument uses the weight of the rain or snow as a means of measuring the amount fallen. The weight is transferred by a series of linkages and a pen-arm system to a recording chart, which in turn converts the weight into inches of fallen liquid.

Freezing precipitation. Freezing precipitation results when droplets freeze on impact with a cold surface. Usually the water is *supercooled,* which simply means that although the temperature of the water is below freezing, the droplet is still in liquid form. Freezing drizzle creates a very light, invisible coating of ice known as *glaze,* which creates many problems for motorists and is quite dangerous to pedestrians. *Freezing rain* ends up as a thick, clear ice coating on all exposed objects. This *thick glaze,* more commonly called an *ice storm,* can snap tree limbs and break power lines (Fig. 28-13).

Solid precipitation. Snow, sleet, and hail are classified as solid precipitation. Although all have the same physical characteristics (solid, frozen, precipitation), they differ from one another in terms of formation, makeup, and effect. *Snow* is formed from the coalescence of many hexagonal ice crystals. These crystals can only exist in temperatures of 32° F (0° C) or less. In such a situation, sublimation and not condensation takes place, and gaseous water vapor becomes solid crystals of ice. It should be understood that the sublimation process is the only way in which ice crystals and hence snowflakes can form. Thus if a snowflake melts the water droplet can never again become a snowflake unless the water droplet evaporates and the process begins anew.

Sleet is really frozen rain. As water droplets leave a cloud and fall toward the earth, they may encounter an

Fig. 28-13. Despite the sparkling beauty of these glaze-covered trees, glaze brings more problems than delights, as the weight of this ice can snap tree branches and break power and telephone lines. (Courtesy United States Department of Commerce, National Oceanic and Atmospheric Administration, Rockville, Md.)

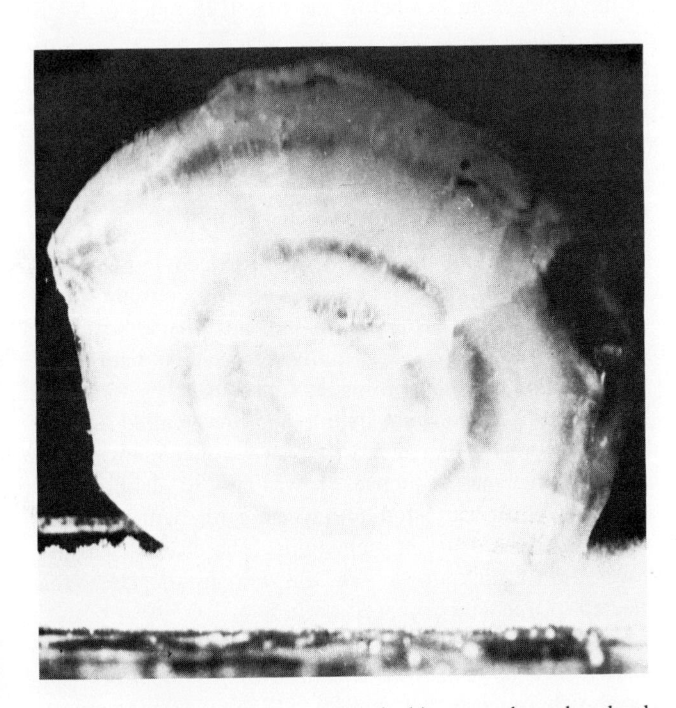

Fig. 28-14. Because of its many vertical journeys through a cloud, a hailstone has a layered structure rather like an onion. (Courtesy United States Department of Commerce, National Oceanic and Atmospheric Administration, Rockville, Md.)

air layer in which the temperature is below the freezing point. The droplets then freeze, becoming small clear pellets of ice, which accounts for the second name, *ice pellets,* given to sleet. Sleet can also form from large snowflakes that have melted on the way down and refrozen in a colder air layer.

The third representative of this group is *hail,* which is a circular piece of ice made up of concentric rings, which indicate that it was formed in layers. Hailstones usually begin as water droplets, which are carried aloft by the strong vertical currents of convective thunderstorm clouds. These clouds are a very favorable medium for hailstone formation because of their strong updrafts, great vertical height above the freezing level, and the enormous quantity of liquid water available. Hence, as the droplet passes above the freezing level, a small pellet of ice forms. This pellet is then forced downward by vicious downdrafts and once more water collects around the outside of the pellet. Then updrafts catch the particle and once again it is swept upward to form a second ring of ice. This process continues until the hailstone falls out of the cloud because its weight is greater than what can be kept aloft by the updrafts. A hailstone cut in half looks just like a divided onion (Fig. 28-14).

29 *the restless air*

Pressure and wind

Tell me how the wind can fare
On his unseen feet of air . . .

To Mother Nature
Frederic Lawrence Knowles

We live at the bottom of a massive ocean of air held close to the earth's surface by gravity. As we have learned in previous chapters, the *atmosphere,* as this ocean is called, extends for hundreds of miles above the earth's surface and is composed of many different gases. Due to the force of gravity, each gas has its own specific weight, which in turn exerts a force on the earth. Since the whole is the sum of all its parts, the combined gases making up the atmosphere must exert a tremendous downward force on the surface of the earth and everything on it.

Atmospheric pressure is expressed as the force per unit area, and every square inch of the earth experiences a pressure related to the weight of the air above it. At sea level this pressure is referred to as the *standard atmospheric pressure*.

It is possible to measure atmospheric pressure. Although the development of an instrument capable of weighing a column of air extending from sea level to the top of the atmosphere seems quite remote, such equipment was finally developed in the middle of the seventeenth century. With this instrument it was found that under normal conditions the pressure of this air column was 14.7 pounds per square inch (psi). This measurement is called the standard atmospheric pressure.

VERTICAL PRESSURE DISTRIBUTION

One of the more important concepts relating to the atmosphere is the *compressibility of air*. In the thermosphere the air is so thin that a cubic foot of space contains only a few molecules; for all practical purposes these are not measureable. At lower altitudes, on the other hand, pressure is great because of both the abundance of air molecules and the weight of the overlying air layers. Since it is known that pressure decreases with height, a discussion as to how this decrease takes place seems the next logical step.

In the lower layers of the troposphere (0 to 7000 ft or 0 to 2135 meters above the earth's surface) the pressure decreases at an almost uniform rate of 1 inch/1000 ft. Above 7000 ft, however, the rate of decrease accelerates rapidly (Table 29). Because of compressibility, one fourth of the atmosphere lies below the first 1.75 miles (2.5 km) of air. Half of the atmosphere is below 3.5 miles (5.5 km), although traversing the second half of the atmosphere would mean an upward journey of several hundred miles. In terms of distance, the vertical variations of pressure are much larger than horizontal pressure differences.

HORIZONTAL PRESSURE VARIATIONS

After inventing the barometer, Torricelli noticed that the level of the liquid did not remain at a constant height above the reservoir. This change indicated that atmospheric pressure must vary horizontally, especially with time. Many years later, meteorologists found that they were able to locate pressure systems by drawing lines (called *isobars*) between points of equal pressure on weather maps. Using daily weather map sequences, it was easy to identify these systems, which wandered around the earth, with pressures at set locations constantly changing. Therefore it was discovered that the atmosphere was a nonstatic body that reacted to imbalances placed on it.

Pressure around the earth varies at any given time, primarily because of the unequal heating of nonuniform surfaces. Other factors that contribute to this pressure variance include moisture content and deflection of air particles due to the earth's rotation. One of the best examples of the effect of differential heating is found near seacoasts. During

Table 29. Pressure changes for various altitudes based on standard atmosphere

Altitude		Pressure		
Kilometers	**Miles**	**Inches**	**Millibars**	**Centimeters**
0.5	0.3	29.92	1013.2	760
1	0.6	26.51	898.8	673
2	1.2	23.45	795.0	596
3	1.9	20.68	701.2	525
4	2.5	18.19	616.6	462
5	3.1	15.94	540.5	405
6	3.7	13.93	472.2	354
7	4.4	12.12	411.0	300
8	5.0	10.52	356.5	267
9	5.6	9.09	308.0	231
10	6.2	7.81	265.0	198
15	9.3	3.57	121.1	90
20	12.4	1.63	55.3	41
25	15.5	0.75	25.5	19
30	18.6	0.35	12.0	9
40	24.8	0.09	2.9	2
50	31.0	0.02	0.8	0.5
60	37.3	0.007	0.23	0.2
70	43.5	0.002	0.06	0.05
80	49.7	Trace	0.01	Trace
90	56.0	Trace	0.002	Trace
100	62.2	Trace	Trace	Trace

the day the land heats up more quickly than the adjacent body of water. The warmer air expands, becomes less dense, and rises. This less dense air exerts less pressure. Exactly the opposite occurs during the evening because the land loses heat more quickly than the water, contracts, and becomes more dense. This results in a higher pressure over the land, while lower pressures are found over the water.

Horizontal pressure patterns

There are two major pressure patterns that are closely related to weather changes. The first is a relatively stationary system over a specific geographic area, while the second pattern includes those systems that meander across the face of the earth.

The stationary systems are zonal in their arrangement; that is, they are arranged according to latitude and are found in both hemispheres. Their presence is the result of more solar radiation in the equatorial regions and less solar energy in the polar areas. Around the equator the hotter, less dense air forms a belt of permanent low-pressure systems known as the *Intertropical Convergence Zone* (where air converges from both north and south). At the poles the lower temperature produces a semipermanent high-pressure area called the *Polar High*. In the middle latitudes the thermal aspect of zonal pressure systems becomes less important, while the earth's rotation and dynamic equilibrium dominate. These mechanisms produce an irregular belt of low pressures at 65° north or south, termed the *Subpolar Lows*. At 30° north or south there is an irregular high-pressure zone called the *Subtropical High*. Fig. 29-1 is an idealized diagram of the earth's zonal sea level pressure.

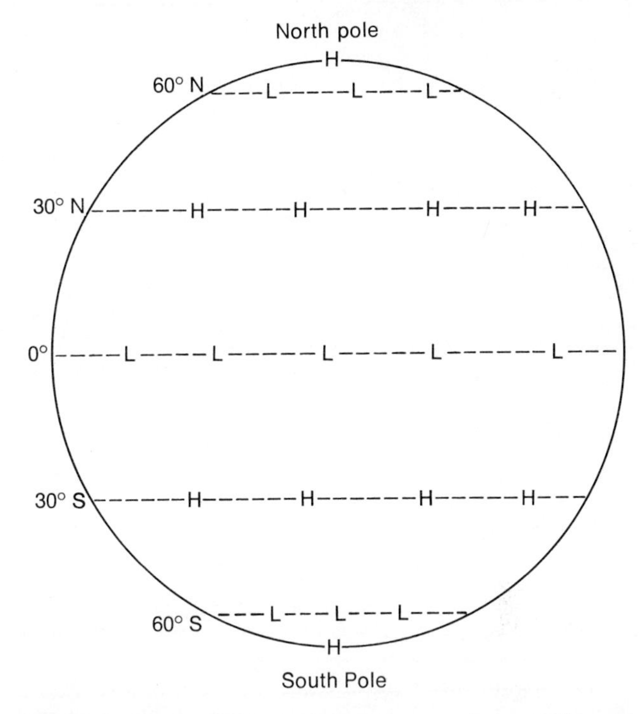

Fig. 29-1. Bands of high and low pressure as they would appear without interfering influences of land masses.

Due to the nonuniform character of the earth's surface, pressure belts in the midlatitudes break up into semipermanent cells that not only change in character but also move slightly during different seasons of the year (Fig. 29-2).

The second type of pressure pattern that influences weather conditions is the migratory system. In the middle latitudes, a series of high- and low-pressure areas constantly moves at varying speeds in a general west-to-east direction. Definite weather patterns accompany these systems and result in constantly changing weather within some of the most populated areas of the world.

By using *isobaric analysis,* which involves drawing lines connecting points of equal pressure on weather maps, systems can be followed and weather forecasts made. (Barometric pressure on such maps is expressed in terms of

A

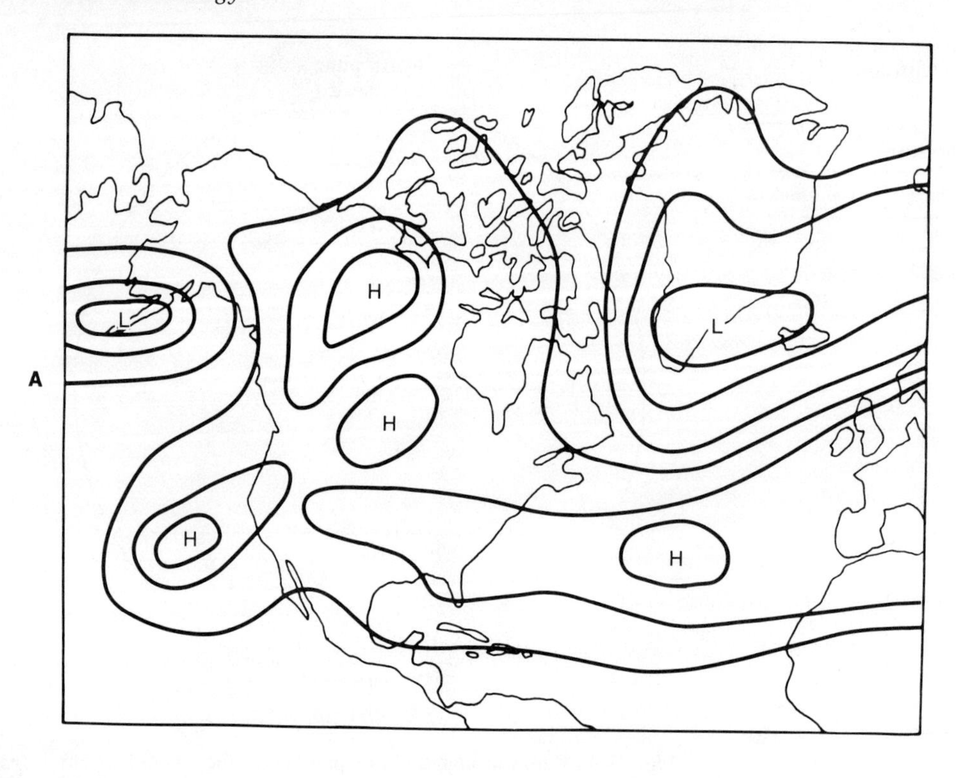

B

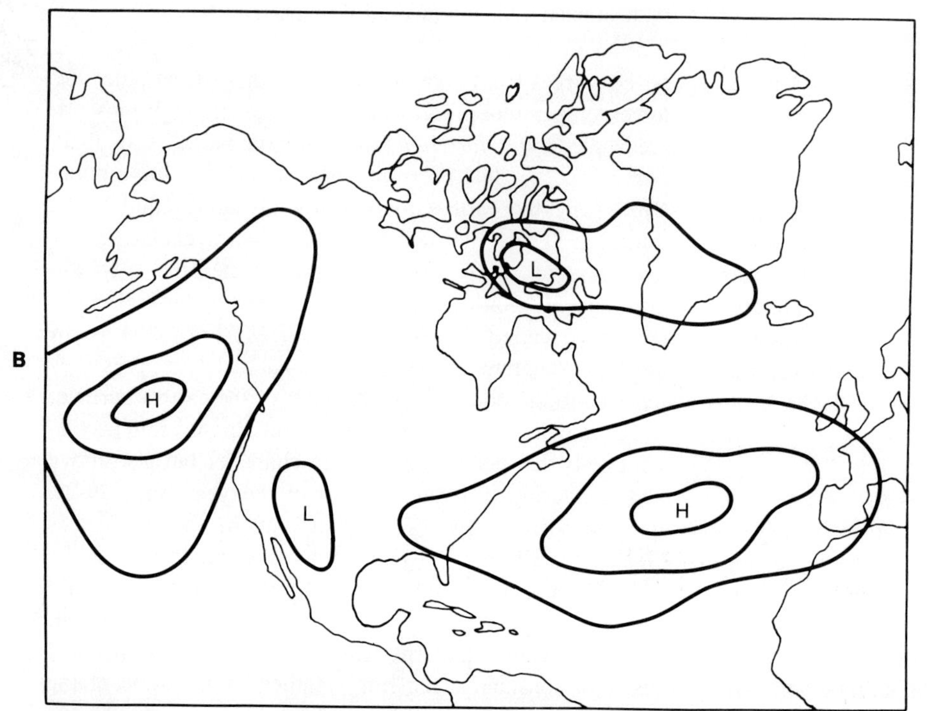

Fig. 29-2. Due to the varying characteristics of land and water, the ideal pressure belts shown in Fig. 29-1 do not exist. Instead, irregularly shaped belts form more-or-less permanent centers of high and low pressure. These centers do not remain constant, however, but migrate with the seasons. **A,** Pressure systems in January; **B,** location of pressure systems in July.

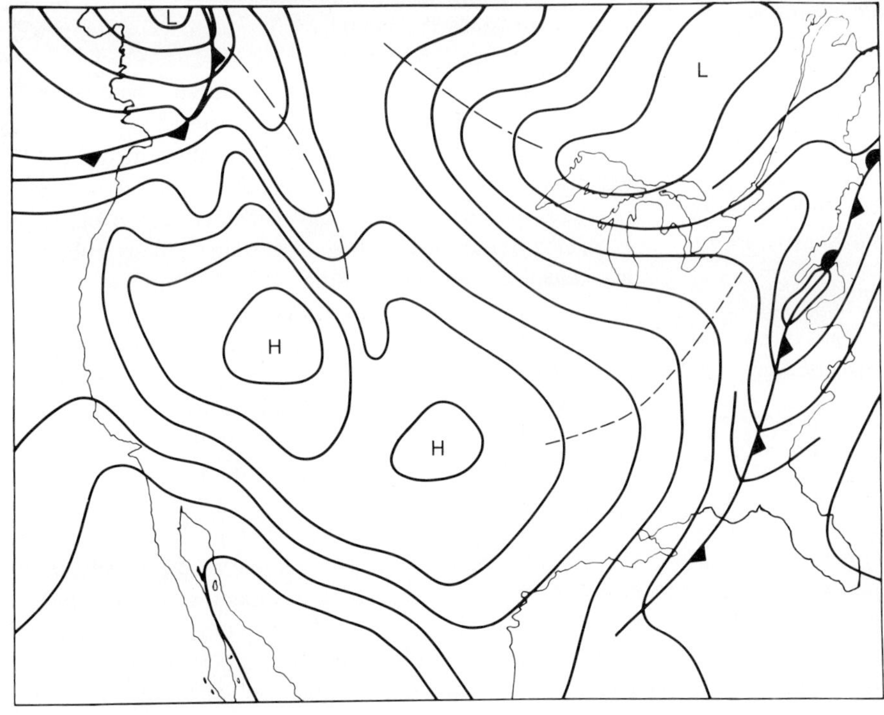

Fig. 29-3. A, Satellite photograph of atmospheric conditions. **B,** Meteorologists transfer this information onto the weather map you see in your daily newspapers. The superimposition of a map of the United States in **A** will help you coordinate the two maps. Notice how the cloud-covered areas generally conform to low-pressure areas, while high-pressure areas are clear. (Courtesy United States Department of Commerce, National Oceanic and Atmospheric Administration, Rockville, Md.)

millibars, a metric system pressure unit.) The following prominent pressure features appear on weather maps (Fig. 29-3).

High-pressure area. A high-pressure area is an elliptically shaped region of air covering hundreds of square miles. The central pressure is higher than that of the surrounding regions. Since the wind circulation around the high-pressure areas is anticyclonic (clockwise), these systems are called *anticyclones*. High-pressure areas are usually associated with pleasant weather.

A typical daily weather map with an accompanying cloud cover pattern taken from a weather satellite is shown in Fig. 29-3, *B*. Two strong high-pressure systems dominate the central and western parts of the United States, and the satellite pictures indicate vividly the clear skies associated with the system.

Low-pressure area. A low-pressure area is an elliptically shaped region of air similar to a high except that it is usually smaller and the central pressure is lower than that in the surrounding region. This *depression,* as it is called, has a *cyclonic* (clockwise) wind pattern and therefore it is commonly called a *cyclone*. Large areas of bad weather are usually associated with low-pressure systems such as those shown over the northeast and northwest sections of the United States in Fig. 29-3.

Trough. A trough is an elongated line of lower pressure coming out of a low-pressure area. It is common to find fronts and resultant cloudiness and bad weather along a trough line. Troughs without associated weather fronts are usually drawn as long dashed lines on weather maps (Fig. 29-3).

Ridge. A ridge is an elongated line of higher pressure coming out of a high-pressure area. Since ridges exhibit characteristics similar to those of highs, there is very little bad weather associated with them. In Fig. 29-3, a ridge indicated by a short dashed line runs from Mississippi to Pennsylvania.

Col. A col is a neutral point between two highs and two lows. Cols can also be the point of intersection of a ridge and a trough. Fig. 29-3 shows a col located between the two troughs in south central Canada.

INSTRUMENTS FOR MEASURING PRESSURE

One of the most accurate pressure-measuring instruments is the *mercurial barometer* (Fig. 29-4). Developed by Evangelista Torricelli in 1644, the barometer works on the principle of air pressure forcing a column of liquid such as mercury up a long, hollow glass tube. As atmospheric pressure changes, the height of the mercury in the tube changes due to the variable weight of the atmosphere. This change continues until the weight of the mercury column exerts a downward pressure equal to the downward force of the air on the reservoir of mercury. At standard atmospheric pressure, the height of this mercury column will be 29.92 inches (760 mm Hg).

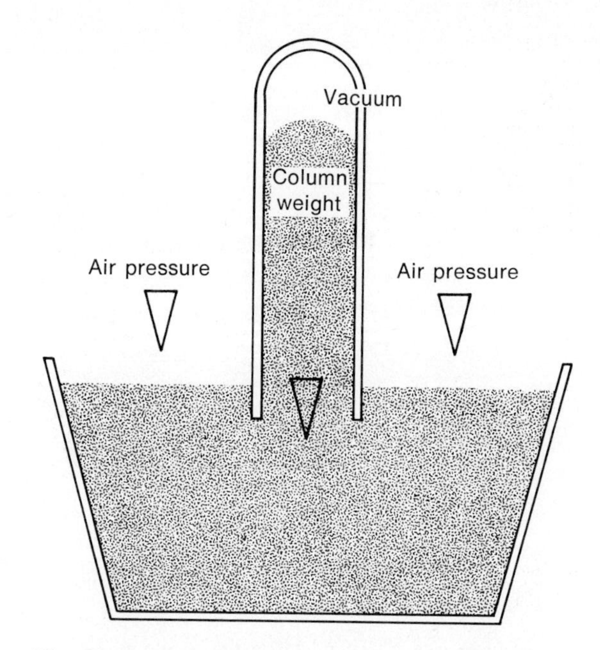

Fig. 29-4. Mercurial barometers work on the principle of balance. The column of mercury in the tube will stand still when the pressure of the air outside the tube on the mercury in the plate exactly balances the tube's own pressure on the mercury in the plate.

Because of its accuracy the mercurial barometer is widely used, but it has several characteristics that make it impractical in many situations. The instrument is both expensive and fragile, and numerous corrections for temperature, altitude, latitude, and manufacturing errors must be made prior to each reading. To eliminate these problems, a mechanical, fluidless instrument called the *aneroid barometer* was invented. This device utilizes a series of evacuated chambers that alter their shape as the surrounding air pressure changes. This change is then transmitted and magnified by a series of levers and pulleys to a pointer that is attached to a dial. The dial is scaled in the same units as a mercurial barometer; after calibration this instrument is fairly accurate (Fig. 29-5).

Special adaptations of the aneroid barometer have successfully been made. An example is the widely used constant recording pressure instrument called the *barograph*. Other than outside design, the only difference between the two instruments is that the barograph uses a recording pen instead of a pointer. This pen traces the

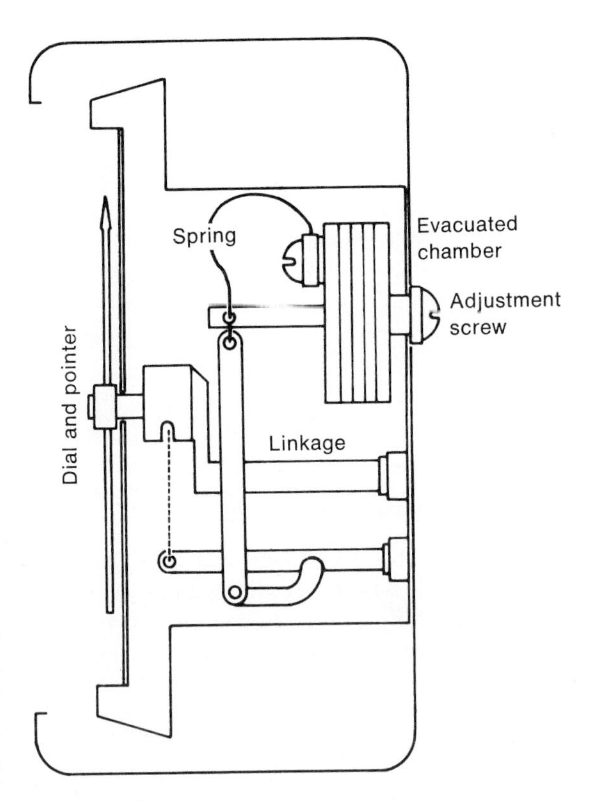

Fig. 29-5. Due to various problems with the mercury barometer, the aneroid barometer was invented. This instrument works on the same principle except that it measures the effects of pressure on the evacuated chamber rather than on a mercury column. The linkages magnify the effect and transfer it to the dial, where an accurate reading can be taken.

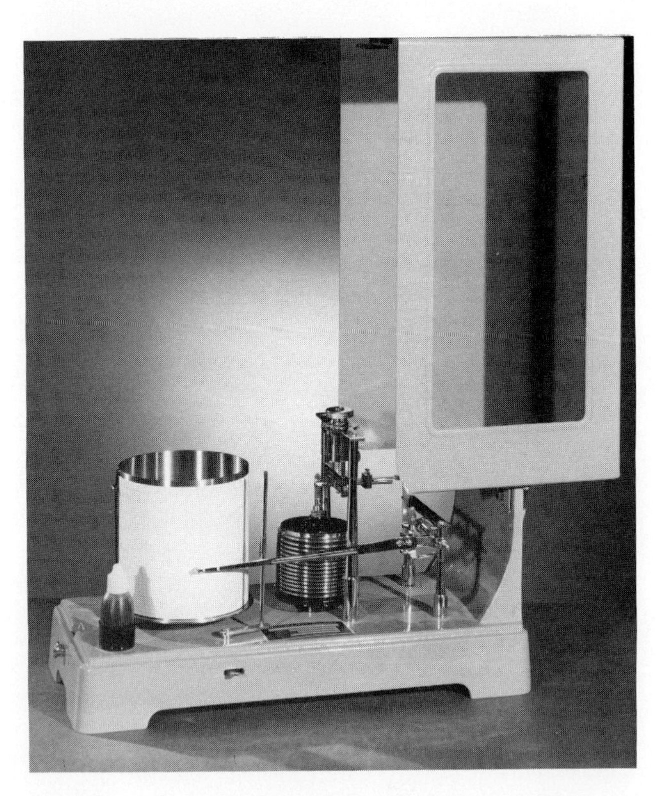

Fig. 29-6. Taking the aneroid barometer one step further, it is possible to construct a barograph like this one, which will not only take pressure readings over a period of time but record them on a paper-covered, rotating drum. (Courtesy WeatherMeasure Corp., Sacramento, Calif.)

pressure changes on a graph attached to a rotating drum (Fig. 29-6) so that a permanent record can be kept.

A second important type of aneroid barometer is the *altimeter*. Other than the altimeter dial, which is calibrated in feet above sea level, there is no difference between the two instruments. Altimeters are extensively used in aircraft and often by mountain climbers to measure altitude. Because of the nature of the instrument, however, slight errors in altitude must be expected, since readings are based on a nonexistent standard decrease in pressure with height.

BASIC CONCEPTS OF WIND FLOW

Due to the many forces acting on it, the atmosphere is constantly in motion. Since these forces tend to create an imbalance, the air's movement is nature's way of restoring equilibrium. Air therefore continues to move until a balance is once again achieved within the atmosphere.

All the physical laws governing the movement of particles can also be applied to the atmosphere. Most important are the three laws of motion developed by Sir Isaac Newton, a noted eighteenth-century physicist. Although these laws were discussed in Chapter 1, a brief review here will serve as an excellent introduction to the behavior of the atmosphere.

Newton's first law, *the law of inertia,* states that a body at rest or in motion will remain that way unless an outside force acts on it. Thus parts of the atmosphere at rest will remain calm unless an outside force acts on the air.

Newton's second law, *the law of acceleration,* asserts that the change in motion of a body relates directly to the force acting to move it. Thus a body will move in a straight line in the same direction as the applied force that caused it to accelerate. In the atmosphere, air particles will move in a straight line if one or more forces produce the motion.

The third law, called *the law of action and reaction,* states that whenever a force is exerted on a body, an equal and opposite force is exerted by that body on the first. Therefore in the atmosphere there cannot exist just one force or one body. There must be at least two forces or bodies in order for air to move.

PRIMARY FORCE AFFECTING WIND

The term "wind" can be defined simply as air in motion relative to the earth's surface. However, this definition implies both horizontal and vertical movements. Meteorologists therefore make a definite distinction regarding air flow. They classify horizontal movement as wind and vertical movement as *convection,* or *turbulence*. Meteorologically, then, wind pertains only to horizontal movements of air.

Although many theories and models of atmospheric motion have been proposed, meteorologists unfortunately do not yet have a full understanding of these complex motions. On the other hand, it is known that there are forces acting on air particles that do indeed result in both wind flow and a general circulation pattern.

The major force responsible for initiating horizontal air movement is called the *pressure gradient force*. (Gravity and heating are responsible for vertical flow.) This force results from differences in atmospheric pressure, which in turn are produced by differential heating (Chapter 28). As land and water areas, dark and light soil, forested and desert areas receive solar energy, they heat up at different rates. This differential heating causes differences in air temperature and ultimately differences in air pressure as well.

As pressure differences occur, air moves in an effort to regain its equilibrium, or pressure balance. Thus air movement is said to begin as the result of an atmospheric pressure force. This force can be better understood by studying air escaping from a punctured tire. The higher pressure in the tire and the lower atmospheric pressure outside are not in equilibrium. As the tire is punctured, the air flows from the area of higher to lower pressure, in an attempt to equalize the difference. The atmosphere acts almost exactly like the tire analogy, as air moves from areas of higher to lower pressure. Wind is the end result (Fig. 29-7).

The atmospheric pressure force responsible for the high-to-low movement of air is also called the *pressure gradient force*. The term "gradient" is not new, for we used it in Chapter 13 to describe stream activity. *Gradient* is a regular rate of change; in the case of the atmosphere, it is used to describe the horizontal change in pressure. If this change is very large, a steep pressure gradient will result, with violent wind speeds such as those found in tornadoes and hurricanes. On the other hand, weak pressure gradients give rise to light, variable winds and calm conditions (Fig. 29-7).

SECONDARY FORCES AFFECTING WIND

Secondary forces are those that come into play only after the air is moving. There are three major secondary forces that cause the air to flow in a curved path and in many different directions. These are the Coriolis force, centrifugal force, and frictional force.

Coriolis force. The Coriolis force is a hypothetical force that appears to turn a moving particle (as of air or water) to the right in the Northern Hemisphere and to the left in the Southern Hemisphere.

If an observer were in space, the wind would appear to follow a straight-line path, and this is exactly what it does. From a moving platform such as the earth, the wind's path appears to curve due to the earth's motion. Another peculiarity of this strange phenomenon is that its effect is greatest at the equator and least at the poles.

A simple experiment can be performed to demonstrate the effect of the Coriolis force. If you try to draw a straight line with a piece of chalk from the center of a rotating record to its outer edge, the result can only be a curved line (Fig. 29-8).

If an observer were motionless in space, the wind would

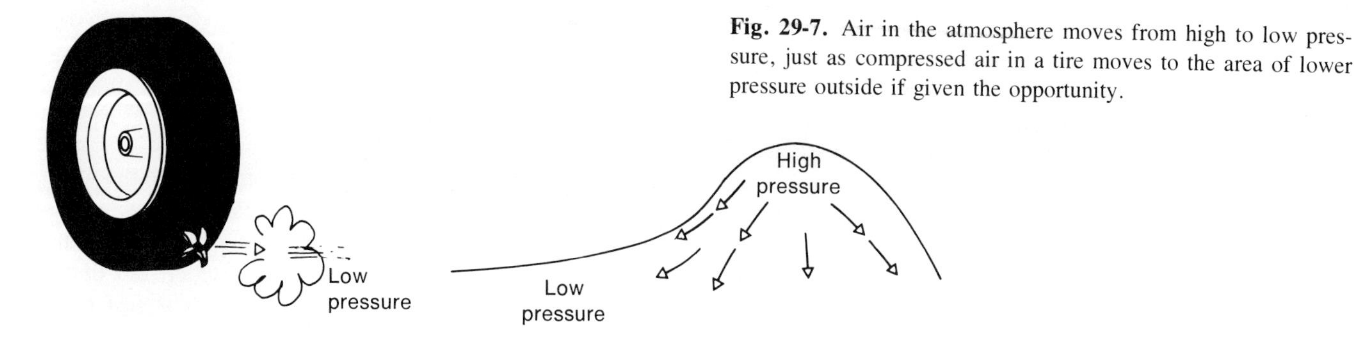

Fig. 29-7. Air in the atmosphere moves from high to low pressure, just as compressed air in a tire moves to the area of lower pressure outside if given the opportunity.

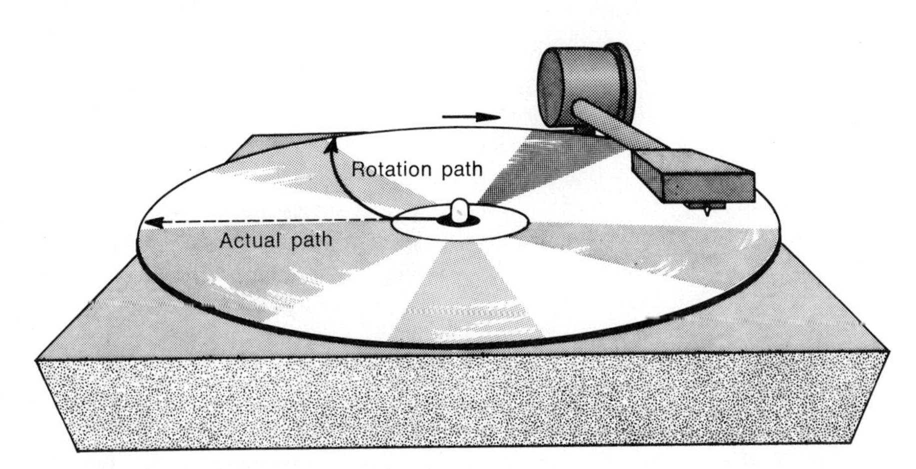

Fig. 29-8. You can see the effects of the Coriolis force for yourself in this experiment. Try to draw a straight line connecting the center and the rim of a rotating record. Even though you may draw a perfectly straight line, the line will end up curved due to the movement of the record.

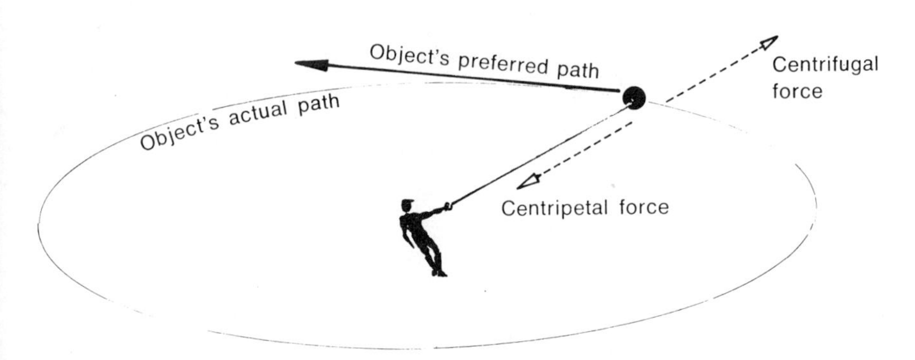

Fig. 29-9. Left alone, an object in motion would fly off on a straight-line course. Affected only by centripetal force, it would swing back and hit the thrower. Affected only by centrifugal force, it would continue indefinitely away from the observer. The combination of all three forces, however, creates the circular orbit this object actually follows.

inscribe a straight-line path because it would be going straight. This curving effect, therefore, only occurs on a rotating object.

To summarize the more important facts about the Coriolis force, it can be stated that (1) it exists only in the minds of observers on a spinning platform, (2) it causes moving particles to appear to curve right in the Northern Hemisphere and left in the Southern Hemisphere, and (3) its effect is greatest at the equator and least at the poles.

Centrifugal force. A rock tied to a string swings in a circle with a constant radius (Fig. 29-9). At any instant the rock's preferred path is a straight line, as explained by Newton's first law of motion. The curved path, however, is maintained by an inward force called the *centripetal force*. If this were the only force acting on the rock, the moving rock would hit the person swinging it on the head. Since this does not happen, there must be an equal and opposite force that keeps the rock on its fixed path. This equal and opposite force (Newton's third law), called centrifugal force, also acts on air moving in a curved path.

In a high-pressure system, for example, air moves outward from the center due to centrifugal force, while low-pressure systems experience an inward flow of air. (In lows, the pressure gradient force cancels out the weaker centrifugal force.)

Frictional force. Frictional force is exerted on any moving object at or near the earth's surface. For all practical purposes, its effect can be ignored at altitudes above 2000 feet (610 meters). Friction tends to retard, or hold back, the movement of air, and the force acts in a direction opposite the direction of flow. Air slows down as the air molecules come in contact with or near the surface roughness of the earth.

A secondary effect also occurs because of friction, for the air not only slows down but also deviates from its original path. This deviation can range anywhere from 10° over smooth water surfaces to 45° over rough topography.

WIND CIRCULATION AROUND PRESSURE SYSTEMS

In the *temperate zones* (30° to 60°) pressure systems constantly migrate, bringing changes in wind speed and direction. The variability of wind in these systems is related to the circulation pattern of each individual pressure area as it passes over a given geographic location.

The wind systems established within any pressure area are the result of an interaction between the primary and secondary forces just discussed.

Anticyclones

In the Northern Hemisphere, the wind of high-pressure areas spirals *clockwise and outward* from the center (Fig. 29-10). In the Southern Hemisphere, on the other hand, a *counterclockwise and outward* motion takes place due to the reverse rotation of moving objects south of the equator. Surface wind speeds in such systems are usually light because air descends from above to spread out gently on the earth's crust.

As discussed previously, fair weather is generally associated with high-pressure systems. These systems generally cover up to several thousand miles, and they are able to transport massive amounts of cold, dry air into an area; this is conducive to fair weather. In the wintertime, huge masses of cold Arctic air funnel down the continents, producing intense cold waves over the interior of the United States and Europe. Very little precipitation develops from these dry, cold air masses.

Another important reason for the absence of bad weather in a high is the descending motion of the air in the center of the system. Subsiding and diverging air such as this is warmed slightly in its descent, thus becoming drier and more stable. Dry, stable air leads to persistent periods of sunny skies with only a few low clouds.

However, this description of what seems to be an ideal weather condition has a hidden flaw. It comes in the form of a *subsidence inversion,* which often is an integral part of a high-pressure system. Remember that the term "inversion" refers to an increase of temperature with increasing height and that, in a high, air warms as it subsides. Therefore it can easily be seen that warmer air frequently lies over colder surface air. With warmer air on top, the colder surface air cannot rise and is trapped. This does not allow pollutants released near the earth to rise to higher levels for dilution and dispersal.

Subsidence inversions are quite common over the northeastern part of the United States in the fall and winter months. One such stagnating anticyclone occurred over the eastern part of the United States on Thanksgiving, 1966, bringing with it one of the more notorious episodes in air pollution history. Since the fall and winter are peak times for burning fossil fuels, sulphur oxides spread quickly over the entire region. Automobiles spewed vast amounts of nitrogen oxides and gaseous hydrocarbons that filled the immobilized air. An increase in the number of deaths due to respiratory diseases was one grim outcome of this incident. Similar experiences have been shared by the Meuse Valley in Belgium, London, England, and Donora, Pennsylvania.

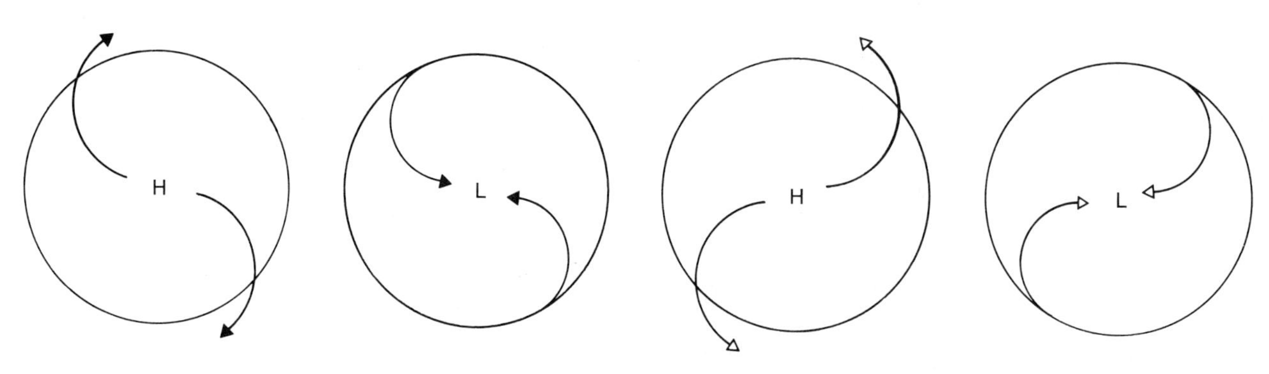

Fig. 29-10. In the Northern Hemisphere, high-pressure systems circle in a clockwise direction, while low-pressure systems move counterclockwise. In the Southern Hemisphere, both systems move in the opposite direction.

Cyclones

In the Northern Hemisphere, air in low-pressure areas flows in a counterclockwise direction. At the earth's surface the air spirals inward toward the center of the system. Wind speeds are usually much higher than those found in high-pressure areas because the pressure gradient is usually greater. Since these systems are so important, they will be discussed separately in Chapter 31.

Low-pressure areas located in the middle and high latitudes are called *extratropical* cyclones. These systems usually move northeastward through the belt of prevailing westerly winds. In the United States many lows migrate toward the Great Lakes area at speeds of up to 25 miles (40 km)/hr. As they move, they show a marked increase in intensity followed by a moderate decay. The bad weather in a low is the product of the convergent surface flow toward the center. As the air meets at the center, it rises; as it rises, it cools to its saturation point. This forms clouds and eventually rain. As a result, large areas of cloudy and rainy weather usually accompany cyclones.

WINDS OF THE WORLD

Meteorologists have known for many years that migrating pressure systems such as highs and lows travel through a more extensive and more stationary wind pattern. These areas of prevailing winds are collectively called the *general circulation pattern*.

It is much easier to understand how these *wind belts* are formed if we start with a very simplified model of the earth. Let us assume that the following conditions are present.

1. The earth does not rotate (there is no day or night).
2. The surface is smooth and of uniform composition (therefore it heats evenly).
3. There is no tilt of the earth's axis (there are no seasonal changes).

If these conditions existed, the equator would be the hottest region on earth. Therefore at the equator the hot air would rise and flow toward both poles (Fig. 29-11). The air at the poles, which would be colder, would sink and flow along the surface toward the equator. (The idea that air flows from areas of high to areas of low pressure still holds, because the cold air at the poles produces high pressure while the warm air at the equator results in lower pressure.) Thus a simple circulation pattern emerges, with northerly surface winds in the Northern Hemisphere and southerly surface winds south of the equator.

This model of a simple circulation system can now be used as a stepping-stone to explain what really happens to our tilted, rotating, nonuniform earth. If we cancel out our previous three assumptions, the equator still remains the hottest region of the earth and the hot air still rises, flowing north and southward toward the poles. (Since the Southern Hemisphere is a mirror image of the Northern Hemisphere, our discussion will be limited to the northern latitudes.) The poleward-moving air is deflected to the right by the Coriolis force (Fig. 29-12), which causes it to pile up at 30° N. Here the air becomes a predominantly eastward-moving, upper-air wind called the Jet Stream, which can reach velocities in excess of 200 miles (322 km)/hr.

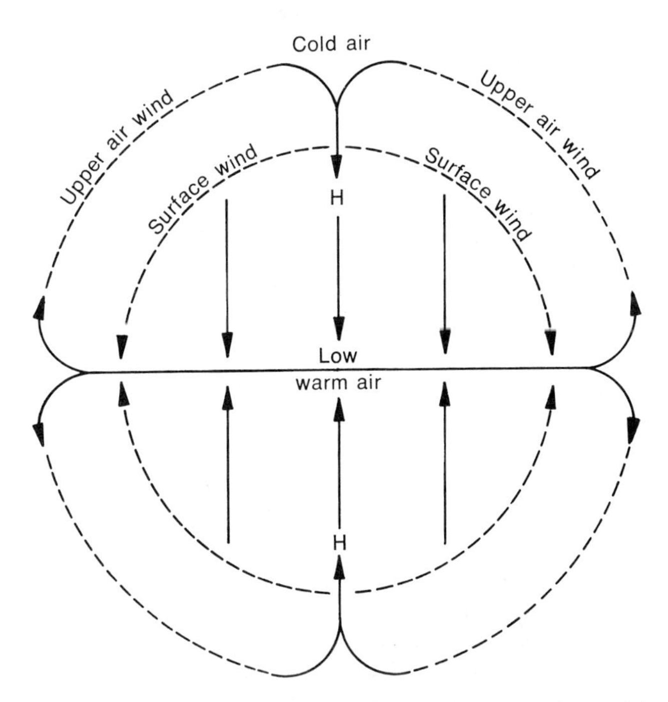

Fig. 29-11. Without complicating features such as rotation, axial tilt, and varying composition, the earth's circulation pattern would look like this. Warm air from the equatorial low would rise, forced up by incoming cooler air flowing from high-pressure areas at the poles toward areas of lower pressure at the equator.

The cold polar air that moves southward from the North Pole is known as the *Polar Easterlies*. These polar winds travel in a region between 90° to 60° N. At 60° N (called the *Polar Front*), the warm air is forced upward. This creates low surface pressure and a second narrow belt of high-speed, upper-air westerly winds called the *Polar Front Jet Stream*.

At 30° N, in the area called the *Horse Latitudes*, cooler air descends, piling up at the surface with a resulting northerly and southerly surface flow. As the air moves southward, it is influenced once again by the Coriolis force, and a northeasterly wind called the *Northeast Trades* is formed. The air moving northward from 30° is deflected

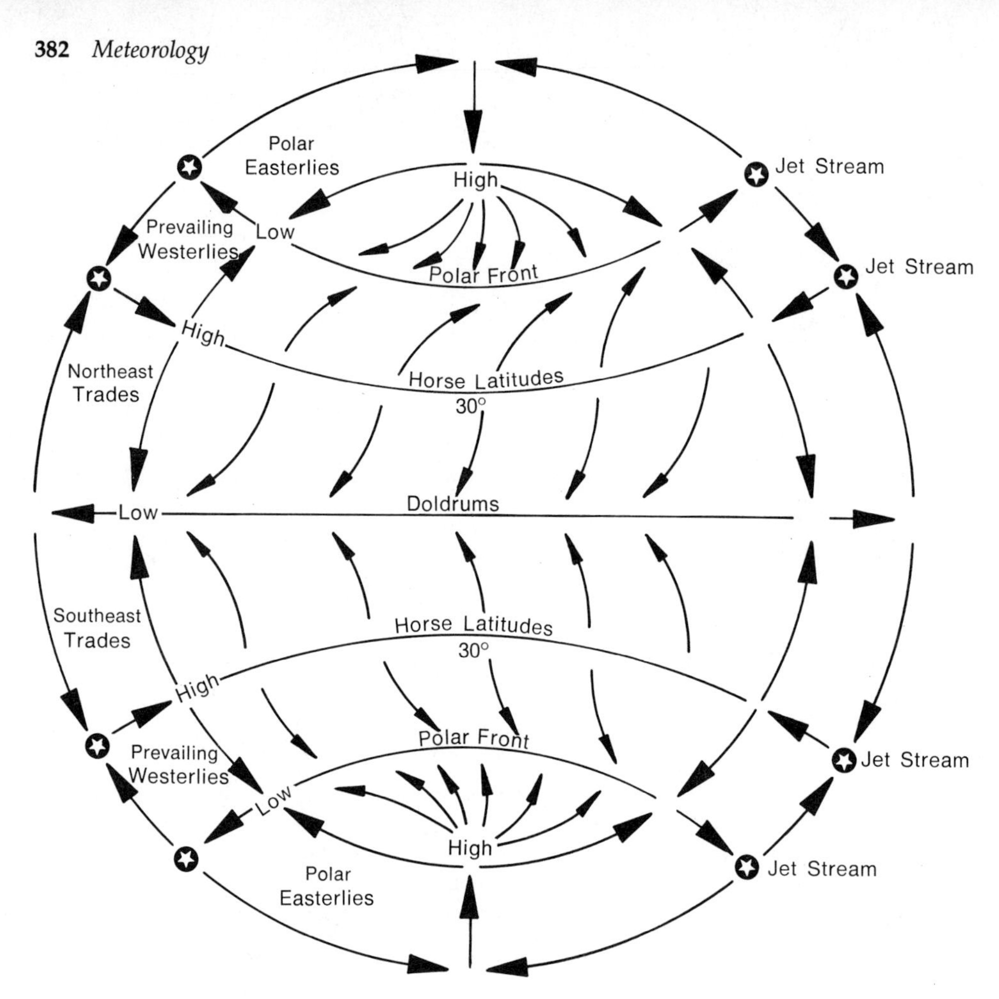

Fig. 29-12. As in our still model (Fig. 29-11), warm, less dense air in the equatorial regions rises and begins to flow north. About the time it reaches 30° N (the Horse Latitudes), it begins to pile up and flow eastward, forced to the right by the Coriolis effect. At the same time polar air flowing south is deflected by the Coriolis force before it reaches the low-pressure area at the equator. The cool air that does penetrate the southern areas passes under the warm Jet Stream in the Horse Latitudes, is swung to the right by the Coriolis force, and approaches the equator as the Northeast Trades.

Table 30. Major wind and pressure systems and related weather

Region	Name	Pressure	Surface winds	Weather
Equator (0°)	Doldrums	Low	Light	Cloudiness, precipitation, breeding ground for hurricanes
0°-30° N and S	Trade Winds	—	Northeast in Northern Hemisphere, southeast in Southern Hemisphere	Pathway for tropical disturbances
30° N and S	Horse Latitudes	High	Light, variable winds	Little cloudiness
30°-60° N and S	Prevailing Westerlies	—	Southwest in Northern Hemisphere, northwest in Southern Hemisphere	Pathway for subtropical high and low pressure
60° N and S	Polar Front	Low	Variable	Stormy, cloudy weather zone
60°-90° N and S	Polar Easterlies	—	Northeast in Northern Hemisphere, southeast in Southern Hemisphere	Cold polar air with very low temperatures
90° N and S	Poles	High	Southerly in Northern Hemisphere, northerly in Southern Hemisphere	Cold, dry air

into a wind coming from the southwest and is called the *Prevailing Westerlies*.

In Fig. 29-12 three distinct, simple circulation systems are shown accompanied by three belts of prevailing surface winds. The weather found in each wind belt is summarized in Table 30.

LOCAL WIND SYSTEMS

Accompanying the general circulation of the earth and the migrating secondary systems is a third wind pattern, which in many cases is much more important than the first two. These are the local winds, which are caused mainly by temperature differences and topography and are confined to much smaller areas.

Land and sea breezes

As their name implies, land and sea breezes occur only in coastal areas. They are produced by the temperature differences between land masses and bodies of water. As we have already noted, the land and the air above it become much warmer during the course of a day than the adjacent water. This produces an area of lower pressure over the land as the warmer, less dense air rises. The colder air over the water has a higher pressure, and a surface wind develops as the air moves from the area of higher to lower pressure. The resultant sea breeze, or *onshore wind,* can be quite strong, especially if the temperature differences are large. Such a wind can penetrate as far as 10 miles (16 km) inland and is usually strongest in the early afternoon (Fig. 29-13).

A reversal of the sea breeze effect occurs at night, when air moves from cooler land (high pressure) toward warmer water (low pressure). This land breeze, or *offshore wind,* is usually not as strong as the sea breeze.

Mountain and valley breezes

Mountain and valley breezes are also thermally induced, although these temperature differences are caused primarily by topographic features. During the day, mountain slopes and the air in contact with them are heated by the sun's rays. As the warm air rises, it appears to flow up and out of a valley; hence the name valley breeze.

At night, the air close to the mountainside is cooled by

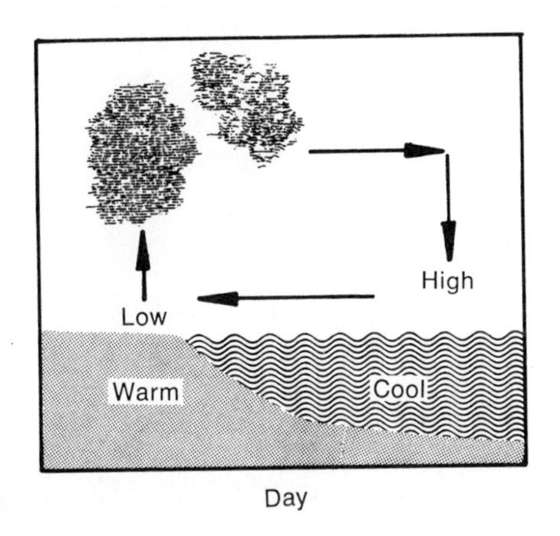

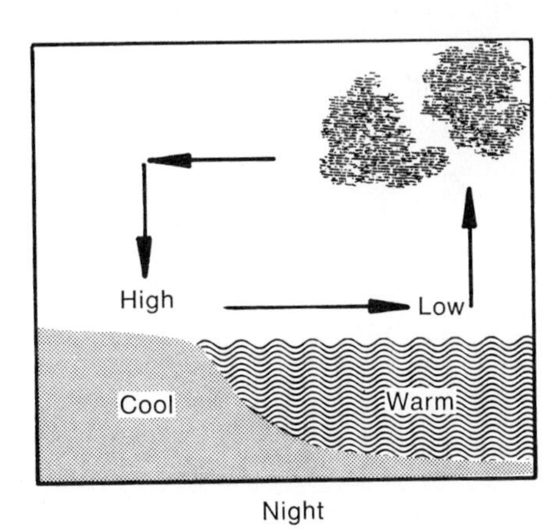

Fig. 29-13. A, Sea breezes are created by the flow of air cooled by contact with the water toward the lower pressure areas created by air warmed by contact with land. **B,** At night, the land cools faster than the water and imparts its coolness to the air above it. As this creates a high-pressure area, the air flows toward the lower pressure area created by warmer air over warmer water.

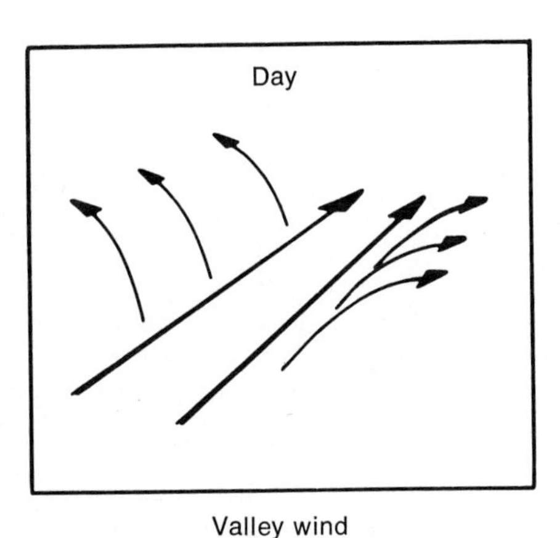

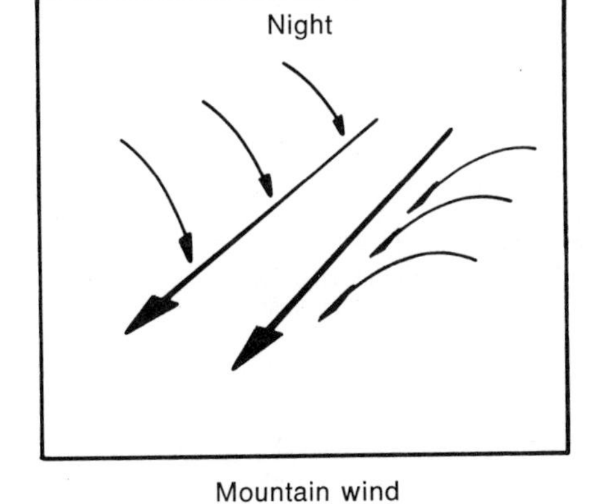

Fig. 29-14. During the day mountain air is warmed by contact with the ground. As it rises it flows up and out of the valley, bringing with it a welcome breeze. At night, the same air cools and falls to the valley floor, rushing down from the mountain. Both breezes are named for their source, the valley wind originating in the valley and the mountain wind on the mountain slopes.

conduction as the slope rapidly loses its heat. The cool air sinks to the valley floor and flows toward still lower levels. This downslope wind is called a mountain breeze (Fig. 29-14).

Foehn winds

Any wind that blows down an incline is termed a *gravity,* or *katabatic,* wind. A special type of gravity wind is the warm, dry foehn (pronounced fern) wind, which is produced by the strong, downward surge on the lee sides of mountains. The air warms adiabatically and dries out rapidly.

Foehn winds are common in the Alps and over the Great Plains of North America. In the United States the foehn is commonly called the *chinook* wind; literally translated, this means "snow eater." If chinooks occur right after a large snowfall, they will cause the snow to disappear by the sublimation process, leaving the ground bare with hardly a trace of wetness.

Whirlwinds

Whirlwinds, or dust devils as they are sometimes called, are miniature storm systems usually caused by strong local convection currents. These convection currents are produced by intense heating of the air just above the earth's surface. The result is a spinning dust column that can attain wind speeds high enough to turn over cars or airplanes and lift small objects high in the air.

• • •

There are many other small local winds that can greatly affect weather conditions of a particular area. Some of the more common types are summarized in Table 31.

Table 31. Local winds around the world

Local name	Location	Type of wind	Weather characteristics
Bora	Northeast shore of Adriatic Sea	Katabatic	Cold and wet
Mistral	Rhone Valley	Katabatic	Cold and dry
Pampero	Central Argentina	Cold air mass	Dusty and rainy
Black Roller	Great Plains, United States	Air mass	Strong dusty wind
Santa Ana	Southern California	Katabatic	Warm and dry

WIND DIRECTION

Wind is classified according to the direction *from which it is blowing*. Therefore a north wind is air coming from the north. It is quite common to use the eight major points of the compass to indicate wind direction. The cardinal points are north, south, east, and west, while the intermediate points are northeast, southeast, southwest, and northwest. In many cases directions are determined to an even finer degree, as in east-northeast.

Although word descriptions of direction are used quite often, meteorologists prefer to report wind directions in degrees measured clockwise from true north, which is called 0°. Using this system, 270° would be used to identify a west wind.

Terms such as *windward* and *leeward* are applied to the windy and calm sides of objects, respectively.

To measure upper-air winds, meteorologists use *pilot balloons*. These balloons are inflated with helium and have a known rate of ascent. The balloons are followed by an instrument called a *theodolite,* which is similar to a surveyor's sextant. The theodolite furnishes the observer with the angular horizontal drift and elevation angle. Using these measurements and orienting to true north, the upper-air wind speed and direction can be mathematically determined.

WIND SPEED

Wind speed is the measure of the average movement of the air. Generally speaking, a 1-minute average is used. Averaging does not include gusts, which are sudden brief increases in the wind speed. The speed of the wind is most often reported in miles per hour, although aviation, military, and seafaring craft prefer to use the nautical mile per hour, or *knot*.*

A simple method of estimating wind speed was developed by Admiral Sir Francis Beaufort in 1805. This system (Table 32) makes use of the wind's apparent effects on natural surroundings. These expressions are used quite often in relation to mountains or seamanship. Other terms such as "backing" (a clockwise change) and "veering" (a counterclockwise change) are used in conjunction with the passage of pressure systems.

Surface wind direction can most simply be measured by a wind vane. You can, for example, act as your own wind vane by orienting yourself toward the north and estimating the direction from which the wind is blowing. Other natural wind vanes include trees blowing in the wind, rising smoke, and even cows standing in an open field. In a strong wind, cows will almost always face downwind with their

*One knot is equivalent to 1.15 miles/hour.

Table 32. Beaufort scale and equivalent values

Beaufort number	Term	Speed		
		Miles/hr	Knots	
0	Calm	Less than 1	Less than 1	Smoke rises vertically
1	Light air	1-3	1-3	Smoke indicates wind direction, wind vane does not move
2	Light breeze	4-7	4-6	Leaves rustle, wind felt on face
3	Gentle breeze	8-12	7-10	Wind extends light flag, dry leaves and paper blow around
4	Moderate breeze	13-18	11-16	Small branches move, dust and paper rise from ground
5	Fresh breeze	19-24	17-21	Crested wavelets form on inland waters, small trees in leaf begin to sway
6	Strong breeze	25-31	22-27	Large branches in motion, telegraph wires whistle
7	Moderate gale	32-38	28-33	Whole trees in motion, walking influenced by wind
8	Gale	39-46	34-40	Twigs break, walking difficult
9	Strong gale	47-54	41-47	Ground littered with branches, slight structural damage
10	Storm	55-63	48-55	Trees uprooted, much structural damage
11	Violent storm	64-75	56-63	Widespread damage
12	Hurricane	Greater than 75	Greater than 63	Severe damage

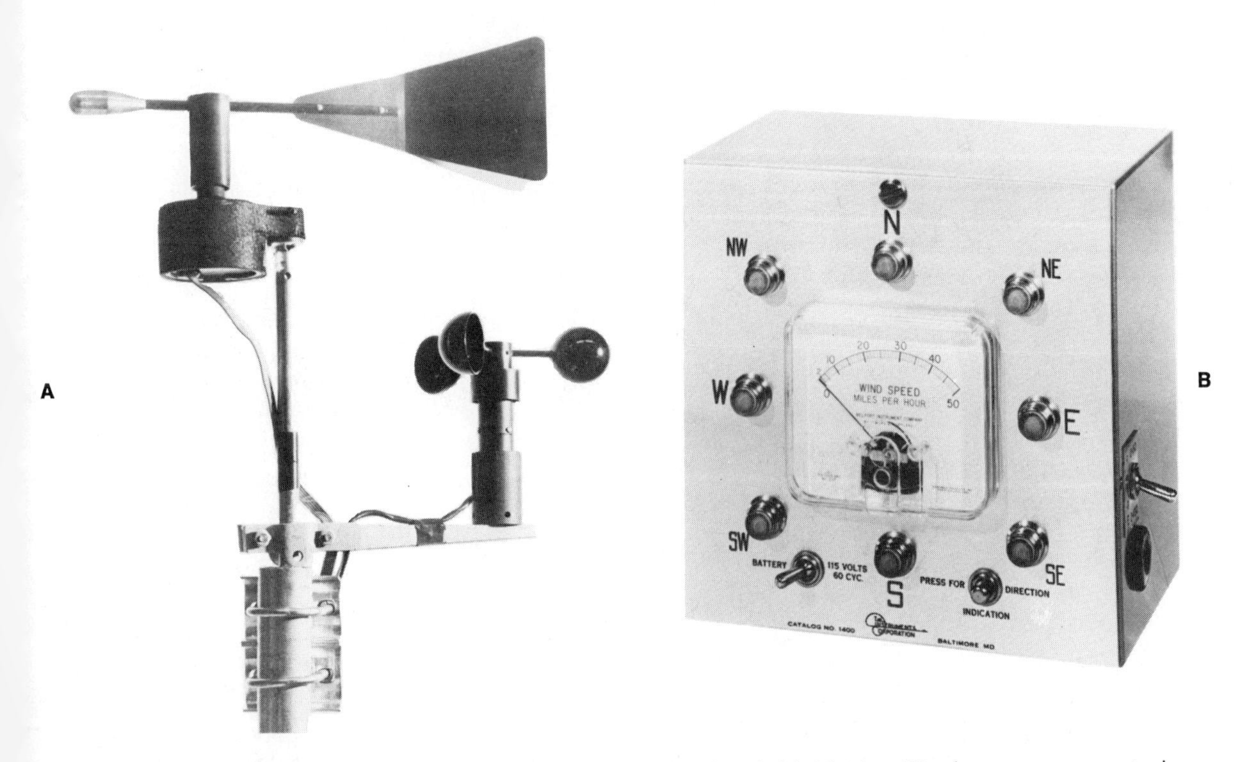

Fig. 29-15. A, This wind vane points in the direction from which the wind is blowing. The three-cup anemometer is used to measure wind speed. **B,** Electronic devices can also be used to measure wind speed and direction. (Courtesy Science Associates, Inc., Princeton, N.J.)

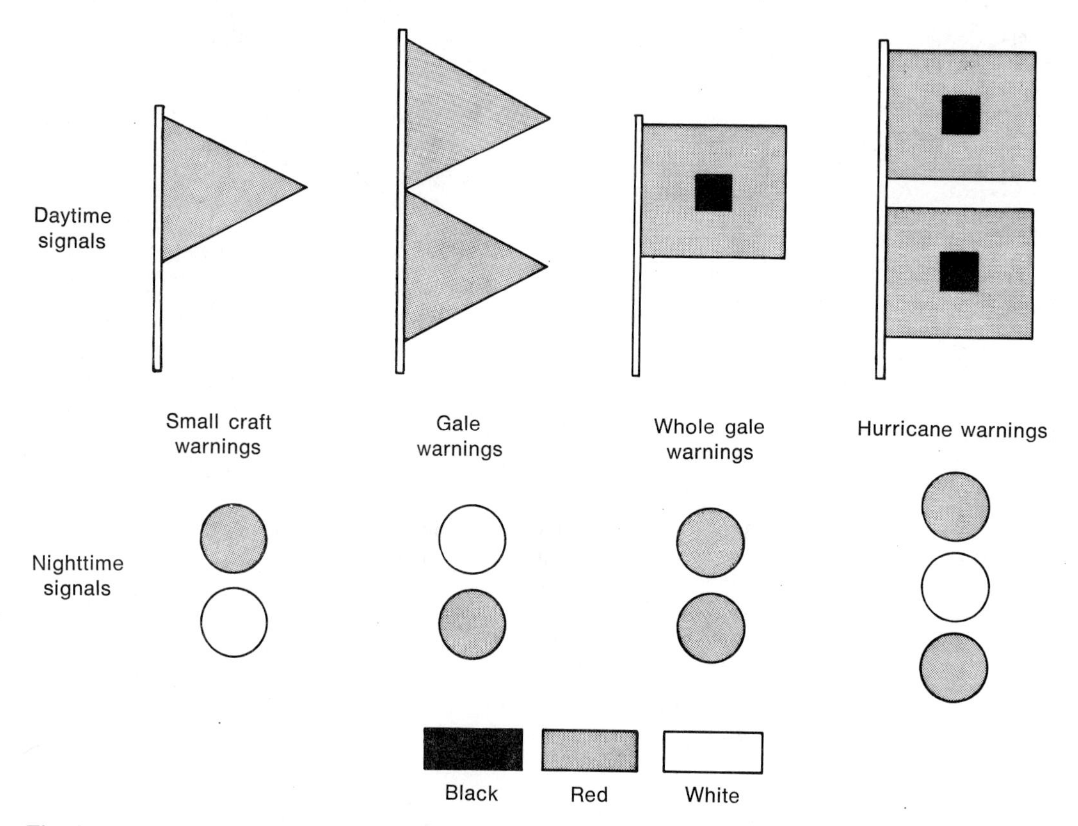

Fig. 29-16. The United States Weather Bureau uses this system of flag and light signals to warn small craft of dangerous winds.

hindquarter pointing toward the direction from which the wind is blowing. There are, of course, many mechanical wind vanes that are more accurate! These operate on a cam system, which activates a switching device to illuminate a directional light that indicates the direction of the wind (Fig. 29-15).

There also are other more sophisticated instruments in use. The *anemometer,* which measures both wind speed and direction, can be set up to yield both instantaneous readings and a permanent record on a revolving chart. This is the most widely used wind speed measuring device, the simplest type being the three-cup type shown in Fig. 29-15. This instrument relates the rotation of the cups to the wind speed by measuring the power it generates. There are other devices such as those utilizing the pressure of the wind exerted on a flat plate and the cooling of a heated wire, but these are much more expensive and limited.

The United States Weather Bureau employs a visual system designed to warn boatsmen of impending dangerous winds. This system utilizes flags during the daylight hours and lights at night (Fig. 29-16).

30 *air in collision*

Air masses and fronts

Every wind has its weather.

Proverbs

AIR MASSES

When the forces exerted on moving air come into balance with one another, the equilibrium that is produced results in a calming of the wind. This *stagnating air* may remain undisturbed for fairly long periods of time as the forces try to gather strength once again. If the air happens to stagnate over a large land mass or body of water, it will take on the temperature and moisture characteristics of that surface. This surface area, called the *source region* of the air, can sometimes stretch for thousands of miles. The air within this source region is called an air mass, which can be defined as an extensive body of air whose temperature and moisture characteristics are horizontally uniform. Even though their temperature or moisture content may vary slightly from place to place, air masses are still quite homogeneous and can therefore be classified as *cold* or *warm, dry* or *wet*.

Source regions

In order for air to take on the characteristic properties of the surface it overlies, the source region must be reasonably uniform. This is especially the case with large land areas, which also must have a smooth topography. The Central Plains area of Canada is an excellent example of the vast, mostly level terrain needed for a source region. The Sahara Desert in North Africa and the desert interior of Asia also represent ideal source locations, as do other deserts, large ocean areas, and vast, ice-covered wastelands. All of these large land and water areas serve as source regions because they represent the main types of surface regions that are uniform enough to impart a uniformity of temperature and moisture to huge quantities of air.

There are six major source regions in North America:

1. Interior Canada, which is covered by ice and snow, and the frozen waters of the Arctic Ocean
2. The large expanse of cold North Atlantic waters eastward from Greenland, Newfoundland, and Labrador
3. The northeastern part of the Pacific Ocean as far as the northwestern coast of the United States
4. The warm ocean waters of the Gulf of Mexico and the Carribean Sea, also including land in parts of Florida and the Yucatan Peninsula of Mexico
5. The warm, narrow region of northern Mexico and the states of Texas, New Mexico, and Arizona, but only during the summer months region
6. Along the lower southwestern coast of the United States and westward into the Pacific Ocean, but only during the winter months

Air mass classification

Tor Bergeron, a Swedish meteorologist, introduced a classification scheme for identifying air masses. Since he wanted to show the weather features of each air mass, the classification involved both temperature and moisture characteristics as well as geographic starting point (source region). Bergeron's classification system is reproduced below.

Geographic classification		Location
T	Tropical air mass	25° N to 25° S
A	Arctic air mass	Close to the North or South Poles
P	Polar air mass	40° to 60° N or S
E	Equatorial air mass	10° N to 10° S
S	Superior air mass	Southwest part of North America

Thermal classification

w Warm air mass
k Cold air mass

Moisture classification

m Maritime air mass (wet)
c Continental air mass (dry)

The geographic classifications are based on the source regions over which the air mass forms. These areas tend to shift slightly with the seasons but still remain within set geographical boundaries.

The temperature, or thermal, characteristics indicate the temperature of the air compared to the temperature of the underlying surface. If the air is colder than the land or water body over which it is traveling, the air mass is classified as cold (k). Thus a cold air mass could have a temperature as high as 60° or 70° F. Warm (w) air masses are also classified in the same way.

Like temperature characteristics, moisture properties are also tied to the source region. If the air starts over land it is classified as *continental* (c), or dry, but if it originated over water it is called a *maritime* (m), or moist, air mass.

By using this classification scheme and a shorthand notation method, meteorologists can identify the major features of each air mass moving over a specific area. This is done by using a combination of the symbols representing moisture, geographical location, and temperature, in that order. Thus a *mTw* air mass (maritime Tropical warm) is an air mass that has formed over the tropical ocean areas and is moist and warm. Being able to identify and understand air mass characteristics enables a weather forecaster to turn the properties of oncoming air into predictable weather changes. The following are other examples of air mass types.

cPk cold, dry Polar air
mPk cold, wet Polar air
cTw warm, dry Tropical air
E warm, wet Tropical air formed near the equator
A cold, dry Arctic air
S hot, dry air

Notice that the letter designations for moisture and temperature of Equatorial, Superior, and Arctic air are omitted. Since cold land masses and ice fields predominate in the Arctic, very little moisture can enter the air. Therefore a moist or warm air mass from this region is very rare. The same is true of the Equatorial air mass that forms mostly over the warm water regions of the equator

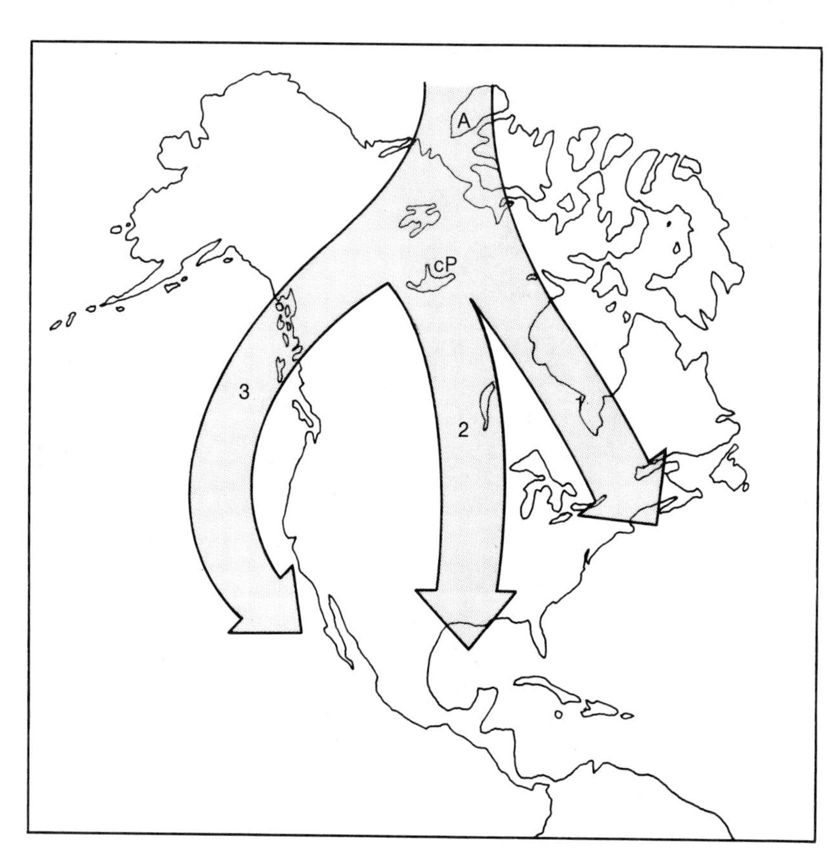

Fig. 30-1. Three major tracks are paths Arctic and Polar air take to the continental United States.

and the Superior warm, dry air mass. Only one letter is therefore needed to identify these types of air masses.

Air masses affecting the United States

Of the six types of air masses just mentioned, all except the equatorial type affect the United States, although the cT air mass influences the daily weather pattern on only rare occasions. The air masses that frequent the North American mainland and their tracks and weather patterns are described in the following discussion.

Arctic (A) and continental polar (cP) air masses

Originating over the snow- and ice-covered higher latitudes of North America, the Arctic and continental Polar air masses, which are composed of extremely cold, dry air, build up enough force to pour down on the heartland of the United States. There are three major wintertime tracks for this Arctic and Polar air, all of which produce frigid conditions in the north and chilly weather in the southern regions of the United States (Fig. 30-1).

The first path takes the P air over the Great Lakes. During the winter, heavy snow showers and gusty winds whip across the lee side of the Great Lakes and continue on to batter the northeastern United States with wind and cold temperatures until the air mass moves offshore.

The second path brings this air to the central part of the United States. This frigid blast of cold air has enough strength to reach across the Gulf of Mexico into Central America, where it is called a *norther*. While traveling southward at high speeds from Canada, the cP or A air leaves in its wake an extreme temperature drop, low relative humidity, and clear skies. When this cold *polar outbreak* is in the making, the citrus growers in central and south Florida start preparing to counteract the expected freezing conditions. They know full well that this polar air has the ability to reach into the very southern portions of their state.

A and cP winds take the third path only rarely, when a very intense high-pressure area is located over central Canada. This high must be strong enough to force the air over the northern Rockies and into northern and central California, bringing cold, rainy weather and occasional snow.

In the summer, cP or A air is much warmer due to a decrease in the number of snow-covered surfaces in the source region. Summer heat waves in the Northeast are normally broken by the invasion of this cP air, which seldom reaches as far south as its winter counterpart.

Fig. 30-2. On those rare occasions when the Coriolis force does not sweep mP air out to sea, it can enter the United States through Canada along this path.

Atlantic maritime Polar (mP) air masses

The Atlantic mP air mass originates in the cold Atlantic waters off the coasts of Newfoundland and Labrador. Since the prevailing wind flow is normally from west to east, it is very rare for mP air to affect the United States. Aided by a wintertime high-pressure area over the North Atlantic, however, this cold wet air can penetrate the coasts of New England, bringing with it freezing temperatures, strong gale winds, high relative humidity, and wintertime precipitation (Fig. 30-2). Along the New England coast these storms are called *northeasters* because of the northeastward direction of the wind.

During the summer an mP-Atlantic air mass is less conspicuous than in the winter. An occasional influx of cool, moderately wet air brings partly cloudy skies, no precipitation, and pleasant temperatures.

Pacific maritime Polar (mP) air masses

Originating over the far northern part of the North Pacific, Pacific mP air becomes a wet air mass that greatly affects the weather of the West Coast. The mP air reaches the North American coast by three distinct paths (Fig. 30-3). The first path originates over Alaska and travels over the relatively short water route southeast to the United States. Accompanying this cold wintertime air flow is a rapid increase in cloudiness, with numerous showers and rain squalls. This is due to the fact that the cold, moist air is forced to rise over the coastal ranges, cooling the air to its saturation point.

The second and third paths are long, over-water trajectories that bring warmer air and high relative humidities. Rain falls along the western coast of the United States but later disappears, leaving sunny skies along the Continental Divide just east of the Rocky Mountains.

In the summer, the influx of mP air from the Pacific generally brings clear skies or only a few low, scattered cumulus clouds.

Atlantic maritime Tropical (mT) air masses

Although Atlantic mT air may appear at any time of the year, its influence is greatest on the eastern half of the United States in the summer (Fig. 30-4). This air mass has one of the highest temperatures and moisture contents of any affecting the United States. As it moves over warmer land, the air is rapidly heated, and the result is showery cloudbursts and violent thunderstorms.

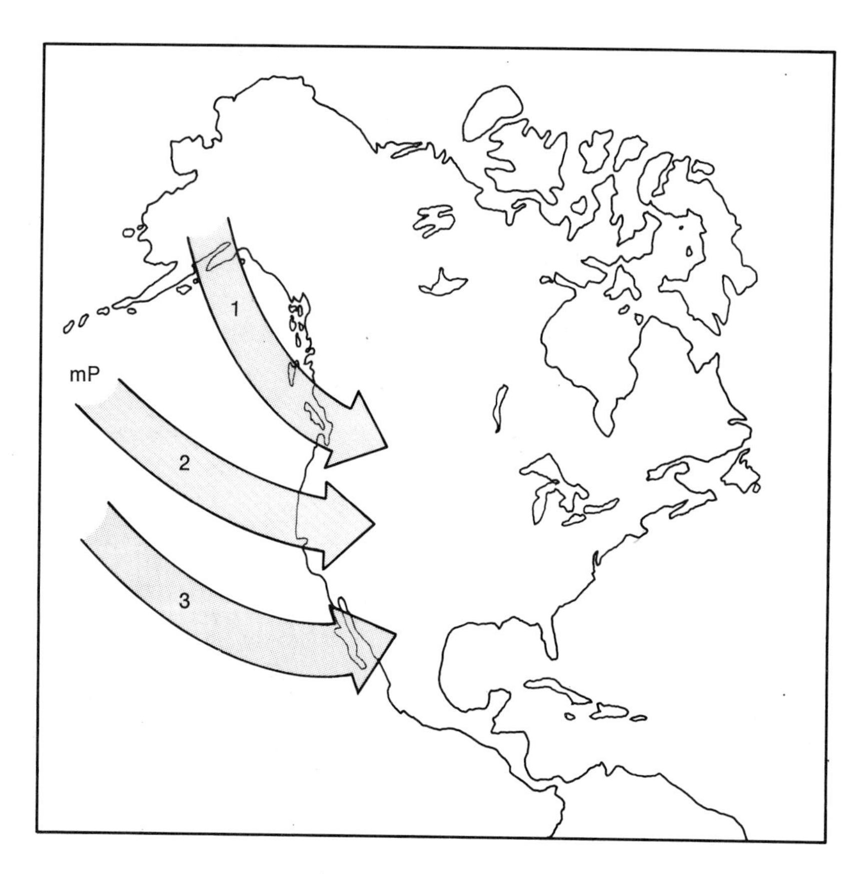

Fig. 30-3. Tracks followed by north Pacific mP air, which greatly influences weather patterns on the West Coast.

If the air mass is particularly strong, it will carry its hot, sultry air northward, creating oppressive heat waves in the densely populated regions of the Northeast. As the mT mass moves off the northeastern coast, it overrides the cooler air over the Labrador currents. This results in advection fog over such areas as the Grand Banks of Newfoundland.

Occasionally a high-pressure area located over Bermuda carries mT air westward over the Gulf of Mexico and into the Southwest. Due to orographic lifting (air is forced to rise to pass over a mountain) huge amounts of rain fall in short periods of time, flooding areas that are usually parched throughout the rest of the year. The local name for this phenomenon is *Sonora weather*.

During the winter the mT air mass is responsible for mild temperatures and high humidities along the South Atlantic and Gulf coasts. Usually night and morning cloudiness give way to fair-weather cumulus clouds during the day. However, when the land has been excessively cooled by colder air, lack of sunshine, or even snowfall, the intrusion of this warm, moist air produces a thick blanket of advection fog.

Superior (S) air masses

Although chiefly an upper air mass, Superior air occasionally reaches the surface. Usually found over the southwestern states, this hot, dry air is thought to have a direct connection with the air in the Horse Latitudes (Chapter 29). Superior air is fairly stagnant and has no definite path when it does move. Therefore only minute amounts of rainfall occur over an area dominated by an S air, and drought-like conditions encourage only sparse vegetation.

Pacific maritime Tropical (mT) air masses

In the winter, mT air affects only the lower coast of California and northwestern Mexico, and then only rarely. It brings with it warm, moist air that, when forced upward by colder air, produces rainfall in record quantities (Fig. 30-5). This air mass does not have any direct influence on the weather of the West Coast in the summer because the dominant Pacific high-pressure area has moved so far south.

Modification of air masses

As an air mass moves out of its source region, it undergoes slow changes in its characteristic properties. These changes are produced by the influence of the different sur-

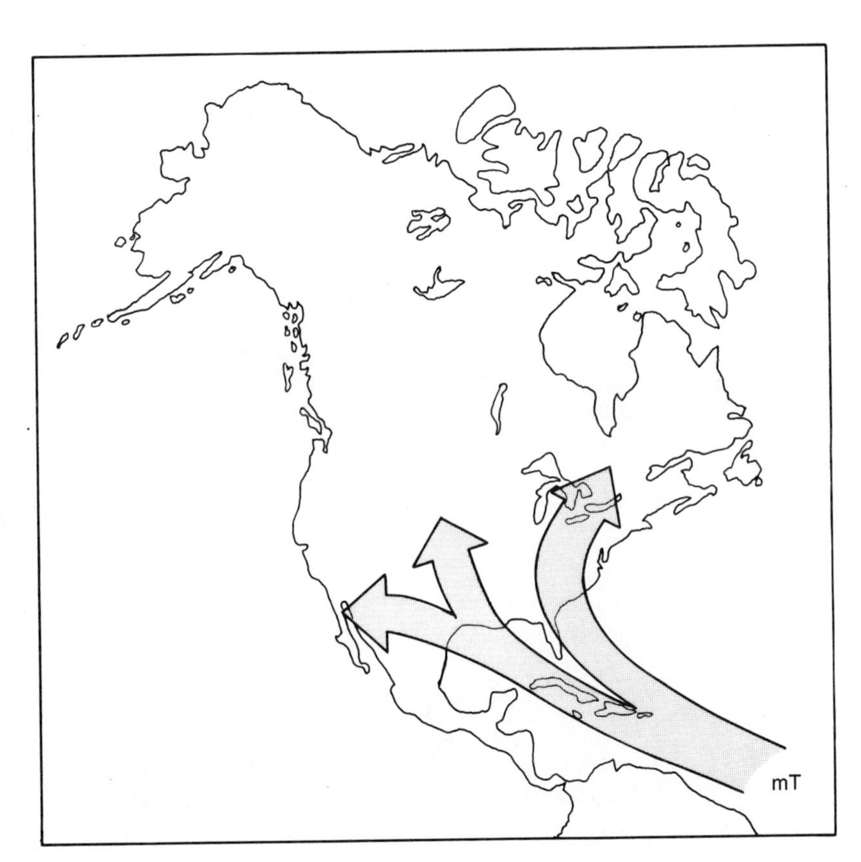

Fig. 30-4. When mT air follows these tracks out of its source area, it brings summer showers and thunderstorms to the East Coast.

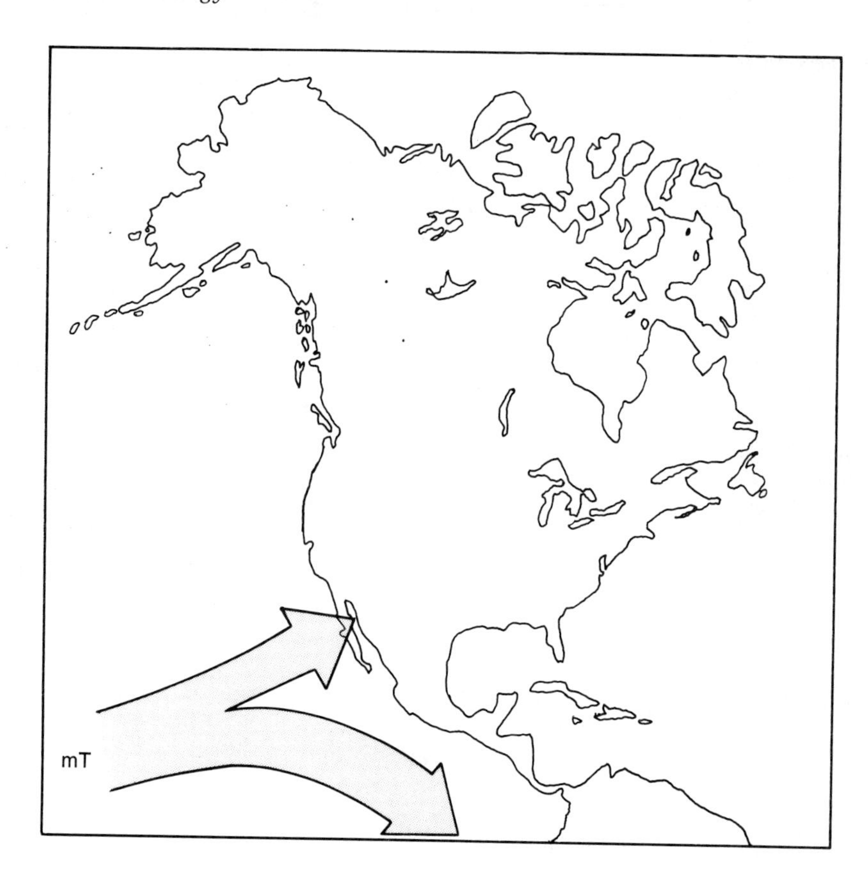

Fig. 30-5. On the rare occasions when mT air swings far enough north to affect the California coast, this is the path it takes.

faces over which the air travels. There are many factors that influence this modification, but the most important are caused by topography, temperature, and moisture.

Topography

Air must move in response to the topography of the ground beneath it. This is nowhere more evident than in mountainous regions, where air is forced to rise on the windward side and then subside on the leeward side of the mountains. The effect of this movement is to remove moisture from the air as it rises and is cooled and to increase the temperature of the air as it descends on the leeward side of the mountain.

Temperature

The temperature of air can be significantly changed by the temperature of the surface over which it is moving. This change ultimately influences associated weather patterns. For example, as warm air from the tropics moves over a cold, snow-covered region, the air in contact with this surface is immediately cooled. As the air moves northward, it is cooled to a greater height, spreading fog throughout the entire area.

Even in winter, Florida seldom receives frigid, Arctic air because of the long path this air must take. Crossing this great distance, the air warms moderately before it reaches the southern limits of the United States.

Moisture

Moisture is added to the atmosphere by evaporation and removed by precipitation. If a dry air mass crosses a large, unfrozen body of water, it takes on additional moisture and become very wet. Any body of water as large or larger than the Great Lakes can act as a good source of moisture. As cold Polar air funnels out of Canada during the winter, it passes over the Great Lakes and picks up large quantities of moisture. On the lee side of the lakes, however, the air is cooled and snow begins to fall. These *snow flurries* can average 3 to 4 inches/day and total as much as 10 to 12 inches without the presence of a storm system.

FRONTS

What would happen if a frigid continental Polar air mass came pouring southward over the heartland of the United States at the same time that a warm, moist Tropical air mass streamed northward from the Gulf of Mexico?

First, they would eventually meet, and second, they would not readily mix, since they have quite different atmospheric characteristics. Therefore there would be a boundary line, or better yet a transition zone that would separate them at various levels in the atmosphere. Such a narrow zone of transition between two air masses whose general characteristics are distinctly different is called a front.

The understanding of a front and its attendant cyclonic storm system was first developed by J. Bjerknes with his publication of *The Polar Front Theory of Cyclones*. This theory, which has been somewhat modified, remains one of the basic guides to modern weather forecasting.

Formation of fronts

Two different air masses are required to produce a front. We must therefore look for areas of the earth where the wind pattern or circulation system is conducive to this prerequisite. At lower latitudes (0° to 30°) the air circulation is such that fronts do not easily form; at higher latitudes (90° to 70°) only cold air is present. In the midlatitudes (30° to 70°), however, there exists a definite battleground between warm Tropical air and cold Polar or Arctic air. This boundary zone is known as the *Polar Front,* a worldwide, discontinuous zone that meanders southward or northward, depending on the intensity of the individual air masses. A second but much weaker frontal zone is the *Arctic Front,* which separates frigid Arctic air from cold Polar air. A third area, the *Intertropical Front* (3° N to 3° S) is nothing more than a convergence zone between the trade winds.

In order for *frontogenesis* (the formation of a front) to occur, two conditions must exist: (1) two air masses of different densities must lie side by side and (2) the wind must be blowing in a direction that brings the air masses into constant contact with one another. A perfect example

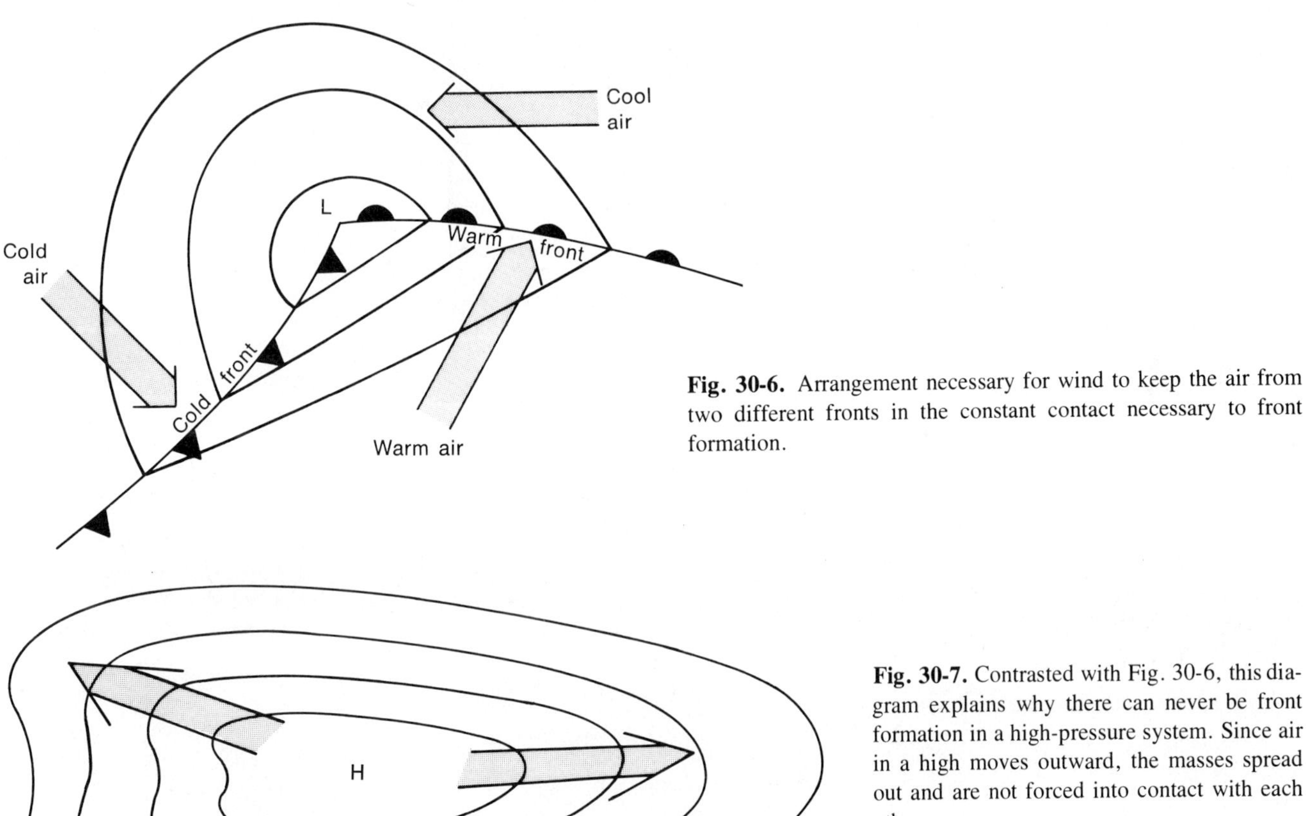

Fig. 30-6. Arrangement necessary for wind to keep the air from two different fronts in the constant contact necessary to front formation.

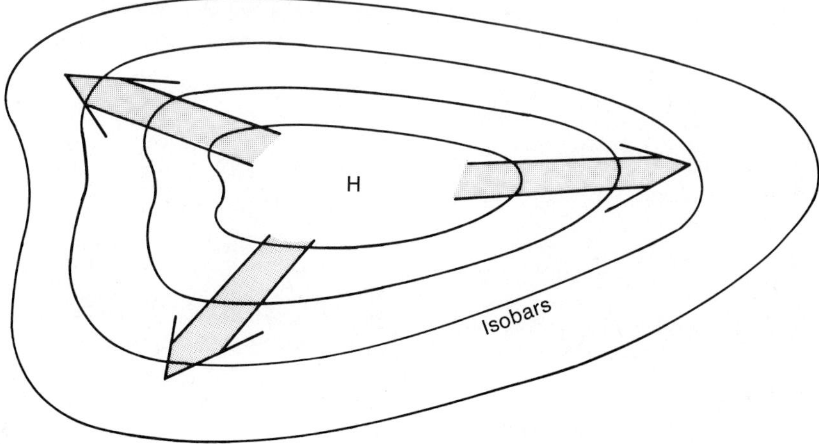

Fig. 30-7. Contrasted with Fig. 30-6, this diagram explains why there can never be front formation in a high-pressure system. Since air in a high moves outward, the masses spread out and are not forced into contact with each other.

of this is the formation of a low, or cyclone. After cyclogenesis (formation of a cyclone), the wind circulation around the low presents the perfect situation for frontal formation. Because of the counterclockwise and inward flow at the surface, cold, northerly air can easily meet warm, southerly air. This can and usually does take place on either side of the low-pressure area (Fig. 30-6).

In an anticyclone, or high-pressure area, the air circulation is outward and clockwise at the surface. Thus air always moves away from a central point so that different air masses can never meet; frontogenesis therefore never takes place in a high (Fig. 30-7).

Frontal classifications

The one important criterion for classifying fronts is the relative strength of the moving air masses. Although certain weather conditions are associated with particular fronts, they are the products of frontal formation and cannot be used for classification on a front itself. Simply put, a front is classified on the basis of which air mass pushes the other out of the way and takes its place.

1. A cold front is a transition zone along which cold air is displacing warm air at the surface.
2. A warm front is a transition zone along which warm air is replacing colder air at the surface.
3. A stationary front is a transition zone between two air masses, neither of which is strong enough to push the other out of the way.
4. An occluded front is the intersection at which a cold front overtakes a warm front and forces the warm air upward.

Cold and warm fronts can also occur above the ground. When this happens they are called upper-air warm or cold fronts.

Cold fronts

When cold air pours southward and replaces warmer air, a cold front develops. This transition zone is three-dimensional in nature and can be visualized as a sloping wedge of cold air underriding warmer air, which is being pushed up and out of the area. The slope of the front depends on the speed at which the cold air is moving. A fast-

Fig. 30-8. This fast-moving squall line is of the type often associated with cold fronts. (Courtesy United States Department of Commerce, National Oceanic and Atmospheric Administration, Rockville, Md.)

moving cold front will have a slope of 1 mile for every 40 to 80 land miles (1:40 to 1:80), while a slow-moving front may have a slope of 1 mile for every 100 land miles (1:100). Since a fast-moving cold front is more important, our discussion will be limited to its characteristics; however, many other types of cold fronts can occur in the atmosphere.

A fast-moving cold front travels at about 25 to 30 miles (40 to 48 km)/hr. The warm air is often pushed violently upward, producing unstable conditions and the formation of cumuliform clouds of great vertical height. Precipitation takes place in a narrow band along and just ahead of the front. In certain instances, a line of violent thunderstorms called a *squall line* develops 100 to 200 miles (161 to 332 km) in advance of the front (Fig. 30-8). Pressure generally falls in advance of a cold front and rises suddenly as the front passes and the colder, denser air arrives. The wind generally veers with the passage of the front, becoming strong and gusty at the frontal surface. Ultimately it comes from a northerly direction, usually from the northwest. The temperature starts to decrease in the warm sector due to the evaporation of falling precipitation (remember, evaporation is a cooling process) and takes a sharp drop behind the front. As the moisture content of the cold air decreases, its relative humidity drops and the storm clears up rapidly. Fig. 30-9 illustrates the cloud and precipitation pattern accompanying a cold front as well as the sloping wedge of the boundary zone.

Warm fronts

Warm fronts develop slowly as warm air replaces a retreating cold air mass. The warm air usually glides slowly over the denser cold air wedge, forming a shallow incline of about 1 mile of elevation for every 150 miles of land surface. This frontal boundary moves at an average speed of between 10 and 20 miles/hr, considerably slower than the typical cold front. The slow advance of both the warm air and the frontal zone results in stable conditions and the formation of stratiform clouds. With the approach of a warm front, all types of clouds can be observed, although cirrus and cirrostratus clouds usually appear miles in advance of the front. Later, lower altostratus clouds appear as the warm air approaches. Finally, nimbostratus or rain clouds appear when the front is about 200 miles away. The precipitation takes the form of steady drizzle, rain, or snow, which at times can be quite heavy. Still closer to the front, stratus clouds and fog make their appearance. After the frontal passage, clearing is very slow (Fig. 30-10). The wind shifts gradually from an east or southeast direction to the west or southwest, and wind speeds are light. The pressure falls as the warm front approaches and levels off

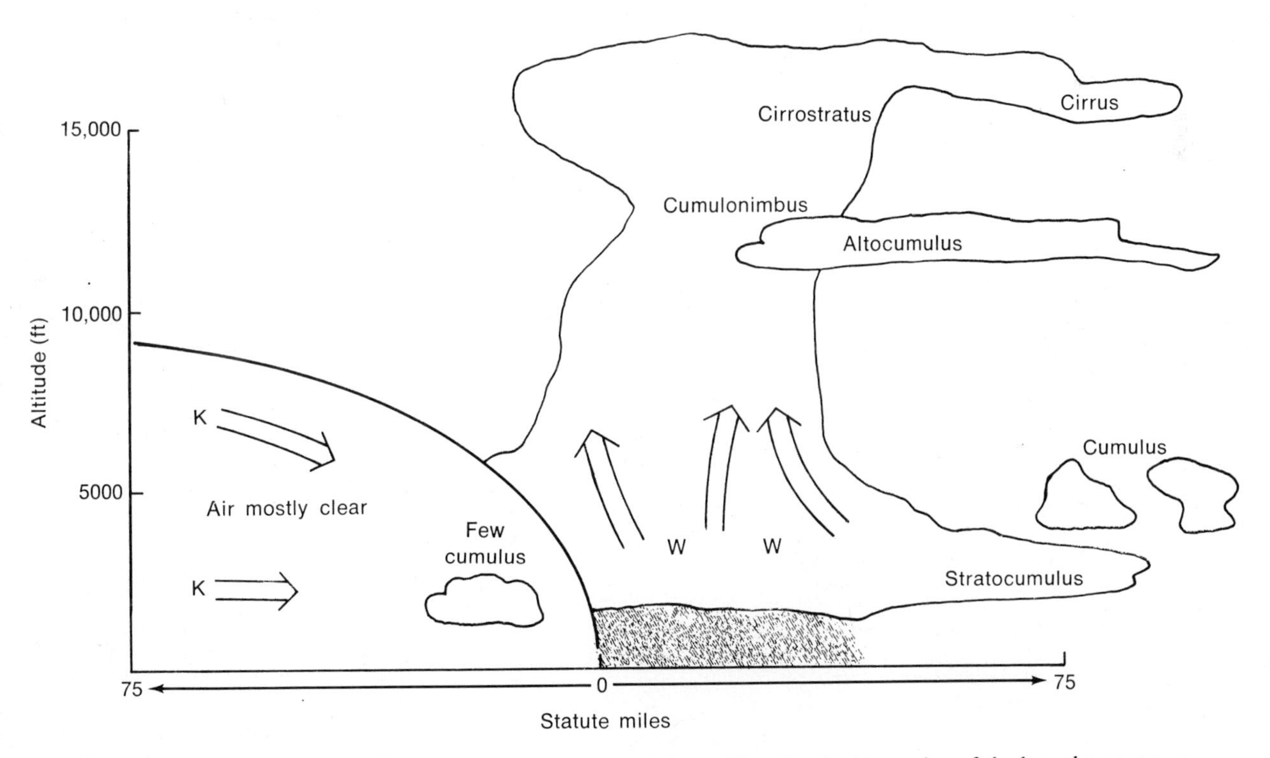

Fig. 30-9. Cloud and precipitation pattern associated with a cold front. Note the sloping wedge of the boundary zone.

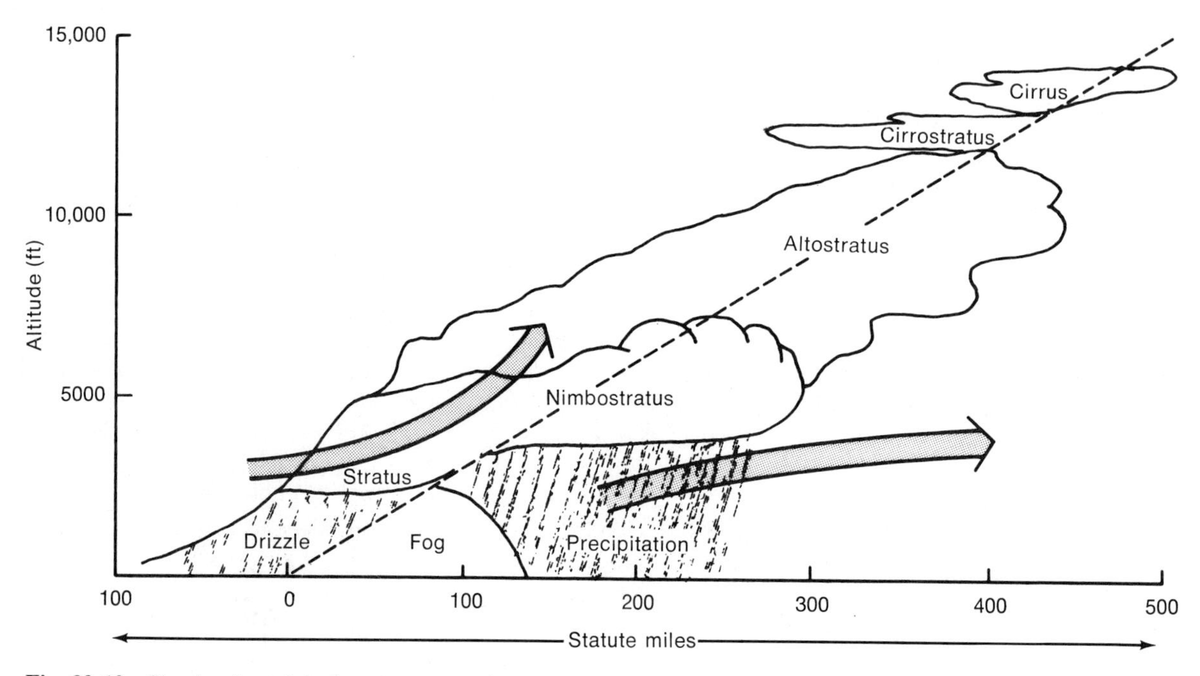

Fig. 30-10. Cloud and precipitation pattern associated with warm fronts. Note the diagonal alignment of the clouds and more gradual transition zone between fronts.

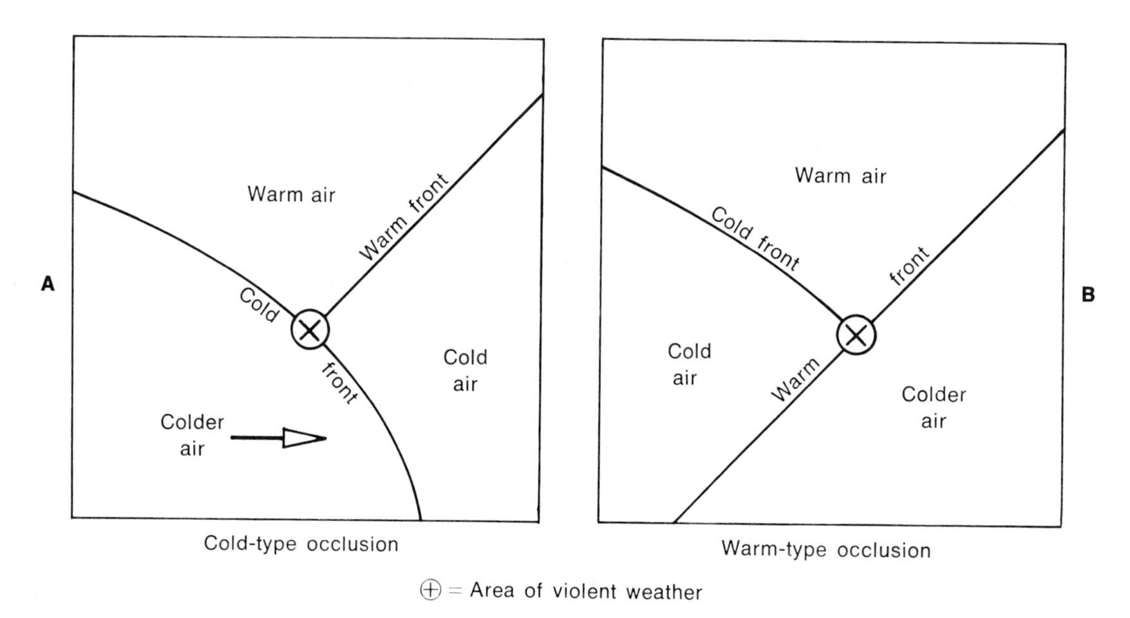

Fig. 30-11. A, Cold-type occlusion involves the circumstances expected when two fronts meet: the cold front slides under the warm front, forcing the warm front upward. **B,** Warm-type occlusion involves an unusual circumstance: the "warm" front is not as "warm" as the "cold" front, so it is the warm front that slides underneath.

after its passage. If the warm air is maritime, humidity is high, leaving lingering, oppressive heat.

Stationary fronts

Since the character and name of a front indicates the air mass that is replacing existing air, a stationary front is one in which there is no replacement. Under such conditions the air masses on both sides of the boundary line move parallel to each other and the front itself does not move.

The weather pattern depends on which sector the observer is in. Warm-front conditions occur in the warm sector and cold front weather is found in the colder part. Eventually the system either disappears as the air masses mix, or one of the two air masses become dominant, forming a cold or a warm front.

Occluded fronts

Cold fronts move faster than warm fronts. Therefore a cold front may overtake a slower-moving warm front, forming an occluded front. When this happens, the cold front generally slips under the warm front because the cold air is denser and therefore heavier (Fig. 30-11, *A*). This is called a *cold-type* occlusion. However, if the cold air in the cold front is warmer than the cold air in the warm front, a *warm-type* occlusion will occur (Fig. 30-11, *B*). Remember that ''cold'' or ''warm'' refers to temperature relative to the ground of the source region, not the absolute temperature.

The weather associated with either type of occlusion is a combination of the very violent weather of a cold front and the widespread cloudiness and steady rain of the warm front. The reason for the odd combination is that the air masses are now mixing and the frontal system will soon start to die out. The worst weather connected with an occlusion, such as violent thunderstorms and tornadoes, occurs principally at the intersection of the three air masses (Fig. 30-12). The skies clear quickly after the occlusion speeds on its way.

Fig. 30-12. Tornadoes such as this usually occur at the intersection of incompatible fronts. (Courtesy United States Department of Commerce, National Oceanic and Atmospheric Administration, Rockville, Md.)

WEATHER MAP SYMBOLS

The United States Weather Bureau utilizes a 6-hour time period for major map analysis of current weather conditions around the world. Meteorologists locate pressure centers on the map by drawing lines between points of equal pressure; these lines are called *isobars*. Then, using known facts such as wind shifts, large temperature differences, bad weather zones, and cloud cover, fronts are accurately pinpointed and drawn. Table 33 summarizes the important frontal symbols found on weather maps.

Table 33. Weather map symbols

Map feature	Symbols	
	Manuscript maps	**Printed maps**
Cold front	Solid blue line	
Warm front	Solid red line	
Stationary front	Alternating red and blue	
Occluded front	Solid purple line	
Cold front aloft	Dashed blue line	
Warm front aloft	Dashed red line	
Stationary front aloft	Dashed alternating red and blue	
Occluded front aloft	Dashed purple line	

31 *the dance of death*

Destructive storms

Lord of the winds: I feel thee nigh,
I know thy breath in the burning sky:
And I wait, with a thrill in every vein,
For the coming of the hurricane.

The Hurricane
José Maria Heredia

Nowhere on earth is there a place safe from some naturally occurring type of disaster. Whether it be a typhoon slamming into the Japanese coastline, a tidal wave wreaking havoc on the Phillippine Islands, or a blizzard racing across the heartland of the United States or Europe, these events usually leave in their wake widespread death and destruction.

Meteorologists classify four major disturbances of the atmosphere as natural disasters. These are (1) estratropical cyclones, (2) thunderstorms, (3) tornadoes, and (4) tropical disturbances (hurricanes).

EXTRATROPICAL CYCLONES

The word "cyclone" literally means "the coil of a snake," and it is applied to any atmospheric system (not necessarily one of great violence) in which air pressure becomes progressively lower approaching the center, with the outer winds coiling inward and upward. These systems, commonly called lows on weather maps, usually are accompanied by bad weather, especially when they develop into tight, spiraling wind systems.

With the coming of autumn, cold, Arctic air begins to dominate the Northern Hemisphere as the warm, tropical air retreats southward. This condition is conducive to the formation of midlatitude cyclones called *blizzards,* which are the winter equivalent of the summer hurricanes.

These storms usually form in the prevailing westerly winds and thus move in a general west-to-east direction. A majority of these storms begin along the boundary of the Polar Front, where cold polar air meets warm tropical air. Here a storm will intensify and begin its sweep over tens of thousands of square miles, deepening into an intense low-pressure area accompanied by raging gale-force winds with speeds greater than 39 miles (63 km)/hr. With the passage of such a storm, temperatures plummet to below freezing and visibility is reduced to near zero. Snowdrifts pile up as high as 30 feet (9 meters), as normal activities come to a standstill (Fig. 31-1).

Death and destruction to man and his property from the great winter storms is vast, but much of the misery comes from automobile accidents, sickness, heart attacks, and falls not directly attributable to the storms. People who live in areas frequented by these storms are always prepared for long sieges of paralyzing weather.

Life cycle of extratropical cyclones

Extratropical cyclonic development is usually associated with frontal activity. A cyclone usually develops on a stationary front and passes through six distinct steps before it dies out, a victim of its own growth. A cyclonic model was first developed by J. Bjerknes of the Norwegian School of Meteorology in 1919 and is still in use today as a basis for weather forecasting. The following is a description of the cycle, which is shown schematically in Fig. 31-2.

Stage 1: inert system. A stationary front separates cold and warm air masses, with winds parallel to the frontal boundary (Fig. 31-2, *A*).

Stage 2: cyclogenesis (birth). Differences in temperature or topography result in a lowering of the pressure along

Fig. 31-1. In the wake of a blizzard only the greatest of efforts keeps our daily routine from coming to a screeching halt. (Courtesy United States Department of Commerce, National Oceanic and Atmospheric Administration, Rockville, Md.)

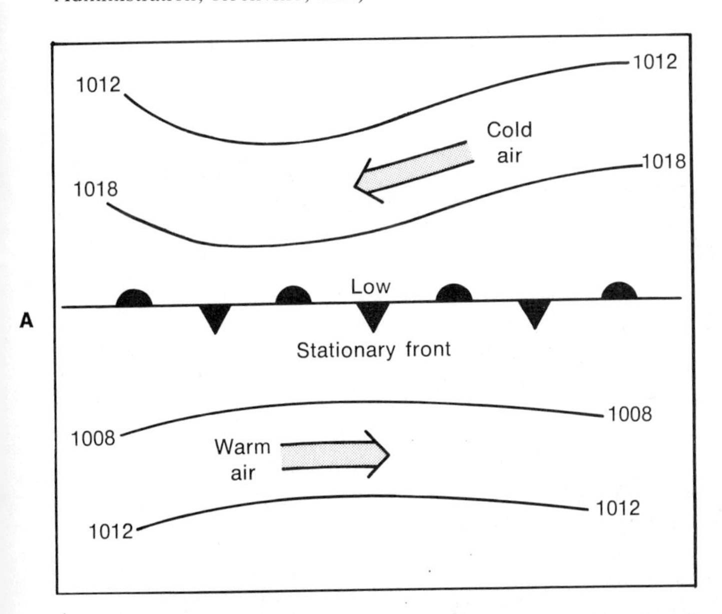

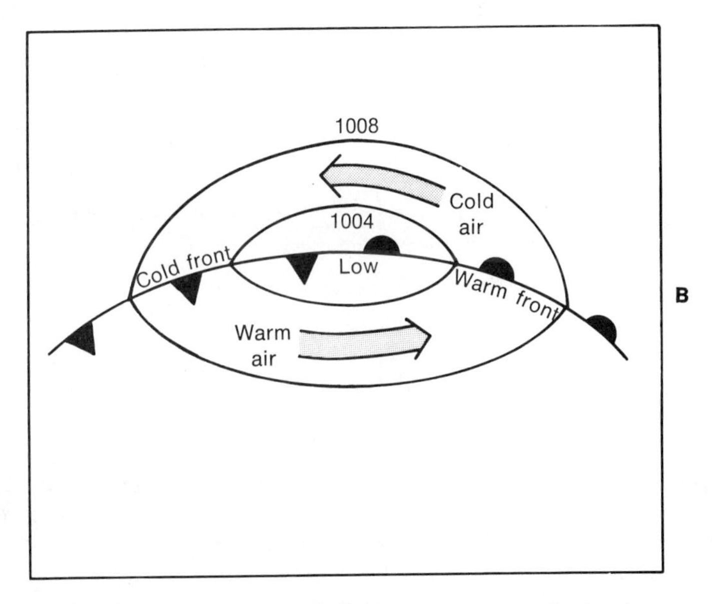

Fig. 31-2. Diagrams will help in tracing the cyclone cycle as described in text. Numbers on the diagrams indicate zones of pressure (isobars). It is the differences in pressure that create the potential for violent storms.

Continued.

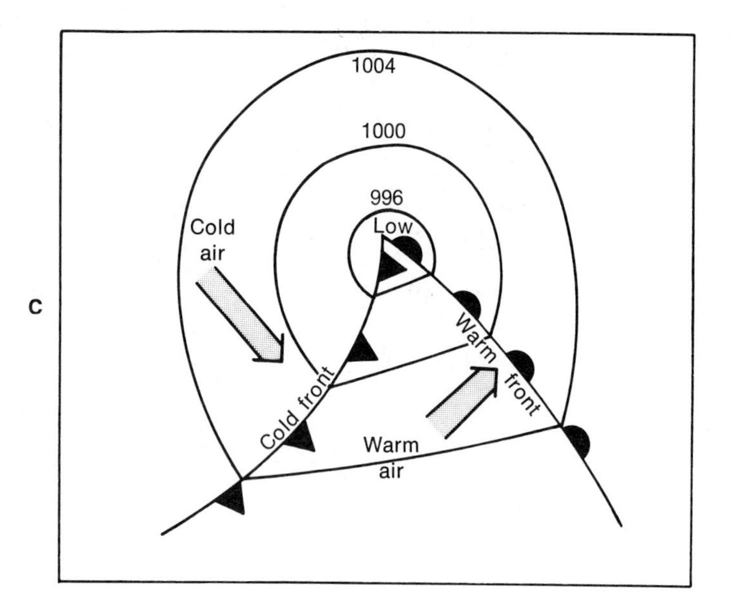

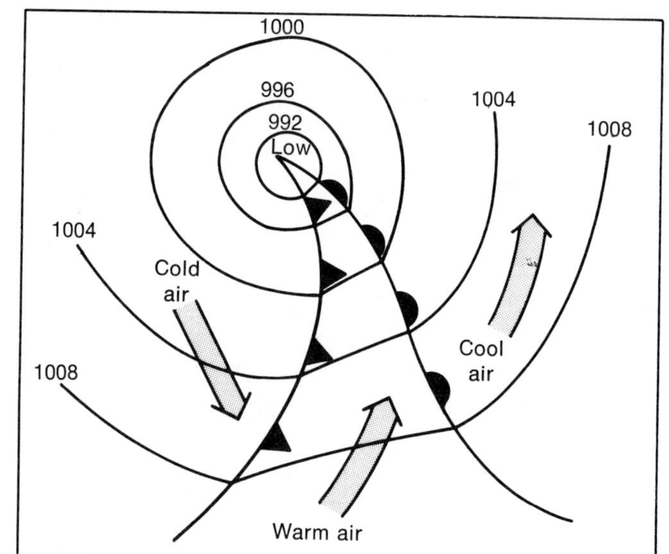

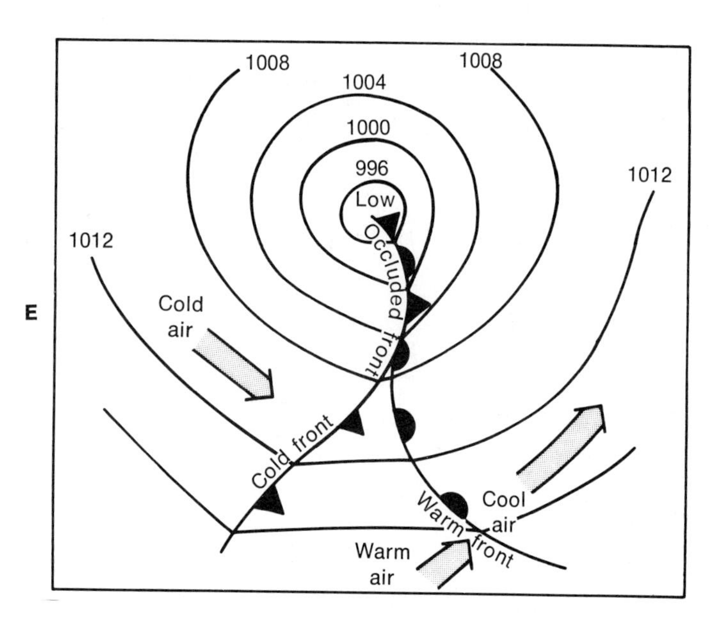

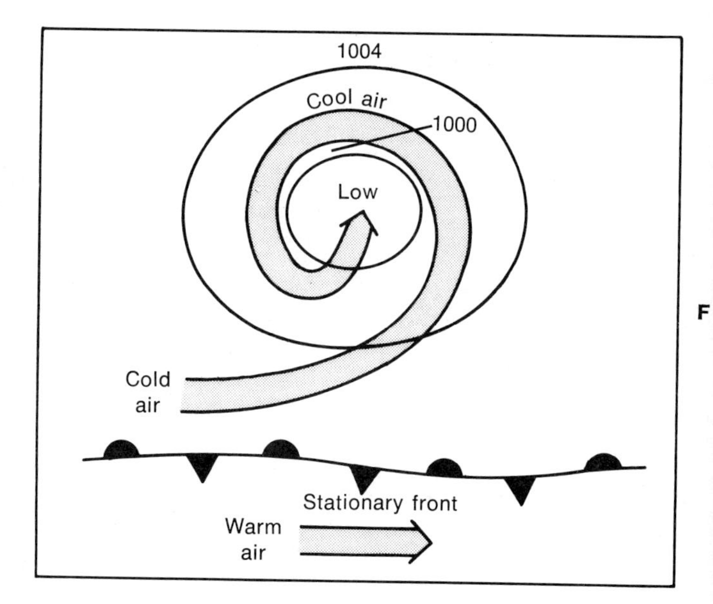

Fig. 31-2, cont'd. For legend see p. 399.

the frontal boundary. The area of lower pressure develops a weak but definite counterclockwise circulation (Fig. 31-2, *B*).

Stage 3: wave formation. As the circulation intensifies, the warm air moves northward and the cold air moves southward. This sets up two distinct fronts in the form of a wave, with the warm front leading the cold front (Fig. 31-2, *C*).

Stage 4: maturity. Pressure continues to fall steeply and the low takes on a circular shape as winds begin spiraling inward and upward, producing continuous cloud cover and widespread precipitation. The cold front, which moves much faster, begins to overtake the slower warm front (Fig. 31-2, *D*).

Stage 5: decline. The sytem reaches maximum development, lasts only a short time, and rapidly breaks up. This process is initiated when the cold front overtakes and mixes with the warm front. This mixing results in an occluded front, and the warm air is lifted off the surface (Fig. 31-2, *E*).

Stage 6: cyclolysis (death). The cold front completely

overtakes the warm front at the surface, and the mixing system deteriorates into a whirl of cold air; a stationary front lies to the south, waiting to be stirred up once again (Fig. 31-2, *F*).

Winter cyclone tracks in the United States

Winter storms pushing across the United States take preferred tracks (Fig. 31-3) that depend on their point of origin or penetration of the mainland. The Rocky Mountains present a formidible obstacle that can rarely be overcome by a majority of storms striking the Pacific coastline. Those storms that are able to cross the mountains arrive in a weakened condition and must redevelop as a Colorado Low (track *3*). Most of these storms, however, move northward to develop into the Alberta Low (track *1*), which, like its southern counterpart, moves toward the Great Lakes region.

Along the eastern seaboard, winter storms usually originate at the North Carolina coast and skirt the coastline, accompanied by gale-force winds and huge snowfalls (track *6*). These are the familiar northeasters, which take their name from the continuous northeast winds whirling ahead of the system.

THUNDERSTORMS

Although not nearly as extensive or long lasting as extratropical or tropical storms, the local cyclonic storm, known as a thunderstorm, is formed wherever unstable conditions occur in the lower atmosphere. Thunderstorms occur in such large numbers each year and are so short lived that they may be nature's most elusive and destructive weather.

Thunderstorms, which are found almost everywhere in the world, are the end product of a wildly growing cumulonimbus cloud. All thunderstorms are marked by these clouds, which at maturity may tower to heights of 70,000 feet (21,350 meters), spread across 1 to 7 miles (1.6 to 11 km) and move at speeds as great as 25 miles (40 km)/hr. Lightning and thunder are determining characteristics of the storm, which may be accompanied by heavy rains, gusty winds, hail, and tornadoes.

Thunderstorms are formed by strong cores of rising air with high relative humidities. As the air is lifted, it eventually reaches saturation, forming small vertical cumulus clouds that continue to grow as the saturated air continues to rise. Individual clouds are called convective cells because of the convective transfer of heat that occurs, and each has a life span of 1 to 3 hours as it clusters with other cells, remains isolated, or lines up in a thick wall called a *squall line*.

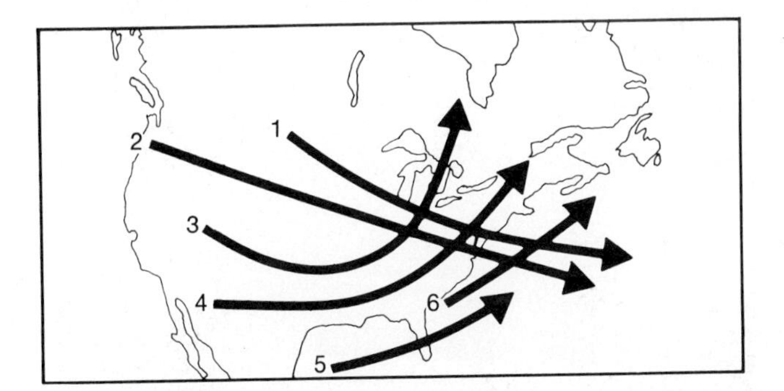

Fig. 31-3. Like the air masses on which they depend, cyclonic storms follow preferred tracks across the continent. See text for discussion.

Phases of thunderstorm development

Thunderstorms develop in three distinct stages; each phase is identified by its own characteristic features.

Cumulus stage. In order for the cumulus stage of a thunderstorm to develop, some type of mechanical lifting

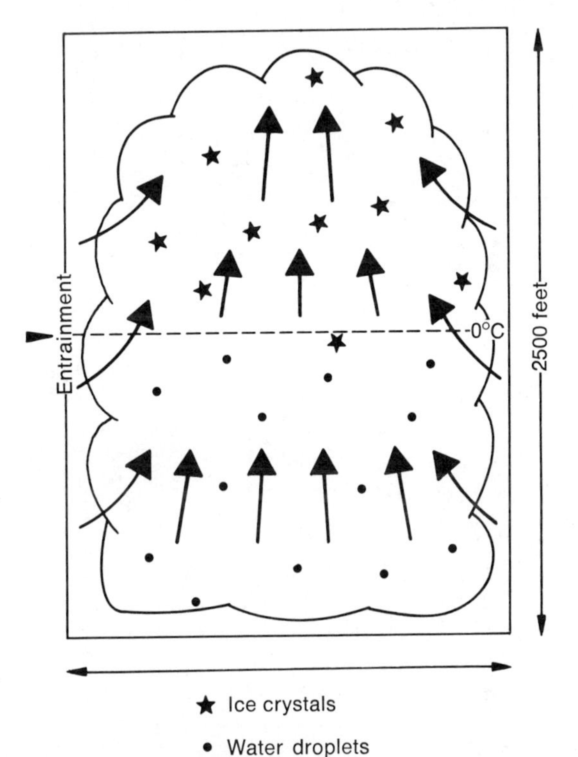

Fig. 31-4. In the first stage of thunderstorm development, two conditions are requisite: something must force the air to rise, and there must be additional surrounding air as well as a force to pull it in to keep the system well supplied.

of air must take place. If the rising air is warmer and less dense than the surrounding air, it will continue to rise of its own accord (unstable air). The rising air eventually becomes saturated and forms small cauliflower-shaped cumulus clouds, which are the distinctive feature of this initial stage. These clouds, called *cumulus humulus* (humble clouds), are formed entirely by updrafts, which in turn pull in air from the surrounding environment to resupply the system with the fresh, moist air needed for growth. This process of pulling new air into the system is called *entrainment* (Fig. 31-4).

In this initial stage cumulus clouds can reach altitudes of 25,000 feet (7625 meters), which is well above the altitude at which freezing occurs, so that ice crystals and large water droplets mix freely, leading ultimately to the formation of the second stage.

Mature stage. In order for this most violent stage of the thunderstorm to begin, downdrafts must form. As raindrops and ice crystals grow in size, the updrafts can no longer overcome the pull of gravity, and the particles fall. Frictional drag and the cooling of the warmer air create these downdrafts; although they are not as strong as updrafts, which reach speeds of 100 miles (160 km)/hr, they can sometimes reach speeds of 50 miles (80 km)/hr). The downdrafts come pouring out of the leading edge of the thunderstorm, creating the familiar strong, cool winds that precede the arrival of the storm. Close behind this initial rush of cool air are heavy rains, gusty winds, possibly hail, and in some cases, tornadoes. The onset of hail usually indicates that the storm's central core has arrived (Fig. 31-5).

HAIL. Hail is a form of precipitation produced only in thunderstorms. These rounded lumps of soft ice are the result of supercooled water growing on ice particles as they constantly circulate through the cumulonimbus cloud. Each layer freezes on ascent in the updrafts, only to have another layer of supercooled water encircle the ice during descent in the downdraft. Thus alternating layers of clear

Fig. 31-5. Summary of the cloud types and spatial relationships involved in a thunderstorm. (Courtesy United States Department of Commerce, National Oceanic and Atmospheric Administration, Rockville, Md.)

and opaque ice are formed. Eventually the pieces of ice become too heavy for the updrafts to support. In the midwestern part of the United States hailstones are frequently the size of golf balls and occasionally baseballs. The largest hailstone ever found fell in Potter, Nebraska, on July 6, 1968; it measured 5½ inches (14 cm) in diameter, 17 inches (43 cm) in circumference, and weighed 1½ pounds (3.8 kg). Although hail does great damage to crops and other property, there have been only a few human deaths directly attributed to it.

LIGHTNING AND THUNDER. Lightning and thunder always accompany the mature stage of a thunderstorm. Lightning is the convective flow of the atmospheric electricity that has accumulated in the cloud. This tremendous surge of charged atoms is capable of causing fires, blowing apart trees and buildings, ionizing the air, and killing. Although the most common type of lightning is the streak form (Fig. 31-6), it can also take the shape of a sheet or ball. Ball lightning is most fascinating because the electrical charges accumulate as a luminous ball about the size of a basketball that floats through the air with a buzzing sound. There are authenticated reports of ball lightning entering a house through a front window, floating through the rooms, and disappearing through an open rear window.

Thunder is an audible by-product of the rapidly expanding gas along the path of discharge. Close to the area of impact it is heard as a sharp clap, but at greater distances it has a characteristic rumbling sound. If an observer wanted to calculate his distance from a lightning stroke, he could use as a rule of thumb 1 mile to 5 seconds between stroke and sound.

Dissipation stage. The last stage of the thunderstorm cycle is called the dissipation stage, which begins when downdrafts become more numerous than updrafts (Fig. 31-7). Eventually the cloud consists entirely of downdrafts, which dry it out. As the air loses its moisture, the cloud rapidly breaks up and light drizzle falls earthward. The final

Fig. 31-6. Examples of the more familiar type of lightning, streak lightning. (Courtesy United States Department of Commerce, National Oceanic and Atmospheric Administration, Rockville, Md.)

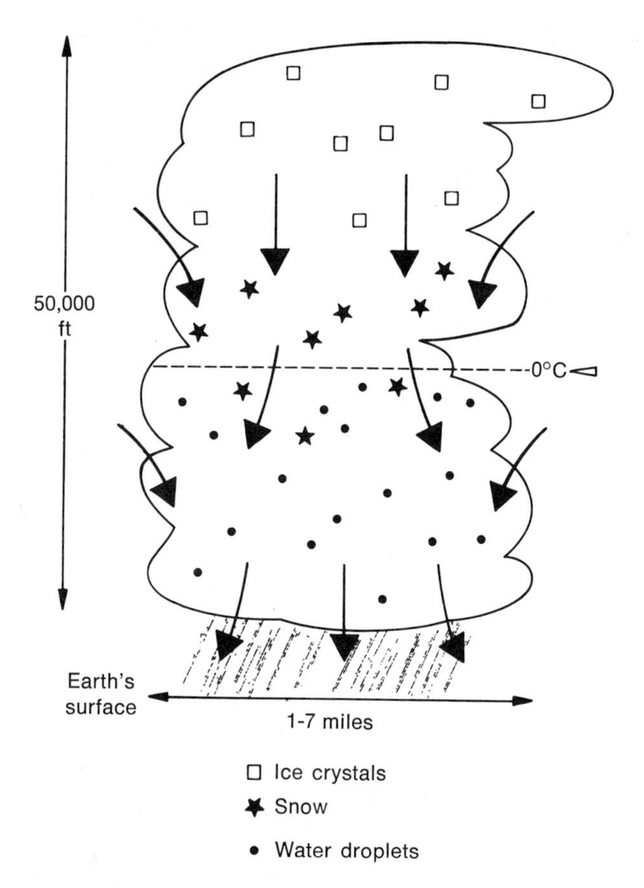

Fig. 31-7. As the cloud's updrafts are outnumbered by growing downdrafts, the cloud breaks up of its own accord and enters the dissipation stage.

Fig. 31-8. Tornadoes, one of the more destructive types of storm and one of those with the fewest redeeming qualities, are likely candidates for weather modification. (Courtesy United States Department of Commerce, National Oceanic and Atmospheric Administration, Rockville, Md.)

process ends with low stratiform-type clouds drifting lazily in an open sky.

Effects of thunderstorms

While thunderstorms, like other cyclonic-type storms, produce great damage and death, they are also beneficial because of their outpouring of precipitation, which is an important source of fresh water. With the exception of Russian hail-suppression studies, little has been done to control these storms. However, the harnessing of lightning for electric power and the modification of thunderstorm-spawned tornadoes are areas of great concern to the meteorologist and are constantly under study.

As with all storms, local or global, precautions and preparedness are the best approaches to the prevention of damage and death. The National Oceanic and Atmospheric Administration (NOAA), through the National Severe Storms Forecast Center in Kansas City, Missouri, keeps track of thunderstorm development and issues warnings whenever severe conditions exist.

TORNADOES

The smallest, most violent, and most feared storm system is the tornado. The word ''tornado'' is derived from the Spanish *tornar,* which means to turn. Even though they are short-lived and cover only a very small area, no other storm on earth can match their destructiveness. It is for this reason that tornadoes merit all our efforts in the area of weather modification (Fig. 31-8).

Tornado occurrence

Tornadoes are invariably associated with severe thunderstorms, although they can accompany other cyclonic storms. They are formed near fast-moving cold fronts, squall lines over water (called waterspouts), and are even embedded in hurricanes. They may develop at any time of the day or night but tend to occur between 4 P.M. and 6 P.M. Although tornadoes have occurred in every month of the year, April, May, and June tend to be the period of maximum frequency. Tornadoes can form anywhere in the world, but the Plains States of the United States can claim the distinction of being the tornado capitals of the world.

Early in the tornado season, usually in February, development in the United States is centered over the Central Gulf states and Florida. Toward summer, the area in which tornadoes develop moves northwestward until it reaches its summer maximum over the Plains area. This movement of the center of maximum tornado occurrence is explained by the shifting of the point at which warm, moist air meets cold, dry air.

Although our knowledge of tornado formation is less than we would like, it appears that the ingredients needed to trigger one are differences in air moisture, temperature, density, and flow. How the vortex is actually formed is still being debated, but it appears to be through a combination of thermal, mechanical, and electrical effects.

Tornado formation

Usually the first sign of funnel formation is the appearance of large undulating clouds attached to a main thunderstorm cloud. A generally warm, humid, and uncomfortable condition accompanies the first visual signs. The appearance of these large hanging clouds indicates severe turbulence and extremely unstable conditions, both prime ingredients for tornado formation. Suddenly a dark extension of the thunderstorm cloud reaches earthward, skipping and jumping back towards its point of origin. Air immediately rises violently around a central vortex as it spins in a spiral motion; air from the region outside the center moves rapidly inward to join the spinning updrafts. This allows large quantities of dust and debris to be carried aloft, making the funnel visible.

The tornado (called a funnel cloud if it does not touch the earth's surface) wavers from side to side as it reaches forward speeds as great as 70 miles (112 km)/hr. The air, still spiraling at over 300 miles (483 km)/hr, causes an extreme pressure drop at the center, which is believed to cause more destruction than the spiraling wind.

Death and destruction

The combined action of low central pressure and high wind speeds are responsible for the death and destruction attributed to tornadoes (Fig. 31-9). As the storm passes over a building, winds around the vortex at speeds of 300 miles (483 km)/hr or more rip at the outside of the struc-

Fig. 31-9. Although tornadoes fortunately follow narrow paths, the destruction left by one's passage is appalling. (Courtesy United States Department of Commerce, National Oceanic and Atmospheric Administration, Rockville, Md.)

ture, while the abrupt pressure change outside causes the air at normal pressure inside to rush outward; this produces the collapse of the structure and the resultant loss of life.

The dramatic results of these forces are well known. Straws have been driven into trees and telephone poles as though they were nails, giant trees have been uprooted like toys, and large objects and people have been hurled through the air and occasionally deposited gently. In 1930, for example, a tornado in Minnesota carried an 83-ton railroad car filled with people 80 feet into the air before dropping it softly into a ditch. In another case, one side of a city street was completely leveled while the other side remained untouched.

Protection and warning

Meteorologists at the National Severe Storms Forecast Center work 24 hours a day, 365 days a year, analyzing atmospheric data from all over the world. These data, acquired from sources such as satellites, radar, upper-air balloon soundings, and weather observers, are collected at a central point and analyzed in order to identify areas favorable to tornado formation (Fig. 31-10). Susceptible areas are put on alert through an efficient warning network, and the warnings are disseminated to the general public by the news media. Usually the *tornado watch,* put into effect in the susceptible area, becomes a tornado warning only after a tornado has been seen or identified as a small hook image on a radarscope. Although there have been numerous schemes proposed for tornado identification, the best source of reliable information is still the United States Weather Bureau.

TROPICAL DISTURBANCES: HURRICANES

On Sunday, August 17, 1969, winds in excess of 200 miles (322 km)/hr slammed into Gulfport, Mississippi. In their wake these hot, sultry winds left more than 200 dead, thousands injured, and property damage of nearly $1.5 billion. The main killer in the storm was the *storm surge,* a series of extremely large waves that inundated the northern coast of the Gulf of Mexico. A 24-foot (7-meter) tide accompanied this most destructive of all storms, the tropical hurricane.

This vivid example, like thousands more that occur each year, show dramatically how people and their property are often at the complete mercy of the weather.

Fig. 31-10. Waterspouts, or tornadoes occurring over water, are carefully watched by the United States Weather Bureau so that warning can be given if one should head for populated areas on land. (Courtesy United States Department of Commerce, National Oceanic and Atmospheric Administration, Rockville, Md.)

Naming tropical systems

The infrequent storm systems of the lower latitudes have many regional names. In the Northern Pacific they are called *typhoons,* while *baguio* is the term used to identify this same disturbance in the Philippines and the China Sea. To most of the population in Australia, a *willy-willy* is exactly like the cyclones of the Indian Ocean and the hurricanes of the West Indies, Caribbean Sea, and Atlantic Ocean. However, the most elaborate name is in use on the west coast of Mexico, where residents call a tropical cyclone *El Cordonazo de San Francisco,* "the last of St. Francis." Regardless of the name they are given, all are storm systems born in the tropics and all are accompanied by high winds, torrential rains, and widespread destruction.

The practice of naming individual storms began during World War II. It was initiated by the United States Weather Bureau but has since spread to many other parts of the world. In the early stages of storm identification the Weather Bureau used a permanent alphabetical list of women's names, retiring only those that caused great damage and death. Women's names were chosen, not because of the unpredictability of the systems (as has been suggested), but because they are often short and easy to remember.

In 1960 the Weather Bureau prepared a semipermanent, alphabetical list of four sets of women's names. Because of the scarcity of names beginning with Q, U, X, Y, and Z, these were omitted. A separate list is used each year, and the first storm is given the first name on the chosen list. Again, names of memorable storms are retired for a 10-year period.

Classifying tropical cyclones

The United States Weather Bureau divides tropical systems into four categories based on criteria that range from light rain and wind to monsoon-type downpours accompanied by destructive winds.

Tropical disturbances. A tropical disturbance is an incipient storm that has only a slight surface wind circulation. It may be just an area of light showers or cloudiness. However, in the tropics such developments are immediately viewed with suspicion as the early stages of possible storm development.

Tropical depressions. If a system acquires one or more closed isobars (an indication that points of equal pressure are forming rings), a rotating wind circulation can develop. As pressure continues to drop near the center, wind speeds increase to speeds of 31 miles (50 km)/hr. Rainfall increases, forming a few narrow bands, some of which are heavy downpours. In the Caribbean Sea a tropical depression is sometimes called an *easterly wave*.

Tropical storms. As the depression continues to develop, the pressure drops drastically, giving rise to a definite circular wind pattern. Large amounts of warm, moist air continue to be pumped into the lower levels by wind speeds, which increase to 72 miles (116 km)/hr. This constant resupply of moisture results in widespread precipitation, with the rain-band effect increasing as the pattern of squall lines develops.

Mature tropical storm: hurricane. The pressure drop within the system is enormous, with a reading at the center of 28 inches (41 cm) of mercury or lower—normal pressure is just about 30 inches (76 cm). Extremely heavy rainfall of the monsoon type occurs in squalls that stop as suddenly as they begin. The wind reaches a force greater than 72 miles (116 km)/hr near the center, with gale-force winds extending outward for hundreds of miles.

Hurricane formation

Hurricanes usually begin as an area of lower pressure embedded in a belt of easterly wind flow, such as the Trade Winds. The easterly wave, as it is called, is believed to originate off the western coast of a large continent. In the case of Atlantic hurricanes, this source area is the desert of North Africa. As the system moves over warmer ocean waters, the air takes up tremendous quantities of water vapor. Since the air is unstable, the water vapor is lifted to altitudes at which it cools and condenses, releasing vast amounts of energy that are the life blood through which tropical storm systems intensify.

Formative stage. The formative stage of tropical cyclone development depends on the temperature of the sea surface. Observations of storms indicate that to maintain a vertical circulation, a steep lapse rate (decrease of temperature) must be maintained. This only occurs when water temperatures are greater than 80° F (27° C). In addition, the rotary circulation so vital to the growth of the tropical system must be maintained by the Coriolis force. Without this force there would be little or no spin of air particles around the low-pressure center and a hurricane would not form. This is the main reason that hurricanes do not form or travel close to the *Doldrum belt* (3° N to 3° S), for here the Coriolis force is zero.

Intensification stage. The tropical storm, or intensification, stage is characterized by a well-developed chimney mechanism whereby air converges toward the center at the surface, spirals upward, and diverges at higher levels. The pressure drops below 1000 millibars as the wind begins to speed in a tight ring, forming a central eye. Immense towering cumuliform clouds spiral outward from the eye wall in narrow bands.

Fig. 31-11. Hurricane Ava, a Pacific hurricane formed off the coast of Salina Cruz, Mexico, on June 2, 1973. Note the distinct eye wall and the spiral banding of clouds as the air spins counterclockwise toward the center. (Courtesy United States Department of Commerce, National Oceanic and Atmospheric Administration, Rockville, Md.)

It is believed that the interaction between high- and low-level winds is the factor that determines whether the system will continue to develop. Theory suggests that if less air is pumped out at higher altitudes than is converging at the surface, the eye will fill and the pressure will rise, breaking up the eye and dissipating the storm.

The eye is a calm center 15 to 40 miles (24 to 64 km) in diameter that is associated with every mature hurricane. The calm of the eye provides a sharp contrast to the wind velocities of 100 miles (161 km)/hr or more found in the *eye wall*. This wall is made up of towering thunderstorm clouds with heights sometimes reaching 80,000 feet (24,400 meters). Updrafts spiral at extreme rates in the central cloud band, while downdrafts of warm, dry air descend gently in the eye. The subsiding warm, dry air is the mechanism responsible for the calmer conditions in the center of the hurricane (Fig. 31-11).

Mature stage. The mature stage of a hurricane is attained when winds reach their maximum speed and pressure hits its low point. The cyclonic circulation is marked by towering bands of cumulonimbus clouds that produce heavy rain in the form of squalls. The strong gale winds reach out to distances as great as 400 miles (644 km), especially in the most dangerous *northeast semicircle* or quadrant. Close to the end of the mature stage, the central pressure and wind speeds level off. By this time, the system has usually swung to the north again and entered the prevailing westerly wind flow, where it moves over colder water (slower evaporation and less energy) or a large land area (no evaporation and no energy) in the middle latitudes. Thus deprived of the energy needed to maintain its strength, the hurricane dries out and loses it power.

Decaying stage. As the decaying stage commences the winds die down, but heavy rain continues, often flooding even inland regions. The system takes on extratropical characteristics and eventually reverts to a simple low-pressure area. However, in certain situations tropical cyclones in the decaying stage may travel off the coast and regenerate into intense storm systems as they once again move over the sea.

Hurricane weather

Far in advance of a hurricane an abundance of cirrus clouds appears. In rapid succession, clouds similar to those seen in the approach of a warm front make their appearance, soon giving way to the low nimbostratus clouds. Bands of dense, dark rain clouds coupled with ugly, thick thunderstorm clouds begin to spiral from the eye (Fig. 31-12).

The precipitation pattern consists of showers and thundershowers starting 50 to 60 miles (80 to 96 km) from the center. These are followed by torrential, wind-driven rains 20 to 30 miles (32 to 48 km) from the eye. The rain is supplemented by water already on the ground, which is whipped into the air and driven horizontally by the wind. This reduces visibility to zero in squalls and makes rainfall measurements almost impossible. Precipitation stops only in the eye, starting again just as rapidly after its passage.

Hurricane tracks

Once they form, hurricanes follow paths that are as varied as the storms themselves (Fig. 31-13). They generally drift eastward shortly after formation; however, sharing the general movement of the easterly wind belt. Soon, as the energy of a storm increases, its movement becomes erratic and unpredictable. Although hurricane movement has not been fully explained, it is believed that their erratic paths are the result of a battle between internal and external forces. The external forces, called *steering*

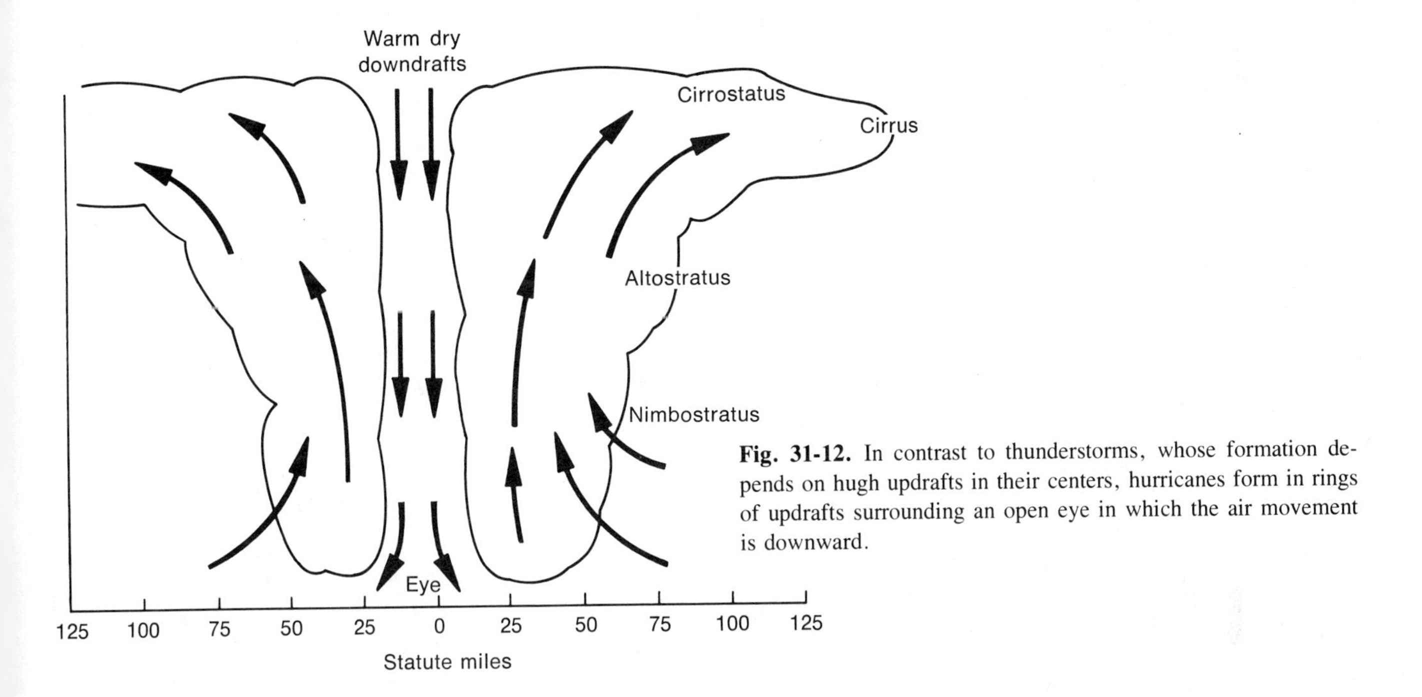

Fig. 31-12. In contrast to thunderstorms, whose formation depends on hugh updrafts in their centers, hurricanes form in rings of updrafts surrounding an open eye in which the air movement is downward.

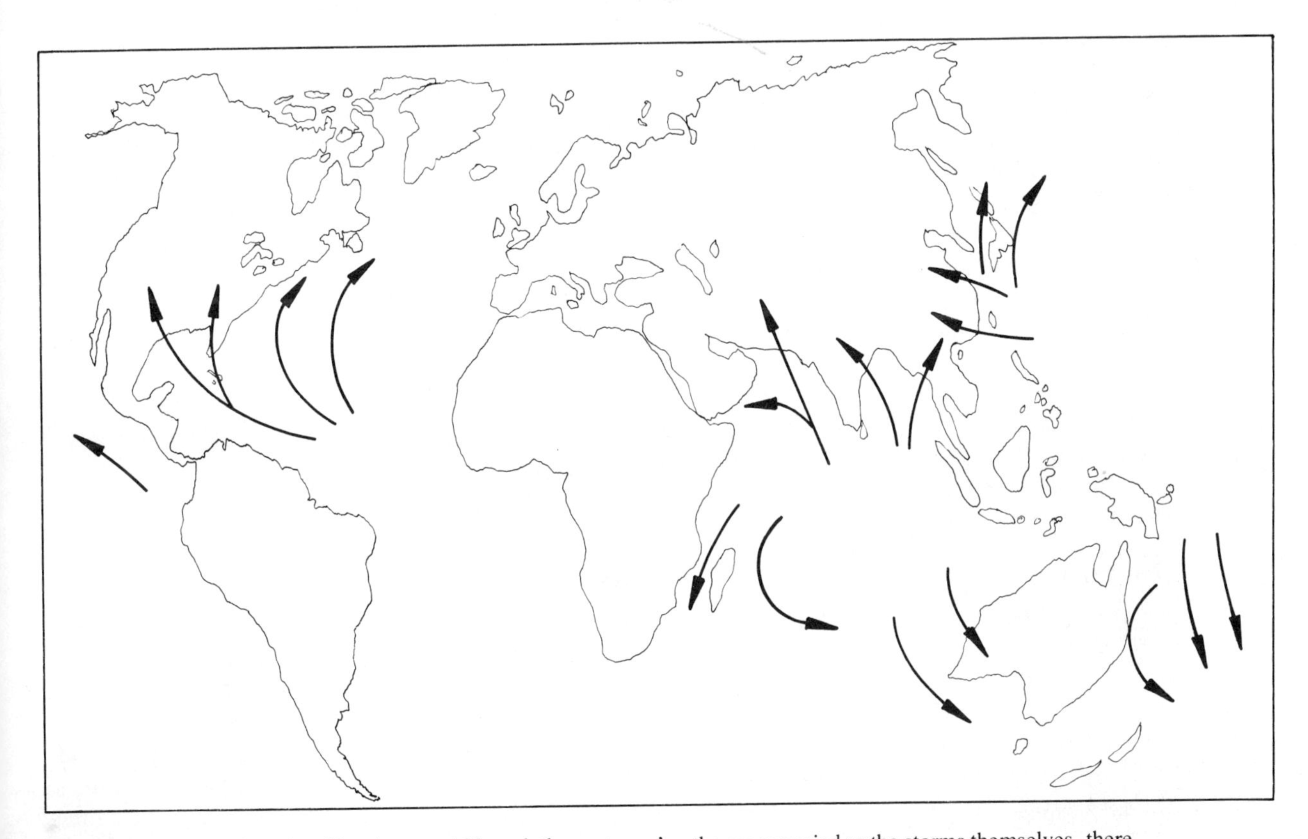

Fig. 31-13. Preferred tracks of hurricanes. Although these storms' paths are as varied as the storms themselves, there does seem to be a certain amount of clustering in certain areas.

currents, are usually dominant. Hurricane tracks often follow a curving clockwise path in the Northern Hemisphere, but too many exceptions exist to make this a rule. Hurricane Inez, for example, doubled back instead of recurving into the North Atlantic, finally expiring in the mountains of Mexico.

The life expectancy of a hurricane that remains over water averages 12 days if it is formed in the month of August and 8 days if the hurricane occurs in the cooler months of July or November. Even after losing its tropical characteristics, the storm may become an extratropical cyclone or combine with an existing low outside the tropics.

The good with the bad

The damage inflicted by hurricanes is well documented. Most of the death and destruction is the result of the force of the wind, flood-producing rains, and the ever-present coastal threat, the storm surge from the ocean.

The wind alone, although at times truly awesome, would have to be considered the smallest threat of the three. It can be lethal, however, as it picks up loose objects and hurls them like missiles or as the wind pressure seeks out flaws in structural design.

A typical hurricane brings from 6 to 12 inches (15 to 30 cm) of rain to the area it strikes. Streams cannot carry this excess volume of water, and floods are the natural result. Hurricane Diane in 1955 flooded Pennsylvania, New York, and the New England states, killing 200 persons and inflicting an estimated $700 million in damage. Hurricane Camille in August, 1969, drowned more than 100 people in Virginia and West Virginia after spending most of its force on the Gulf Coast.

In coastal areas the greatest danger comes from the ocean turned killer. Hurricane-force winds can whip up waves as high as 50 feet. These storm waves, often coinciding with high tide, can destroy low-lying coastal areas both by flooding and by the tremendous impact of the waves themselves. The toll in human life and property from storm surges has been staggering. Besides the 500,000 dead in East Pakistan in 1970, 300,000 were killed in India in 1737, and another 50,000 lost their lives in the same region in 1864. In Florida in 1928, 2000 were killed as a hurricane turned Lake Okeechobee into a nightmare. In 1932, 2500 people lost their lives in Cuba.

Almost lost in the statistics of destruction is a benefit the storms provide—rainfall. It is estimated that one fourth of the annual rainfall of the southeastern United States comes as a result of hurricanes or their related weather patterns. Without this rain this area would certainly not become a desert, but over a period of years significant changes would occur in the flora and fauna. Crop-supporting characteristics of this agricultural region would change, and there would be a noticeable drop in the water table.

Notable hurricanes in recent history

The most deadly hurricane of the twentieth century hit Galveston, Texas, on September 8 and 9, 1900. The storm brought with it a 5-foot storm surge that hit a city already deluged by heavy rains. Of a population of 38,000, more than 6000 drowned.

Another important storm was the San Felipe hurricane in 1928. It lashed across Puerto Rico after demolishing the islands of Guadeloupe and St. Kitts. Although it lost much of its force crossing the mountains of Puerto Rico, it quickly regained its strength and spread its large tentacles over the United States mainland. Moving inland at Palm Beach, Florida, it raced northward up the Atlantic seaboard. The death toll of this killer storm was 300 in the West Indies and more than 1500 in the United States. Most of the loss of life in the United States occurred in the inland regions of Florida, especially in the vicinity of Lake Okeechobee, where losses from wind and flooding exceeded $75 million.

The most costly hurricane and one of the strongest storms ever to hit the United States was 1969's Camille. Approximately $1.5 billion in property damage was caused by continuous flooding along the Mississippi and Louisiana coasts and in the interior of Virginia and West Virginia. The low death toll (200) was attributed to warnings given by the Weather Bureau and heeded by the populace. Otherwise, a tragedy of the first magnitude could have occurred.

Forecasts

If people do not intervene, the great storms will continue to ravage the earth. Intervention by man in many areas of nature's realm must be considered failures, however, so the watchwords for future hurricane control must be "research with caution."

The National Oceanic and Atmospheric Administration (NOAA) and the United States Navy have for the past several years been engaged in a joint research program to explore methods of modifying or neutralizing hurricanes. This joint effort is known as Project Stormfury. Whenever a hurricane with suitable characteristics comes within a designated area, attempts are made to modify it.

It is known that the power of a hurricane is so vast that no energy source at the disposal of man, with the possible exception of a hydrogen bomb, can neutralize it. Project Stormfury is attempting to probe ways in which these storms can be made to destroy themselves. There are four approaches under consideration.

1. Prevent the formation of the storms by introducing competitive points of preferred convection.
2. Drain off the inflowing air in the outer parts of the storm before it has a chance to rise.
3. Cut off evaporation of water from the ocean surface in the area where a storm appears ready to form.
4. Alter the condition of the clouds in the eyewall of the storm to (a) force the incoming air up prematurely, before formation of the deadly central vortex of low pressure, and (b) encourage precipitation, so that moisture output exceeds intake.

Thus far only the last technique has been tried directly. Hurricanes Esther in September, 1961, Beulah in August, 1963, Debbie in 1969, and Ginger in 1971 were seeded with silver iodide crystals. Although the results were encouraging, they were also inconclusive, for the minor alterations that occurred could have been the result of normal changes within the storms.

Until the day when the intensity of the great storms can be diminished without loss of the beneficial rains they bring to the earth, we must keep our eyes turned to the sea in a state of watchful awareness. We must listen to the forecasts and advisories issued by the weather bureau and be prepared for prompt action should the worst come.

Hurricanes, like other weather elements, know no boundaries, for weather is international. Their study therefore raises many serious political, economic, social, and legal questions. Through well-planned weather modification, however, our lot on this planet will be improved and secured.

EPILOGUE

Thine individual being, shalt thou go
To mix forever with the elements,
To be a brother to the insensible rock.

Thanatopsis
William Cullen Bryant

Although there are other planets in our solar system and probably other planets orbiting other stars in our galaxy or other galaxies elsewhere in the universe, the earth is unique, as far as we know, as a sustainer of life. Its primordial elements have formed everything on this planet—land, water, atmosphere, even life itself. Except for small amounts of meteoric material that fall to the earth each day from space, the earth is a closed environment. As a result, we must live with the raw materials we have, for we will get no more.

We continue to believe that when the earth as an environment is spoiled and can no longer be inhabited because of the problems of overpopulation, depleted natural resources, air unfit to breathe, or water too dirty to drink, we can move on to another planet. With our present space technology or with any technology even remotely within our reach we cannot transport masses of people from the earth to another planet, even if we knew of such a planet that could support human life. The distances of space are too great and times required for travel are too long. Before such technologies could be developed we would, at our present rate of environmental destruction, follow the dinosaur and passenger pigeon to extinction. Of all the creatures ever to live on earth, we alone have the power to destroy ourselves and every living thing that inhabits the planet.

Therefore we must stop raping our environment and start caring for it gently as though our lives depend on it, for in fact they do. This care must reach from the minuscule to the gross, from our own backyards to the larger picture, the earth itself. We must not only understand the parts of the earth but also how they are interrelated. We must work with the planet rather than against it, and we must realize there are some things on earth that we cannot change and others that we must. As an umbilical cord is to an astronaut in space, so we are inexorably linked to the earth environment.

This is the legacy we must leave to those who come after us, or soon there will be no others. It is not yet too late, but time is running short. We are naturally nearsighted to problems that are larger than ourselves, but our vision must be improved. We, eternal optimists, must become pessimists about our earth. The earth is a beautiful planet. It is the only home we have. It can, however, be destroyed just as surely as if some mysterious force were to rend it into billions of useless pieces.

APPENDICES

appendix A

The constellations*

Constellation	English name or description	Abbreviation	Approximate position of constellation's center	
			Right ascension (hours)	Declination (degrees)
Andromeda	Princess of Ethiopia	And	1	+40
Antlia	Air pump	Ant	10	−35
Apus	Bird of Paradise	Aps	16	−75
Aquarius	Water bearer	Aqr	23	−15
Aquila	Eagle	Aql	20	+5
Ara	Altar	Ara	17	−55
Aries	Ram	Ari	3	+20
Auriga	Charioteer	Aur	6	+40
Boötes	Herdsman	Boo	15	+30
Caelum	Graving tool	Cae	5	−40
Camelopardus	Giraffe	Cam	6	−70
Cancer	Crab	Cnc	9	+20
Canes Venatici	Hunting dogs	CVn	13	+40
Canis Major	Big dog	CMa	7	−20
Canis Minor	Little dog	CMi	8	+5
Capricornus	Sea goat	Cap	21	−20
Carina†	Keel of Argonauts' ship	Car	9	−60
Cassiopeia	Queen of Ethiopia	Cas	1	+60
Centaurus	Centaur	Cen	13	−50
Cephus	King of Ethiopia	Cep	22	+70
Cetus	Sea monster (whale)	Cet	2	−10
Chamaeleon	Chameleon	Cha	11	−80
Circinus	Compasses	Cir	15	−60
Columba	Dove	Col	6	−35
Coma Berenices	Berenice's hair	Com	13	+20
Corona Australis	Southern crown	CrA	19	−40
Corona Borealis	Northern crown	CrB	16	+30
Corvus	Crow	Crv	12	−20
Crater	Cup	Crt	11	−15
Crux	Cross (southern)	Cru	12	−60
Cygnus	Swan	Cyg	21	+40
Delphinus	Porpoise	Del	21	+10
Dorado	Swordfish	Dor	5	−65
Draco	Dragon	Dra	17	+65
Equuleus	Little horse	Equ	21	+10
Eridanus	River	Eri	3	−20
Fornax	Furnace	For	3	−30
Gemini	Twins	Gem	7	+20
Grus	Crane	Gru	22	−45

*Modified from Abell, G.: Exploration of the universe, New York, 1969, Holt, Rinehart & Winston, Inc.

†The four constellations Carina, Puppis, Pyxis, and Vela originally formed the single constellation of Argo Navis.

Constellation	English name or description	Abbreviation	Approximate position of constellation's center	
			Right ascension (hours)	Declination (degrees)
Hercules	Hercules, son of Zeus	Her	17	+30
Horologium	Cloak	Hor	3	−60
Hydra	Sea serpent	Hya	10	−20
Hydrus	Water snake	Hyi	2	−75
Indus	Indian	Ind	21	−55
Lacerta	Lizard	Lac	22	+45
Leo	Lion	Leo	11	+15
Leo Minor	Little lion	LMi	10	+35
Lepus	Hare	Lep	6	−20
Libra	Balance	Lib	15	−15
Lupus	Wolf	Lup	15	−45
Lynx	Lynx	Lyn	8	+45
Lyra	Lyre or harp	Lyr	19	+40
Mensa	Table Mountain	Men	5	−80
Microscopium	Microscope	Mic	21	−35
Monoceros	Unicorn	Mon	7	−5
Musca	Fly	Mus	12	−70
Norma	Carpenter's level	Nor	16	−50
Octans	Octant	Oct	22	−85
Ophiuchus	Holder of serpent	Oph	17	0
Orion	Orion, the hunter	Ori	5	+5
Pavo	Peacock	Pav	20	−65
Pegasus	Pegasus, the winged horse	Peg	22	+20
Perseus	Perseus, hero who saved Andromeda	Per	3	+45
Phoenix	Phoenix	Phe	1	−50
Pictor	Easel	Pic	6	−55
Pisces	Fishes	Psc	1	+15
Piscis Austrinus	Southern fish	PsA	22	−30
Puppis†	Stern of the Argonauts' ship	Pup	8	−40
Pyxis (= Malus)†	Compass on the Argonauts' ship	Pyx	9	−30
Reticulum	Net	Ret	4	−60
Sagitta	Arrow	Sge	20	+10
Sagittarius	Archer	Sgr	19	−25
Scorpius	Scorpion	Sco	17	−40
Sculptor	Sculptor's tools	Scl	0	−30
Scutum	Shield	Sct	19	−10
Serpens	Serpent	Ser	17	0
Sextans	Sextant	Sex	10	0
Taurus	Bull	Tau	4	+15
Telescopium	Telescope	Tel	19	−50
Triangulum	Triangle	Tri	2	+30
Triangulum Australe	Southern triangle	TrA	16	−65
Tucana	Toucan	Tuc	0	−65
Ursa Major	Big bear	UMa	11	+50
Ursa Minor	Little bear	UMi	15	+70
Vela†	Sail of the Argonauts' ship	Vel	9	−50
Virgo	Virgin	Vir	13	0
Volans	Flying fish	Vol	8	−70
Vulpecula	Fox	Vul	20	+25

appendix B

Sidereal time equivalent to 8 P.M. (2000 hours) local mean time*

Day	Jan.	Feb.	March	April	May	June	July	Aug.	Sept.	Oct.	Nov.	Dec.	Day
1	2^h45^m	4^h48^m	6^h38^m	8^h40^m	10^h39^m	12^h41^m	14^h39^m	16^h41^m	18^h44^m	20^h42^m	22^h44^m	0^h42^m	1
2	2^h49^m	4^h52^m	6^h42^m	8^h44^m	10^h43^m	12^h45^m	14^h43^m	16^h45^m	18^h48^m	20^h46^m	22^h48^m	0^h46^m	2
3	2^h53^m	4^h56^m	6^h46^m	8^h48^m	10^h46^m	12^h49^m	14^h47^m	16^h49^m	18^h51^m	20^h50^m	22^h52^m	0^h50^m	3
4	2^h57^m	5^h00^m	6^h50^m	8^h52^m	10^h50^m	12^h53^m	14^h51^m	16^h53^m	18^h55^m	20^h54^m	22^h56^m	0^h54^m	4
5	3^h01^m	5^h04^m	6^h54^m	8^h56^m	10^h54^m	12^h57^m	14^h55^m	16^h57^m	18^h59^m	20^h58^m	23^h00^m	0^h58^m	5
6	3^h05^m	5^h07^m	6^h58^m	9^h00^m	10^h58^m	13^h01^m	14^h59^m	17^h01^m	19^h03^m	21^h02^m	23^h04^m	1^h02^m	6
7	3^h09^m	5^h11^m	7^h02^m	9^h04^m	11^h02^m	13^h05^m	15^h03^m	17^h05^m	19^h07^m	21^h06^m	23^h08^m	1^h06^m	7
8	3^h13^m	5^h15^m	7^h06^m	9^h08^m	11^h06^m	13^h08^m	15^h07^m	17^h09^m	19^h11^m	21^h09^m	23^h12^m	1^h10^m	8
9	3^h17^m	5^h19^m	7^h10^m	9^h12^m	11^h10^m	13^h12^m	15^h11^m	17^h13^m	19^h15^m	21^h13^m	23^h16^m	1^h14^m	9
10	3^h21^m	5^h23^m	7^h14^m	9^h16^m	11^h14^m	13^h16^m	15^h15^m	17^h17^m	19^h19^m	21^h17^m	23^h20^m	1^h18^m	10
11	3^h25^m	5^h27^m	7^h18^m	9^h20^m	11^h18^m	13^h20^m	15^h19^m	17^h21^m	19^h23^m	21^h21^m	23^h24^m	1^h22^m	11
12	3^h29^m	5^h31^m	7^h22^m	9^h24^m	11^h22^m	13^h24^m	15^h22^m	17^h25^m	19^h27^m	21^h25^m	23^h27^m	1^h26^m	12
13	3^h33^m	5^h35^m	7^h25^m	9^h28^m	11^h26^m	13^h28^m	15^h26^m	17^h29^m	19^h31^m	21^h29^m	23^h31^m	1^h30^m	13
14	3^h37^m	5^h39^m	7^h29^m	9^h32^m	11^h30^m	13^h32^m	15^h30^m	17^h33^m	19^h35^m	21^h33^m	23^h35^m	1^h34^m	14
15	3^h41^m	5^h43^m	7^h33^m	9^h36^m	11^h34^m	13^h36^m	15^h34^m	17^h37^m	19^h39^m	21^h37^m	23^h39^m	1^h38^m	15
16	3^h45^m	5^h47^m	7^h37^m	9^h40^m	11^h38^m	13^h40^m	15^h38^m	17^h40^m	19^h43^m	21^h41^m	23^h43^m	1^h41^m	16
17	3^h49^m	5^h51^m	7^h41^m	9^h43^m	11^h42^m	13^h44^m	15^h42^m	17^h44^m	19^h47^m	21^h45^m	23^h47^m	1^h45^m	17
18	3^h53^m	5^h55^m	7^h45^m	9^h47^m	11^h46^m	13^h48^m	15^h46^m	17^h48^m	19^h51^m	21^h49^m	23^h51^m	1^h49^m	18
19	3^h56^m	5^h59^m	7^h49^m	9^h51^m	11^h50^m	13^h52^m	15^h50^m	17^h52^m	19^h55^m	21^h53^m	23^h55^m	1^h53^m	19
20	4^h00^m	6^h03^m	7^h53^m	9^h55^m	11^h54^m	13^h56^m	15^h54^m	17^h56^m	19^h58^m	21^h57^m	23^h59^m	1^h57^m	20
21	4^h04^m	6^h07^m	7^h57^m	9^h59^m	11^h57^m	14^h00^m	15^h58^m	18^h00^m	20^h02^m	22^h01^m	0^h03^m	2^h01^m	21
22	4^h08^m	6^h11^m	8^h01^m	10^h03^m	12^h01^m	14^h04^m	16^h02^m	18^h04^m	20^h06^m	22^h05^m	0^h07^m	2^h05^m	22
23	4^h12^m	6^h14^m	8^h05^m	10^h07^m	12^h05^m	14^h08^m	16^h06^m	18^h08^m	20^h10^m	22^h09^m	0^h11^m	2^h09^m	23
24	4^h16^m	6^h18^m	8^h09^m	10^h11^m	12^h09^m	14^h12^m	16^h10^m	18^h12^m	20^h14^m	22^h13^m	0^h15^m	2^h13^m	24
25	4^h20^m	6^h22^m	8^h13^m	10^h15^m	12^h13^m	14^h15^m	16^h14^m	18^h16^m	20^h18^m	22^h16^m	0^h19^m	2^h17^m	25
26	4^h24^m	6^h26^m	8^h17^m	10^h19^m	12^h17^m	14^h19^m	16^h18^m	18^h20^m	20^h22^m	22^h20^m	0^h23^m	2^h21^m	26
27	4^h28^m	6^h30^m	8^h21^m	10^h23^m	12^h21^m	14^h23^m	16^h22^m	18^h24^m	20^h26^m	22^h24^m	0^h27^m	2^h25^m	27
28	4^h32^m	6^h34^m	8^h25^m	10^h27^m	12^h25^m	14^h27^m	16^h26^m	18^h28^m	20^h30^m	22^h28^m	0^h31^m	2^h29^m	28
29	4^h36^m		8^h29^m	10^h31^m	12^h29^m	14^h31^m	16^h30^m	18^h32^m	20^h34^m	22^h32^m	0^h34^m	2^h33^m	29
30	4^h40^m		8^h32^m	10^h35^m	12^h33^m	14^h35^m	16^h33^m	18^h36^m	20^h38^m	22^h36^m	0^h38^m	2^h37^m	30
31	4^h44^m		8^h36^m		12^h37^m		16^h37^m	18^h40^m		22^h40^m		2^h41^m	31

*Courtesy Unitron Scientific, Inc., Newton Highlands, Mass.

appendix C

Catalog of selected celestial bodies

Messier's Catalog no.	New General Catalog no.	Right ascension (hours and minutes)*	Declination (degrees and minutes)*	Apparent visual magnitude	Diameter (seconds of arc)	Distance from sun (parsecs, kiloparsecs, megaparsecs)	Constellation	Description
1	1952	5 31.5	+22 00	11.3	6 × 4	1050 pc	Taurus	Crab nebula
2	7089	21 30.9	−1 02	6.4	12	16 kpc	Aquarius	Globular cluster
3	5272	13 39.8	+28 38	6.3	19	14 kpc	Canes Venatici	Globular cluster
4	6121	16 20.6	−26 24	6.5	23	2.3 kpc	Scorpio	Globular cluster
5	5904	15 16.0	+2 17	6.1	20	8.3 kpc	Serpens	Globular cluster
6	6405	17 36.8	−32 10	5.3	26	630 pc	Scorpio	Open cluster
7	6475	17 50.7	−34 48	4.0	50	250 pc	Scorpio	Open cluster
8	6523	18 00.6	−24 23	6.0	90 × 40	1.5 kpc	Sagittarius	Lagoon nebula
9	6333	17 16.3	−18 28	8.0	6	7.9 kpc	Ophiuchus	Globular cluster
10	6254	16 54.5	−4 02	6.7	12	5.0 kpc	Ophiuchus	Globular cluster
11	6705	18 48.4	−6 20	6.3	12	1.7 kpc	Scutum	Open cluster
12	6218	16 44.7	−1 52	7.1	12	5.8 kpc	Ophiuchus	Globular cluster
13	6205	16 39.9	+36 33	5.9	23	6.9 kpc	Hercules	Globular cluster
14	6402	17 35.0	−3 13	8.5	7	7.2 kpc	Ophiuchus	Globular cluster
15	7078	21 27.5	+11 57	6.4	12	15 kpc	Pegasus	Globular cluster
16	6611	18 16.1	−13 48	6.4	8	1.8 kpc	Serpens	Open cluster (nebular)
17	6618	18 17.9	−16 12	7.0	46 × 37	1.8 kpc	Sagittarius	Swan (Omega) nebula
18	6613	18 17.0	−17 09	7.5	7	1.5 kpc	Sagittarius	Open cluster
19	6273	16 59.5	−26 11	7.4	5	6.9 kpc	Ophiuchus	Globular cluster
20	6514	17 59.4	−23 02	9.0	29 × 27	1.6 kpc	Sagittarius	Trifid nebula
21	6531	18 01.6	−22 30	6.5	12	1.3 kpc	Sagittarius	Open cluster
22	6656	18 33.4	−23 57	5.6	17	3.0 kpc	Sagittarius	Globular cluster
23	6494	17 54.0	−19 00	6.9	27	660 pc	Sagittarius	Open cluster
24	6603	18 15.5	−18 27	4.6	4	5.0 kpc	Sagittarius	Open cluster
25	4725†	18 28.7	−19 17	6.5	35	600 pc	Sagittarius	Open cluster
26	6694	18 42.5	−9 27	9.3	9	1.5 kpc	Scutum	Open cluster
27	6853	19 57.5	+22 35	8.2	8 × 4	200 pc	Vulpecula	Dumbbell nebula‡
28	6626	18 21.4	−24 53	7.6	15	4.6 kpc	Sagittarius	Globular cluster
29	6913	20 22.2	+38 21	7.1	7	1.2 kpc	Cygnus	Open cluster
30	7099	21 37.5	−23 24	7.7	9	13 kpc	Capricornus	Globular cluster
31	224	0 40.0	+41 00	3.5	160 × 40	700 kpc	Andromeda	Galaxy (spiral)
32	221	0 40.0	+40 36	8.2	3 × 2	700 kpc	Andromeda	Elliptical galaxy

*Data based on 1950 coordinates.
†International Catalog Number.
‡Denotes planetary nebulae.
§Visual magnitudes are 9.0 and 9.3.

Continued.

Messier's Catalog no.	New General Catalog no.	Right ascension (hours and minutes)*	Declination (degrees and minutes)*	Apparent visual magnitude	Diameter (seconds of arc)	Distance from sun (parsecs, kiloparsecs, megaparsecs)	Constellation	Description
33	598	1 31.0	+30 24	5.8	60 × 40	700 kpc	Triangulum	Spiral galaxy
34	1039	2 38.8	+42 35	5.5	30	440 pc	Perseus	Open cluster
35	2168	6 05.7	+24 21	5.3	29	870 pc	Gemini	Open cluster
36	1960	5 33.0	+34 04	6.3	16	1.3 kpc	Auriga	Open cluster
37	2099	5 49.1	+32 33	6.2	24	1.3 kpc	Auriga	Open cluster
38	1912	5 25.3	+35 47	7.4	18	1.3 kpc	Auriga	Open cluster
39	7092	21 30.4	+48 13	5.2	32	250 pc	Cygnus	Open cluster
40	—	12 20.0	+59 00	§	—	400 kpc	Ursa Major	Close double star
41	2287	6 44.9	−20 41	4.6	32	670 pc	Canis Major	Loose open cluster
42	1976	5 32.9	−5 25	4.0	66 × 60	460 pc	Orion	Orion nebula
43	1982	5 33.1	−5 19	9.0	—	460 pc	Orion	Northeast part of M42
44	2632	8 37.0	+20 10	3.7	90	158 pc	Cancer	Praesepe open cluster
45	—	3 44.5	+23 57	1.6	—	126 pc	Taurus	Pleiades open cluster
46	2437	7 39.5	−14 42	6.0	27	1.8 kpc	Puppis	Open cluster
47	2478	7 52.4	−15 17	5.2	25	548 pc	Puppis	Loose group of stars
48	—	8 11.0	−1 40	5.5	35	480 pc	Hydra	"Cluster of stars"
49	4472	12 27.3	+8 16	8.5	4 × 4	11 mpc	Virgo	Elliptical galaxy
50	2323	7 00.6	−8 16	6.3	16	910 pc	Monoceros	Loose open cluster
51	5194	13 27.8	+47 27	8.4	12 × 6	2 mpc	Canes Venatici	Whirlpool spiral galaxy
52	7654	23 22.0	+61 20	7.3	13	2.1 kpc	Cassiopeia	Loose open cluster
53	5024	13 10.5	+18 26	7.8	14	20 kpc	Coma Berenices	Globular cluster
54	6715	18 51.9	−30 32	7.8	6	15 kpc	Sagittarius	Globular cluster
55	6809	19 36.8	−31 03	6.2	15	5.8 kpc	Sagittarius	Globular cluster
56	6779	19 14.6	+30 05	8.7	5	14 kpc	Lyra	Globular cluster
57	6720	18 51.7	+32 58	9.0	1 × 1	550 pc	Lyra	Ring nebula‡
58	4579	12 35.2	+12 05	9.6	4 × 3	11 mpc	Virgo	Spiral galaxy
59	4621	12 39.5	+11 56	10.0	3 × 2	11 mpc	Virgo	Spiral galaxy
60	4649	12 41.1	+11 50	9.0	4 × 3	11 mpc	Virgo	Elliptical galaxy
61	4303	12 19.3	+4 45	9.6	6	11 mpc	Virgo	Spiral galaxy
62	6266	16 58.0	−30 02	7.3	6	6.9 kpc	Ophiuchus	Globular cluster
63	5055	13 13.5	+42 17	8.6	8 × 3	4 mpc	Canes Venatici	Spiral galaxy
64	4826	12 54.2	+21 57	8.5	8 × 4	6 mpc	Coma Berenices	Spiral galaxy
65	3623	11 16.3	+13 22	9.4	8 × 2	9 mpc	Leo	Spiral galaxy
66	3627	11 17.6	+13 16	9.0	8 × 2	9 mpc	Leo	Spiral galaxy
67	2682	8 48.4	+12 00	6.1	18	830 pc	Cancer	Open cluster
68	4590	12 36.8	−26 29	8.2	9	12 kpc	Hydra	Globular cluster
69	6637	18 28.1	−32 24	8.0	4	7.2 kpc	Sagittarius	Globular cluster
70	6681	18 40.0	−32 30	8.1	4	20 kpc	Sagittarius	Globular cluster
71	6838	19 51.5	+18 39	9.0	6	5.5 kpc	Sagitta	Globular cluster
72	6981	20 50.7	−12 45	9.3	5	18 kpc	Aquarius	Globular cluster
73	6994	20 56.2	−12 50	9.0	3	75 kpc	Aquarius	Open cluster
74	628	1 34.0	+15 32	9.3	8	8 mpc	Pisces	Spiral galaxy
75	6864	20 03.1	−22 04	8.6	5	24 kpc	Sagittarius	Globular cluster
76	650	1 39.1	+51 19	11.4	2 × 1	2.5 kpc	Perseus	Planetary nebula
77	1068	2 40.1	−0 12	8.9	2	16 mpc	Cetus	Spiral galaxy
78	2068	5 44.2	+0 02	8.3	8 × 6	500 pc	Orion	Small emission nebula

Messier's Catalog no.	New General Catalog no.	Right ascension (hours and minutes)*	Declination (degrees and minutes)*	Apparent visual magnitude	Diameter (seconds of arc)	Distance from sun (parsecs, kiloparsecs, megaparsecs)	Constellation	Description
79	1904	5 22.1	−24 34	7.5	8	13 kpc	Lepus	Globular cluster
80	6093	16 14.0	−22 52	7.5	5	11 kpc	Scorpio	Globular cluster
81	3031	9 51.7	+69 18	7.0	16 × 10	3.0 mpc	Ursa Major	Spiral galaxy
82	3034	9 51.9	+69 56	8.4	7 × 2	3 mpc	Ursa Major	Irregular galaxy
83	5236	13 34.2	−29 37	8.3	10 × 8	4 mpc	Hydra	Spiral galaxy
84	4374	12 22.6	+13 10	9.4	3	11 mpc	Virgo	Elliptical galaxy
85	4382	12 22.8	+18 28	9.3	4 × 2	11 mpc	Coma Berenices	Elliptical galaxy
86	4406	12 23.6	+13 13	9.2	4 × 3	11 mpc	Virgo	Elliptical galaxy
87	4486	12 28.2	+12 40	8.7	3	11 mpc	Virgo	Elliptical galaxy
88	4501	12 29.4	+14 42	9.5	6 × 3	11 mpc	Coma Berenices	Spiral galaxy
89	4552	12 33.1	+12 50	10.3	2	11 mpc	Virgo	Elliptical galaxy
90	4569	12 34.3	+13 26	9.6	6 × 3	11 mpc	Virgo	Spiral galaxy
91	—	12 35.0	+14 02	—	—	Unknown	Virgo	Planetary nebula?
92	6341	17 15.6	+43 12	6.4	12	11 kpc	Hercules	Globular cluster
93	2447	7 42.4	−23 45	6.0	18	1.1 kpc	Puppis	Open cluster
94	4736	12 48.6	+41 24	8.3	5 × 4	6 mpc	Canes Venatici	Spiral galaxy
95	3351	10 41.3	+11 58	9.8	3	9 mpc	Leo	Barred spiral galaxy
96	3368	10 44.1	+12 05	9.3	7 × 4	9 mpc	Leo	Spiral galaxy
97	3587	11 12.0	+55 17	11.1	3	800 pc	Ursa Major	Owl nebula‡
98	4192	12 11.2	+15 11	10.2	8 × 2	11 mpc	Coma Berenices	Spiral galaxy
99	4254	12 16.3	+14 42	9.9	4	11 mpc	Coma Berenices	Spiral galaxy
100	4321	12 20.4	+16 06	9.4	5	11 mpc	Coma Berenices	Spiral galaxy
101	5457	14 01.4	+54 36	7.9	22	3 mpc	Ursa Major	Spiral galaxy
102	—	— —	— —	—	—	Unknown	Unknown	Unknown (omitted)
103	581	1 29.9	+60 26	7.4	6	2.6 kpc	Cassiopeia	Open cluster
104	4594	12 37.4	−11 21	8.3	7 × 2	4.4 mpc	Virgo	Spiral galaxy
105	3379	10 45.2	+13 01	9.7	2 × 2	9 mpc	Leo	Elliptical galaxy
106	4258	12 16.5	+47 35	8.4	20 × 6	6 mpc	Canes Venatici	Spiral galaxy
107	6171	16 29.7	−12 57	9.2	8	7.0 kpc	Ophiuchus	Globular cluster
108	—	11 08.7	+55 57	10.7	8 × 2	3 mpc	Ursa Major	Spiral galaxy
109	—	11 08.0	+53 39	10.8	7	3 mpc	Ursa Major	Elliptical galaxy

Planetary nebulae and other celestial bodies not listed in Messier's Catalog

Right ascension (hours and minutes)	Declination (degrees and minutes)	Diameter (seconds of arc)	Apparent visual magnitude	Constellation	Description
4 08.4	+36 17	145 × 40	4.0	Perseus	California nebula
5 10.6	+19 25	Variable	Variable	Taurus	Hind's nebula
17 42.3	−69 04	20 × 20	? (dark)	Dorado	Tarantula nebula
06 29.6	+04 40	64 × 61	? (diffuse)	Monoceros	Rosette nebula
06 26.4	+08 46	Variable	Variable	Monoceros	Hubble's nebula
06 38.2	+09 57	60 × 30	4.7	Monoceros	Cone nebula
20 13.6	+30 32	70 × 6	? (diffuse)	Cygnus	Cirrus nebula
20 46.0	+31 30	40 × 20	? (diffuse)	Cygnus	Cirrus nebula
20 46.9	+44 11	85 × 75	1.3	Cygnus	Pelican nebula
20 44.3	+31 30	78 × 8	? (diffuse)	Cygnus	Cirrus nebula
20 57.0	+44 08	120 × 100	1.3	Cygnus	North America nebula

The selected celestial bodies described in Appendix C can be located on this chart. Numbers on the chart are Messier's Catalog numbers.

Degree of declination

Hours of right ascension

- Diffuse nebulae
- Galaxies
- Planetary nebulae
- Globular clusters
- Open clusters

appendix D

Symbols for the constellations of the zodiac

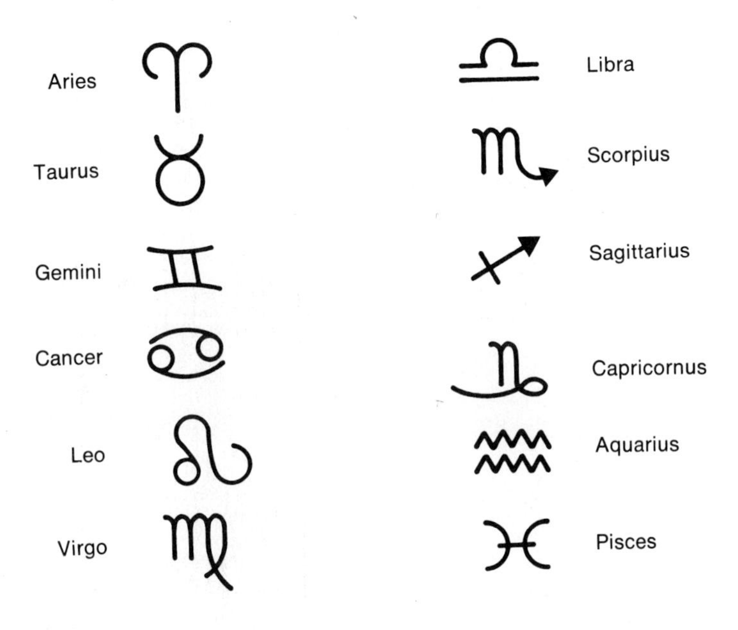

appendix E

The twenty brightest stars*

Star	Constellation	Right ascension (hours and minutes)†	Declination (degrees and minutes)†	Distance (parsecs)‡	Proper motion (seconds of arc)	Spectra of components§		
						A	B	C
Sirius	Canis major	6 42.9	−16 39	2.7	1.32	A1V	wd	
Canopus	Carina	6 22.8	−52 40	30	0.02	F0Ib-II		
α Centauri	Centaurus	14 36.2	−60 38	1.3	3.68	G2V	K5V	M5eV
Arcturus	Boötes	14 13.4	+19 27	11	2.28	K2IIIp		
Vega	Lyra	18 35.2	+38 44	8.0	0.34	A0V		
Capella	Auriga	5 13.0	+45 57	14	0.44	GIII	MIV	M5V
Rigel	Orion	5 12.1	−8 15	250	0.00	B8Ia	B9	
Procyon	Canis minor	7 36.7	+5 21	3.5	1.25	F5IV-V	wd	
Betelgeuse	Orion	5 52.5	+7 24	150	0.03	M2Iab		
Achernar	Eridanus	1 35.9	−57 29	20	0.10	B5V		
β Centauri	Centaurus	14 00.3	−60 08	90	0.04	B1III	?	
Altair	Aquila	19 48.3	+8 44	5.1	0.66	A7IV-V		
α Crucis	Crux	12 23.8	−62 49	120	0.04	B1IV	B3	
Aldebaran	Taurus	4 33.0	+16 25	16	0.20	K5III	M2V	
Spica	Virgo	13 22.6	−10 54	80	0.05	B1V		
Antares	Scorpius	16 26.3	−26 19	120	0.03	M1Ib	B4eV	
Pollux	Gemini	7 42.3	+28 09	12	0.62	K0III		
Fomalhaut	Pisces Austrinus	22 54.9	−29 53	7.0	0.37	A3V	K4V	
Deneb	Cygnus	20 39.7	+45 06	430	0.00	A2Ia		
β Crucis	Crux	12 44.8	−59 24	150	0.05	B0.5IV		

*Modified from Abell, G.: Exploration of the Universe, New York, 1969, Holt, Rinehart & Winston, Inc.

†Data are based on 1950 coordinates.

‡Distances to the more remote stars have been estimated from their spectral types and apparent magnitudes and are only approximate.

§Several of the components listed are themselves spectroscopic binaries. A "v" after magnitude indicates that the star is variable; in such cases the magnitude at median light is given. A "p" after spectral type indicates that the spectrum is peculiar. An "e" after spectral type indicates that emission lines are present. When the luminosity classification is uncertain a range is given.

Visual magnitudes of components			Absolute visual magnitudes of components		
A	B	C	A	B	C
−1.42	+8.7		+1.4	+11.5	
−0.72			−3.1		
−0.01	+1.4	+10.7	+4.4	+5.8	+15
−0.06			−0.3		
+0.04			+0.5		
+0.05	+10.2	+13.7	−0.7	+9.5	+13
+0.14	+6.6		−6.8	−0.4	
+0.38	+10.7		+2.7	+13.0	
+0.41v			−5.5		
+0.51			−1.0		
+0.63	+4		−4.1	−0.8	
+0.77			+2.2		
+1.39	+1.9		−4.0	−3.5	
+0.86	+13		−0.2	+12	
+0.91v			−3.6		
+0.92v	+5.1		−4.5	−0.3	
+1.16			+0.8		
+1.19	+6.5		+2.0	+7.3	
+1.26			−6.9		
+1.28v			−4.6		

appendix F

Brightest stars in the major constellations

The brightest stars in the major constellations are shown for the four seasons of the year. Maps include the sky from ±50° declination. The circumpolar star maps fill in the gaps from ±50° to ±90°. The symbols along the bottom of the maps are the zodiac signs through which the sun appears to move during each season. The months listed along the top and bottom of the maps are the months when you can go outside and see the appropriate stars (based on 8 P.M. LST). By holding the maps over your head with north facing north and east facing east, the view will be faithfully reproduced. For example, about 8 P.M. on March 15, you would see from north to south Lynx, Gemini, Canis Minor, Monoceros, Canis Major, etc.

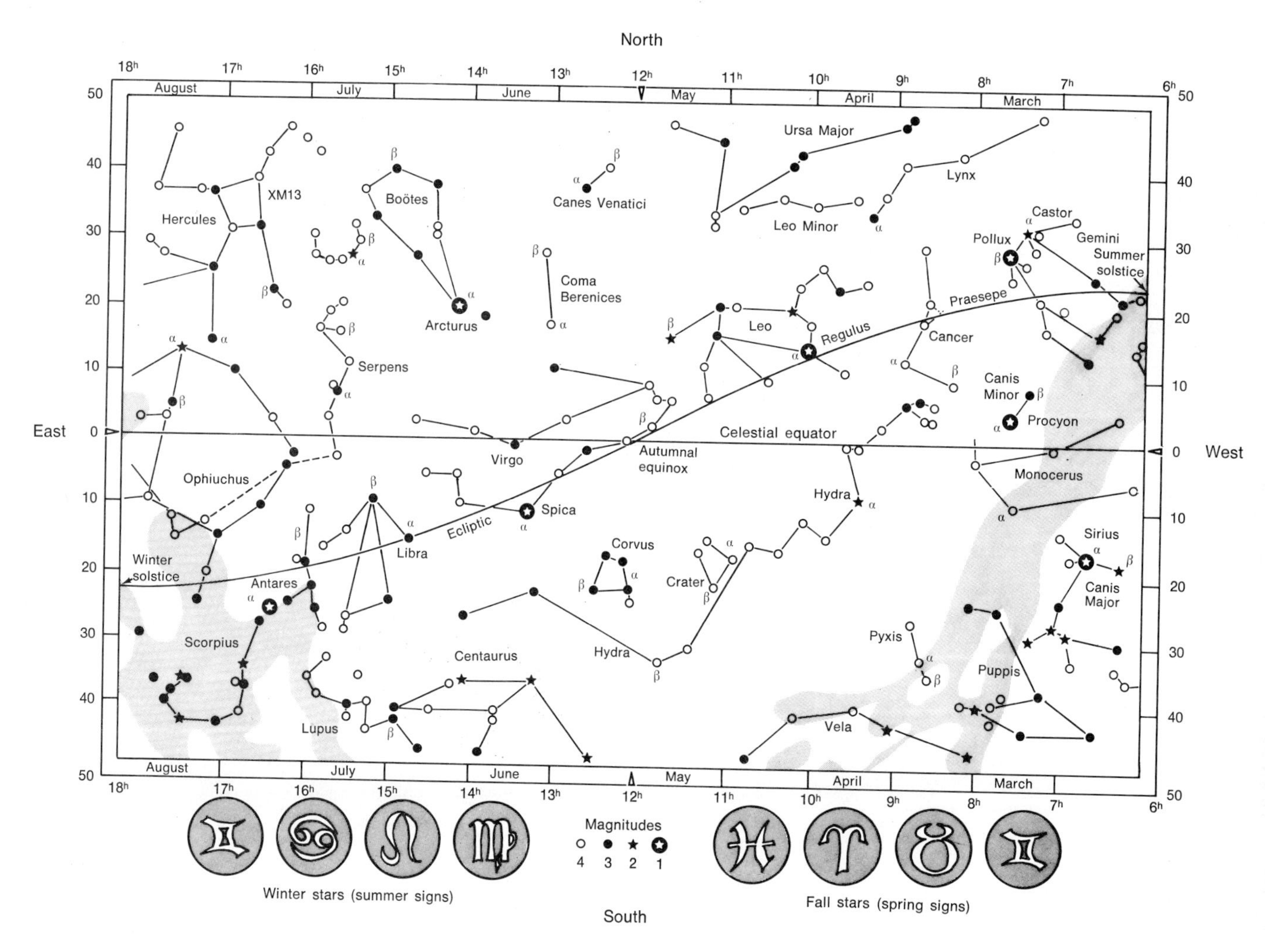

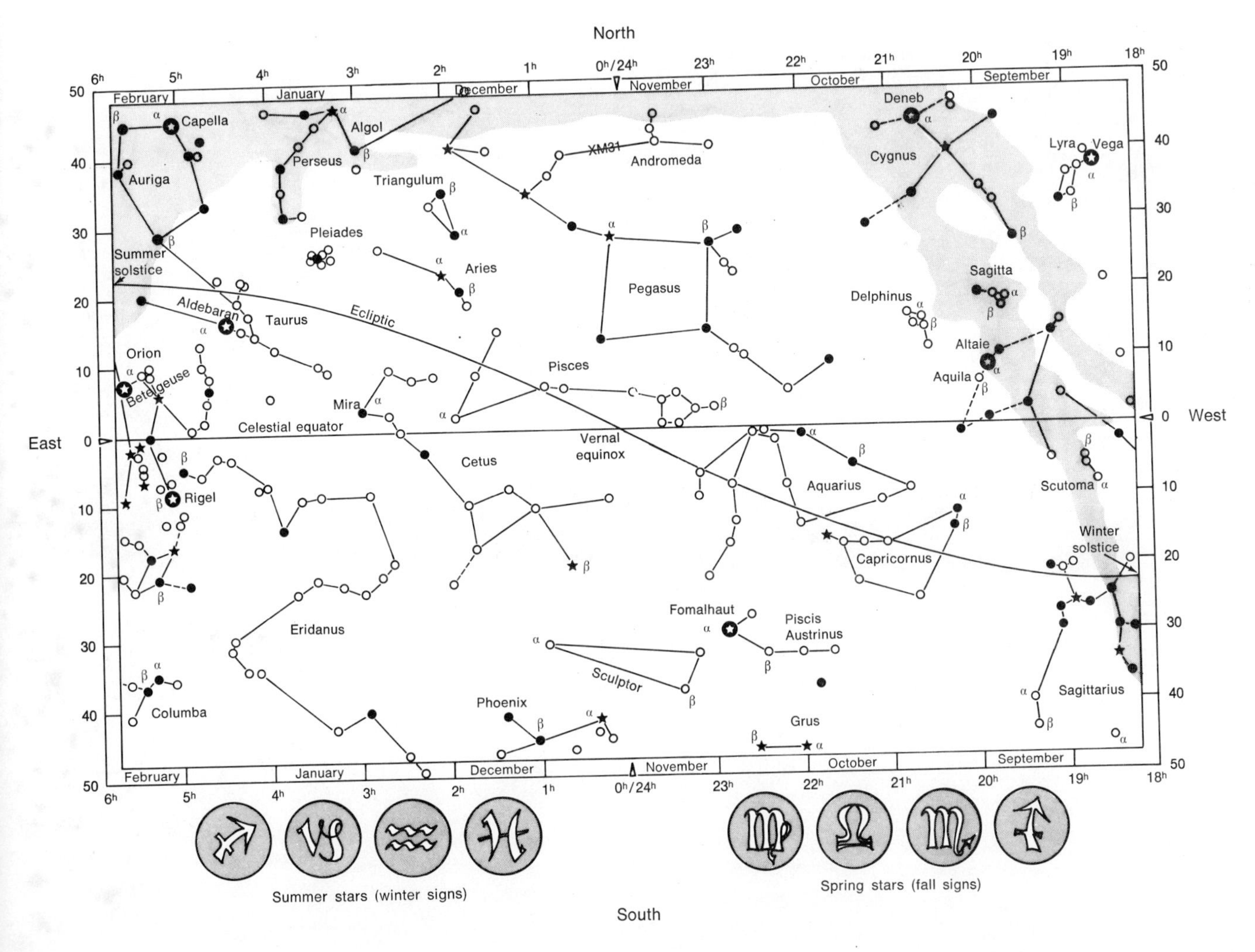
North
South
East
West
Celestial equator
Ecliptic
Vernal equinox
Summer solstice
Winter solstice
Capella
Auriga
Perseus
Algol
Triangulum
Andromeda
Cygnus
Deneb
Lyra
Vega
Pleiades
Aries
Pegasus
Delphinus
Sagitta
Aldebaran
Taurus
Orion
Betelgeuse
Rigel
Mira
Cetus
Pisces
Aquila
Altaie
Aquarius
Capricornus
Scutoma
Sagittarius
Eridanus
Columba
Phoenix
Sculptor
Fomalhaut
Piscis Austrinus
Grus
February
January
December
November
October
September
Summer stars (winter signs)
Spring stars (fall signs)

appendix G

Circumpolar stars

The south celestial pole region is never seen from areas on the 40° N latitude or farther north. Similarly, the north celestial pole and surrounding stars are not visible from below 40° S latitude. In these extreme latitudes, however, the stars surrounding their own poles appear to be circumpolar; that is, they rotate around the poles, but never set. The north celestial pole is found in the constellation of Ursa Minor; the south celestial pole is in the constellation of Octans.

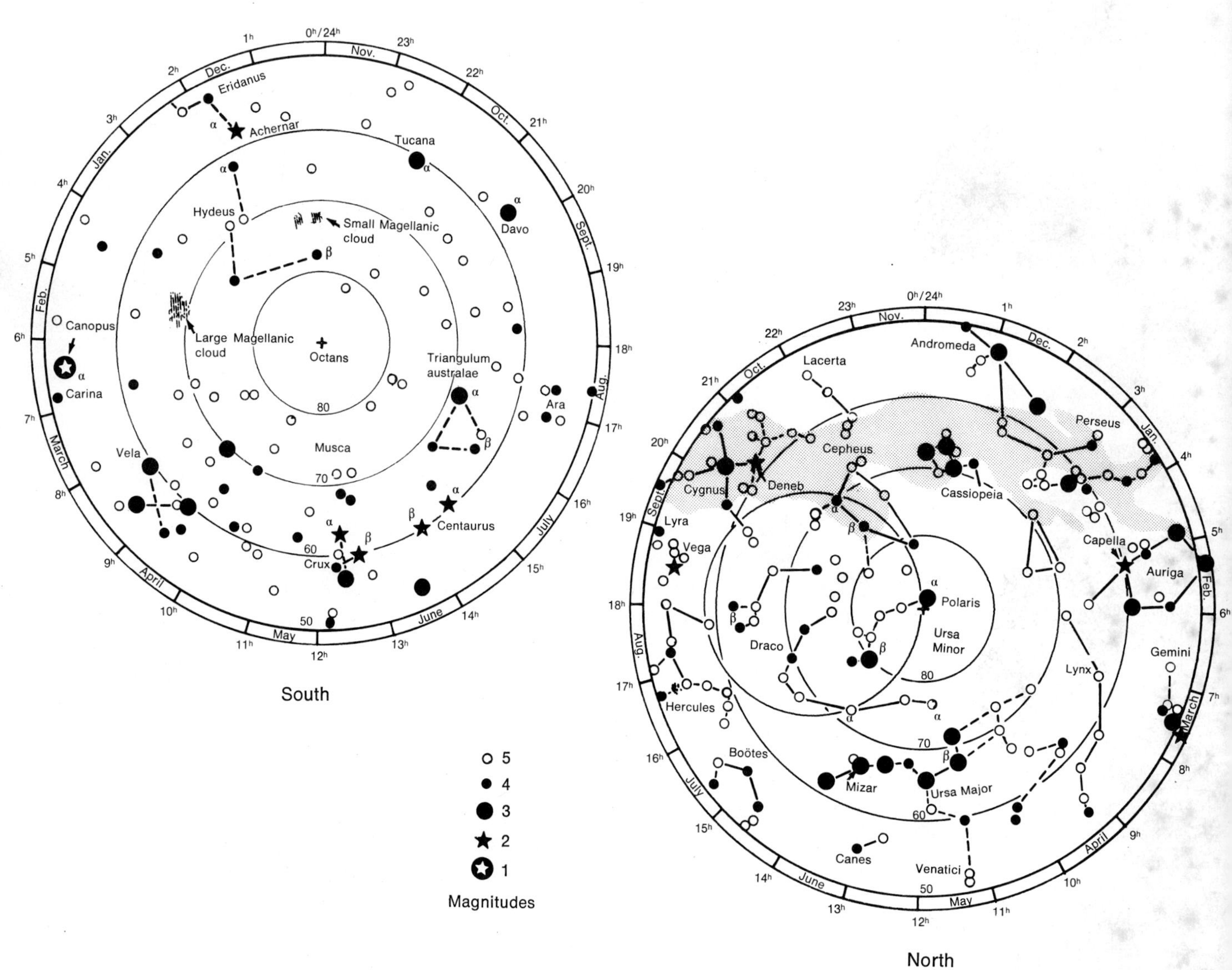

Index

D

Q

R

T